Biological Monitoring of Rivers

Applications and Perspectives

Water Quality Measurements Series

Series Editor

Philippe Quevauviller
European Commission, Brussels, Belgium

Published Titles in the Water Quality Measurements Series

Hydrological and Limnological Aspects of Lake Monitoring
Edited by Pertti Heinonen, Giuliano Ziglio and Andre Van der Beken

Quality Assurance for Water Analysis
Authored by Philippe Quevauviller

**Detection Methods for Algae, Protozoa and Helminths in Fresh and
Drinking Water**
Edited by Andre Van der Beken, Giuliano Ziglio and Franca Palumbo

Analytical Methods for Drinking Water: Advances in Sampling and Analysis
Edited by Philippe Quevauviller and K. Clive Thompson

Biological Monitoring of Rivers: Applications and Perspectives
Edited by Giuliano Ziglio, Maurizio Siligardi and Giovanna Flaim

Forthcoming Titles in the Water Quality Measurements Series

Wastewater Quality Monitoring
Edited by Philippe Quevauviller, Olivier Thomas and Andre Van der Berken

Biological Monitoring of Rivers

Applications and Perspectives

GIULIANO ZIGLIO
Università di Trento, Trento, Italy

MAURIZIO SILIGARDI
APPA, Trento, Italy

GIOVANNA FLAIM
Istituto Agario San Michele all'Adige, Trento, Italy

John Wiley & Sons, Ltd

Other Wiley Editorial Offices

John Wiley & Sons Inc., 111 River Street, Hoboken, NJ 07030, USA

Jossey-Bass, 989 Market Street, San Francisco, CA 94103-1741, USA

Wiley-VCH Verlag GmbH, Boschstr. 12, D-69469 Weinheim, Germany

John Wiley & Sons Australia Ltd, 42 McDougall Street, Milton, Queensland 4064, Australia

John Wiley & Sons (Asia) Pte Ltd, 2 Clementi Loop #02-01, Jin Xing Distripark, Singapore 129809

John Wiley & Sons Canada Ltd, 22 Worcester Road, Etobicoke, Ontario, Canada M9W 1L1

Wiley also publishes its books in a variety of electronic formats. Some content that appears
in print may not be available in electronic books.

Library of Congress Cataloguing-in-Publication Data

Biological monitoring of rivers : applications and perspectives / [edited by]
    Giuliano Ziglio, Maurizio Siligardi, Giovanna Flaim.
       p.   cm. – (Water quality measurements series)
    Based on a workshop held at Istituto Agraio of S. Michele TN, Italy,
in June 1998.
    Includes bibliographical references and index.
    ISBN-13: 978-0-470-86376-3 (cloth : alk. paper)
    ISBN-10: 0-470-86376-5 (cloth : alk. paper)
    1. Water quality biological assessment.   2. Stream ecology.
3. Environmental monitoring.   I. Ziglio, G.   II. Siligardi, Maurizio.
III. Flaim, Giovanna.   IV. Series.

QH96 .8.B5B5473 2006
577 .6′427 – dc22   2005020013

British Library Cataloguing in Publication Data

A catalogue record for this book is available from the British Library

ISBN-13: 978-0-470-86376-3 (HB)
ISBN-10: 0-470-86376-5 (HB)

Typeset in 10.5/12.5pt Times New Roman by TechBooks, New Delhi, India
Printed and bound in Great Britain by Antony Rowe, Chippenham, Wiltshire
This book is printed on acid-free paper responsibly manufactured from sustainable forestry
in which at least two trees are planted for each one used for paper production.

Contents

Series Preface

Water is a fundamental constituent of life and is essential to a wide range of economic activities. It is also a limited resource, as we are frequently reminded of by the tragic effects of drought in certain parts of the world. Even in areas with high precipitation, and in major river basins, over-use and mis-management of water have created severe constraints on availability. Such problems are widespread and will be made more acute by the accelerating demand on freshwater arising from trends in economic development.

Despite the fact that water-resource management is essentially a local, river-basin-based activity, there are a number of areas of action that are relevant to all or significant parts of the European Union and for which it is advisable to pool efforts for the purpose of understanding relevant phenomena (e.g. pollution and geochemical studies), developing technical solutions and/or defining management procedures. One of the keys for successful co-operations aimed at studying hydrology, water monitoring, biological activities, etc. is to achieve and ensure good water quality measurements.

Quality measurements are essential to demonstrate the comparability of data obtained worldwide and they form the basis for correct decisions related to the management of water resources, monitoring issues, biological quality, etc. Besides the necessary quality control tools developed for various types of physical, chemical and biological measurements, there is a strong need for education and training related to water quality measurements. This need has been recognized by the European Commission which has funded a series of training courses on this topic, covering aspects such as monitoring and measurements of lake recipients, measurements of heavy metals and organic compounds in drinking and surface water, use of biotic indexes, and methods to analyse algae, protozoa and helminths. In addition, a number of series of research and development projects have been or are being developed.

This book series will ensure a wide coverage of issues related to water quality measurements, including the topics of the above mentioned courses and the outcome of recent scientific advances. In addition, other aspects related to quality control tools (e.g. certified reference materials for the quality control of water analysis) and the monitoring of various types of waters (river, wastewater, groundwater, etc.) will also be considered.

This book on 'Biological Monitoring of Rivers: Applications and Perspectives' is the fifth one of the series. It has been written by leading scientific experts in river monitoring and offers the reader an updated and integrated view of river ecology, the application of biotic indices using the more common biological indicators and the interpretation and future development of river monitoring in different parts of the world.

The Series Editor – Philippe Quevauviller

Preface

This book was derived as output from the Techware Workshop, *A Comparison Among Four Different European Biotic Indices (IBE, BBI, BMWP and RIVPACS) for River Quality Evaluation*, held at the Istituto Agraio of S. Michele TN, Italy, in June 1998. The objectives of this course were the study of the aquatic environment in a comprehensive way and to promote the use of biotic methods in evaluating the quality of running waters. The uncommon opportunity to evaluate, as a scientific exercise, the performance of different methods based on the same or similar biological/environmental rationale, was regarded as an exciting experience by the main developers of the tested biotic indices – who were the teachers in the course – and stimulated exchange of experiences at all levels.

The possibility given to us by John Wiley & Sons, Ltd to publish the lectures presented at this workshop and other contributions pertaining to river ecology and monitoring suggested by an editorial panel, is greatly appreciated.

The workshop anticipated the aims and strategies of the European Water Framework Directive (WFD) in as much as it focused on methodological questions related to a common understanding of the technical and scientific implications.

In the time which has elapsed between the end of the workshop and the publication of this book, some relevant changes have occurred in technical and legislative fields related to the topics covered. For these reasons, this text has also embraced other aspects of river monitoring, including new tools and strategies for river ecology evaluation, relevant for a complete coverage. The contents of this book offer scientists and all workers involved in river monitoring the possibility to have an updated and integrated view on river ecology, the application of biotic indices using the more common biological indicators and the interpretation and future development of river monitoring in different parts of the world.

We are grateful to all of the authors for their contributions and for their efforts in achieving the objectives of this book.

Giuliano Ziglio, Maurizio Siligardi and Giovanna Flaim

List of Contributors

Javier Alba-Tercedor Departamento de Biologia Animal y Ecologie, Facultad de Ciencias, Universidad de Granada, E-18071 Granada, Spain

Patrick D. Armitage Centre for Ecology and Hydrology (CEH), Winfrith Technology Centre, Winfrith Newburgh, Dorchester, Dorset, DT2 8ZD, UK

Michael T. Barbour Center for Ecological Studies, Tetra Tech, Inc., 10045 Red Run Boulevard, Suite 110, Owings Mills, MD 21117, USA

Tim Burton Department of Geography, University of Durham, Durham, DH1 3LE, UK

Cristina Cappelletti Dipartimento Risorse Naturali ed Ambientali, Istituto Agario, via Mach 2, I-38010 San Michele all'Adige, Trento, Italy

Ana Cristina Cardoso European Commission Joint Research Centre, Ispra Institute for Environment and Sustainability, I-21020 Ispra-Varese, Italy

Daren M. Carlisle Water Resources Discipline, US Geological Survey, 413 National Center, 12201 Sunrise Valley Drive, Reston, VA 20192, USA

James l. Carter US Geological Survey, 345 Middlefield Road, MS465, Menlo Park, CA 94025, USA

Stefano Cataudella Laboratory of Experimental Ecology and Aquaculture, Department of Biology, University of Rome, 'Tor Vergata', via della Ricerca Scientifica, I-00133 Rome, Italy

Francesca Ciutti   Dipartimento Risorse Naturali ed Ambientali, Istituto Agario, via Mach 2, I-38010 San Michele all'Adige, Trento, Italy

William H. Clements   Department of Fishery and Wildlife Biology, Colorado State University, Fort Collins, CO 80523, USA

John Davy-Bowker   Centre for Ecology and Hydrology (CEH), Winfrith Technology Centre, Winfrith Newburgh, Dorchester, Dorset, DT2 8ZD, UK

Niels De Pauw   Laboratory of Environmental Toxicology and Aquatic Ecology, Department of Applied Ecology and Environmental Biology, Ghent University, J. Plateaustraat 22, B-9000 Ghent, Belgium

Stefano De Toni   Department of Civil and Environmental Engineering, University of Trento, via Mesiano 77, I-38050 Trento, Italy

Corrado Diamantini   Department of Civil and Environmental Engineering, University of Trento, via Mesiano 77, I-38050 Trento, Italy

Martin Dokulil   Institute for Limnology, Austrian Academy of Sciences, Mondseestrasse 9, A-5310 Mondsee, Austria

Barbara J. Downes   School of Anthropology, Geography and Environmental Studies, The University of Melbourne, 221 Bouverie Street, Melbourne, Victoria 3010, Australia

Giovanna Flaim   Dipartimento Risorse Naturali ed Ambientali, Istituto Agario, via Mach 2, I-38010 San Michele all'Adige, Trento, Italy

Nikolai Friberg   Department of Streams and Riparian Areas, National Environmental Research Institute, Vejlsøvej, PO Box 314, DK-8600 Silkeborg, Denmark

Wim Gabriels   Laboratory of Environmental Toxicology and Aquatic Ecology, Department of Applied Ecology and Environmental Biology, Ghent University, J. Plateaustraat 22, B-9000 Ghent, Belgium

Jereon Gerritsen   Center for Ecological Studies, Tetra Tech, Inc., 10045 Red Run Boulevard, Suite 110, Owings Mills, MD 21117, USA

Peter L. M. Goethals Laboratory of Environmental Toxicology and Aquatic Ecology, Department of Applied Ecology and Environmental Biology, Ghent University, J. Plateaustraat 22, B-9000 Ghent, Belgium

Konstantinos C. Gritzalis Hellenic Centre for Marine Research, Institute of Inland Waters, 46.7 km Athens-Sounion Avenue, 190 13 Anavyssos, Attica, Greece

Bruna Gumiero Department of Biology, University of Bologna, via Selmi 3, I-40126 Bologna, Italy

Jacques Haury UMR Agrocampus – INRA Ecobiologie et Qualité des Hydrosystèmes Continentaux, 65 Rue de Saint Brieuc, CS 84215, I-35042 Rennes Cedex, France

Jan Helešic Laboratory of Running Waters Biology, Department of Zoology and Ecology, Faculty of Science, Masaryk University, Kotláská 2, CZ-611 37, Brno, Czech Republic

Georg Janauer Department for Limnology and Hydrobotany, University of Vienna, Althanstrasse 14, A-1090 Vienna, Austria

Laura Mancini National Institute of Health, Viale Regina Elena 299, I-00161 Rome, Italy

Catia Monauni Agenzia Provinciale per la Protezione dell'Ambiete della Provincia Autonoma di Trento (APPA), via Mantova 14, I-38100 Trento, Italy

John F. Murphy Centre for Ecology and Hydrology (CEH), Winfrith Technology Centre, Winfrith Newburgh, Dorchester, Dorset, DT2 8ZD, UK

Gilles Pinay Centre d'Ecologie Fonctionnelle and Evolutive, CNRS, 1919 Route de Mende, F-34293 Montpellier Cedex 5, France

Guido Premazzi European Commission Joint Research Centre, Ispra Institute for Environment and Sustainability, I-21020 Ispra-Varese, Italy

Jean Prygiel Agence de l'Eau Artois-Picardie, Centre Tertiaire de l'Arsenal, 200 Rue Marceline, BP 818, F-59508 Douai Cedex, France

Vincent H. Resh Environmental Science, Policy and Management, 201 Wellman Hall, University of California, Berkeley, CA 94702, USA

Trefor B. Reynoldson Acadia Centre for Estuarine Research, Box 115, Acadia University, Wolfville, NS B4P 2P8, Canada

David M. Rosenberg Fisheries and Oceans Canada, Freshwater Institute, 501 University Crescent, Winnipeg, MB RST 2N6, Canada

Leonard Sandin Department of Environmental Assessment, Swedish University of Agricultural Science, PO Box 7050, SE-750 07, Uppsala, Sweden

Michele Scardi Laboratory of Experimental Ecology and Aquaculture, Department of Biology, University of Rome, 'Tor Vergata', via della Ricerca Scientifica, I-00133 Rome, Italy

Paolo Scotton Department of Civil and Environmental Engineering, University of Trento, via Mesiano 77, I-38050 Trento, Italy

Maurizio Siligardi Agenzia Provinciale per la Protezione dell'Ambiete della Provincia Autonoma di Trento (APPA), via Mantova 14, I-38100 Trento, Italy

Angelo Giuseppe Solimini European Commission Joint Research Centre, Ispra Institute for Environment and Sustainability, I-21020 Ispra-Varese, Italy

Lorenzo Tancioni Laboratory of Experimental Ecology and Aquaculture, Department of Biology, University of Rome, 'Tor Vergata', via della Ricerca Scientifica, I-00133 Rome, Italy

Piet F. M. Verdonschot Alterra Green World Research, Freshwater Ecology, PO Box 47, Droevendaalsesteeg 3, 6700 AA Wageningen, The Netherlands

Section 1
The River Environment

1.1

Floodplains in River Ecosystems

Gilles Pinay, Tim Burt and **Bruna Gumiero**

1.1.1   FLOODPLAINS AS OPEN SYSTEMS

Strictly speaking, the riparian zone includes only vegetation along the bed and banks of the river channel (Tansley, 1911). However, the definition has been extended in recent years to include a wider strip alongside the channel; very often, the floodplain is taken to be consistent with this more broadly defined area. The riparian landscape is unique among environments because it is a terrestrial habitat strongly affecting and affected by aquatic environments (Malanson, 1993). Pinay *et al.* (1990) identify the salient process elements – water, nutrient and sediment fluxes – in relation to transverse and longitudinal structure of the riparian zone. They, like Naiman and Décamps (1990) highlight the role of the riparian zone as a terrestrial-aquatic ecotone between the terrestrial and the aquatic ecosystems, whereas Forman and Godron (1986) prefer to view the riparian zone as a corridor, thus emphasizing longitudinal rather than transverse fluxes.

Biological Monitoring of Rivers   Edited by G. Ziglio, M. Siligardi and G. Flaim
© 2006 John Wiley & Sons, Ltd.

Schumm (1977) divided the channel network into three zones, as follows. Headwater streams, first- to third-order links within the channel network, which are the source regions for water, sediment and dissolved load (Figure 1.1.1). In Schumm's terminology, this is the *production* zone. Here, streams are closely coupled to hillslopes. There may be narrow floodplains between slope and channel, but often steep slopes connect directly to the channel. Given the inherent nature of the channel network, most of the channel length – and hence most of the basin area – is to be found in the headwater tributaries, and hence the emphasis on slope–channel coupling. The middle or *transfer* zone, channel links of fourth to sixth order, represents a transition. The floodplain becomes wider and transfers from water to land become relatively much more important when compared to fluxes in the other direction. By implication, longitudinal fluxes increase in importance given the dependence of (downstream) floodplains on fluxes from (upstream) headwater regions. In its lower regions, the floodplain is a *storage* zone, or sink, in which channel-to-floodplain transfers dominate.

The nature of the riparian zone will vary, depending on its location within the channel network and on regional climatic conditions (Décamps *et al.*, 2004). In the headwater tributaries, the delivery of water, sediment and solutes from slope to channel is most important. In the middle section, transfers from slope to channel can remain important but the channel also becomes a significant input source to the floodplain. Wide, lowland floodplains tend to be isolated from the surrounding hillslopes; they receive significant inputs from the channel and themselves become important source areas, especially during the flood recession and periods of low flow.

Given their location and topography, floodplains are likely to form wetlands, temporarily if not permanently. Even where the floodplain sediments are permeable, the combination of width and low gradient helps to maintain a high water table, with this control being accentuated as the floodplain width increases or where the alluvium is more fine-grained. Even above the water table, the soil is likely to remain close to saturation because of the capillary fringe effect (Gillham, 1984). Accepting the transverse nature of fluxes, as defined by Pinay *et al.* (1990), the floodplain water balance may be defined as follows.

Inputs: (a) overland flow from upslope (UOF); (b) subsurface flow from upslope (USSQ); (c) precipitation directly onto the floodplain (RF); (d) groundwater discharge from local aquifers into the floodplain (GW); (e) seepage from the river though the channel bank (BS); (f) overbank flooding from the channel to inundate the floodplain surface (OBI).

Outputs: (a) overland flow from floodplain to channel (FOF); (b) subsurface flow from the floodplain sediments to the river (FSSQ); (c) evaporation from the floodplain surface (ET); (d) percolation from the floodplain into local aquifers (PERC).

An imbalance between inputs and outputs must, by definition, involve a change of water storage within the floodplain (ΔS). The floodplain water balance may therefore be expressed as follows:

$$\text{UOF} + \text{USSQ} + \text{RF} + \text{GW} + \text{BS} + \text{OBI} - \text{FOF} - \text{FSSQ} - \text{ET} - \text{PERC} \pm \Delta S = 0$$

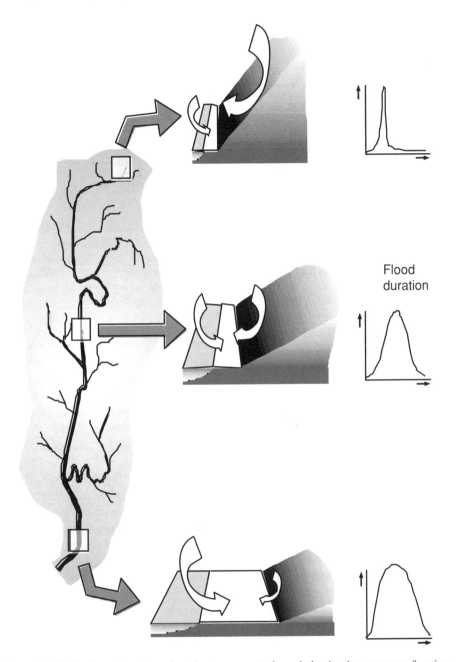

Figure 1.1.1 Preferential water and nutrient movements through the riparian zones as a function of their location within the drainage basin. Arrows symbolize the main water and associated suspended and dissolved matter transfer between upland and stream via the riparian zone. The riparian zones are shown in white, the rivers in light grey and the upland catchments in dark grey. Along small streams, most of the water and associated nutrients flow from the upland via the riparian zone whenever it exists, while along large streams the main flow direction is from the stream towards the floodplain (mainly during flood events) (adapted from Tabacchi *et al.*, 1998)

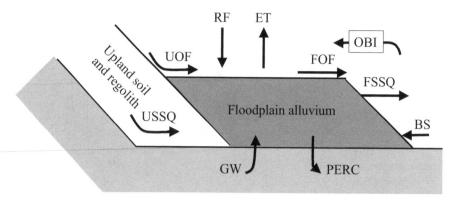

Figure 1.1.2 Water balance in the floodplain: UOF, overland flow from upslope; USSQ, subsurface flow from upslope; RF, precipitation directly onto the floodplain; GW, groundwater discharge from local aquifers into the floodplain; OBI, overbank flooding from the channel to inundate the floodplain surface; BS, see page from the river through the channel bank; FOF, overland flow from floodplain to channel; FSSQ, subsurface flow from the floodplain sediments to the river; ET, evaporation from the floodplain surface; PERC, percolation from the floodplain into the local aquifers. Reproduced by permission of Haycock Associates Limited from Burt, T. P., 1997, 'The hydrological role of floodplain within the drainage basin system', in *Buffer Zones: Their Processes and Potential in Water Protection*, Haycock, N. E., Burt, T. P., Goulding, K. W. T. and Pinay, G. (Eds), Quest Environmental Publications, Harpenden, UK, pp. 21–32

   The floodplain water balance equation highlights the point that inputs to the floodplain can originate from both the adjacent hillslopes and from the river channel (Figure 1.1.2). As noted above, the relative importance of these runoff sources alters along the channel network. In headwater tributaries, riparian zone hydrology is closely controlled by topography. Assuming that there is no floodplain present and that steep slopes abut the channel, the river has only very limited influence on the water table within the riparian zone. High water tables may develop but these depend on drainage from upslope and will be temporary. If a floodplain is present, high water tables will be a much more regular occurrence, and both river and hillslope can influence water table height within the floodplain (Burt *et al.*, 2002b). As one moves down the channel network, and floodplain width tends to increase, so the river becomes a relatively more important source of water to the floodplain. During winter floods, the floodplain can remain inundated for long periods of time; only during the flood recession is the dominance of hillslope sources re-established (Burt *et al.*, 2002a).

   River systems are open systems, dynamically linked in three dimensions by hydrological and geomorphologic processes which largely control their existence and maintenance via the timing and duration of floods and low-flow events. Process rates vary spatially in relation to the floodplain's location within the channel network.

   Floodplains tend to widen downstream; this tends to be associated with higher water tables and a greater relative importance of river inputs when compared to hillslope sources. The operation of any particular process can therefore be expected to vary in relation to location within the river basin; further downstream, longitudinal fluxes will tend to become more important when compared to transverse fluxes.

Fluxes of solutes and sediments into and out of the floodplain are, of course, very largely dependent on the water fluxes. Floodplains may act as either a conduit or a barrier for all of these fluxes (Burt and Pinay, 2005). Where the floodplain acts as a barrier, or buffer zone, the internal storage and process mechanisms operating within the floodplain assume great importance. For example, the micromorphology and sediment stratigraphy of floodplains depends on a combination of channel and overbank deposition. These, in turn, influence flow paths and soil wetness across the floodplain, important controls on a process like denitrification, for instance, where process rates depend on a combination of nitrate input, soil carbon content and the development of anaerobic conditions.

In this hydrogeomorphic context, we define three fundamental principles driving the nitrogen cycle in river systems (Pinay *et al.*, 2002). In general, these principles should apply to other nutrients (Hedin *et al.*, 1998). The three fundamental principles that regulate the cycling and transfer of nitrogen in rivers are as follows: the mode of nitrogen delivery affects ecosystem functioning; the degree of contact between water and soil or sediment increases nitrogen retention and processing; the role of floods and periods of low flow in influencing pathways of nitrogen cycling.

1.1.2 THE MODE OF NITROGEN DELIVERY AFFECTS RIVER ECOSYSTEM FUNCTIONING

The first principle is related to the delivery patterns of nitrogen inputs along river corridors. River systems and their riparian zones can be viewed as open ecosystems, dynamically linked longitudinally, laterally and vertically by hydrologic and geomorphic processes (Ward, 1989). In small, forested headwater streams, particulate organic nitrogen is the main form of nitrogen transferred to the aquatic system, primarily as litter fall from the adjacent riparian vegetation (Cummins *et al.*, 1983; Minshall *et al.*, 1983). Nitrogen-fixing plants such as alder (*Alnus* spp.) are often found in riparian forests. They contribute large amounts of nitrogen rich organic matter, which can reach several kg of dry matter per m^2 (Chauvet, 1987). Eventually, these particulate inputs contribute to the export of dissolved organic nitrogen via surface and subsurface pathways after degradation and recycling processes have occurred (Newbold *et al.*, 1981; Elwood *et al.*, 1983; McClain *et al.*, 1997; Clark *et al.*, 2000; Stepanauskas *et al.*, 2000). Due to their location along the edge of rivers, riparian forests also receive, recycle and transfer large amounts of sediments and nutrients to streams (for nitrogen, it is mainly as nitrate by subsurface flow) from upslope ecosystems (Peterjohn and Correll, 1984; Lowrance *et al.*, 1995). Fortunately, riparian zones can efficiently utilize and retain nitrate inputs from upslope as long as the subsurface water flow intercepts roots and microorganisms. Therefore, riparian zones deliver nitrogen to streams mainly as particulate organic matter.

In the floodplains of most large rivers, the main inputs of nutrients, sediment and organic matter are mainly via surface flow from upstream (Figure 1.1.1). Indeed, significant amounts of these materials are deposited during floods (Brinson *et al.*, 1983; Schlosser and Karr, 1981; Lowrance *et al.*, 1986; Grubaugh and Anderson, 1989; Brunet *et al.*, 1994). River floodplains are recognized as important storage sites for sediments and associated nutrients mobilized from upstream catchments during floods (He and Walling, 1997). The transfer and storage of materials in flood-plains are largely under the control of flood duration, frequency and magnitude that, collectively, create a mosaic of geomorphic surfaces influencing the spatial pat-tern and successional development of riparian vegetation (Salo *et al.*, 1986; Roberts and Ludwig, 1991). The fluxes of matter via flood deposits are responsible for the high nutrient-cycling capacity of floodplain soils, as compared to upland ecosystems (Brinson *et al.*, 1984). The significantly higher fertility of floodplain soils is illustrated in an agricultural example from Bangladesh. Historically, flood-mediated sediment and nutrient deposits on the Ganges and Bramaputhra river floodplains supported up to three crops of rice per year without fertilizer addition, while upland soils only sustained one crop a year (Mathab and Karim, 1992; Haque and Zaman 1993).

1.1.3   INCREASING CONTACT BETWEEN WATER AND SOIL OR SEDIMENT INCREASES NITROGEN RETENTION AND PROCESSING

The second basic principle is that the area of water–substrate interface (i.e. water–sediment or wetland–upland length of contact) is positively correlated with the effi-ciency of nitrogen retention and use in river ecosystems. This occurs both instream and in the riparian and floodplain zones. The nitrogen cycle is driven by processes that occur on or at the interface of particulate material such as stones, soils, sediments or algal mats (Ponnamperuma, 1972; Hill, 1979; Jones and Holmes, 1996; Valett *et al.*, 1996). Hence, increased length of contact between water and these substrates increase the biological use and thereby the total amount of nitrogen processed.

Riparian wetlands provide a large contact area between water and soils that pro-mote nitrogen retention and processing, thereby regulating fluxes from uplands to streams. The wetland–upland contact zone can be envisioned, as a first approxima-tion, as the contact zone between the riparian wetland and the uplands. This zone of contact varies both in depth and width as a function of the river's geomorphology and hydrologic regime. During high water periods, the extension of the saturated area increases away from the stream and extends further upslope while, during low-water periods, the saturated area shrinks in extent and decreases in length upslope (Beven and Kirkby, 1979).

There has been considerable interest in restoring or promoting the use of ripar-ian wetlands to mitigate diffuse nutrient pollution (Peterjohn and Correll, 1984; Lowrance *et al.*, 1985; Pinay and Labroue, 1986, Haycock *et al.*, 1997). Surprisingly, several attempts to relate the percentage area of riparian wetlands with nutrient fluxes at the outlet of drainage basins have failed (Osborne and Wiley, 1988; Tufford *et al.*, 1998). One of the major reasons for the lack of correlation is that riparian efficiency is driven by hydraulic connection with upland inputs since nitrate is often a limiting nutrient (Groffman and Hanson, 1997; Lowrance *et al.*, 1997). Riparian zones represent a mosaic of physical and functional units whose patterns are shaped by long-term geomorphic development of the floodplain. These biophysical units can be connected or disconnected hydrologically from each other and from the upland catchment (Brinson, 1993). Therefore, the efficiency of a riparian zone in regulating nitrogen fluxes is not a function of the surface area of the riparian zone but rather a function of the hydrological length of contact between the riparian zone and the upland drainage basin (Haycock and Pinay, 1993; Matchett, 1998, Burt *et al.*, 1999).

This can be illustrated by comparing first-order streams (i.e., small perennial streams without any tributaries) to larger rivers. First-order streams represent more than 50 % of the entire length of the river network, while higher-order rivers represent only a few percentage of the total length in a given catchment (Naiman, 1983). As a consequence, riparian zones associated with small-order stream develop a more intimate wetland–upland interface than riparian zones along high order rivers (Brinson, 1993), and better contribute to mitigating diffuse pollution from the catchment (Peterson *et al.*, 2001). Moreover, for a given surface area of riparian zone in a catchment, small streams are more efficient in retaining upland nutrients than larger streams because of the close proximity of water to sediments or soils.

Throughout the world, many upland streams have been subjected to human modifications such as channelization, impoundment or removal of riparian vegetation. All anthropogenic impacts tend to reduce both the spatial extent of saturated areas and the duration of riparian soil saturation (Worrall and Burt, 1998). Moreover, straightening river channels, dredging riverbeds, or clogging of interstitial spaces by fine sediments, reduce the size of the hyporheic zone and its exchange rates with surface water, thereby affecting the nutrient recycling capacity of the stream. As a consequence, these human impacts reduce the efficiency of the river network to mitigate diffuse nutrient pollution.

1.1.4 FLOODS AND DROUGHTS ARE NATURAL EVENTS THAT STRONGLY INFLUENCE PATHWAYS OF NITROGEN CYCLING

The third principle is related to the role of floods in shaping the characteristics of nitrogen cycling. Changes to the water regime, either through alterations to the frequency, duration, period of occurrence and intensity of water levels, directly affect nitrogen cycling in alluvial soils by controlling the duration of oxic and anoxic phases (Ponnamperuma, 1972; Keeney, 1973; Patrick, 1982). Flooding duration is controlled by local topography; low areas are flooded more often and longer than higher ones, producing variations in biogeochemical patterns at the metre scale (Pinay *et al.*, 1989; Pinay and Naiman, 1991). Biogeochemical processes, especially for nitrogen, are sensitive to the redox status of the soil. For instance, ammonification of organic nitrogen can be realized both under aerobic and anaerobic conditions but the nitrification process, which requires oxygen, can only occur in aerated soils or sediments. For instance, Hefting *et al.* (2004) found in a pan-European study that water table elevation was the prime determinant of the nitrogen dynamics and its endproducts. Three consistent water table thresholds were identified. In sites where the water table is within −10 cm of the soil surface, ammonification is the main process and ammonia accumulates in the topsoils. Average water tables between −10 and −30 cm favour denitrification and therefore reduce the nitrogen availability in soils. In drier sites, where the water table is below −30 cm, nitrate accumulates as a result of high net nitrification.

Short-term periodicity of aerobic–anaerobic conditions through groundwater level movements allows all nitrogen cycling processes to occur simultaneously at the same location in accordance with the level of soil-water saturation. Several studies have demonstrated that alternating aerobic and anaerobic conditions enhance organic matter decomposition and nitrogen loss through denitrification (e.g. Reddy and Patrick, 1975; Groffman and Tiedje, 1988; Pinay *et al.*, 1993). Moreover, it has been shown in an experimental study that the rate of nitrogen mineralization is much greater during a flooded period than during a non-flooded one (Neill, 1995). Overall, natural water table fluctuations in floodplains are key drivers of soil fertility with changes to the natural flood regime often decreasing productivity. However, drier periods are also important since they allow mineralization of more complex organic matter structures (e.g. hemicellulose and lignin) and contribute to soil fertility by providing another inorganic nitrogen source (ammonia and nitrate) to plant and microbes which otherwise would be sequestered as organic residues.

The flood regime also indirectly affects nutrient cycling in floodplain soils by influencing soil structure and texture through the deposition of sediment. The alluvial soil grain size mosaic and the proportion of different grain size deposits varies spatially and temporally following extreme flooding (Petts and Maddock, 1996; Richter and Richter, 2000). At small scales, geomorphic and hydrologic processes influence the sorting of sediment deposits on a grain-size basis. This creates a mosaic of soils of different textures. It is the soil or sediment texture that influences denitrification rates, as well as other biogeochemical cycles (Pinay *et al.*, 2000). Fine structures such as clay develop large surface areas per unit of weight or volume, which provide greater chemical adsorption sites (Paul and Clark, 1996) and microbial habitats (Ranjard *et al.*, 2000). For instance, the fastest denitrification rates are measured in soils with fine texture. Below a threshold of ∼ 65 % of silt and clay, floodplain soils do not show any significant denitrification. Above that threshold, denitrification increases linearly. In fine-textured floodplain soils, the denitrification rates are of the same order of magnitude as nitrogen mineralization, with annual denitrification representing up to 70 % of the nitrogen deposited during floods (Pinay *et al.*, 1995). Thus, floodplains contribute to the regulation of nitrogen fluxes by sorting sediments mobilized during floods and recycling nitrogen deposited during a flood.

1.1.5 NEW CHALLENGES

There are several new scientific challenges which need urgent attention. Quantifying the cumulative effects of water regime changes on the functioning of floodplains is an important scientific challenge. For instance, a reduction of the high-water period duration along low-order streams would lead to downstream movement of the dry–wet interface into the riparian zone, thereby favouring further intrusion of upland allochthonous nitrates within the riparian zone (Burt, 1997). Moreover, it is expected that this will also lead to further intrusion of pesticides in riparian soils since these xenobiotic molecules are often found together with nitrate. To what extent does this intrusion reduce bacterial denitrification of newly contaminated soils where pesticide decomposition would be less effective since the bacterial population was not adapted to repeated pesticide application (Abdelhafid *et al.*, 2000)? A second challenge concerns the scale of appraisal of water regime change effects on nutrient cycling. Depending on the spatial and temporal monitoring scale, the consequences of water regime changes on gaseous nitrogen end products varies (Bodelier *et al.*, 2000). Contrasting field and laboratory results on the respective importance of different processes on end products appears to result from the difference between space and time scale at which the studies are conducted. Another issue is related to the determination of the respective roles of human impacts and natural changes on river ecosystem functions (Vitousek *et al.*, 1997). In most cases, pristine ecosystems are in different ecoregions than human impacted ones, hence limiting the ability to extrapolate process rates and trends from natural to modified systems (Tol and Langen, 2000). However, in both cases, the main principles regulating the cycling and transfer of nitrogen in river ecosystems remain valid despite differences in process rates. Therefore, sustainable management practices should be assessed according to their impact on (1) the delivery mode of nitrogen to river ecosystems, (2) the length of contact between water, soil and sediment, and (3) the timing, rate and duration of floods and drought.

REFERENCES

Abdelhafid, R., Houot, S. and Barriuso, E., 2000. 'Dependence of atrazine degradation on C and N availability in adapted and non-adapted soils'. *Soil Biology Biochemistry*, **32**, 389–401.

Beven, K. and Kirkby, M. J., 1979. 'A physically based, variable contributing area model of basin hydrology'. *Hydrological Sciences Bulletin*, **24**, 43–69.

Bodelier, P. L. E., Hahn, A. H., Arth, I. R. and Frenzel, P., 2000. 'Effects of ammonium-based fertilisation on microbial processes involved in methane emission from soil planted with rice'. *Biogeochemistry*, **51**, 225–257.

Brinson, M. M., 1993. 'Changes in the functioning of wetlands along environmental gradients'. *Wetlands*, **13**, 65–74.

Brinson, M. M., Bradshaw, H. D. and Holmes, R. N., 1983. 'Significance of floodplain sediments in nutrient exchange between a stream and its floodplain'. In: *Dynamics of Lotic Ecosystems*, Fontaine III, T. D. and Bartel, S. M. (Eds). Ann Arbor Science: Ann Arbor, MI, USA, pp. 199–220.

Brinson, M. M., Bradshaw, H. D. and Kane, E. S., 1984. 'Nutrient assimilative capacity of an alluvial floodplain swamp'. *Journal of Applied Ecology*, **21**, 1041–1057.

Brunet, R. C., Pinay, G., Gazelle, F. and Roques, L., 1994. 'The role of floodplain and riparian zone in suspended matter and nitrogen retention in the Adour River, southwest France'. *Regulated Rivers: Research and Management*, **9**, 55–63.

Burt, T. P., 1997. 'The hydrological role of floodplain within the drainage basin system'. In: *Buffer Zones: Their Processes and Potential in Water Protection*, Haycock, N. E., Burt, T. P., Goulding, K. W. T. and Pinay, G. (Eds). Quest Environmental Publications: Harpenden, UK, pp. 21–32.

Burt T. P. and Pinay, G. 2005. 'Linking hydrology and biogeochemistry in complex landscapes'. *Progress in Physical Geography*, **29**, 297–316.

Burt, T. P., Matchett, L. S., Goulding, K. W. T., Webster, C. P. and Haycock, N. E., 1999. 'Denitrification in riparian buffer zones: the role of floodplain sediments'. *Hydrological Processes*, **13**, 1451–1463.

Burt, T. P., Bates, P. D., Stewart, M. D., Claxton, A. J., Anderson, M. G. and Price, D. A., 2002a. 'Water table fluctuations within the floodplain of the River Severn, England'. *Journal of Hydrology*, **262**, 1–20.

Burt, T. P., Pinay, G., Matheson, F. E., Haycock, N. E, Butturini, A., Clement, J. C., Danielescu, S., Dowrick, D. J., Hefting M. M., Hillbricht-Ilkowska, A. and Maitre, V., 2002b. 'Water table fluctuations in the riparian zone: comparative results from a pan-European experiment'. *Journal of Hydrology*, **265**, 129–148.

Chauvet, E., 1987. 'Changes in the chemical composition of alder, poplar and willow leaves during decomposition in a river'. *Hydrobiologia*, **148**, 35–44.

Clark, G. M., Mueller, D. K. and Mast, M. A. 2000. 'Nutrient concentrations and yields in undeveloped stream basins of the United States'. *Journal of the American Water Resources Association*, **36**, 849–860.

Cummins, K. W., Sedell, J. R., Swanson, F. J., Minshall, G. W., Fisher, S. G., Cushing, C. E., Peterson, R. C. and Vannote, R. L., 1983. 'Organic matter budgets for stream ecosystems: problems in their evaluation'. In: *Stream Ecology*, Barnes, J. R. and Minshall, G. W. (Eds). Plenum Press, New York, NY, USA, pp. 299–353.

Décamps, H., Pinay, G., Naiman, R. J., Petts, G. E., McClain, M. E., Hillbricht-Ilkowska, A., Hanley, T. A., Holmes, R. M., Quinn, J., Gibert, J., Planty-Tabacchi, A. M., Schiemer, F., Tabacchi, E. and Zalewski, M., 2004. 'Riparian zones: where biogeochemistry meets biodiversity in management practice'. *Polish Journal of Ecology*, **52**, 3–18.

Elwood, J. W., Newbold, J. D., O'Neill, R. V. and Winkle, W. V., 1983. 'Resource spiraling: an operational paradigm for analysing lotic ecosystems'. In: *Dynamics of Lotic Ecosystems*, Fontaine III, T. D. and Bartell, S. M. (Eds). Ann Arbor Science, Ann Arbor, MI, USA, pp. 3–27.

Forman, R. T. T. and Godron, M., 1986. *Landscape Ecology*. Wiley: New York, N. Y., USA.

Gillham, R.W, 1984. 'The capillary fringe and its effect on water-table response'. *Journal of Hydrology*, **67**, 307–324.

Groffman, P. M. and Tiedje, J. M., 1988. 'Denitrification hysteresis during wetting and drying cycles in soil'. *Soil Science Society of America Journal*, **52**, 1626–1629.

Groffman, P. M. and Hanson, G. C., 1997. 'Wetland denitrification: influence of site quality and relationships with wetland delineation protocols'. *Soil Science Society of America Journal*, **61**, 323–329.

Grubaugh, J. W. and Anderson, R. V., 1989. 'Upper Mississippi River: seasonal and floodplain forest influences on organic matter transport'. *Hydrobiologia*, **174**, 235–244.

Haque, C. E. and Zaman, M. Q., 1993. 'Human responses to riverine hazards in Bangladesh: a proposal for sustainable floodplain development'. *World Development*, **1**, 93–107.

Haycock, N. E. and Pinay, G., 1993. 'Groundwater nitrate dynamics in grass and poplar vegetated riparian buffer strips during the winter'. *Journal of Environmental Quality*, **22**, 273–278.

Haycock, N. E., Burt, T. P., Goulding, K. W. T. and Pinay, G. (Eds), 1997. *Buffer Zones: Their Processes and Potential in Water Protection*, Quest Environmental Publications: Harpenden, UK.

He, Q. and Walling, D. E., 1997. 'Spatial variability of the particle size composition of overbank floodplain deposits'. *Water, Air and Soil Pollution*, **99**, 71–80.

Hedin, L. O., Vonfischer, J. C., Ostrom, N. E., Kennedy, B. P., Brown, M. G. and Robertson, G. P., 1998. 'Thermodynamic constraints on nitrogen transformations and other biochemical processes at soil–stream interfaces'. *Ecology*, **79**, 684–703.

Hefting, M., Clément, J. C., Dowrick, D., Cosandey, A. C., Bernal, S., Cimpian, C., Tatur, A., Burt, T. P. and Pinay, G., 2004. 'Water table elevation controls on soil nitrogen cycling in riparian wetlands along a European climatic gradient'. *Biogeochemistry*, **67**, 113–134.

Hill, A. R. 1979. 'Denitrification in the nitrogen budget of a river ecosystem'. *Nature*, **281**, 291–292.

Jones, J. B. and Holmes, R. M., 1996. 'Surface–subsurface interactions in stream ecosystems'. *Trends in Ecology and Evolution*, **11**, 239–242.

Keeney, D. R., 1973. 'The nitrogen cycle in sediment-water systems'. *Journal of Environmental Quality*, **2**, 15–29.

Lowrance, R., Leonard, R. and Sheridan, J. M., 1985. 'Managing riparian ecosystems to control nonpoint pollution'. *Journal of Soil and Water Conservation*, **40**, 87–91.

Lowrance, R., Sharpe, K. and Sheridan, J. M., 1986. 'Long term sediment deposition in the riparian zone of a coastal plain watershed'. *Journal of Soil and Water Conservation*, July/August, 266–271.

Lowrance, R., Altier, L. S., Newbold, J. D., Schnabel, R. R., Groffman, P. M., Denver, J. M., Correll, D. L., Gilliam, J. W., Robinson, J. L., Brinsfield, R. B., Staver, K. W., Lucas, W. C. and Todd, A. H., 1995. *Water Quality Functions of Riparian Forest Buffer Systems in the Chesapeake Bay Watersheds*, EPA 903-R-95-004, US Environmental Protection Agency, Chesapeake Bay Program Report, US EPA Region 3 Chesapeake Bay Program Office, Annapolis, MA, USA.

Lowrance, R., Altier, L. S., Newbold, J. D., Schnabel, R. R., Groffman, P. M., Denver, J. M., Correll, D. L., Gilliam, J. W., Robinson, J. L., Brinsfield, R. B., Staver, K. W., Lucas, W. and Todd, A. H., 1997. 'Water quality functions of riparian forest buffers in Chesapeake Bay watersheds'. *Environmental Management*, **21**, 687–712.

Malanson, G. P., 1993. *Riparian Landscapes*, Cambridge University Press: Cambridge, UK.

Matchett, L. S., 1998. *Denitrification in Riparian Buffer Zones*, PhD Dissertation, University of Oxford: Oxford, UK.

Mathab, F. U. and Karim, Z., 1992. 'Population and agricultural land use: towards a sustainable food production system in Bangladesh'. *Ambio*, **21**, 50–55.

McClain, M. E., Richey, J. E., Brandes, J. A. and Pimentel, T. P., 1997. 'Dissolved organic matter and terrestrial–lotic linkages in the central Amazon basin of Brazil'. *Global Biogeochemical Cycles*, **11**, 295–311.

Minshall, G. W., Petersen, R. C., Cummins, K. W., Bott, T. L., Sedell, J. R., Cushing, C. E. and Vanotte, R. L., 1983. 'Interbiome comparison of stream ecosystem dynamics'. *Ecological Monographs*, **53**, 1–25.

Naiman, R. J., 1983. 'The annual pattern and spatial distribution of aquatic oxygen metabolism in boreal forest watersheds'. *Ecological Monographs*, **53**, 73–94.

Naiman, R. J. and Décamps, H. (Eds), 1990. *The Ecology and Management of Aquatic–Terrestrial Ecotones*, Man and the Biosphere Series, Vol. 4., UNESCO: Paris, France.

Neill, C., 1995. 'Seasonal flooding, nitrogen mineralization and nitrogen utilization in a prairie marsh'. *Biogeochemistry*, **30**, 171–189.

Newbold, J. D., Elwood, J. W., O'Neill, R. V. and Van Winkle, W., 1981. "Measuring nutrient spiralling in streams'. *Canadian Journal of Fisheries and Aquatic Sciences*, **38**, 860–863.

Osborne, L. L. and Wiley, M. J., 1988. 'Empirical relationships between land use cover and stream water quality in an agricultural watershed'. *Journal of Environmental Management*, **26**, 9–27.

Patrick, W. J., 1982. 'Nitrogen transformations in submerged soils'. *Agronomy Monographs*, **22**, 449–765.

Paul, E. A. and Clark, F. E., 1996. *Soil Microbiology and Biochemistry*, Academic Press: London, UK.

Peterjohn, W. T. and Correll, D. L., 1984. 'Nutrient dynamics in an agricultural watershed: Observations on the role of a riparian forest'. *Ecology*, **65**, 1466–1475.

Peterson, B. J., Wollheim, W. H., Mulholland, P. J., Webster, J. R., Meyer, J. L., Tank, J. L., Marti, E., Bowden, W. B., Valett, H. M., Hershey, A. E., McDowell, W. H., Dodds, W. K., Hamilton, S. K., Gregory, S. and Morrall, D. J., 2001. 'Control of nitrogen export from watersheds by headwater streams'. *Science*, **292**, 86–90.

Petts, G. E. and Maddock, I. 1996. 'Flow allocation for in-river needs'. In: *River Restoration*, Petts, G. E. and Calow, P. P. (Eds), Blackwell Sciences: London, UK, pp. 60–79.

Pinay, G. and Labroue, L. 1986. 'Une station d'Èpuration naturelle des nitrates transportÈs dans les nappes alluviales: l'aulnaie glutineuse'. *Comptes Rendus de l'Académie des Sciences de Paris*, série III, **302**, 629–632.

Pinay, G. and Naiman, R. J., 1991. 'Short term hydrologic variations and nitrogen dynamics in beaver created meadows'. *Archiv für Hydrobiologie*, **123**, 187–205.

Pinay, G., Décamps, H., Arles, C. and Lacassin-Seres, M., 1989. 'Topographic influence on carbon and nitrogen dynamics in riverine woods'. *Archiv für Hydrobiologie*, **114**, 401–414.

Pinay, G., Decamps, H., Chauvet, E. and Fustec, E., 1990. 'Functions of ecotones in fluvial systems'. In: *The Ecology and Management of Aquatic–Terrestrial Ecotones*, Man and the Biosphere Series, Vol. 4, Naiman, R. J. and Décamps, H. (Eds), UNESCO, Paris, France.

Pinay, G., Roques, L. and Fabre, A., 1993. 'Spatial and temporal patterns of denitrification in a riparian forest'. *Journal of Applied Ecology*, **30**, 581–591.

Pinay, G., Ruffinoni, C. and Fabre, A., 1995. 'Nitrogen cycling in two riparian forest soils under different geomorphic conditions'. *Biogeochemistry*, **4**, 1–21.

Pinay, G., Black, V. J., Planty-Tabacchi, A. M., Gumiero, B. and Décamps, H., 2000. 'Geomorphic control of denitrification in large river floodplain soils'. *Biogeochemistry*, **30**, 9–29.

Pinay, G., Clement, J. C. and Naiman, R. J., 2002. 'Basic principles and ecological consequences of changing water regimes on nitrogen cycling in flurial systems.' *Environmental Management*, **30**, 481–491.

Ponnamperuma, F. N., 1972. 'The chemistry of submerged soils'. *Advances in Agronomy*, **24**, 29–96.

Ranjard, L., Poly, F., Combrisson, J., Richaume, A., Gourbière, F., Thioulouse, J. and Nazaret, S., 2000. 'Heterogeneous cell density and genetic structure of bacterial polls associated with various soil microenvironments as determined by enumeration and DNA fingerprinting approach (RISA)'. *Microbial Ecology*, **39**, 263–272.

Reddy, K. R. and Patrick, W. H., 1975. 'Effect of alternate aerobic and anaerobic conditions on redox potential, organic matter decomposition and nitrogen loss in a flooded soil'. *Soil Biology and Biochemistry*, **7**, 87–94.

Richter, B. D. and Richter, H. E., 2000. 'Prescribing flood regimes to sustain riparian ecosystems along meandering rivers'. *Conservation Biology*, **1**, 1467–1478.

Roberts, J. and Ludwig, J. A., 1991. 'Riparian vegetation along current-exposure gradients in floodplain wetlands of the River Murray, Australia'. *Journal of Ecology*, **79**, 117–127.

Salo, J., Kalliola, R., Häkkinen, J., Mäkinen, Y., Niemelä, P., Puhakka, M. and Coley, P. B., 1986. 'River dynamics and the diversity of Amazon lowland forest'. *Nature*, **332**, 254–258.

Schlosser, I. J. and Karr, J. R., 1981. 'Riparian vegetation and channel morphology impact on spatial patterns of water quality in agricultural watersheds'. *Environmental Management*, **5**, 233–243.

Schumm, S. A., 1977. *The Fluvial System*, Wiley: New York, NY, USA.

Stepanauskas, R., Laudon, H. and Jorgensen, N. O. G., 2000. 'High DON bioavailability in boreal streams during a spring flood'. *Limnology and Oceanography*, **45**, 1298–1307.

Tabacchi, E., Correll, D. L., Hauer, R., Pinay, G., Planty-Tabacchi, A. M. and Wissmar, R. C. 1998. 'Development, maintenance and role of riparian vegetation in the river landscape'. *Freshwater Biology*, **40**, 497–516.

Tansley, A. G., 1911. *Types of British Vegetation*, Cambridge University Press: Cambridge, UK.

Tol, R. S. J. and Langen, A., 2000. 'A concise history of Dutch river floods'. *Climate Change*, **46**, 357–369.

Tufford, D. L., McKellar, Jr. H. N. and Hussey, J. R. 1998. 'In-stream non point source nutrient prediction with land-use proximity and seasonality'. *Journal of Environmental Quality*, **27**, 100–111.

Valett, H. M., Morrice, J. A., Dahm, C. N. and Campana, M. E. 1996. 'Parent lithology, surface–groundwater exchange and nitrate retention in headwater streams'. *Limnology and Oceanography*, **41**, 333–345.

Vitousek, P. M., Aber, J. D., Howarth, R. W., Likens, G. E., Matson, P. A., Schindler, D. W., Schlesinger, W. H. and Tilman, D. G., 1997. 'Human alteration of the global nitrogen cycle: sources and consequences'. *Ecological Applications*, **7**, 737–750.

Ward, J. V., 1989. 'The four-dimensional nature of lotic ecosystems'. *Journal of the North American Benthological Society*, **8**, 2–8.

Worrall, F. and Burt, T. P., 1998. 'Decomposition of river nitrate time-series. Comparing agricultural and urban signals'. *The Science of the Total Environment*, **210/211**, 153–162.

1.2

Instream and Bankside Habitat in Rivers

Patrick D. Armitage

1.2.1 INTRODUCTION

The main concerns of water managers in the past were water supply, pollution, power generation and flood control. However, with increasing knowledge of the complexities of the environment and cause and effect relationships between catchment disturbances and the condition of the river, a more holistic ecological view of river management has emerged. This has resulted in truly interdisciplinary and collaborative studies, particularly in the field of flow regulation, where since 1979 there has been a triennial series of symposia dedicated to the amalgamation of

Biological Monitoring of Rivers   Edited by G. Ziglio, M. Siligardi and G. Flaim
© 2006 John Wiley & Sons, Ltd.

science with management issues (Lane, 2001). In addition, several publications have appeared specifically addressing the application of scientific knowledge to environmental management of rivers (Gore and Petts, 1989; Boon *et al.*, 1992; Calow and Petts, 1992a,b; Rosenberg and Resh, 1993; Harper and Ferguson, 1995; Petts and Amoros, 1996). A theme common to most of this work is the recognition that impacts and disturbances affect not only the water quality but also the condition and distribution of riverine habitat.

Management interest in habitat has arisen from a number of separate sources. First, the relation between discharge and wetted usable habitat is fundamental to the setting of *ecologically acceptable flows* and this relationship has been explored in the development of the Instream Flow Incremental Methodology (IFIM) (Bovee, 1995). This system has focused attention on the relationship between stream hydrology and hydraulics and ecology and given rise to a new area of collaborative studies 'ecohydraulics' (see Petts, 2003). Secondly, the restoration and rehabilitation of river sections for recreational, environmental and aesthetic purposes (Brookes and Shields, 1996; Boon and Raven, 1998; Lane, 2001) requires knowledge of the relationship between channel geomorphology and habitat distribution. Thirdly, the United Nations Convention on Biological Diversity demands a commitment to the conservation of biodiversity, a goal which requires detailed information on the habitat requirements of a wide range of biotic communities. This, in conjunction with EEC regulations, particularly the Habitats (92/43/EEC) and Water Framework (2000/60/EC) Directives, has focused interest on the role of habitat in resource management (Logan and Furse, 2002).

This chapter will focus on the importance of habitat in determining river biocoenoses with particular reference to macroinvertebrate communities. The specific objectives are as follows:

- To describe instream and bankside habitat in relation to the hierarchical structure of rivers.

- Provide examples of how this may be altered or impacted by natural and anthropogenic means and briefly consider current habitat assessment methodologies and outline future developments and needs.

1.2.2   HABITAT HIERARCHIES

The hierarchical organization of rivers has long been recognized by geomorphologists and several classification systems have been developed. For the present purposes, the system devised by Frisell *et al.* (1986) forms a convenient framework for a discussion of ecological habitat.

Lotic systems incorporate a number of levels that are nested at successively smaller spatio–temporal scales. These range from the entire catchment, with a high temporal stability to the microhabitat (stick and leaf, detritus, sand over cobbles, etc.), that may last as little as one month or even less. When this conceptual approach is blended with 'River Continuum' theory (Vannote *et al.*, 1980), which stresses the role of the downstream gradient of physical factors as a control on biological strategies and dynamics of river systems, we have a broad holistic view of river structure and function. Further developments incorporating anthropogenic influences and a historical perspective have led to the view that rivers ('fluvial hydrosystems'), (Amoros *et al.*, 1987, Petts and Amoros, 1996) are four-dimensional systems which are dependent on longitudinal, lateral and vertical transfers of energy, material and biota over a range of time-scales (Ward, 1989).

It is useful to simplify the complex array of environmental variables that determine the occurrence and distribution of river biota, by classifying sections of rivers into major types or *functional sectors* (Amoros *et al.*, 1987). Figure 1.2.1 presents a

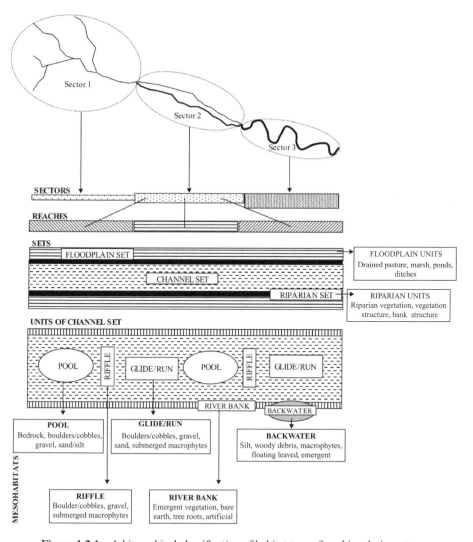

Figure 1.2.1 A hierarchical classification of habitat types found in a lotic system

schematic showing habitat hierarchies partially adapted from the 'Fluvial Hydrosys-tem' approach with the main emphasis on the instream component. *Sectors* may be defined as lengths of river within which there is a uniformity of major physi-cal characteristics such as geology and slope. Sector boundaries are usually related to a major change in the flow regime, water chemistry and/or channel form. They usually show predictable changes with increasing distance from source, from small upland relatively unmanaged sectors to lowland regions that may be intensively man-aged. Each sector type is composed of a series of *reaches* which are lengths of river within which there is a uniformity of local habitat features. Reach boundaries may relate to localized features such as weirs, which create ponded reaches upstream and faster-flowing gravel reaches downstream.

Alternatively the boundaries may be associated with local changes in land use from open pasture to woodland, or local changes in channel management (dredg-ing or channel straightening) (Petts and Bickerton, 1997). Reaches themselves are made up of *sets* of habitats associated with particular flow velocity and substratum combinations and bounded by bankside and floodplain habitats. This classification provides a useful framework to view habitat relationships with each other and with their position along the stream. In addition, when information is available from all components, this approach provides a comprehensive description of the water system which allows the integration of basic and applied knowledge in a manner suitable for use by water managers.

1.2.3   HABITAT TYPES

Figure 1.2.1 identifies a range of habitats found throughout a lotic system. Some large mobile mammals and fish may cover a wide range of habitat types. Invertebrates, in contrast, although widely dispersed in running water, have a more focused habitat range which is best considered at the 'mesoscale', intermediate between microhabitat (leaf tips, individual stones, etc.) and reach level.

There are many accounts of the types of habitat to be found in rivers (see Harper *et al.* (1995) for a review) and their occurrence and relative proportions at a site will reflect both sector and reach characteristics which are themselves largely controlled by catchment hydrology. A generalized and condensed list based on the UK River Habitat Survey Methodology (Raven *et al.*, 1997) is presented below.

Instream mineral habitats:
Bedrock, boulders, cobbles, pebbles, gravel, sand, silt, clay, hyporheic sediments
Instream vegetation habitats:
Macrophytes
        Emergent and marginal (reeds, rushes, grasses and sedges)
        Emergent broad-leaved (*Apium*, *Rorippa* spp.)
        Submerged broad-leaved (*Elodea* spp., *Callitriche* spp.)
        Submerged fine-leaved (*Ranunculus* spp., *Myriophyllum* spp.

Floating-leaved and rooted (*Nuphar lutea*)
Free-floating (*Lemna* spp.)
Submerged mosses/liverworts/lichens
Filamentous algae
Other categories
Tree roots
Aggregations of coarse particulate organic matter (detritus)
Submerged wood
***Riparian and floodplain habitats (not covered by the above categories)*:**
Bare earth banks
Vegetated river bank
Cattle drinks
Backwaters (connected and isolated) and side channels
Ditches
Artificially modified (iron pilings, concrete, rip rap, etc.)

The river ecosystem is composed of combinations of these basic units which may themselves be modified by prevailing flow conditions Superimposed on this are the temporal dynamics of habitat change, such as siltation of coarse substrata during seasonal low-flow periods and the growth and senescence of vegetative substrata. In addition, the habitat requirements of instream fauna may change with time. Examples are seen in the pupation of riffle beetles (Elmidae) in the wet banks of streams (Holland, 1972) and the temporal shifts in the physical habitat of the crayfish, *Orconectes neglectus* (Faxon), throughout its development (Gore and Bryant, 1990). Therefore, although different types of habitat are relatively easy to recognize it is important to be aware that the communities of biota they support are dynamic entities responding to seasonal changes in discharge and hydraulics, vegetation growth and intrinsic biotic interactions. These in turn, are subject to the modifying effects of natural and anthropogenic disturbances in the catchment area and in the channel itself.

1.2.4 HABITAT/FAUNA ASSOCIATIONS

1.2.4.1 Choice of scale

The association of faunal communities with habitat patches has been a common area for study by freshwater ecologists but it is only relatively recently that emphasis has been placed on habitat availability – a feature central to the IFIM methodology and of major interest to environmental management. Habitat availability has referred in the main to habitat for single species (usually fish) but there is merit in extending this concept to habitat of faunal assemblages. By placing emphasis on the proportions of habitat and the faunal communities which live in them, we have a tool that can be applied in assessment of biological quality, conservation, and classification.

The scale of study chosen for ecological investigations will have a major influence on the results obtained and their interpretation (Armitage and Cannan, 1998). In river systems, the importance of scale in research studies has been addressed with respect to classification systems (Naiman *et al.*, 1992) and 'pattern and process' (Cooper *et al.*, 1998). It is recognized that at very large scales, assemblages of lotic biota are predictable and persistent in the absence of any catastrophic natural or anthropogenic disturbance. Observation of faunal assemblages at a microscale will in contrast be less predictable and more susceptible to environmental change. An intermediate scale of study is required which will provide reliable data that can be used in management decisions.

1.2.4.2 Reach level and mesoscale

Fish biologists have placed importance on investigations of available habitat at reach level (Plafkin *et al.*, 1989; Kershner *et al.*, 1992; Parasiewicz, 2001) but macroin-vertebrate studies have in general not followed this research route, due mainly to the high cost of data processing and different research needs (assessment of water quality). There is little in the published literature on the application of macroinvertebrate studies at reach level, except in broad longitudinal zonation studies (Hawkins, 1984; Statzner and Higler, 1986; Palmer *et al.*, 1991). A description of a river in terms of the distribution of habitats along it and their faunal associations has, however, been central to studies on the Rhône (Amoros *et al.*, 1987; Statzner *et al.*, 1994; Petts and Amoros, 1996).

Research efforts into macroinvertebrate/habitat associations at the reach scale in other countries have been less intensive and floodplain issues and historical aspects of change have not been much studied. Instead, the stimulus for ecological instream habitat investigations with application to management has come mainly from flow regulation, conservation and rehabilitation issues.

Studies of flow-related impacts on rivers, such as impoundment, abstraction and natural low flows, have shown that it is habitat which is being affected by these disturbances (Armitage, 1984; Armitage and Petts, 1992; Gore, 1994). Information on the faunal assemblages of specific habitats was required because flow modifications affected the proportion and distribution of these habitats and hence the faunal characteristics of the river.

In one series of studies (Armitage *et al.*, 1995; Armitage and Pardo, 1995), visually distinct areas of river bottom (weed beds intermixed with patches of gravel, sand and silt) were identified from the bank. These were referred to as mesohabitats to introduce a scalar dimension, which the term biotope does not have, and to distinguish them from microhabitats such as a leaf tip or stone surface. A mesohabitat may be defined as **'a medium-scale habitat that arises through the interactions of hydrological and geomorphological forces which may include instream macrophyte growth'**. This choice of habitat unit was deliberately anthropocentric because changes in mesohabitat or their proportion are easily recorded. The question was asked – do mesohabitats support specific faunal communities that remain distinct throughout the seasons? If so, information on their distribution in relation to hydrological/hydraulic variables will provide a way of assessing disturbances that affect the river bottom. This study and its application to a regulated reach and a channelized stream (Armitage and Pardo, 1995; Tickner *et al.*, 2000) have shown that the mesohabitat approach can be used to assess disturbance in cases where most standard biotic indices would be unsuitable.

Similar work at this scale, where the habitat units are called 'functional habitats', has been carried out to provide ecological guidance to those whose work affected the river environment (Smith *et al.*, 1991). The work has now developed into a methodology for providing advice on conservation of river habitat and may also provide a means of assessing river restoration works and have application in both riparian zones and the floodplain (Harper *et al.*, 1995).

Despite the varied nomenclature, mesohabitats (introducing a scale aspect), functional habitats and biotopes, are all essentially the same thing where river management is concerned. That is, areas of the river bottom or river system which are usually visually distinct. They may be defined by resident hydraulic conditions, macrophyte growth, or deposition of woody debris or combinations of these, and may support specific biotic communities. Work at this scale (Newson and Newson, 2000; Beisel *et al.*, 2000; Buffagni *et al.*, 2000; Brunke *et al.*, 2002; Usseglio-Polatera and Beisel, 2002) has provided information appropriate to the management of rivers and as part of a nested set of habitats (Figure 1.2.1) (Armitage and Cannan, 1998), can be applied throughout the catchment.

1.2.4.3   Spatial and temporal aspects

Habitats exist in time as well as space and it is therefore important that data on habitat characteristics are gathered throughout the year, or at least seasonally. The species assemblages of mesohabitats in a river in southern England (Pardo and Armitage, 1997) were observed in spring, summer and autumn. Macrophyte mesohabitats showed seasonal variations and were not discrete units throughout the year. '*Ranunculus*' constituted a distinct habitat in spring but as discharge decreased through summer and autumn this mesohabitat became similar faunistically to two marginal macrophyte habitats, '*Nasturtium*' and '*Phragmites*'. In contrast, 'gravel' mesohabitats showed a more persistent species structure although this also showed some seasonal variation. It was concluded that species assemblages could only be descriptors of mesohabitats if the seasonal distinctiveness is taken into account.

Seasonality is of major importance when considering **bankside habitat** (Armitage *et al.*, 2001). The interactions between plants and their effects on flow hydraulics, and the associated fauna of invertebrates, fish, birds and mammals, are most highly developed in bankside habitat. In certain rivers, the bankside may contribute most of the faunal diversity (Cogerino *et al.*, 1995). Production and organic matter retention may be several times higher in the littoral than in mid-channel (Chauvet and

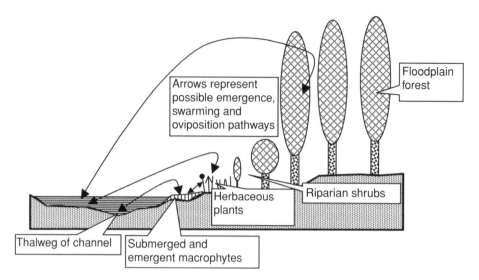

Figure 1.2.2   Bankside habitats and their relationship to the channel and floodplain

Jean-Louis, 1988) and bankside vegetation will furnish cover for both adult and juvenile fish (Schiemer and Zalewski, 1992). Bankside habitat will also provide refugia for macroinvertebrates and fish during periods of high flow (Cogerino *et al.*, 1995). In addition, the bankside habitat provides an important link between instream fauna and the terrestrial environment (Figure 1.2.2). Species migrate from the main channel to marginal areas to complete their life-histories (Holland, 1972; Gore and Bryant, 1990) and use marginal vegetation for emergence and oviposition and the riparian vegetation as refugia for adults and as swarm markers (Gibbons and Pain, 1992; Armitage, 1995).

Generally speaking, this physically dynamic zone, which lies between the riparian strip and instream habitats proper, has been neglected, particularly in regard to management issues. Comprehensive ecological study of macroinvertebrate assemblages has been concentrated on the Rhône (Cogerino *et al.*, 1995). Armitage *et al.* (2001) found that bank profile and structure were major determinants of macroinvertebrate assemblages. Total abundances were five to six times greater in shallow-sloping vegetated sites compared with steeply sloped and artificial banks. Highest abundances were found in spring but the greatest biodiversity was observed in August and September, both findings having particular relevance to management issues.

1.2.5   FACTORS AFFECTING STREAM HABITAT

Any disturbances to stream structure and/or function will affect the habitat available to riverine biota. The natural annual hydrological cycle will result in seasonal modifications to depth, velocity, bed hydraulics and channel morphometry to which the

Table 1.2.1 Major disturbances and their primary effects on stream habitat

Disturbance	Impact	Primary effect
Flow regulation	Increased flow	Substratum instability
	Reduced flow	Loss of wetted area, sedimentation
	Constant flow	Substratum stability, growth of macrophytes, loss of floodplain habitat
	Variable flow	Substratum instability
	Flood relief channel	Reduction of flushing flows in main channel, loss of floodplain habitat
Engineering	Bridge construction	Substratum disturbance
	Channelization	Bank structure, substratum, floodplain disturbance, loss of riparian habitat
	Dredging	Substratum disturbance, sedimentation
	Dam construction	Altered flow regime, sedimentation
	Run-of-river diversions	Altered flow regime, substratum modifications
Water supply and disposal	Abstraction	Reduced wetted area, sedimentation
	Transfer	Flow patterns, water chemistry
	Pollution from effluents	Increased nutrient load, algal growth, subtratum hydraulics
Catchment activities	Agriculture	Nutrients, pesticides, sedimentation, loss of riparian habitat
	Flood relief / drainage	Loss of riparian and wetland habitats
	Land clearance	Sedimentation, riparian modification
	Forestry	Sedimentation
	Quarrying and mining	Sedimentation
	Construction work	Sedimentation
	Urbanization and industrial	Channelization, altered hydrology, pollution
	Catastrophic events (vulcanism, earthquakes)	Total environmental disturbance

biotic communities are adapted. Extreme natural events such as floods and droughts, together with anthropogenic disturbances, will further modify habitat conditions.

Major disturbances and their **primary effects** on stream habitat are listed in Table 1.2.1. These disturbances will induce a variety of responses, depending on their intensity, the river type in question and whether they are acting alone or with other impacting agents. Most disturbances have some effect, either direct or indirect, on riverine habitat and many of the resultant changes are deleterious to primary producers, macroinvertebrates and fish populations (see Wood and Armitage (1997) for a review of these impacts)

1.2.6 HABITAT ASSESSMENT METHODS

It is clear that most stresses/disturbances on the lotic system involve changes, either directly or indirectly, to the quantity and quality of instream and marginal habitat.

Habitat assessment and classification methodologies have arisen from the need to protect and conserve the environment by monitoring the effects of disturbances and predicting their possible impacts. These methods range from basic descriptions of the environment incorporating instream and riparian features (River Habitat Survey (Raven *et al.*, 1997); SERCON – System for Evaluating Rivers for Conservation (Boon *et al.*, 1997)) to those which relate habitat availability or characteristics to the composition and distribution of biotic communities (Rapid Bioassessment Protocols (Plafkin *et al.*, 1989); HABSCORE (Milner *et al.*, 1985); RIVPACS (Wright *et al.*, 1993)). Lastly, there are systems such as the Instream Flow Incremental Methodology (IFIM) and Physical Habitat Simulation Modeling (PHABSIM) (Bovee, 1995) which attempt to predict habitat availability in relation to flow conditions.

These systems can be used to classify and assess the quality and quantity of river habitats (Boon and Raven, 1998), provide tools for maintaining a sustainable management regime for rivers (Raven and Boon, 2002) and present a means of assessing the ecological integrity of running waters (Jungwirth *et al.*, 2000). However, it is important when evaluating environmental quality that habitat modification is considered alongside water quality. At present, it is quite possible to record top water quality conditions in a river whose habitat has been severely modified.

1.2.7   FUTURE DIRECTIONS

The overwhelming importance of habitat in determining river biocoenoses is clear. A wealth of data are now available to demonstrate the need to maintain a diversity of habitat in the face of increasing pressures arising from catchment disturbance and the demand for water for domestic and industrial purposes. What remains to be investigated are the mechanisms that give rise to habitat and the particular features of that habitat to which the benthic communities respond. Recent symposia have addressed the need to maintain 'environmental flows' (Petts, 2003) and it is the relationship between these flows, available habitat and the communities they support that requires more study.

In order to improve the use of the mesohabitat/habitat unit approach, it is important that habitat occurrence is predictable. That is to say, the physical conditions that give rise to a specific mesohabitat need to be defined in a range of channel and river types. A number of authors have related the occurrence and distribution of pools runs and riffles to geomorphological and hydrological features (Jowett, 1993; Cohen *et al.*, 1998; Rowntree and Wadeson, 1996) and near-bed hydraulics have been used to develop a model (CASIMIR) which can be used to simulate the impact of different minimum flow regulations (Jorde and Bratrich, 1998). Further research of this type, together with approaches for linking habitats, flow types and species requirements (Newson *et al.*, 1998), are necessary requirements to understand fully community/habitat interactions and facilitate river management.

In recent years, there has been an increasing interest in habitat mapping and modeling, mainly in relation to environmental flow assessment (Tharme, 2003). Studies may be divided into three basic types – those mainly concerned with mapping habitat over small and large scales, those whose prime interest is in the hydraulics of habitat types, and those whose main aim is to model habitat response to change (see Hardy (1998) for a review of this subject area). The rapid advances in mapping technology and the quantification of riverine habitats using airborne multispectral imagery (Whited *et al.*, 2002) now provide a means of describing river habitat along its entire length and in relation to landscape features in both large and small water courses.

The ability to predict changes in the relative proportions of habitat units in a reach is central to the further development of successful tools for environmental management because such changes will affect the balance of faunal communities and provide a comprehensive assessment of environmental disturbance (Armitage and Cannan, 2000). Mapping and modelling approaches, in conjunction with habitat-specific faunal studies, will eventually provide the information needed to predict the occurrence and distribution of habitat units and assess comprehensively ecological conditions in the whole river system.

Much current research tends to be fragmented, whereas management requires cohesive long-term data which will build towards comprehensive data bases (Armitage *et al.*, 2001). It is hoped that funding organizations will encourage the development of studies which will help build knowledge of habitat/biota interactions in a systematic way.

ACKNOWLEDGEMENTS

I am grateful to my colleagues at the Centre for Ecology and Hydrology for providing information and useful discussions, and to Marty Gurtz (US Geological Survey) and the Environment Agency (UK) for supplying relevant reports. This work was partly funded by the Natural Environment Research Council.

REFERENCES

Amoros, C., Roux, A. L., Reygrobellet, J. L., Bravard, J. P. and Pautou, G., 1987. 'A method for applied ecological studies of fluvial hydrosystems'. *Regulated Rivers: Research and Management*, **1**, 17–36.

Armitage, P. D., 1984. 'Environmental changes induced by stream regulation and their effect on lotic macroinvertebrate communities'. In: *Regulated Rivers – Proceedings of the 2nd International Symposium on Regulated Stream Limnology*, Lillehammer, A. and Saltveit, S. J. (Eds). Norwegian University Press: Oslo, Norway, pp. 139–164.

Armitage, P. D., 1995. 'Behaviour and ecology of adults'. In: *The Chironomidae – Biology and Ecology of Non-Biting Midges*, Armitage, P. D., Cranston, P. S. and Pinder, L. C. V. (Eds). Chapman & Hall: London, pp. 194–224.

Armitage, P. D. and Cannan, C. E., 1998. 'Nested multi-scale surveys in lotic systems – tools for management'. In: *Advances In River Bottom Ecology*, Bretschko, G. and Helesic J. (Eds). Backhuys Publishers: Leiden, The Netherlands, pp. 293–314.

Armitage, P. D. and Cannan, C. E., 2000. 'Annual changes in summer patterns of mesohabitat distribution and associated macroinvertebrate assemblages'. *Hydrological Processes*, **14**, 3161–3179.

Armitage, P. D. and Pardo, I., 1995. 'Impact assessment of regulation at the reach level using mesohabitat information'. *Regulated Rivers: Research and Management*, **10**, 147–158.

Armitage, P. D. and Petts, G. E., 1992. 'Biotic score and prediction to assess the effects of water abstractions on river macroinvertebrates for conservation purposes'. *Aquatic Conservation: Marine and Freshwater Ecosystems*, **2**, 1–17.

Armitage, P. D., Pardo, I. and Brown, A., 1995. 'Temporal constancy of faunal assemblages in mesohabitats. Application to management'. *Archiv für Hydrobiologie*, **133**, 367–387.

Armitage, P. D., Lattmann, K., Kneebone, N. and Harris, I., 2001. 'Bank profile and structure as determinants of macroinvertebrate assemblages – seasonal changes and management'. *Regulated Rivers: Research and Management*, **17**, 543–556.

Beisel, J. N., Usseglio-Polatera, P. and Moreteau, J. C., 2000. 'The spatial heterogeneity of a river bottom: a key factor determining macroinvertebrate communities'. *Hydrobiologia*' **422**, 163–171.

Boon, P. J., and Raven, P. J. (Eds), 1998. The Application of Classification and Assessment Methods to River Management in the UK, Special Issue, *Aquatic Conservation: Marine and Freshwater Systems*, **8**, 383–644.

Boon, P. J., Petts, G. and Calow, P. (Eds), 1992. *River Conservation and Management*. Wiley: Chichester, UK.

Boon, P. J., Holmes, N. T. H., Maitland, P. S., Rowell, T. A. and Davies, J., 1997. 'A system for evaluating rivers for conservation (SERCON): development, structure and function', In: *Freshwater Quality: Defining the Indefinable*, Boon, P. J. and Howell, D. L. (Eds). HNSO: Edinburgh, UK, pp. 299–326.

Bovee, K. D. (Ed.), 1995. *A Comprehensive Overview of the Instream Flow Incremental Methodology*. National Biological Service: Fort Collins, CO, USA.

Brookes, A. and Shields, Jr, F. D. (Eds), 1996. *River Channel Restoration: Guiding Principles for Sustainable Projects*. Wiley: Chichester, UK.

Brunke, M., Hoffmann, A. and Pusch, M., 2002. 'Association between invertebrate assemblages and mesohabitats in a lowland river (Spree, Germany): A chance for predictions?'. *Archiv Für Hydrobiologie*, **154**, 239–259.

Buffagni, A., Crosa, G. A., Harper, D. M. and Kemp, J. L., 2000. 'Using macroinvertebrate species assemblages to identify river channel habitat units: an application of the functional habitats concept to a large, unpolluted Italian river (River Ticino, northern Italy)'. *Hydrobiologia*, **435**, 213–225.

Calow, P. and Petts, G. E. (Eds), 1992a. *The Rivers Handbook*, Vol. 1. Blackwell Scientific: Oxford, UK.

Calow, P. and Petts, G. E. (Eds), 1992b.*The Rivers Handbook*, Vol. 2. Blackwell Scientific: Oxford, UK.

Chauvet, E. and Jean-Louis, A. M., 1988. 'Production de litière de la ripisylve de la Garonne et apport au fleuve'. *Acta Oecologica, Oecologia Generalis*, **9**, 265–279.

Cogerino, L., Cellot, B. and Bournaud, M., 1995. 'Microhabitat diversity and associated macroinvertebrates in aquatic banks of a large European river'. *Hydrobiologia*, **304**, 103–115.

Cohen, P., Andriamahefa, H. and Wasson, J.-G., 1998. 'Towards a regionalization of aquatic habitat: distribution of mesohabitats at the scale of a large basin'. *Regulated Rivers: Research & Management*, **14**, 391–404.

Cooper, S. D., Diehl, S., Kratz, K. and Sarnelle, O., 1998. 'Implications of scale for patterns and processes in stream ecology'. *Australian Journal of Ecology*, **23**, 27–40.

Frisell, C. A., Liss, W. J., Warren, C. E. and Hurtley, M. D., 1986. 'A hierarchical framework for stream classification: viewing streams in a watershed context'. *Environmental Management*, **10**, 199–214.

Gibbons, D. W. and Pain, D., 1992. 'The influence of river flow rate on the breeding behaviour of *Calopteryx* damselflies'. *Journal of Animal Ecology*, **61**, 283–289.

Gore, J. A., 1994. 'Hydrological change'. In: *The Rivers Handbook*, Vol. 2, Calow, P. and Petts, G. E. (Eds). Blackwell Scientific Publications: Oxford, UK, pp. 33–54.

Gore, J. A. and Bryant, Jr, R. M., 1990. 'Temporal shifts in physical habitat of the crayfish, *Oronectes neglectus* (Faxon)'. *Hydrobiologia*, **199**, 131–142.

Gore, J. A. and Petts, G. E. (Eds), 1989. *Alternatives in Regulated River Management*. CRC Press, Boca Raton, FL, USA.

Hardy, T. B., 1998. The future of habitat modeling and instream flow assessment techniques. *Regulated Rivers: Research & Management*, **14**, 405–420.

Harper, D. and Ferguson, A. (Eds), 1995. *The Ecological Basis For River Management*. Wiley: Chichester, UK.

Harper, D., Smith, C., Barham, P. and Howell, R., 1995. 'The ecological basis for the management of the natural river environment'. In: *The Ecological Basis for River Management*, Harper, D. and Ferguson, A. (Eds). Wiley: Chichester, UK, pp. 59–78.

Hawkins, C. P., 1984. 'Substrate associations and longitudinal distributions in species of Ephemerellidae (Ephemeroptera: Insecta) from Western Oregon'. *Freshwater Invertebrate Biology*, **3**, 181–188.

Holland, D. G., 1972. *A Key to the Larvae, Pupae and Adults of the British Species of Elminthidae*, Scientific Publication No. 26. Freshwater Biological Association: Ambleside, pp. 1–58.

Jorde, K. and Bratrich, C., 1998. 'River bed morphology and flow regulations in diverted streams: effects on bottom shear stress patterns and hydraulic habitat'. In: *Advances In River Bottom Ecology*, Bretschko, G. and Helesic, J. (Eds). Backhuys Publishers: Leiden, The Netherlands, pp. 47–63.

Jowett, I. G., 1993. 'A method for objectively identifying pool, run, and riffle habitats from physical measurements'. *New Zealand Journal of Marine and Freshwater Research*, **27**, 241–248.

Jungwirth, M., Muhar, S. and Schmutz, S. (Eds), 2000. Assessing the Ecological Integrity of Running Waters, Special Issue, *Hydrobiologia*, **442/443**, 1–487.

Kershner, J. L., Snider, W. M., Turner, D. M. and Moyle, P. B., 1992. 'Distribution and sequencing of mesohabitats: are there differences at the reach scale?'. *Rivers*, **3**, 179–190.

Lane, S., 2001. 'Foreword: Eighth International Symposium on Regulated Streams'. *Regulated Rivers: Research and Management*, **17**, 301.

Logan, P. and Furse, M. T., 2002. 'Preparing for the European Water Framework Directive – making the links between habitat and aquatic biota'. *Aquatic Conservation: Marine and Freshwater Ecosystems*, **12**, 425–437.

Milner, N. J., Hemsworth, R. J. and Jones, B. E., 1985. 'Habitat evaluation as a fisheries management tool'. *Journal of Fish Biology*, Supplement A, **27**, 85–108.

Naiman, R. J., Lonzarich, D. G., Beechie, T. J. and Ralph, S. C., 1992. 'General principles of classification and the assessment of conservation potential in rivers'. In: *River Conservation and Management*, Boon, P. J., Calow, P. and Petts, G. E. (Eds). Wiley: Chichester, UK, pp. 93–123.

Newson, M. D. and Newson, C. L., 2000. 'Geomorphology, Ecology and River Channel Habitat: Mesoscale Approaches To Basin-Scale Challenges'. *Progress in Physical Geography*, **24**, 195–217.

Newson, M. D., Harper, D. M., Padmore, C. L., Kemp, J. L. and Vogel, B., 1998. 'A Cost-Effective Approach For Linking Habitats, Flow Types And Species Requirements'. *Aquatic Conservation: Marine and Freshwater Ecosystems*, **8**, 431–446.

Palmer, C. G., O'Keeffe, J. H. and Palmer, A. R., 1991. 'Are macroinvertebrate assemblages in the Buffalo River, southern Africa, associated with particular biotopes?'. *Journal of the North American Benthological Society*, **10**, 349–357.

Parasiewicz, P., 2001. 'MesoHABSIM: A concept for application of instream flow models in river restoration planning'. *Fisheries*, **26**(9), 6–13.

Pardo, I. and Armitage, P. D., 1997. 'Species assemblages as descriptors of mesohabitats'. *Hydrobiologia*, **344**, 111–128.

Petts, G. E., 2003. Editorial, Environmental Flows for River Systems, Special Issue, *River Research and Applications*, **19**.

Petts, G. E. and Amoros, C. (Eds), 1996. *Fluvial Hydrosystems*. Chapman & Hall: London, UK.

Petts, G. E. and Bickerton, M. A., 1997. *River Wissey Investigations: Linking Hydrology and Ecology*, Manual for using macroinvertebrates to assess in-river needs, Report to the Environment Agency, Anglian Region , OI\526\3\A., Environment Agency, Bristol, UK, pp. 43.

Plafkin, J. L., Barbour, M. T., Porter, K. D., Gross, S. K. and Hughes, R. M., 1989. *Rapid Bioassessment Protocols for Use in Streams and Rivers: Benthic Macroinvertebrates and Fish*, EPA/444/4-89–001. Office of Water, USEPA: Washington, DC, USA.

Raven, P. J. and Boon, P. J. (Eds), 2002. 'Sustainable river basin management in the UK: Needs and opportunities'. Special Issue, *Aquatic Conservation: Marine and Freshwater Ecosystems*, **12**, 327–483.

Raven, P. J., Fox, P., Everard, M., Holmes, N. T. H. and Dawson, F. H., 1997. 'River Habitat Survey: a new system for classifying rivers according to their habitat quality'. In: *Freshwater Quality: Defining the indefinable*, Boon, P. J. and Howell, D. L. (Eds). HMSO, Edinburgh, UK, pp. 215–234.

Rosenberg, D. M. and Resh, V. H. (Eds), 1993. *Freshwater Biomonitoring and Benthic Macroinvertebrates*. Chapman & Hall: New York, NY, USA.

Rowntree, K. M. and Wadeson, R. A., 1996. 'Translating channel morphology into hydraulic habitat: application of the hydraulic biotope concept to an assessment of discharge related habitat changes'. In: *Ecohydraulics 2000*, Leclerc, M., Capra, H., Valentin, S., Boudreault, A. and Côté, Y. (Eds), Proceedings of the second IAHR Symposium on Habitats' Hydraulics, International & Association of Hydraulics Research, Quebec, 281–292.

Schiemer, F. and Zalewski, M., 1992. 'The importance of riparian ecotones for diversity and productivity of riverine fish communities'. *Netherlands Journal of Zoology*, **42**, 323–335.

Smith, C. D., Harper, D. M. and Barham, P. J., 1991. *Physical Environment for River Invertebrate Communities*, Project Report, National Rivers Authority (Anglian Region) Operational Investigation **A13–38A**, National Rivers Authority, Peterborough: 1–101.

Statzner, B. and Higler, B., 1986. 'Stream hydraulics as a major determinant of benthic invertebrate zonation patterns'. *Freshwater Biology*, **16**, 127–139.

Statzner, B., Resh, V. H. and Dolédec, S. (Eds), 1994. 'Ecology of the Upper Rhône River: a test of habitat templet theories', *Freshwater Biology*, **31**, 253–556.

Tharme, R. E., 2003. 'A global perspective on environmental flow assessment: emerging trends in the development and application of environmental flow methodologies for rivers'. *River Research and Applications*, **19**, 397–441.

Tickner, D. P., Armitage, P. D. Bickerton, M. A. and Hall, K. A., 2000. 'Assessing stream quality using information on mesohabitat distribution and character'. *Aquatic Conservation: Marine and Freshwater Ecosystems*, **10**, 179–196.

Usseglio-Polatera, P. and Beisel, J. N., 2002. 'Longitudinal changes in macroinvertebrate assemblages in the Meuse River: Anthropogenic effects versus natural change'. *River Research and Applications*, **18**, 197–211.

Vannote, R. L., Minshall, G. W., Cummins, K. W., Sedell, J. R. and Cushing, C. E., 1980. 'The river continuum concept'. *Canadian Journal of Fisheries and Aquatic Sciences*, **37**, 130–137.

Ward, J. V., 1989. 'The four dimensional nature of lotic ecosystems'. *Journal of the North American Benthological Society*, **8**, 2–8.

Whited, D., Stanford, J. A. and Kimball, J. S., 2002. 'Application of airborne multispectral digital imagery to quantify riverine habitats at different base flows', *River Research and Applications*, **18**, 583–594.

Wood, P. J. and Armitage, P. D., 1997. 'Silt and siltation in the lotic environment'. *Environmental Management*, **21**, 203–217.

Wright, J. F., Furse, M. T. and Armitage, P. D., 1993. 'RIVPACS – a technique for evaluating the biological quality of rivers in the UK'.*European Water Pollution Control*, **3**(4), 15–25.

1.3

Main Features of Watercourses' Hydrodynamics

Paolo Scotton, Stefano De Toni and **Catia Monauni**

1.3.1   INTRODUCTION

Until a few years ago, studies on water courses were performed, from the engineering point of view, in a way almost completely independent of all aspects relevant to water biology. At the same time, relations between engineers and biologists were somewhat occasional, both because of substantially different languages and a different way to face and solve problems. The present unavoidable necessity to evaluate environmental impacts of civil engineering works is producing, as a consequence, a tendency to a deeper reciprocal attention. From this arises the necessity to express concepts and communicate experiences, belonging to different professions, in a shared technical language.

This present brief communication is an attempt to introduce, in a simple way, some of the most important concepts of the hydrodynamics of watercourses.

Biological Monitoring of Rivers   Edited by G. Ziglio, M. Siligardi and G. Flaim
© 2006 John Wiley & Sons, Ltd.

1.3.2   SOME GENERAL ASPECTS

It is clear that gravity force is responsible for water movement to the sea. It appears to be equally evident that this cannot be the only force that acts on water molecules. In fact, in absence of other forces, transformation of potential energy into kinetic energy would induce water to move at a velocity of 140 m s^{-1}, after travelling a difference in elevation of 1000 m. Actually, the largest part of the energy, possessed by water at the beginning of its movement to the sea, is dissipated or, better, transformed into heat.

Energy transformation into heat depends on the characteristics of watercourses, which can be imagined as being constituted by a 'content' and a 'container'. The *content* is formed by the fluid and solid material dragged downstream by the fluid, while the *container* is represented by the territorial portion that circumscribes the content.

Therefore, energy dissipation depends on the characteristics of content and container. Parameters which mainly characterize the content are density and viscosity of the fluid, and density, shape and dimensions of the moving solid material. Parameters which characterize the container are its shape, together with the shape and dimensions of the material representing the separation surface between the container and content.

In general, the bottom of a watercourse is formed by loose material, vegetation and regulation works, which actively interact with the fluid. The hydrodynamic pressures, exerted by the fluid on the bottom material, are able, under appropriate conditions, to move and drive it downstream.

In the case of normal discharge values, it appears quite easy to separate container and content, with the former being the solid part of the watercourse, while the latter is the fluid part. The same distinction appears to be much more difficult under high discharge values, when large sediment transport and important phenomena of erosion and deposition occur.

It is well known that water is made up of a large amount of molecules that gives the fluid its macroscopic appearance of a continuum matter. Nevertheless, the behaviour of single particles influences in a very important way the motion as a whole. Whenever the water particles have a large momentum, interaction with other particles produces frequent and large variations in velocity direction and value. On the contrary, to a little momentum corresponds a small variation in the velocity vector.

The fluid motion in watercourses can be described as the movement of parallel layers at a different mean velocity. The fluid velocity (Figure 1.3.1) of the layer which is very near to the bottom is practically zero, while the velocity increases approaching the free surface.

The maximum velocity occurs at a little below free surface. Near to the bottom, the velocity is low and the chance of large fluctuations is inhibited by the presence of the bottom itself. The particles tend to remain in their layer, thus leading to the so-called 'laminar motion' (Figure 1.3.2(a)). In the remaining volume occupied by the water, the particles are able to move from one layer to another and the motion is defined as

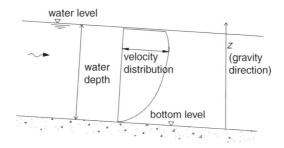

Figure 1.3.1 Velocity distribution along the depth in a watercourse

'turbulent' (Figure 1.3.2(b)). From the hydraulic point of view, the tendency of the fluid to produce a turbulent rather than a laminar motion is estimated by the Reynolds number, which represents the ratio between the inertia and viscous force acting on a water particle (Hinze, 1975). In natural watercourses, the Reynolds number is usually quite high and the motion is turbulent almost everywhere.

Resistance exerted by the bottom to water movement depends on the relationship between the thickness of the laminar layer and the dimensions of solid bodies forming the bottom (Figure 1.3.3). If the thickness of the layer in laminar motion is large enough to cover the bottom roughness, the bottom surface is defined as being 'smooth' and the resistance depends on the Reynolds number, i.e. on the water viscosity coefficient. If the bottom roughness is large in comparison with the bottom laminar layer, the bottom surface is defined as being 'rough' and the motion resistance depends on the dimensions of the bottom material.

Mountain watercourses, characterized by a large mean velocity and large dimensions of the bottom material, have rough surfaces. Lowland watercourses, characterized by a lower mean velocity and smaller dimensions of the bottom material, can exhibit smooth surfaces.

The properties described above have an important influence on the biotic activity in watercourses. The intensity of velocity fluctuations of the water particles is responsible for the mixing processes of organic and inorganic solid components driven by the water. The shape and dimensions of macro-benthos, populating the bottom of watercourses, should be strongly influenced by local motion characteristics.

It is worth briefly mentioning here some of the macroscopic aspects of watercourses (Henderson, 1989).

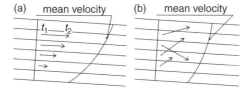

Figure 1.3.2 (a) Laminar motion, where water particles remain in their layer. (b) Turbulent motion, where water particles move widely

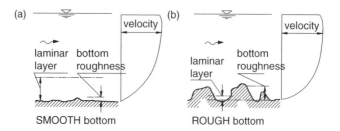

Figure 1.3.3 Turbulent motion for (a) a hydraulically smooth and (b) a rough bottom

The way in which perturbations, produced inside a watercourse, propagate is particularly important because it enables us to evaluate which zones can be affected by activities carried out inside it.

The definition of the Froude number (*Fr*) of the current allows the evaluation of these aspects, and gives a numerical expression of the ratio between inertia and gravitational forces acting on the fluid particles.

If the inertia forces are larger than the gravitational forces (Froude numbers larger than 1), the hydraulic effects of any kind of perturbation, for instance, an engineering work, are not able to move upstream; they can only have an influence downstream (Figure 1.3.4). This situation is common in mountainous reaches of watercourses and is defined as 'super-critical flow' or 'fast-flow'.

On the contrary, if the inertia forces are small compared to the gravitational forces (Froude numbers smaller than 1), perturbations can experience their influence, not only downstream, but also upstream, sometimes travelling long distances. This case is common in lowland rivers and the motion conditions are defined as 'sub-critical flow' or 'slow-flow'.

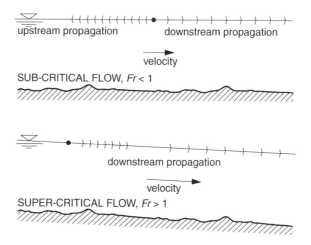

Figure 1.3.4 The ways in which hydraulic perturbations propagate inside a watercourse (*Fr* is the Froude number)

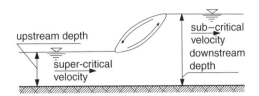

Figure 1.3.5   Transition from super-critical flow to sub-critical flow – the hydraulic jump

Another very remarkable hydraulic phenomenon is the so-called 'hydraulic jump' (Figure 1.3.5). Its importance is due to an associated large energy dissipation and to considerable variation in all of the hydraulic properties which are produced (in particular, of the water depth and local velocity).

A water stream can show a gradual change from sub-critical to super-critical flow conditions (flowing over the notch of a weir, crossing through a narrowing, etc.), whereas transition from fast to slow flow is only possible through the above mentioned phenomenon. The latter can be easily observed in natural mountain watercourses and is frequently used by engineers as an energy dissipator. It shows itself in different shapes according to the Froude number of the upstream current.

Water depth variation can be very large – for strong jumps, corresponding to Froude numbers larger than 6, of the order of magnitude of 6 to 7 times the upstream water depth.

Similarly, energy dissipation can be considerable, reaching 80 % of the energy possessed upstream. The length of a hydraulic jump is of an order of magnitude of 5 to 6 times the downstream water depth.

In the upper part of watercourses (in mountain regions), the presence of large boulders distributed on the bottom is responsible for frequent and local transitions from super-critical to sub-critical flow by means of hydraulic jumps where large energy dissipation occurs. In the medium and low reaches, watercourses show hydraulic jumps which involve the entire wetted section, often caused by engineering works.

1.3.3   SEDIMENT TRANSPORT AND BED MORPHOLOGY

Another very important aspect of watercourses' hydrodynamics is relevant to sediment transport.

In the very upper part of mountain catchments, huge sediment transport is caused by a phenomenon known as 'debris flows' (Figure 1.3.6) (Iverson, 1997; Scotton and Deganutti, 1997), three-phase mixtures of water, sediment and air, responsible for the formation of alluvial fans. This impulsive phenomenon shows very different properties, depending on the geometric characteristics of the solid material and the type of triggering rainfalls. The movement is accompanied with intense processes

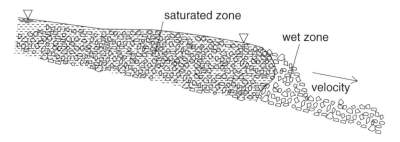

Figure 1.3.6 Schematic representation of a debris flow surge. This aspect is assumed when the mixture is rich in coarse and poor in fine material

of bottom erosion and deposition. They are normally generated in very small basins, whose dimensions are of the order of magnitude of some square kilometres, although they can exhibit a high bulk density (roughly twice the water density), a high velocity (up to 10–20 m s^{-1}) and very large discharges (some hundreds of cubic metres per second).

While in the case of debris flow, the mixture is forced to move downstream directly by gravity, in the case of ordinary sediment transport, gravity acts on the fluid mass and hydrodynamic forces, generated on the bottom material, are responsible for bed material movement.

In ordinary sediment transport (Figure 1.3.7), the basic physical parameters are the friction exerted on the bottom by the water, the water depth, the physical properties of the fluid (density and viscosity) and the physical properties of the bed material (density, shape and dimensions).

The initial motion conditions can be described by the Shields theory (van der Berg and de Vries, 1979), which has been extended in order to take account of a bottom slope which is different from zero, a 'hiding' effect (in the case where the bottom is formed of grains of different dimensions) and low submergence (in the case where the water depth is of the same order of magnitude of the dimensions of the bottom material).

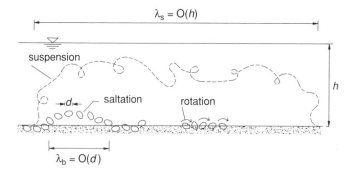

Figure 1.3.7 Ordinary sediment transport, illustrating the ways in which the bottom material moves inside a watercourse

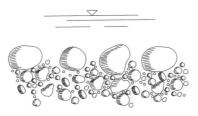

Figure 1.3.8 Under normal discharge conditions, the bed of the watercourse appears 'armoured'

In the literature, many theories are available for describing the value of the sediment discharge as a function of the parameters cited above. However, the applicability range is often very different. In addition, in the case where they have a common range of applicability, very different values of sediment discharge are found (differences of the order of 50 % are not surprising).

Under normal discharges, mobile bed watercourses appear 'armoured' (Figure 1.3.8). This means that the superficial layer of the bottom is constituted by material having a mean diameter larger than the mean diameter of the material at deeper layers.

The largest part of the bottom changes, due to sediment transport, occur in flood conditions, while the stabilizing effect due to bed vegetation is reduced because erosion around the root apparatus is much larger than that of its breakage.

The morphological evolution of a watercourse is normally analysed numerically. Simulations applied to real cases are strongly influenced by boundary conditions whose determination, when possible, is very demanding.

1.3.4 BED MORPHOLOGY

Under flood conditions significant phenomena of erosion and deposition may occur, both in natural conditions and due to engineering works.

The beds of watercourses show different shapes, depending on topographical and sediment characteristics and on hydrodynamic properties. Instability phenomena of the plane shape of the bottom are often responsible for the formation of 'peculiar' bed forms.

With low values of Froude numbers, much lower than 1, small sand waves are generated, known as 'ripples' (Figure 1.3.9). The latter are only observed when the bottom sand grains have dimensions smaller than 0.5–0.6 mm. The dimensions of the ripples depend on the thickness of the laminar layer at the bottom. Their wavelengths are of the order of 30–50 cm, with magnitudes of just a few centimetres for their amplitudes.

With larger Froude numbers, but still lower than 1, bottom shapes of larger dimensions, known as 'dunes', are observed. Their shape is very similar to those of the ripples, although the longitudinal dimensions and amplitudes are comparable to that of the current depth. These dunes slowly move downstream. With Froude numbers larger than 1, 'anti-dunes' are generated. The geometrical shapes are similar to those of dunes, but they slowly migrate upstream.

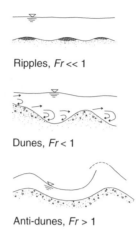

Figure 1.3.9   Illustrations of some one-dimensional bed forms inside watercourses (*Fr* is the Froude number)

Other bottom structures show typically two-dimensional properties. Among them, 'bars' (Figure 1.3.10) have a geometrical scale much larger than the above-mentioned shapes. Their longitudinal dimensions are in relation to the width of the watercourse, while the vertical scales are of the order of magnitude of the water depth. Bars can be 'free' inside a straight channel or forced by a bend, an inlet or an engineering work. They form because of instability phenomena of the bottom plane shape. Forced bars are practically stationary, while free bars migrate downstream at a velocity of metres per day.

Bank erosion phenomena are responsible for the formation of 'meanders' and play a very important role in the evolution of 'braiding rivers'.

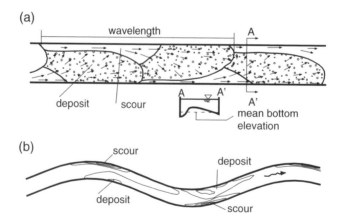

Figure 1.3.10   Illustrations of two-dimensional bed forms inside watercourses: (a) free bars; (b) forced bars

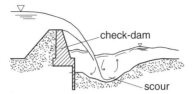

Figure 1.3.11 Sediment transport control by means of a check-dam, leading to strong modification of the velocity field and creation of longitudinal bottom discontinuity

1.3.5 ENGINEERING WORKS

Engineering works are built inside a watercourse in order to control water resource and sediment transport and to stabilize bed and banks (Ferro *et al.*, 2004). They contribute, in a very important way, to alter the natural behaviour of watercourses (Figure 1.3.11).

Check-dams and sills (Figure 1.3.12) tend to stabilize the bottom level, sometimes creating vertical discontinuities. 'Ramps' carried out by big boulders (Ferro *et al.*, 2004) behave in the same way, reducing environmental impact.

Groynes (Figure 1.3.13) are cross-sectional works which stabilize banks, naturally under erosion, deeply modifying the water velocity field. Bank revetments, of different types, fulfil the same purpose, so directly increasing bank resistance to erosion.

1.3.6 A LINKAGE BETWEEN ENGINEERING AND HYDROBIOLOGY: THE RIM PROJECT

Some years ago, the Department of Civil and Environmental Engineering of the University of Trento (Italy) started a co-operation (River Management and Macrobenthos, RIM project) with the Agricultural Institute of San Michele all'Adige (TN, Italy) and the Provincial Office for Stream Control, in order to assess the

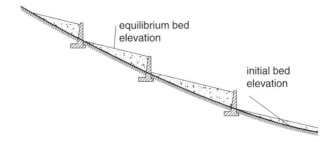

Figure 1.3.12 Restoration of a torrent by means of check-dams – bottom evolution

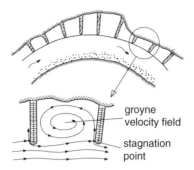

Figure 1.3.13   Stabilization of the outer bend of a watercourse by means of groynes, leading to a strong modification of the velocity field

influence of the variation of some hydraulic parameters on the macroinvertebrate and periphyton communities, with the aim of obtaining some information useful for river management and rehabilitation.

An experimental apparatus has been located inside the riverbed of the Fersina torrent along the right bank, about 15 km from the source (580 m above sea level), near Canezza, in the Trentino Province (Italy) (Figure 1.3.14) (Armanini and Scotton, 1996). By means of a water reservoir (located just downstream from a check dam), a feeding channel (delivering 0.5 $m^3 s^{-1}$ at low water conditions) and a head tank, five flumes can be fed simultaneously with torrent water. The discharge of each flume is kept reasonably constant in time by means of a spillway located in the head tank.

The five steel flumes are each 20 m in length and have a square cross-section with dimensions of 30 cm. The slope of each flume is adjustable, ranging from 0 to 5 %.

Figure 1.3.14   Experimental apparatus installed inside the Fersina torrent (Trento, Italy) for the study of the evolution of benthos population

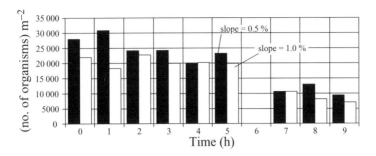

Figure 1.3.15 Macroinvertebrate abundance at two different slopes during the turbidity experiment

The bottom of each channel is formed by fifty sampling boxes filled with gravel in order to simulate natural conditions.

As an example, we report a case study carried out in 1996 to evaluate the effects of exposition to turbidity on macrobenthos, at different slopes (Ciutti *et al.*, 1999).

This study has been carried out by using two flumes, with the former placed at a slope of 0.5 % (water velocity of 0.4 m s^{-1}) and the latter at a slope of 1 % (water velocity of 0.6 m s^{-1}). Turbidity has been realized by adding a controlled mixture of torrent water and clay, for a 10 h period.

Samples of macrobenthos were taken every hour, with the results obtained being shown in Figure 1.3.15.

Turbidity effects on the macrobenthic community are evident after a period of 6 h; moreover, the effects seem to be more evident at larger slopes.

Some other experiments have been performed in order to evaluate the effects of shadowing (Siligardi *et al.*, 1998), slope (Ciutti *et al.*, 2004), colonization rate (Siligardi *et al.*, 1999; Siligardi *et al.*, 1996) and substrate size (Ciutti *et al.*, 2004) on the macrobenthonic community.

Shadowing effects have been tested by using four different artificial coverings, simulating different natural situations of riparian canopy. Macroinvertebrates, periphytic–epilithic communities and leaf-pack degradation have been studied. The shadowing condition that gave a better response on the investigated communities (larger diversity and abundance of individuals and better organic material degradation) seems to be a black net with openings, which simulates the canopy produced by a thick riparian vegetation.

The macrobenthic colonization rate has been investigated by using five flumes at different slopes. Samples from each flume were taken for a period of 43 days. Sample analysis show that low slopes (0.1 %) seem to favour the establishment of a macrobenthic community with low diversity.

The influence of bottom roughness and current mean velocity on the composition of the macrobenthic community has also been investigated. Two samples of bottom material were used (dimensions of 20 and 50 mm), mixed in five different compositions (100–0 %, 75–25 %, 50–50 %, 25–75 % and 0–100 %, by volume). In addition, the water velocity has been set at three different values, i.e. 0.4, 0.6 and 1.0 m s^{-1}.

The current velocity influenced the composition and diversity of macroinvertebrates, showing a tendency to preferentially colonize substrata when the current velocity is higher.

Substrate composition also acted as a variability factor, both in terms of composition and heterogeneity. Substrates with heterogeneous particle sizes, in particular, those with a predominance of gravel, seem to be colonized by higher numbers of individuals. Within the same channel, substrata with homogeneous composition (100–0 % and 0–100 %) do not differ in terms of macroinvertebrate density.

1.3.7   CONCLUSIONS

Nowadays, it is generally accepted that altering the structure of rivers and torrents, in order to increase flow capacity, provide protection against erosion or regulate discharge may seriously impoverish river habitats, drastically reducing biodiversity. From the engineering point of view, it is often possible to take into consideration various design solutions to modify a natural phenomenon which is considered to be potentially dangerous. The best solution should probably take account of all aspects of a particular phenomenon. Sensibility to non-strictly hydrodynamic aspects can give engineers the chance to take into consideration different point of views and to interact with different professional skills.

REFERENCES

Armanini, A. and Scotton, P., 1993. *Criteri di Dimensionamento e di Verifica delle Stabilizzazioni di Alveo e di Sponda con Massi Sciolti e Massi Legati*, Studio Sperimentale dell'Influenza dei Parametri Idrodinamici sullo sviluppo dei Macroinvertebrati, Conv. N. 3263 Provincia, Autonoma di Trento. Department of Civil and Environmental Engineering, University of Trento: Trento, Italy (in Italian).

Ciutti, F., Siligardi, M., Giordani, V., Cappelletti, C. and Monauni, C., 1999. 'Effetti della torbidità sulla comunità macrobentonica'. In: *Atti del Seminario di Studi 'I Biologi e l'Ambiente.....Oltre il Duemila'*. Venice, Italy, 22–23 November, 1996, Baldaccini, G. N. and Sansoni, G. (Eds). Edizioni CISBA: Reggio Emilia, Italy, pp. 351–355 (in Italian).

Ciutti, F., Cappelletti, C., Monauni, C. and Siligardi, M., 2004. 'Influence of substrate composition and current velocity on macroinvertebrates in a semi-artificial system'. *Journal of Freshwater Ecology*, **19**, 455–460.

Ferro, V., Dalla Fontana, G., Pagliata, S. Pugliesi, S. and Scotton, P., 2004. *Opere di Sistemazione Idraulico-Forestale a Basso Impatto Ambientale*. McGraw-Hill: Milano, Italy (in Italian).

Henderson, F. M., 1989. *Open Channel Flow*, Macmillan Series in Civil Engineering. Macmillan: New York, NY, USA.

Hinze, J. O., 1975. *Turbulence*. McGraw-Hill: New York, NY, USA.

Iverson, R. M., 1997. *The Physics of Debris Flows*. US Geological Survey: Cascades Volcano Observatory, Vancouver, WA, USA.

Scotton, P. and Deganutti, A., 1997. 'Phreatic line and dynamic impact in laboratory debris flow experiments'. In: *Debris-Flow Hazards Mitigation: Mechanics, Prediction and Assessment*, Proceedings of the 1st International Conference, San Francisco, CA, USA, August 7–9, 1997, Chen C. -L. (Ed.). ASCE: New York, NY, USA, pp. 777–786.

Siligardi, M. and Ciutti, F., 1996. 'Studio del processo di colonizzazione macrobenthonico in condizioni di diverso regime idraulico'. In: *Atti del Seminario 'Dalla Tossicologia alla Ecotossicologia'*, Pordenone, Italy, 16–17 September, 1994, Azzoni, R., De Marco, N. and Sansoni, G. (Eds). Edizioni CISBA, Reggio Emilia, Italy, pp. 217–224 (in Italian).

Siligardi, M., Ciutti, F., Cappelletti, C. and Monauni, C., 1998. 'Effects of shadowing on macroinvertebrates and periphyton communities in an artificial stream'. In: *Assessing the Ecological Integrity of Running Waters Symposium*, Vienna, Austria, 9–11 November, 1998, Book of Abstracts, pp. 75–76.

Siligardi, M., Ciutti, F., Cappelletti, C., Monauni, C. and Giordani, V., 1999. 'Il progetto RIM – regimazione idraulica e macrobenthos'. In: *Atti del Seminario di Studi 'I Biologi e l'Ambiente.....Oltre il Duemila'*, Venice, Italy, 22–23 November, 1996, Baldaccini, G. N. and Sansoni, G. (Eds). Edizioni CISBA: Reggio Emilia, Italy, pp. 357–360 (in Italian).

Van der Berg, J. and de Vries, M., 1979. *Principles of River Engineering*. Pitman: London, UK.

1.4

Riverine Fish Assemblages in Temperate Rivers

Lorenzo Tancioni, Michele Scardi and **Stefano Cataudella**

1.4.1   INTRODUCTION

Fish are the most abundant class of Vertebrates, including about 25 000 species overall (Nelson, 1994; Maitland, 1995) and 10 000 freshwater species. They are very diverse in shape and functioning, and potentially able to colonize most aquatic environments, including those deeply disturbed by human impacts. Most freshwater fish species occur in stream and rivers and laterally connected floodplains.

Fish diversity and, in particular, species composition of fish assemblages in temperate rivers are controlled by a number of biotic and abiotic factors (Zalewski and Naiman, 1985). Geological factors, as well as thermal regimes, hydrological regimes and interactions among fishes, play a major role in determining fish fauna composition (Poff *et al.*, 2001; Matthews, 1998). Past events, such as Pleistocenic glaciations, may be considered as the 'screens' that shaped present fish assemblages and induced spatial distribution patterns in fish fauna, which significantly varies in terms of species richness in different ecoregions (Matthews, 1998; Moyle, 1994).

Biological Monitoring of Rivers   Edited by G. Ziglio, M. Siligardi and G. Flaim
© 2006 John Wiley & Sons, Ltd.

Because of these events, for instance, European fish fauna includes less riverine species than its North American counterpart. Independently of the effects of large-scale processes, fish species richness also varies within ecoregions. At a local scale, each fish assemblage can be considered as the end result of the selection of a potential fish fauna operated by several 'natural filters' (i.e. environmental turbulence of Pleistocene, zoogeographic barriers, physiological factor, natural disturbance, etc.), which acted from continental to local scale and from the Pleistocene to the present age (Tonn, 1990; Moyle, 1994).

In most regions of the world, anthropogenic disturbances induced changes in the 'natural' factors that regulated species diversity and fish assemblage composition at different spatial and temporal scales. Impacts of human activities and other disturbances such as point source pollution from industrial and domestic wastes, diffuse pollution from agriculture, loss of habitat due to damming and canalization of rivers, and introduction of non-native species often resulted in alteration of the natural structure and function of biotic communities in lotic systems.

The effect of environmental perturbations involves different levels of biological organization in fish, as in other aquatic organisms, with responses ranging from sub-cellular to community level (Whitfield and Elliott, 2002).

Studies on fish populations and assemblages can be useful for the assessment of biotic integrity (Karr, 1981, 1991) of riverine systems since they tend to be sensitive to both chemical water pollution and physical habitat degradation. Changes in expected population and assemblage parameters (reference conditions), such as number and occurrence of species, dominance, number of trophic links, population size, age structure and growth rates (Hughes *et al.*, 1987; Fausch *et al.*, 1990; Karr, 1991; Angermeier and Karr, 1994; Lafaille *et al.*, 1999; Oberdorff and Hughes, 1992; Oberdorff *et al.*, 2001b) can be interpreted as responses to ecosystem perturbations.

Moreover, fish populations and assemblages can represent a useful tool to assess the socio-economic impacts of pollution. In fact, fish populations and assemblages are often exploited as renewable resources by game and commercial fishery (FAO, 1995, 1997) and therefore their economic and aesthetic values can be easily assessed (Kahn and Buerger, 1994). Therefore, changes in these values provide an estimate of the costs associated to water pollution, which is not as straightforward in the case of other groups of organisms. Thus, fish assemblages are among the most widely used bioindicators of environmental and ecological change (Karr, 1981; Karr and Dudley, 1981; Fausch *et al.*, 1984; Karr *et al.*, 1986; Angermeier and Karr, 1986; Plafkin *et al.*, 1989; Oberdorff and Hughes, 1992; Karr and Chu, 1997, 1999; Simon, 1999; Oberdorff *et al.*, 2002), and responses at different levels of biological organisation, from sub-organism level (biomarkers) to individual, population and community level (bioindicators), have been considered (Peakall and Fournier, 1992; Paller *et al.*, 1996; Kedwards *et al.*, 1999; Whitfield and Elliott, 2002; Van der Oost *et al.*, 2003; Lawrence and Hemingway, 2003; Scardi *et al.*, 2006 (Chapter 2.2 in this text)).

This chapter is aimed at providing a basic description of fish assemblages in temperate rivers, with particular emphasis on the main effects of pollution. Here, pollution is intended in a broad sense as 'any man-made or man-induced alteration of the physical, chemical, biological or radiological integrity of water' (US Clean Water Act, 1987, in Karr and Chu, 1999) and according to the European Water Framework Directive (European Union, 2000), which focuses on biotic effects (e.g. on changes in fish populations and assemblages), as chemical and hydromorphological variables are only considered as supporting the interpretation of biotic ones (European Union, 2000).

1.4.2 RELEVANT FEATURES OF RIVERINE FISH POPULATIONS AND ASSEMBLAGES

A set of populations of different fish species that co-exist in a given site at a given time can be considered as a *fish assemblage* that in turn is a subset of a broader community. In many cases, the term fish assemblage has been used 'to describe all the species in a defined area irrespective of whether they interact or not' (Wootton, 1990). Matthews (1998) defined fish assemblage as 'fishes that occur together in a single place, such that they have at least a reasonable opportunity for daily contact with each other'.

Some descriptors of fish populations and assemblages are particularly relevant to the structure and function of communities in fluvial systems. For example, Maitland (1995) considered characteristics such as: discreteness (a fish population is often confined into a single fluvial system and vulnerable to the effects of pollution and habitat alterations); migration (life cycles of many, mainly diadromous, cyclostoms and teleosts are characterized by migration periods during which they are vulnerable – i.e. eels, salmons, sturgeons, sea and fluvial lampreys – which have to pass through the lower river stretch from and to the sea); life cycles/longevity (large slow-growing species – i.e. sturgeons – and small short-lived species may be highly vulnerable to human pressures); loss of habitats (if the species life cycle requirements are not found within a fluvial system, species become either migratory or do not establish permanent populations); communities (fish assemblages are key components of aquatic communities and food webs, and thus the characteristics of fish populations and structure and functions of fluvial systems can be strongly modified by habitat alteration, pollution and especially by the introduction of non-native predator–competitor species).

According to Matthews (1998), studies on fish assemblage structure should take into account the following attributes: (1) number of species, (2) number of families, (3) numbers of prey species versus piscivorous species, (4) proportional composition of the assemblage by trophic groups or by functional groups; (5) distribution of abundance of species; (6) body-size patterns for the whole assemblage; (7) distribution of 'trophic potential' (mouth size) for the whole assemblage.

Deep past and more recent environmental events determined the zoogeographic distributions of fish fauna in streams and rivers in different geographic areas and, within the latter, in different ecoregions. Therefore, fish fauna is arranged in space as a complex patchwork and significant differences in fish assemblages may be observed, even between river basins that are located in the same sub-ecoregion. Thus, it can be very difficult to develop fish-based bioassessment methods that can be generalized and applied at the ecoregional or continental scale.

In order to simplify the analysis of fish assemblages, it is possible to group fish species according to the concept of ecological/functional guild (Austen *et al.*, 1994). This can be done, for instance, by classifying species according to their preferences for water quality, habitat, specific requirement for breeding, feeding, growth, recruitment and survival (Noble and Cowx, 2002). Wootton (1990) reported a definition of guild as formerly stated by Root (1967), i.e. 'a group of species in a community which exploit the same class of resources in a similar way'.

However, classification of fish species into groups or guilds may not be an easy task. In fact, it could be difficult to assign a fish species to guilds, for instance, because of the flexibility of many riverine species. This is particularly evident when trying to distinguish trophic groups in fish assemblages, mainly because of changes in trophic ecology during fish ontogeny. Most species eat planktonic and microbenthic taxa at the larval or postlarval stage, increase the macroinvertebrate consumption as juveniles, and widen the range of food items in an opportunistic way when adult. Gerking (1994) reviewed several trophic classifications for freshwater fish and pointed out that all of them were not effective, mainly because of trophic adaptability of fish species, thus concluding that guilds lacked 'maturity as a concept'.

Trophic classification can be aimed at distinguishing either 'type of food', reflecting what species eat (e.g. crustaceans), or 'feeding habitat', reflecting where fish species find their food (e.g. water columns) (Keenleyside, 1979; Wootton, 1990). The most widely used system for fish trophic classification for North American and European streams and rivers is probably the one proposed by Karr *et al.* (1986), who distinguished as different trophic categories invertivores, piscivores, herbivores, omnivores, and planktivores. Differences in relative abundance of trophic groups, in samples collected in four streams in Midwestern United States, highlighted an upland–low-gradient pattern, with many more insenctivores in permanently flowing upland streams than in low-gradient/slow-flowing streams, where omnivores and piscivores were more abundant (Brown and Matthews, 1995).

Another important criterion for classification of fish species is reproduction. An example of a classification scheme based on reproductive guilds of teleost fishes is the one proposed by Balon (1975), who classified fish into 33 reproductive guilds on the basis of ontogeny, spawning behaviour and substratum for egg deposition. In order to simplify this complex reproductive classification, particularly for using the reproductive guild as a metric in biotic indices, simplified systems of reproductive grouping were also developed (Oberdorff and Hughes, 1992; Cowx and Welcomme, 1998; Cowx, 2001).

The guild approach has been used (Karr *et al.*, 1986; Karr and Chu, 1997, 1999) for the development of fish-based, multi-metric indices, which are useful for assessing the ecological status of fluvial ecosystems, and may actually be considered as one of the keystones in these approaches. For instance, in the framework of an European research programme (FAME), which was aimed at developing a 'standardized fish-based assessment method for the ecological status of European rivers', the following sets of criteria were proposed for classifying European fish species into guilds and functional groups: (1) trophic (planktivores, herbivores, detritivores, omnivores, insectivores/invertivores, benthivores, piscivores, parasite.); (2) reproductive (lithophils, phytophils, phytolithophils, psammophils, ostracophils, pelagophils, lithopelagophils, ariadnophlis, speleophils, viviparous, polyphils); (3) habitat 'rheophily' (reophilic, eurytopic, limnophilic); (4) feeding habitat (water column, benthic); (5) residency/migration (short, intermediate, long anadromous/catadromous); (6) tolerance, as 'combined tolerance to water quality, chemical acidification and habitat degradation' (tolerant, intermediate, intolerant); (6) longevity and maturation (short-lived, < 5 years; intermediate, 5–15 years, long-lived, > 15 years); (7) sentinel species and population structure (key species and zero+ or older/length data) (Noble and Cowx, 2002).

An example of taxonomic and guild classification is shown in Table 1.4.1. This includes feeding habitat, type of food and reproductive classification of fish species collected between 1998 and 2004 in the urban reach of the Tiber River (Tancioni *et al.*, 2001a,b; Tancioni and Cataudella unpublished data).

One of the most important attributes of a fish assemblage, as for the other biotic assemblages, is the species diversity. Several indices have been used as measures of species diversity, but the most widely accepted are probably the *Shannon–Wiener Index* (Shannon and Weaver, 1949) and the *Evennes Index* (Pielou, 1966). These indices are useful as measures of alpha diversity (*sensu* Whittaker, 1960), i.e. of local-scale diversity. In case comparisons between different fish assemblages are to be made, similarity or dissimilarity indices, such as the *Jaccard similarity coefficient* (Jaccard, 1908) or the *Rogers–Tanimoto dissimilarity index* (Rogers and Tanimoto, 1960) can be used.

Demographic studies and, in particular, those involving fish population dynamics might play a more relevant role in biomonitoring programmes in streams and rivers, especially where the number of fish species is very low, as in many temperate head water reaches. In such situations, it is common to find only one or two taxa (in most cases, belonging to the genera *Salmo* spp. and *Cottus* spp.). A similar problem is also commonplace in northern European streams and rivers, where the fish fauna is poorer than in central and southern Europe, mainly because of the effects of the last glaciations (Noble and Cowx, 2002). By the way, this is also the reason why fish fauna in temperate northern American rivers is richer than its European counterpart (Matthews, 1998).

A fish population is defined as a group of individual fish belonging to the same species that live in a given (and, in theory, well-defined) area at a given time. Spatial limits of populations may be difficult to set and in many cases they are arbitrarily defined by fisheries biologists, who also rely on standardized sample units while collecting data about population and assemblage structure (e.g. a five hundred metres stretch in a river that is fifty metres in width). Fishery ecologist and other professionals involved in fisheries management frequently use the term *stock* for an exploited fish population (Wootton, 1990).

Table 1.4.1 Taxonomic and guild classification of cyclostomes and teleosts collected in the urban reach of the Tiber River (Rome, Italy) between 1998 and 2004 (Tancioni and Cataudella, unpublished data)

Taxon	Common name	Origin in the basin[a]	Feeding habitat[b]	Type of food[c]	Reproductive[d]
CYCLOSTOMATA					
PETROMYZONTIFORMES					
Petromyzontidae					
Lampetra fluviatilis (Linnaeus, 1758)	Fluvial lamprey	D		Pa	L
Petromyzon marinus (Linnaeus, 1758)	Sea lamprey	D		Pa	L
OSTEICHTHYES					
ANGUILLIFORMES					
Anguillidae					
Anguilla anguilla (Linnaeus, 1758)	Eel	M	B	Iv	
CLUPEIFORMES					
Clupeidae					
Alosa agone (Scopoli, 1786)	Twaite shad	D	W	O	L
CYPRINIFORMES					
Cyprinidae					
Abramis brama (Linnaeus, 1758)	Common bream	I	B	O	PhL
Alburnus alburnus alborella (DeFilippi, 1844)	Bleak	N	W	O	PhL
Barbus plebejus (Bonaparte, 1839)	Italian barbel	I	B	O	L
Barbus tyberinus (Bonaparte, 1839)	Tyberian barbel	N	B	O	L
Barbus barbus (Linnaeus, 1758)	Barbel	I	B	O	L
Carassius auratus (Linnaeus, 1758)	Golden fish	I	B	O	PhL
Carassius carassius (Linnaeus, 1758)	Crucian carp	I	B	O	PhL
Cyprinus carpio (Linnaeus, 1758)	Common carp	I	B	O	Ph
Leuciscus cephalus (Linnaeus, 1758)	Chub	N	W	O	L
Rutilus rubilio (Bonaparte, 1837)	Mid-Italian roach	N	B/W	O	PhL
Scardinius erytrophthalmus (Linnaeus, 1758)	Rudd	N	B/W	O	Ph
Tinca tinca (Linnaeus, 1758)	Tench	N	B	O	Ph

Table 1.4.1 *Continued*

Taxon	Common name	Origin in the basin[a]	Feeding habitat[b]	Type of food[c]	Reproductive[d]
Pseudorasbora parva (Temmink and Schlegel, 1846)	Stone moroko	I	W/B	Iv	PhL
Rutilus rutilus (Linnaeus, 1758)	Roach	I	W	O	PhL
SILURIFORMES					
Ictaluridae					
Ictalurus melas (Rafinesque, 1820)	Black bull-head	I	B	Iv	L
Ictalurus punctatus (Rafinesque, 1818)	Channel catfish	I	B	Iv	L
Silurus glanis (Linnaeus, 1758)	Wels catfish	I	B	P	PhL
ESOCIFORMES					
Esocidae					
Esox lucius (Linnaeus, 1758)	Pike	I	W	P	Ph
CYPRINODONTIFORMES					
Poecilidae					
Gambusia holbrooki (Girard, 1859)	Mosquito fish	I	W	Iv	V
GASTEROSTEIFORMES					
Gasterosteidae					
Gasterosteus aculeatus (Linnaeus, 1758)	Three-spined stickelback	N	W	Iv	A
PERCIFORMES					
Percidae					
Perca fluviatilis (Linnaeus, 1758)	Perch	I	W	P	Ph
Sander lucioperca (Linnaeus, 1758)	Pikeperch	I	B	P	PhL
Gymnocephalus cernuus (Linnaeus, 1758)	Ruffe	I	B	O	PhL
Centrarchidae					
Lepomis gibbosus (Linnaeus, 1758)	Pumpkinseed	I	W	Iv	L
Micropterus salmoides (Lacépède, 1802)	Black bass	I	W	P	PhL
Mugilidae					
Mugil cephalus (Linnaeus, 1758)	Grey mullet	M	W	D	
Liza ramada (Risso, 1826)	Thin-lipped mullet	M	W	D	

(*Continued*)

Table 1.4.1 *Continued*

Taxon	Common name	Origin in the basin[a]	Feeding habitat[b]	Type of food[c]	Reproductive[d]
Serranidae					
Dicentrarchus labrax (Linnaeus, 1758)	Seabass	M	W	P	
Blennidae					
Salaria fluviatilis (Asso, 1801)	Freshwater blenny	N	B	Iv	L
Gobiidae					
Knipowitschia panizzae (Verga, 1841)	Panizza's gobid	I	B	Iv	L

[a] D, Diadromous; I, Introduced; N, Native; M, Marine.
[b] B, Benthic; W, Water column.
[c] D, Detrivore; Iv, Invertivore; O, Omnivore; P, Piscivore; Pa, Parasite.
[d] A, Ariadnophil; Ph, Phytophil; PhL, Phytolithophil; L, Lithophil; V, Viviparous.

In the last few decades, morphological analysis and genetic studies of the species have increasingly been used to identify populations that have distinct genetic characteristics as well as to identify genetic changes that affect whole populations, like, for instance, adaptation to pollution (Hauser *et al.*, 2003).

Fish populations may vary in abundance and production because of the effects of environmental factors. Thus, the variability in abundance of a fish population can be regarded as the overall effect of a change in the survival probabilities and in the reproductive success for individual fish. The dynamics of fish populations can be studied by means of different analyses, such as: the analysis of age classes and length–frequency distributions; the assessment of fish population abundance (e.g. mark-recapture, catch per unit effort; virtual population analysis, acoustic methods); the assessment of mortality at different life-stages (e.g. eggs and yolk-sac larvae); the fecundity analysis (Ricker, 1975; Gulland, 1977; Wootton, 1990).

Growth rates, mortality and fecundity may be dependent on density (density-dependent growth, mortality, natality), and theoretically this allows us to reach an equilibrium population density at which death and birth rates balance each other (Wootton, 1990). If the population dynamics are influenced by unnatural factors, such as pollution, the fish populations are usually density-independent (Elliott *et al.*, 2003). In this case, the birth and death rates per individual fish are not directly linked to population size (Hasting, 1997).

One of the most important variables in fish population dynamics is length. For instance, total length and weight data are usually recorded in order to define length–weight relationships according to the $W = aTL^n$ formula (Ricker, 1975), where W is the body weight in g and TL is the total body length in cm. The constants a and n are determined empirically on the basis of a linear regression on log–log transformed data and can vary between 2,5 and 3,5 (Weatherly and Gills, 1987).

Age-structured and length-converted catch curves have been utilized to compare different populations of riverine fish (Pauly *et al.*, 1995).

Another common fish population descriptor, which has been used as an indicator in stream assessment systems (Siligato and Böhmer, 2001, 2002) is the condition factor K (Ricker, 1975), which is a measure for body fatness (plumpness), calculated as $K = W \times 100/TL^3$, where W is the body weight in g and TL is the total body length in cm.

One of the most widely used models in growth analysis is the von Bertalanffy function, $L_t = L_\infty \left\{ 1 - e^{[-k(t-t_0)]} \right\}$, in which L_∞ is the asymptotic length, that is the mean length the fish of a given stock would reach if they grow indefinitely, K is the rate (in units of 1/time) at which L_∞ is approached, and t_0 is the 'age of the fish at zero length' if they had always grown according to the theoretical function in the manner described by the equation (t_0 is generally negative) (Bertalanffy, 1938; Pauly, 1981; Gayanilo and Pauly, 1997).

1.4.3 DISTRIBUTION OF RIVERINE FISH FAUNA: PATTERNS AND CONCEPTS

Patterns in fish species diversity and in fish assemblage structure have been studied in lotic systems at different scales and taking into account different theoretical frameworks (Oberdorff *et al.*, 1997, 1998, 2001a; Lafaille *et al.*, 1999), such as: (1) the 'species–area hypothesis' (McArthur and Wilson, 1967) according to which fish species richness should increase as power function of river basin area; (2) the 'historical hypothesis' (Whittaker, 1977), which considers fish species richness as the outcome of fish re-colonization and dispersion processes that happened in fluvial ecosystems after the last glaciation; (3) the 'species–energy hypothesis' (Wright, 1983), which states that species richness is correlated with energy availability.

Moreover, as (allopatric) speciation requires geographic isolation for long periods, species richness is generally higher in regions of the world that have provided suitable habitats for fishes during previous adverse periods (Poff *et al.*, 2001). Thus, for example, freshwater fish fauna of North America includes 1033 native species (Williams and Miller, 1990), while the European one consists of 267 native species of teleosts and cyclostoms (Noble and Cowx, 2002). Moreover, more than 1000 fish species have been recorded in freshwater systems in China (Liu and Chen, 1999). Among those, 361 species and subspecies (177 endemic) have been recorded in the Yangtze River basin, the third longest river in the world (Fu *et al.*, 2003). In northern American lotic systems, fish fauna consists mainly of darters (Percidae, i.e. *Etheostoma* spp.), minnows (Cyprinidae, i.e. *Phoxinus* spp., *Notropis* spp., *Pimephales* spp., *Campostoma* spp.), suckers (Catastomidae, i.e. *Catastomus commersoni*, *Carpioides carpio*, *Moxostoma* spp.) and sunfish (Centrarchidae, i.e. *Lepomis* spp.).

In Europe, riverine fish usually include species belonging to the family of Ciprinidae (i.e. *Barbus* spp., *Leuciscus* spp., *Rutilus* spp., *Abramis* spp.), which are mostly found in lowland reaches, corresponding to potamon, while mainly Salmonidae (i.e. *Salmo trutta* and in some basins *Thymallus thymallus*) occupy mountain and foothill reaches, partially corresponding to rhithron. In riverine systems of Central Europe, this longitudinal distribution pattern of fish assemblages was first observed at the beginning of the last century and was used as the basis for a river classification system based on 'fish zone' (Thieneman, 1925a,b; Huet 1949, 1959).

In general, many species of riverine fish could be considered as 'non-primary' in the northern regions of the temperate areas, because of paleoclimatic reasons (snow and ice cover during the last ice age). These species, which have been selected throughout the extreme Pleistocenic environmental disturbances, are in most cases very flexible and not specialized (Moyle, 1994). This complex framework has to be carefully taken into account when trying to predict the characteristics of lotic fish assemblages, such as structure, stability and resilience.

According to Matthews (1998), local diversity and fish assemblage structure depends on the following factors: zoogeographic factors (e.g. geological phenomena, river basin boundaries, historical biogeography, glaciations) and evolution of fishes; local abiotic phenomena (e.g. longitudinal zonation, size of habitat, such as stream width, habitat structure, microhabitat phenomena such as dissolved oxygen concentrations, temperature gradient, hydrology characteristics); autoecology of individual species (e.g. habitat preference, physico-chemical tolerances, resistance to disturbance, feeding habits, reproduction); biotic interactions among fish (e.g. competition, predation, resource partitioning).

Thus, the main factors that regulate species diversity and fish assemblages from continental down to regional and local scales include geomorphology, temperature, history and hydrology (Poff. *et al.*, 2001). Temperature and hydrology regimes, together with geomorphology and riparian vegetation are the bases for the habitat template (*sensu* Southwood, 1977), which includes all of the main factors controlling fish species diversity at local and regional scales (Poff and Ward, 1990). Hildrew and Townsend (1987) modified the habitat template concept on the basis of the organization of freshwater benthic communities, and postulated that the frequency of disturbance is the main factor controlling productivity and diversity in lotic communities (Giller and Malmqvist, 1998). For instance, flow variations may be considered as one of most important disturbances that affect the physical environment, which, in turn, influences the overall nature of fish assemblages in lotic systems (Poff and Allen, 1995). Moreover, according to the *intermediate disturbance hypothesis*, first proposed by Connell (1978) and extended to aquatic communities by Ward and Stanford (1983), the highest species richness is attained under environmental conditions that are not completely undisturbed. Of course, the intermediate disturbance hypothesis is also relevant to biomonitoring programmes, and particularly to those relying on biotic indices based on species richness (see Scardi *et al.*, 2006 (Chapter 2.2 in this text)).

Riverine fish assemblages, as any other component of biotic communities, may be influenced by processes that occur at different spatial and temporal scales in lotic ecosystems. For instance, local-scale landscape changes may have an impact on lotic systems at different lower spatial levels due to hierarchical organization of these ecosystems (Frissell *et al.*, 1986; Schlosser, 1990; Hawkins *et al.*, 1993).

However, a broader set of concepts, representing the foundation of lotic ecology (Minshall, 1988; Ward *et al.*, 2002), has to be taken into account for classifying spatial patterns, including fish assemblages distribution. In fact, classification can be considered as a fundamental step in any environmental evaluation procedures. Thus, to classify patterns in fish species diversity and fish assemblages in lotic systems, from large-scale ones, i.e. ecoregion (Hughes *et al.*, 1987; Hughes and Larsen, 1988) or catchment, to local-scale ones, i.e. reach or macrohabitat, it is also important to consider concepts such as: hierarchical classification of rivers based on stream order (Horton, 1945; Strahler, 1957); stream zonation (Huet, 1949, 1959; Illies and Botosaneanu, 1963); river *continuum* (Vannote *et al.*, 1980); flood pulse (Junk *et al.*, 1989); patch dynamics (Pickett *et al.*, 1985; Pringle *et al.*, 1988); serial discontinuity (Ward and Stanford, 1995); aquatic–terrestrial ecotones (Naiman and Decamps, 1990); four-dimensional perspective of lotic ecosystems (Ward, 1989).

Although it is not possible to summarize here the rationale supporting all of the above-mentioned concepts, it is certainly useful to briefly elucidate those that are more frequently taken into account in classification and evaluation schemes for fish assemblages in temperate lotic systems (Karr *et al.*, 2000; Iversen *et al.*, 2000).

In temperate rivers, where unidirectional flow is the major forcing function, it is obvious that detecting gradients, from headwater to the lower reaches, has been a recurrent research topic (Ward *et al.*, 2002). Longitudinal fish zones (Huet, 1959), for instance, are still widely considered in Europe. This concept combines longitudinal changes in fish assemblages with morphodynamic characteristics (i.e. river width, slope, etc.) for classifying central European river reaches into four fish zones, namely trout zone, grayling zone, barbel zone and bream zone. More recently, changes in riverine communities along clinal rather then zonal gradient have been also considered. For instance, the river *continuum* concept (Vannote *et al.*, 1980) has been frequently used as a basis for elucidating the most relevant biological links between different stages of lotic environments, as organic and inorganic matter availability as well as energy flows vary longitudinally in rivers according to several physical variables (e.g. width, shading, etc.), thus inducing changes in the structure and function of fish assemblages.

In river basins with large floodplains, a major pathway in energy fluxes is often found in laterally linked floodplains. The ecological relevance of floodplains supports the flood-pulse concept (Junk *et al.*, 1989; Tockner *et al.*, 2000). Several studies on fishery ecology of floodplain rivers showed that these zones play a relevant role in reproduction, feeding and growth of larvae and juveniles (nursery areas) of many fish species (Welcomme, 1979).

Lotic systems have been also considered, from a different viewpoint, as mosaics of patches (Pringle *et al.*, 1988). Thus, unifying concepts that put the emphasis on longitudinal linkages, such as river *continuum* (Vannote *et al.*, 1980) and nutrient spiralling (Newbold *et al.*, 1981), as well as those focusing on lateral linkages, such as flood pulse (Junk *et al.*, 1989), may be simultaneously taken into account by studying changes in the characteristics of patches. For instance, a patch perspective has been adopted in some fish studies in North American streams (Argermeier and Karr, 1984).

More recently, the new concept of *ecohydrology* has been also considered in fluvial ecology and fisheries management (Zalewski, 1998; Zalewski *et al.*, 1997).

1.4.4   EFFECTS OF POLLUTION ON RIVERINE FISH ASSEMBLAGES

Fluvial hydrosystems in boreal temperate regions have been influenced and, in many cases, shaped by human civilizations since their earliest stages. Therefore, their present ecological properties, as well as those that have been recorded in the past, cannot be considered as pristine. This situation, in conjunction with the introduction of exotic species since the Roman age, makes it difficult, if not impossible, to define the undisturbed structure of fish assemblages in many river reaches.

The influence of pollution and habitat degradation in temperate fluvial systems varies considerably among different regions. In general, the highest human pressure on fish assemblages has been recorded in regions where heavy industrial activities are associated with ineffective waste-control technologies and the lack of a specific legislation on the sustainable management of water resources. This has been the case, for instance, of many eastern European countries after the Second World War (Giller and Malmqvist, 1998). However, human impacts on fluvial ecosystems may also affect river catchments very far from the most polluted regions, as a consequence of global processes and phenomena (e.g. the greenhouse effect, ozone layer depletion, acid rains, deposition of pesticides). Thus, 77 % of river systems in Europe, in the countries of the former Soviet Union and in North America have been considered as strongly modified (Cowx and Welcomme, 1998). Moreover, heavy habitat alterations, such as, for instance, those induced by river damming, and management of fisheries in riverine impounded areas, often based on the introduction of alien species, have affected lotic ecosystems and their fish assemblages also in many developing countries. Among the latter, China can be considered the most relevant case because of the scale and of the number of river reaches that have been recently affected by deep modifications, such as, for instance, those related to major dams. According to McCully (1996), the construction and management of large dams – about 50 000 wordwide – have impacted river systems at a very large spatial scale, extending far from the rivers themselves (e.g. production of greenhouse gases by reservoirs, impacts on estuarine–costal fisheries, flood plains destruction), with multiple implications at ecological (e.g. changes in fish assemblages and loss of biodiversity because of river fragmentation), socio-economic (e.g. loss of commercial fisheries, loss of agricultural activities because of the salinization of soils, etc.) and socio-political reasons (e.g. forced displacement of millions of people for the construction of storage reservoirs, potential conflicts between States, etc.).

Thus, human activities and the changes they induced in lotic systems have modified the natural factors that regulate fish assemblages (Cowx and Welcomme, 1998), both because of direct impacts on fish and their habitats (e.g. introduction of alien species, rearrangement of river channels and/or floodplains, etc.) or indirect ones (e.g. changes in land use, increasing urbanization, presence of manufacturing industries). However, the main anthropogenic factors affecting fish population and assemblages are probably habitat degradation, fragmentation and water pollution (Jobling, 1995; Maitland, 1995; Cowx and Welcomme, 1998; Baldigo and Lawrence, 2001; Poff *et al.*, 2001; Northcote and Hartman, 2004) (Table 1.4.2).

Table 1.4.2 The main anthropogenic disturbances (impacts) affecting riverine fish populations and assemblages

Impacts	Effects
Global changes in climate and land use (regional scale)	Fish populations and assemblages' adaptation through strong selection; shift in the original species distribution. Potential extinction of some southern fish populations because of global warming
Acid deposition	Freshwater acidification has an impact that may be dramatic on sensitive species (e.g. Salmonids), especially in carbonate-poor river catchments. Moreover, reduction of pH by 'acid rain' may facilitate leaching of ions. For instance, aluminium is toxic to fish at a water pH of 5
Land use – farming and forestry, including logging	Eutrophication, mainly induced by nitrogen and phosphorus diffuse sources, may cause dissolved oxygen depletion during night and therefore death or displacement of intolerant species. Poor forestry or farming management may influence sedimentation patterns and suspended solids' concentration in rivers, which in turn has an impact on fish gills. Spawning grounds for lithophil species may be also affected. Other effects include acidification and increase in pollutants
Extraction industries – mining for metal and extraction of gravel and sand	Acid discharge and toxic salts of heavy metals mainly released through the mine water discharges. In cases of high concentration, these salts can be directly toxic to fish. Moreover, the extraction industries can alter suspended solid levels and change stream-bed topography
Manufacturing industries – Whole range of industrial process	Deposition of pollutants such as pesticides and radioactive materials. Toxic discharges and organic matter may trigger oxygen depletion and production of hydrogen sulphide, i.e. environmental conditions that are only suitable for very tolerant species. Waste heat may play a similar role
Urbanization – sewage and run-off from paved surfaces	Organic pollution, usually from sewage or other point sources, may induce oxygen depletion and therefore disturbance on fish. Oestrogenic or hormonal-like substances, which are often present in sewage water, may cause permanent changes in male fish (feminizing) with inter-sex effects

Table 1.4.2 *Continued*

Impacts	Effects
Physical habitat degradation – river obstructions and impoundment (e.g. dams and barrages); infilling, drainage and canalization; water abstraction; fluctuating water levels	Structural effects such as blocking of upstream migration (reduction or extinction of diadromous species), loss of fry during 'passive' passage through dams, flooding of spawning grounds downstream; physical and chemical alterations such as changes in discharge and water quality patterns; flow fluctuations; unnatural pulses of poor-quality water; changes in water temperature patterns; reduced availability of spawning grounds for lithophil species in impounded areas. Loss of habitat and transfer of species
Fish farming, angling, fishery management, introduction of new species	Eutrophication, overfishing, genetic change, competition between native and introduced competitive species, elimination of native species, introduction of new parasites and diseases

As already mentioned, impacts of human activities on fish may involve changes at different levels of biological organization, which range from hormonal to community level (Figure 1.4.1).

In general, the response of individual fishes to changes in environmental conditions due to anthropogenic activities (e.g. pollution) results in adaptation to new conditions, or avoidance of the latter, to reduce exposure to toxics, or extinction, in case specific tolerance limits are exceeded (lethal effects). Thus, heavy disturbances may result in almost instantaneous mortality, and long-term exposure to less severe stressors may also result in the death of some individuals, while the population may eventually undergo acclimatization and genetic adaptation. Short-term exposures or sub-lethal stressors induce changes or disorders that can lead to impairment of embryo and larval development, individual growth, reproduction and recruitment success, and disease resistance, which in turn may have consequences for populations and fish assemblages. (Jobling, 1995; Howells, 1994; Müller and Lloyd, 1994).

Thus, the effects of pollution on fish could be detected by studying patterns in different types of responses, such as, for instance: changes in hormone levels; changes in cell structure, enzyme induction, immunology; physiology, respiration, osmoregulation; individual mortality, growth, behaviour, reproduction; population growth rate, reproduction, recruitment; community structure and function.

Biological response signatures (Yoder and Rankin, 1995) at the population and community level can be considered the most relevant to ecosystem and fisheries conservation, while effects at lower levels of biological organisation (e.g. cellular or individual) may be more relevant to a basic understanding of the physiological processes related to pollution.

In particular, toxicological tests have been mainly carried out on single fish species exposed to single toxicants (or to a simple mixture). Nevertheless, they have been

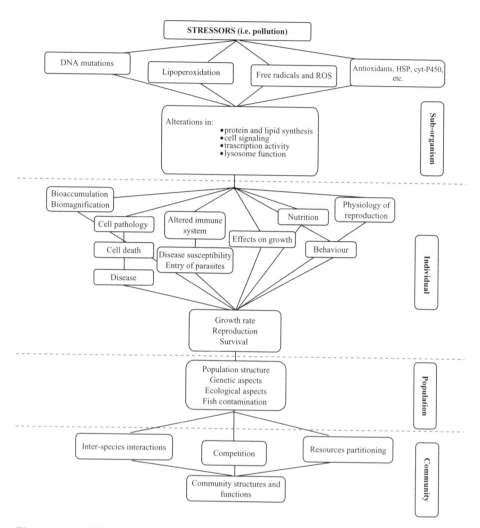

Figure 1.4.1 Different levels of fish response to environmental stresses from the molecular level (biomarkers) to individual, population and community levels (bioindicators)

frequently used as water quality criteria, and detailed information can be easily found in a vast set of papers and technical literature (e.g. Hynes, 1960; Alabaster and Lloyd, 1982 ; Hellawell, 1986; Haslam, 1990; Howells, 1994, Müller and Lloyd, 1994, etc.).

1.4.5   FINAL REMARKS

Recently, a growing interest in a more holistic approach to ecological and human risk assessment has emerged and was applied, for instance, in evaluating the effects of contaminants on fish health at multiple levels of biological organization (Adams *et al.*, 2000). Moreover, Knopper and Siciliano (2002) suggested that focusing on pollution-induced community tolerance might play a role in biomonitoring activities.

In many monitoring programmes aimed at assessing the effect of pollution on fish, there has been a growing consensus about concentration–effect studies on fish (Müller and Lloyd, 1994). Nevertheless, this approach cannot be really exhaustive, as it takes into account only a limited number of chemical disturbances and only a few fish species. Thus, according to Karr and Chu, (1999) and Simon (2003), chemical criteria based on dose–response curves for single toxicants cannot be considered as a sound basis for studying interactions among several contaminants in aquatic environments and, according to Elliott *et al.* (2003), there are no available studies which demonstrate that a change in fish assemblage is caused by specific xenobiotics. Moreover, chemical and toxicological approaches completely failed in analysing other anthropogenic threats to fish, such as physical habitat alteration and regulation of flow regimes, which are probably the most important source of disturbance for fish populations and assemblages.

On the contrary, the biotic integrity concept (Karr, 1991; Angermeier and Karr, 1994; Karr and Chu, 1999) is based on a more general perspective and has rapidly become a keystone in biomonitoring. According to this concept, an undisturbed biotope must support a 'biota-balanced, integrated, adaptive system having the full range of elements (genes, species, assemblages) and processes (mutation, demography, biotic interactions, nutrient and energy dynamics, metapopulation, or fragmented population, processes) that are expected in the region's natural environment' (Karr, 1991; Angermeier and Karr, 1994). In this framework, five classes of lotic systems' features that can be affected by human disturbances were pointed out, and their effects on fish assemblages were also considered. These features included water quality (e.g. temperature, dissolved oxygen, turbidity, toxic substances); habitat structure (e.g. substrate, current velocity, water depth); flow regime (e.g. flux variability); energy sources (e.g. amount and particle size of organic matter entering water courses); biotic interactions (e.g. competition and predation) (Karr, 1991; Karr and Chu, 1999).

Obviously, tolerance to specific disturbances varies among fish species. Therefore, the response of the latter can be used as a basis for evaluating environmental quality, as, for instance, in the *Index of Biotic Integrity* (Karr, 1981) and in many other indices that are more or less related to it. In particular, ranking tolerances of fish species to specific disturbances allowed to select indicator species, which have been taken into account for developing these indices.

Finally, some attributes of fish assemblages have been also proposed as indicators of environmental alteration and they have been taken into account in biotic indices. The attributes that tend to decrease under disturbance include, for instance (Karr, 1981; Fausch *et al.*, 1990), native species richness and species richness of single guilds, number of intolerant species, proportion of trophic specialists such as insectivores and top carnivores, abundance of reproductive guilds requiring silt-free coarse substrate for spawning, etc. On the other hand, environmental disturbances may cause an increase in the proportion of tolerant species and trophic generalists, such as omnivores, in the occurrence of parasites and DELT (deformities, eroded fins, lesions, tumours) anomalies (Karr *et al.*, 1986).

The role of these attributes of fish assemblages in biomonitoring is obviously of paramount importance and is thoroughly discussed in other sections of this book.

REFERENCES

Adams, S. M., Greeley, M. S., Jr and Ryon, M. G., 2000. 'Evaluating effects of contaminants of fish health at multiple levels of biological organization. Extrapolating from lower to higher levels'. *Human and Ecological Risk Assessment*, **6**, 15–27.

Alabaster, J. S. and Lloyd, R., 1982. *Water Quality Criteria for Freshwater Fish*, 2nd Edition, Food and Agriculture Organization of the United Nations. Butterworths: London, UK.

Angermeier, P. L. and Karr, J. R., 1984. 'Relationships between eroding debris and fish habitat in a small stream'. *Transactions of the American Fisheries Society*, **113**, 716–726.

Angermeier, P. L. and Karr, J. R., 1986. 'Applying an Index of Biotic Integrity Based on Stream-Fish Communities: Considerations in Sampling and Interpretation'. *North American Journal of Fisheries Management*, **6**, 418–429.

Angermeier, P. L. and Karr, J. R., 1994. 'Biological integrity versus biology diversity as policy directives. Protecting biotic resources'. *BioScience*, **44**, 690–697.

Austen, D. J., Bayley, P. B. and Menzel, B. W., 1994. 'Importance of the guild concept to fisheries research and management'. *Fisheries*, **19**(6), 12–20.

Baldigo, B. P. and Lawrence, G. B., 2001. 'Effects of stream acidification and habitat on fish populations of a North American river'. *Aquatic Science*, **63**, 196–222.

Balon, E. K., 1975. 'Reproductive guides of fishes: a proposal and definition'. *Journal of the Fish Research Board of Canada*, **32**, 821–864.

Bertalanffy, L. von, 1938. 'Untersuchungen über die Gesetzlinchkeiten des Wachstumjs. 1. Algemeine Grundlangen der Theorie'. *Roux' Arch Entwincklungsmech Org*, **131**, 613–653.

Brown, A. and Matthews, W. J., 1995. 'Streams ecosystems of the central United States'. In: *Ecosystems of the World*, Vol. 22, *River and Stream Ecosystems*, Cushing, C. E., Cummins, K. W. and Minshall, G. W. (Eds). Elsevier: Amsterdam, The Netherlands, pp. 89–116.

Connell, J. H., 1978. 'Diversity in tropical rain forests and coral reefs'. *Science New York*, **199**, 1302–1309.

Cowx, I. G., 2001. *Factors Influencing Coarse Fish Populations in Rivers: A Literature Review*, Research and Development Publication, Vol. 18. Environmental Agency: Bristol, UK.

Cowx, I. G. and Welcomme, R. L. (Eds), 1998. *Rehabilitation of Rivers for Fish*, FAO – Fishing News Books. Blackwell Science: Oxford, UK.

Elliott, M., Hemingway, K. L., Krueger, D., Thiel, R., Hylland, K., Arukwe, A., Förlin, L. and Sayer, M., 2003. 'From the individual to the population and community responses to pollution'. In: *Effects of Pollution on Fish: Molecular Effects and Population Responses*, Lawrence, A. J. and Hemingway, K. L. (Eds). Blackwell Science: Oxford, UK, pp. 221–255.

European Union, 2000. Parliament and Council Directive 2000/60/EC of 23rd October 2000. Establishing a Framework for Community Action in the Field of Water Policy. *Official Journal PE-CONS 3639/1/00 REV 1*. European Union: Brussels, Belgium.

FAO, 1995. *Code of Conduct for Responsible Fisheries*. Food and Agriculture Organization of the United Nations: Rome, Italy.

FAO, 1997. *Inland Fisheries* 'FAO Technical Guidelines for Responsible Fisheries, No. 6. Fisheries Department, Food and Agriculture Organization of the United Nations: Rome, Italy.

Faush, K. D., Karr, J. R. and Yant, P. R., 1984. 'Regional application of an Index of Biotic Integrity based on stream fish communities'. *Transactions of the American Fisheries Society*, **113**, 39–55.

Fausch, K. D., Lyons, J., Karr, J. R. and Angermeier, P. L., 1990. 'Fish communities as indicators of environmental degradation'. *American Fisheries Society, Symposium*, **8**, 123–144.

Frissell, C. A., Liss, W. L., Warren, C. E. and Hurley, M. D., 1986. 'A hierarchical framework for stream habitat classification: viewing streams in a watershed context'. *Environmental Management*, **10**, 199–214.

Fu, C., Wu, J., Chen, J., Wu, Q. and Lei, G., 2003. 'Freshwater fish biodiversity in the Yangtze River basin of China: patterns, threats and conservation'. *Biodiversity and Conservation*, **12**, 1649–1685.

Gayanilo, F. C., Jr and Pauly, D. (Eds), 1997. *FAO-ICLARM Stock Assessment Tools (FISAT)*, Reference Manual, FAO Computerized Information Series (Fisheries), No. 8. 262 p. Food and Agriculture Organization of the United Nation: Rome, Italy.

Gerking, S. D., 1994. *Feeding Ecology of Fish*. Academic Press: San Diego, CA, USA.

Giller, P. and Malmqvist, B. (Eds), 1998. *The Biology of Streams and Rivers*. Oxford University Press: New York, NY, USA.

Gulland, J. A., 1977. *Fish Population Dynamics*. Wiley: Chichester, UK.

Haslam, S. M., 1990. *River Pollution: An Ecological Perspective*. Wiley: Chichester, UK.

Hasting, A., 1997. *Population Biology*. Springer-Verlag: Berlin, Germany.

Hauser, L., Hemingway, K. L., Wedderburn, J. and Lawrence, A. J., 2003. 'Molecular/cellular processes and the population genetics of a species'. In: *Effects of Pollution on Fish: Molecular Effects and Population Responses*, Lawrence, A. J. and Hemingway, K. L. (Eds), Blackwell Science: Oxford, UK, pp. 256–288.

Hawkins, C. J., Kerschner, J. L., Bisson, P. A., Bryant, M. D., Decker, L. M., Gregory, S. V., McCullough, D. A., Overton, C. K., Reeves, G. H., Steedman, R. J. and Young, M. K., 1993. 'A hierarchical approach to classifying stream habitat features'. *Fisheries (Bethesda, MD)*, **18**(6), 3–11.

Hellawell, J. M., 1986. *Biological Indicators of Fresh Water Pollution and Environmental Management*. Elsevier Applied Science: London, UK.

Hildrew, A. G. and Townsend, C. R., 1987. 'Organization in benthic communities'. In: *Organization of Communities: Past and Present*, Gee, J. H. R. and Giller, P. S. (Eds). Blackwell Science: Oxford, UK, pp. 347–371.

Horton, R. E., 1945. 'Erosional development of streams and their drainage basins: hydrophysical approach to quantitative morphology'. *Geological Society of America Bulletin*, **56**, 275–370.

Howells, G. (Ed.), 1994. *Water Quality for Freshwater Fish: Further Advisory Criteria*, Environmental Topics, ISSN 1046-5294, V. 6. Gordon and Breach: Amsterdam, The Netherlands. [Most of the material published in this book first appeared as European Inland Fisheries Advisory Commission Reports T/37, T/43, T/45 and T/46. Chapters 3, 4 and 5 later appeared in *Chemistry and Ecology*, **3**(1 and 3) and **4**(3)].

Huet, M., 1949. 'Aperçu des relations entre la pente et les population piscicoles dans les eaux courantes'. *Revue Suisse Hydrolopie.*, **11**, 332–351.

Huet, M., 1959. 'Profiles and biology of Western Europe streams as related to fisheries management.' *Transactions of the American Fisheries Society*, **88**, 155–163.

Hughes, R. M. and Larsen, D. P., 1988. 'Ecoregions: an approach to surface water protection'. *Journal of the Water Pollution Control Federation*, **60**, 486–493.

Hughes, R. M., Rexstad, E. and Gammon, J. R., 1987. 'The relationship of aquatic ecoregions, river basin and physiographics provinces to the ichthyogeographic regions of Oregon'. *Copeia*, **87**, 423–432.

Hynes, H. B. N., 1960. *The Biology of Polluted Waters*. Liverpool University Press, Liverpool, UK.

Illies, J. and Botosaneanu, L., 1963. 'Problems et methodes de la classification et de la zonation ecologique des eaux courantes, considerees surtout du point de vue faunistique'. *Mitteilungen der internationale Vereinigung für theoretische und angewaudte Limnologie*, **12**, 1–57.

Iversen, T. M., Madsen, B. L. and Bøgestrand, J., 2000. 'River conservation in the European Community, including Scandinavia'. In: *Global Perspectives on River Conservation: Science, Policy and Practice*, Boon, P. J., Davies, B. R. and Petts, G. E. (Eds). Wiley: Chichester, UK, pp. 79–103.

Jaccard, P., 1908. 'Nouvelles recherches sur la distribution florale'. *Bulletin Société Vadoise des Sciences Naturelles*, **44**, 223–270.

Jobling, M., 1995. 'Human impacts on aquatic environments'. In: *Environmental Biology of Fishes*, Jobling, M. (Ed.). Chapman & Hall: London, pp. 415–436.

Junk, W. J., Bayley, P. and Sparks, R. E., 1989. 'The flood pulse concept in river-floodplain systems'. *Canadian Special Publication of Fishery and Aquatic Science*, **106**, 110–127.

Karr, J. R., 1981. 'Assessment of biotic integrity using fish communities'. *Fisheries*, **6**, 21–27.

Karr, J. R., 1991. 'Biological integrity: a long-neglected aspect of water resource management'. *Ecological Applications*, **1**, 66–84.

Karr, J. R. and Dudley, D. R., 1981. 'Ecological perspectives on water quality goals'. *Environmental Management*, **5**, 55–68.

Kahn, J. R. and Buerger, R. B., 1994. 'Valuation and consequences of multiple sources of environmental deterioration: the case of the New York striped bass fishery'. *Journal of Environmental Management*, **40**, 257–273.

Karr, J. R. and Chu, E. W., 1997. 'Biological monitoring: essential foundation for ecological risk assessment'. *Human and Ecological Risk Assessment*, **3**, 993–1004.

Karr, J. R., and Chu, E. W., 1999. *Restoring Life in Running Waters – Better Biological Monitoring*. Island Press: Washington, DC, USA.

Karr, J. R., Fausch, K. D., Angermeier, P. L., Yant, P. R. and Schlosser, I. J., 1986. *Assessing Biological Integrity in Running Waters: a Method and its Rationale*, Illinois Natural History Survey Special Publication, Vol. 5. Illinois Natural History Survey: Champaign, IL, USA.

Karr, J. R., Allan, J. D. and Benke, A. C., 2000. 'River conservation in the United States and Canada'. In: *Global Perspective on River Conservation: Science, Policy and Practice*, Boon, P. J., Davies, B. R. and Petts, G. E.(Eds). Wiley: Chichester, UK, pp. 3–39.

Kedwards, T. J., Maund, S. J. and Chapman, P. F., 1999. 'Community level analysis of ecotoxicological field studies: I. Biological monitoring'. *Environmental Toxicology and Chemistry*, **18**, 149–157.

Keenleyside, M. H. A., 1979. *Diversity and Adaptation in Fish Behaviour*. Springer-Verlag: Berlin, Germany.

Knopper, L. D. and Siciliano, S. D., 2002. 'A Hypothetical Application of the Pollution-Induced Community Tolerance Concept in Megafaunal Communities Found at Contaminated Sites'. *Human and Ecological Risk Assessment*, **8**, 1057–1066.

Lafaille, P., Lek, S. and Oberdorff, T., 1999. 'Fish ecology and use for monitoring in Europe'. In: *PAEQANN* (Predicting Aquatic Ecosystem Quality using Artificial Neural Networks), Deliverable 3, EVK-1999-00125. Review Publications, pp. 43–46.

Lawrence, A. J. and Hemingway, K. L. (Eds), 2003. *Effects of Pollution on Fish: Molecular Effects and Population Responses*. Blackwell Science: Oxford, UK.

Liu, H. and Chen, Y., 1999. 'Resources of aquatic organisms and conservation'. In: *Advanced Hydrobiology*, Liu, J. (Ed.). Science Press: Beijing, China, pp. 362–375.

Maitland, P. S., 1995. 'The conservation of freshwater fish: past and present experience'. *Biological Conservation*, **72**, 259–270.

Matthews, W. J. (Ed.), 1998. *Patterns in Freshwater Fish Ecology*. Chapman and Hall: New York, NY, USA.

McArthur, R. H. and Wilson, E. O., 1967. *The Theory of Island Biogeography*. Princeton University Press: Princeton, NJ, USA.

McCully, P., 1996. *Silenced Rivers: The Ecology and Politics of Large Dams*. Zed Books: London. In: *Global Perspectives on River Conservation: Science, Policy and Practice*, Boon, P. J. Davies, B. R. and Petts, G. E. (Eds). (2000). Wiley: Chichester, UK, pp. xi–xvi.

Minshall, G. W., 1988. 'Stream ecosystem theory: a global perspective '. *Journal of the North American Benthological Society*, **7**, 263–288.

Moyle, P. B., 1994. 'Biodiversity, biomonitoring, and the structure of stream fish communities '. In: *Biological Monitoring of Aquatic Systems*, Loeb, S. L. and Spacie, A. (Eds). Lewis Publishers: Boca Raton, FL, USA, pp. 171–186.

Müller, R. and Lloyd, R. (Eds), 1994. *Sublethal and Chronic Effects of Pollutants on Freshwater Fish*, published by arrangement with the Food and Agriculture Organization of the United Nations by Fishing News Books. Blackwell Science: London, UK.

Naiman, R. H. and Decamps, H. (Eds), 1990. *The Ecology and Managment of Aquatic–Terrestrial Ecotones*. Parthenon Publishers: Carnforth, UK.

Nelson, J. S., 1994. *Fishes of the World*, 3rd Edition. Wiley: New York, NY, USA.

Newbold, J. D., Elwood, J. W., O'Neill, V. O. and Van Winkle, W., 1981. 'Measuring nutrient spiralling in streams '. *Canadian Journal of Fisheries and Aquatic Sciences*, **38**, 860–863.

Noble, R. A. A. and Cowx, I. G., 2002. 'Compilation and harmonisation of fish species classification', Work Package 1b. *Development, Evaluation and Implementation of a Standardised Fish-based Assessment Method for the Ecological Status of European Rivers – A Contribution to the Water Framework Directive (FAME)*. Contract No., EVK1-CT-2001-00094. (www.fame.boku.ac).

Northcote, T. G. and Hartman, G. F. (Eds), 2004. *Fishes and Forestry – Worldwide Watershed Interactions and Management*. Blackwell Science: Oxford, UK.

Oberdorff, T. and Hughes, R. M., 1992. 'Modification of an index of biotic integrity based on fish assemblages to characterize rivers of the Seine Basin, France'. *Hydrobiologia*, **228**, 117–130.

Oberdorff, T., Hugueny, B. and Guégan, J., 1997. 'Is there an influence of historical events on contemporary fish species richness in rivers? Comparisons between Western Europe and North America '. *Journal of Biogeography*, **24**, 461–467.

Oberdorff, T., Hugueny, B., Compin, A. and Belkessam, D., 1998. 'Non interactive fish communities in the coastal streams of North-Western France'. *Journal of Animal Ecology*, **67**, 472–484.

Oberdorff, T., Hugueny, B. and Vigneron, T., 2001a. 'Is assemblage variability related to environmental variability? An answer for riverine fish'. *Oikos*, **93**, 419–428.

Oberdorff, T., Pont, D., Hugueny, B. and Chessel, D., 2001b. 'A probabilistic model characterizing fish assemblages of French rivers: a framework for environmental assessment'. *Freshwater Biology*, **46**, 399–415.

Oberdorff, T., Pont, D., Hugueny B. and Porcher, J. P., 2002. 'Development and validation of a fish-based index for the assesment of 'river health' in France'. *Freshwater Biology*, **47**, 1720–1734.

Paller, M., Reichert, M. J. and Dean J. N., 1996. 'Use of fish communities to assess environmental impacts in South Carolina coastal plain streams'. *Transactions of the American Fisheries Society*, **125**, 633–644.

Pauly, D., 1981. 'The relationship between gill surface area and growth performance in fish: a generalization of Von Bertalanffy's theory of growth'. *Meeresforschung*, **28**, 251–282.

Pauly, D., Moreau, J. and Abad, N., 1995. 'Comparisons of age-structured and length-converted catch curves of brown trout *Salmo trutta* in two French Rivers'. *Fisheries Research*, **22**, 197–204.

Peakall, D. B. and Fournier, M. (Eds), 1992. *Animal Biomarkers as Pollution Indicators*. Chapman and Hall: London, UK.

Pickett, S. T. A. and White, P. S. (Eds), 1985. *The Ecology of Natural Disturbance and Patch Dynamics*. Academic Press: Orlando, FL, USA.

Pielou, E. C., 1966. 'The measurement of diversity in different types of biological collections'. *Journal of Theoretical Biology*, **13**, 131–144.

Plafkin, J. L., Barbour, M. T., Porter, K. D., Gross S. K. and Hughes, R. M., 1989. *Rapid Bioassessment Protocols for Use in Streams and Rivers. Benthic Macroinvertebrates and Fish*, EPA 440/4-891001. US Environmental Protection Agency: Washington, DC.

Poff, N. L. and Allan, D., 1995. 'Functional organisation of stream fish assemblages in relation to hydrological variability'. *Ecology*, **76**, 606–627.

Poff, N. L. and Ward, J. V., 1990. 'Physical habitat template of lotic systems: recovery in the context of historical pattern of spatiotemporal heterogeneity'. *Environmental Management*, **14**, 629–645.

Poff, N. L., Angermeier, P. L., Cooper, S. D., Lake, P. S., Fausch, K. D., Winemiller, K. O., Mertes, L. A. K., Oswood, M. W., Reynolds, J. and Rahel, F. J., 2001. 'Fish diversity in streams and rivers'. In: F. S Chapin III, Osvaldo Etuart Sala, Elisabeth Huber-Sannwald (eds.),. *Global Biodiversity in a Changing Environment: Scenarios for the 21st Century*, Stuart Chapin III, F., Sala, O. E. and Huber-Sannwald, E. (Eds). Springer-Verlag: New York, NY, USA, pp. 315–340.

Pringle, C. M., Naiman, R. J., Bretschko, G., Karr, J. R., Oswood, M. W., Webster, J. R., Welcomme, R. L. and Winterbourn, M. J., 1988. 'Patch dynamics in lotic systems: the stream as a mosaic'. *Journal of the North American Benthological Society*, **7**, 503–524.

Ricker, W. E., 1975. 'Computation and interpretation of biological statistics of fish population'. *Bulletin of the Fish Research Board of Canada*, **191**, 1–382.

Rogers, D. J. and Tanimoto., T. T., 1960. 'A computer program for classifying plants'. *Science*, **132**, 1115–1118.

Root, R. B. 1967. 'The niche exploitation pattern of the blue–gray gnatcatcher'. *Ecological Monographs*, **37**, 317–350.

Scardi, M., Tancioni, L. and Cataudella, S. (2006). 'Monitoring methods based on fish'. In: *Biological Monitoring of River: Applications and Perspectives*, Ziglio, G. Siligardi, M. and Flaim, G. (Eds). Wiley: Chichester, UK, pp. 135–153.

Schlosser, I. J., 1985. 'Flow regime, juvenile abundance, and the assemblage structure of stream fishes'. *Ecology*, **66**, 1484–1490.

Schlosser, I. J., 1990. 'Environmental variation, life history attributes, and community structure in stream fishes: implications for environmental management and assessment'. *Environmental Management*, **14**, 621–628.

Shannon, C. E. and Weaver, W., 1949. *The Mathematical Theory of Communication*. University of Illinois Press: Urbana, IL, USA.

Siligato, S. and Böhmer, J., 2001. 'Using indicators of fish health at multiple levels of biological organization to assess effects of stream pollution in southwest Germany'. *Journal of Aquatic Ecosystem Stress Recovery*, **8**, 371–386.

Siligato, S. and Böhmer, J., 2002. 'Evaluation of biological integrity of a small urban stream system by investigating longitudinal variability of the fish assemblage'. *Chemosphere*, **47**, 777–788.

Simon, T. P. (Ed.), 1999. *Assessing the Sustainability and Biological Integrity of Water Resource Using Fish Communities*. Lewis Press: Boca Raton FL, USA.

Simon, T. P. (Ed.), 2003. *Biological Response Signatures – Indicator Patterns Using Aquatic Communities*. CRC Press: Boca Raton, FL, USA.

Southwood, T. R. E., 1977. 'Habitat, the templet for ecological strategies?'. *Journal of Animal Ecology*, **46**, 337–365.

Strahler, A. N., 1957. 'Quantitative analysis of watershed geomorphology'. *American Geophysical Union Transactions*, **38**, 913–920.

Tancioni, L., Baldari, F., Ferrante, I., Scardi, M. and Mancini, L., 2001a. 'Feeding habits of some native and introduced fish species in the low stretch of Tiber River (Central Italy)'. *Journal of Freshwater Biology*, **30**, 159–162.

Tancioni, L., Cecchetti, M., Costa, C., Eboli, A. and Di Marco, P., 2001b. 'Aspects of reproductive biology of the barbel, *Barbus tyberinus* (Bonaparte, 1839)'. *Journal of Freshwater Biology*, **30**, 155–158.

Thienemann, A., 1925a. 'Die Süßwasserfische Deutschlands – Eine tiergeographische Skizze'. In: *Handbuch der Binnenfischerei Mitteleuropas*, Demoll, R. and Maier H. N. (Eds). Bd. IIIA: Stuttgart, Germany, pp. 1–32 (in German).

Thienemann, A., 1925b. 'Die Binnengewasser Mitteleuropas'. *Die Binnengewasser*, **1**, 54–83.

Tockner, K., Malard, F. and Ward, J. V., 2000. 'An extension of the Flood Pulse Concept'. *Hydrological Processes*, **14**, 2861–2883.

Tonn, W. M., 1990. 'Climate change and fish communities: A conceptual framework'. *Transactions of the American Fisheries Society*, **119**, 337–352.

Van der Oost, R., Beyer, J. and Vermeulen, N. P. E., 2003. 'Fish bioaccumulation and biomarkers in environmental risk assessment: a review'. *Environmental Toxicology and Pharmacology*, **13**, 57–149.

Vannote, R. L., Minshall, G. W., Cummins, K. W., Sedell, J. R and Cushing, C. E., 1980. 'The river continuum concept'. *Canadian Journal of Fisheries and Aquatic Science*, **37**, 130–137.

Ward, J. V., 1989. 'The four-dimensional nature of lotic ecosystems'. *Journal of the North American Benthological Society*, **8**, 2–8.

Ward, J. W. and Standford, J. A., 1983. 'The intermediate-disturbance hypothesis: an explanation for biotic diversity patterns in lotic ecosystems'. In: *Dynamics of Lotic Ecosystems*, Fontaine III, T. D. and Bartell, S. M. (Eds). Ann Arbor Science: MI, USA, pp. 347–356.

Ward, J. W. and Standford, J. A., 1995. 'The serial discontinuity concept: extending the model to floodplain rivers'. *Regulatory and Rivers Research Management*, **11**, 105–119.

Ward, J. V., Robinson, C. T. and Tockner, K., 2002. 'Applicability of ecological theory to riverine ecosystems'. *Verhandlungeu Internationale Vereinigung für theoretishe und angewandte Limnologie*, **28**, 443–450.

Weatherley, A. H. and Gills, H. S., 1987. *The Biology of Fish Growth*, Vol. XII. Academy Press: London, UK.

Welcomme, R., 1979. *Fishery Ecology of Floodplain Rivers*. Longman: New York, NY, USA.

Welcomme, R., 1985. *River Fisheries*, FAO Fisheries Technical Paper, No. 262. Food and Agriculture Organization of the United Nations: Rome, Italy.

Whitfield, A. K. and Elliott M., 2002. 'Fishes as indicators of environmental and ecological changes within estuaries: a review of progress and some suggestions for the future'. *Journal of Fish Biology*, **61** (supplement A), 229–250.

Whittaker, R. H., 1960. 'Vegetation of the Siskiyou Mountains, Oregon and California'. *Ecological Monographs*, **30**, 279–338.

Whittaker, R. H., 1977. 'Evolution of species diversity in land communities'. *Evolutionary Biology*, **10**, 1–67.

Williams, J. E. and Miller, R. R., 1990. 'Conservation status of the North American fish fauna in fresh water'. *Journal of Fish Biology*, **37** (supplement A), 79–85.

Wootton, R. J. 1990. *Ecology of Teleost Fishes*, Fish and Fisheries, series. Chapman and Hall: London, UK.

Wright, D. H., 1983. 'Species energy theory: an extension of species–area theory'. *Oikos*, **41**, 495–506.

Yoder, C. O. and Rankin, E. T., 1995. 'Biological response signature and the area of degradation value: new tools for interpreting multimetric data', In: *Biological Assessment and*

Criteria: Tools for Water Resource Planning and Decision Making, Davis, W. S. and Simon, T. P. (Eds). Lewis Publishers: Boca Raton, FL, USA, pp. 263–286.

Zalewski, M., 1998. 'Ecohydrology and fisheries management'. *Italian Journal of Zoology*, **65**, 501–506.

Zalewski, M. and Naiman, R. J., 1985. 'The regulation of riverine fish communities by a continuum of abiotic–biotic factors'. In: *Habitat Modification and Freshwater Fisheries*, Alabaster, J. S. (Ed.). Butterworths: London, UK, pp. 3–9.

Zalewski, M., Janauer, G. A. and Jolánkai, G., 1997. *Ecohydrology. A New Paradigm for the Sustainable Use of Aquatic Resources*, Technical Document in Hydrology, IHP. UNESCO: Paris, France.

1.5

Aquatic Macroinvertebrates

Javier Alba-Tercedor

1.5.1   INTRODUCTION

The term 'macroinvertebrate' does not respond to a taxonomical concept but to an artificial delimitation of part of the groups of invertebrate animals. In running waters, we generally consider as macroinvertebrates those organisms large enough to be able

Biological Monitoring of Rivers   Edited by G. Ziglio, M. Siligardi and G. Flaim
© 2006 John Wiley & Sons, Ltd.

to be caught with a mesh size of 250 μm, and thus observed at first sight. In fact, most of them are larger than 1 mm in size.

Despite the general definition given above, some animal groups that could fit in it are never considered as macroinvertebrates (i.e. Protozoa and Tardigrada), while other groups (i.e. Nematoda, Nematomorpha, Cladocera, Copepoda, etc.) are not taken into account by methodologies based on macroinvertebrates.

At the beginning of this century, Kolwitz and Marsson (1902) clearly formulated the relationship of aquatic organisms to the purity and pollution of water. Since then, many methods to assess biological water quality, using different organisms (virus, bacteria, fungi, lichens, algae, plants, protozoa, macroinvertebrates and fishes) have been performed. However, most of the methodologies are based on macroinvertebrates (Hellawell, 1986; De Pauw *et al.*,1992; Rosenberg and Resh, 1993; Ghetti, 1997).

1.5.2 MACROINVERTEBRATE GROUPS

Most of the macroinvertebrate groups (Table 1.5.1) are arthropods, and the insects represent the great majority (Figure 1.5.1). To identify these groups, there are several general texts permitting identification, at least to the family, and in many cases to the generic level. Thus, for instance, for the European fauna there are general publications written in different languages, i.e. Catalan (Puig *et al.*, 1999), French (Tachet *et al.*, 2000), Italian (Campaioli *et al.*, 1994; Sansoni, 1988), Danish (Dall and Lingegaard, 1995), Dutch (De Pauw and Vannevel, 1990), English (Fitter and Manuel, 1986; De Pauw *et al.*, 1996; Puig *et al.*, 1999) and Spanish (Puig *et al.*, 1999). There exist many publications dedicated to each particular group and numerous websites are now appearing all over the world, hence permitting easy

Table 1.5.1   Most common groups of benthic macroinvertebrates

Arthropods	
Insecta	Crustacea
Plecoptera	Amphipoda
Ephemeroptera	Isopoda
Trichoptera	Decapoda
Odonata	
Heteroptera	
Coleoptera	
Megaloptera	
Diptera	
Non-arthropods	
Turbellaria	
Mollusca	
Oligochaeta	
Hirudinea	

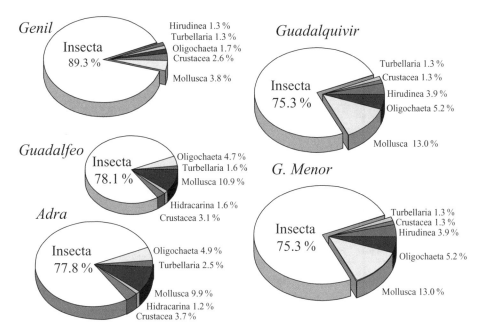

Figure 1.5.1 Macroinvertebrate compositions in different rivers in southern Spain

identification to family or genus level. However, an identification to the species level is more complex, requiring an interaction with taxonomists, and in many cases it is not possible to identify the immature aquatic forms, simply because they are unknown.

1.5.3 ADAPTATIONS TO SURVIVE IN A FLOWING HABITAT

Water current is the most important characteristic of running waters. Thus, aquatic macroinvertebrates have needed to develop anatomical and behavioral strategies in order to survive in this ecosystem to avoid being swept away.

According to Hynes (1970), the most common anatomical adaptations can be summarized as follows.

1.5.3.1 Flattening of the body

In running waters, the highest velocities are found at the surface and near to the centre of the channel, where the friction is lower (in deep rivers, due to the friction with the atmosphere current the speed is highest just below the surface). The velocity decreases, from the surface to the bottom as a function of the logarithm of the depth

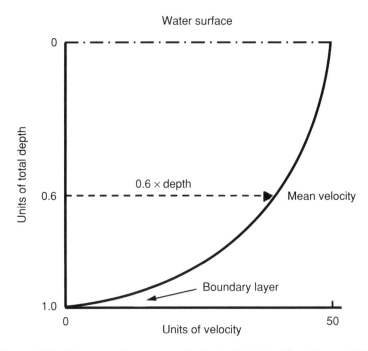

Figure 1.5.2   Decrease of current speed with depth (adapted from Hynes, 1970)

(Figure 1.5.2), and there is a *boundary layer* right on the bed in which it declines very rapidly to zero (Hynes, 1970; Allan, 1995).

   Some animals develop their activities in stony substrata and are flattened to be able to pass through the narrow pathways existing between pebbles and stones. Moreover, others, such as mayflies of the genera *Rhithrogena, Ecdyonurus* or *Epeorus*, feed while on the stones, and thus they would be exposed to the possibility of being swept away. However, they have solved this problem by developing flat enough bodies which allow them to live in the relative still boundary layer (Figure 1.5.3).

1.5.3.2   Streamlining (hydrodynamic shapes)

The advantages of a hydrodynamic shape, which decreases the resistance to the water, are well known in fishes, and one might expect it to be common among stream inhabitants. However, it is rather rare and almost the only animals that display it to perfection are the baetid mayflies (Figure 1.5.4).

1.5.3.3   Reduction of projecting structures

In general, projecting structures increase the resistance to the current and animals living in the current, where such structures are at risk of being swept away. Therefore,

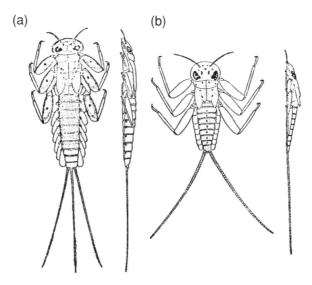

Figure 1.5.3 Flattening of the body in genera of mayflies of the family Heptageniidae: (a) *Rhithrogena*; (b) *Epeorus*. Reproduced by permission of the Blackburn Press from Hynes, H. B. N., 1979, *The Ecology of Running Waters*, Liverpool University Press, Liverpool, UK

Figure 1.5.4 Mayflies of the family Baetidae: (a) *Baetis maurus*; (b) *Rhodani*; (c) *Allainites muticus*

what has been observed in some animals seems logical. For instance, some mayflies, such as *Epeorus* or some species of *Baetis*, typical inhabitants of high velocities zones, reduce the terminal filament (Figure. 1.5.4).

1.5.3.4   Fixative or anchorage structures

Suckers

True hydraulic suckers can be found in leeches (Hirudinea) and in Diptera of the family Blephariceridae (Figure 1.5.5).

From a functional point of view, the broad feet of the gastropod molluscs are quite similar to suckers; in fact, the limpet-like Ancylidae are very characteristic of stony substrata (Figure 1.5.6), which support strong current velocities.

Hooks

Most of the stream-inhabitant arthropods present well-developed tarsal claws, hence permitting them to hold on to the 'roughnesses' on stone surfaces. Trichopteran larvae present conspicuous posterior claws. Circles of hooks help to fix the animals, and can be found, for instance, on the prolegs on some dipteran families, as in the case of the Blephariceridae (see Figure 1.5.5) and at the base of the body of the Simuliidae (Figure 1.5.7). Some pupae shown hook-attaching structures, for instance, the dipteran *Limnophora* shows posterior prolegs transformed in to hook

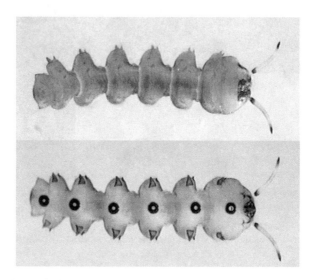

Figure 1.5.5   Views of a larva of Blephariceridae (Diptera), showing the characteristic ventral suckers which permit it to live in currents with high velocities

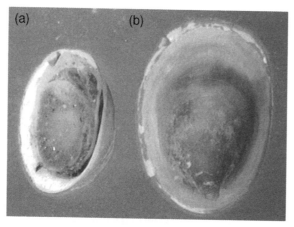

Figure 1.5.6 Views of the common mollusc gastropod species *Ancylus fluviatilis*: (a) ventral; (b) dorsal

structures (Figure 1.5.7), while some nymphs of mayflies (i.e. *Serratella ignita*) and stoneflies (i.e. *Taenyopteryx* sp.) (Figure. 1.5.7) have dorsal abdominal spines, thus helping them to anchor to the vegetation.

Sometimes, i.e. larvae of caddisflies, simulids and chironomids, secrete silk that helps to attach the individuals to the substrate, so enhancing at the same time the use of the hook structures.

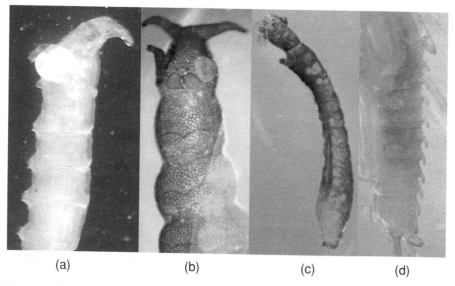

| (a) | (b) | (c) | (d) |

Figure 1.5.7 (a) Larva and pupa of *Limnophora* (Diptera), (b) larva of Simuliidae (Diptera), (c) nymph of *Serratella ignite* (Ephemeroptera) and (d) latero-abdominal view of the nymphs of *Taeniopteryx* sp. (Plecoptera)

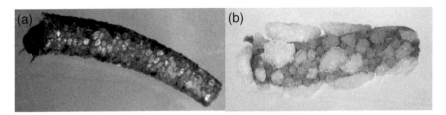

Figure 1.5.8   Caddisflies: (a) Sericostomatidae; (b) cases of the family Goeridae

Increasing the weight (ballasting)

Many caddisflies build their cases with large-sized sand grains ('little stones') that make them heavier and difficult to be swept out, while the Goeridae include additional 'large' pieces of sand (Figure 1.5.8).

   Among the mollusks, it is also noticeable that running-water species of the family Unionidae (*Anodonta* sp., *Unio* sp.) are often massive, as opposed to still-water forms (Figure 1.5.9).

Friction-pads and marginal contact

In many insects, the flattened ventral surface, or some structure around the edge of the animal, helps to maintain the body in close contact with the substratum, so increasing the frictional resistance and thus avoiding the possibility that the animal be dragged out by the current. In this sense, the soft flexible periostracum of the Ancylidae (see Figure. 1.5.6) fits closely to the irregularities of the surface of the substratum: the larvae of the beetles of the genus *Elmis* are flattened and have a peripheral complex of bristles which fit the surface and seal off the ventral side of the larva (Figure 1.5.10).

   The ventral gills of nymphs of some heptagenid genera of mayflies (*Rhithrogena* and *Epeorus*) are disposed, forming a ventral ring in contact with the substratum, hence increasing the area of marginal contact and reducing the possibility that water flows under the animals (Figure 1.5.11).

Figure 1.5.9   Bilvalvia molluscs of the family Unionidae: (a) *Anodonta* sp.; (b) *Unio* sp.

Figure 1.5.10 Larva of a beetle of the family Elmidae (*Elmis*)

Among the molluscs, it is also noticeable that running water species of the family Unionidae (*Anodonta, Unio* and *Margaritifera*) are often massive, as opposed to still-water forms.

1.5.3.5 Small size

It is clear that animals with small body size can inhabit the boundary layer, thus avoiding being dragged out. In this sense, we can find examples within the first instars of many insects, e.g. beetles of the family Elmidae (Figures 1.5.10 and 1.5.12), Hydracarina, etc.

Figure 1.5.11 Mayflies of the genera *Rhithrogena* (a, b) and *Epeorus* (c), where (b) and (c) are ventral views showing the friction pads formed by the abdominal gills

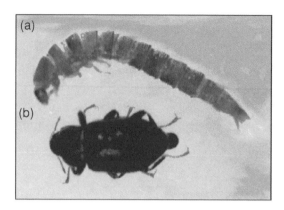

Figure 1.5.12   (a) Larva and (b) adult of an elmid Coleoptera of the genus *Riolus*

1.5.4   THE DRIFT

Despite their adaptations, stream-dwelling animals have high probabilities of being dragged out by the current when they are moving. The phenomenon of being swept down stream is known as 'drift'. There is a 'normal' daily drift that affects quite a large number of organisms, according to a circadian rhythm, and this generally increases during the night (Figure 1.5.13) (see Hynes, 1970 (in Allan, 1995)).

Different stress situations (floods, drought, pollution, etc.) produce 'catastrophic' drifts and in fact this phenomenon can be used to measure disturbances.

To compensate the drift, in many macroinvertebrate a positive rheotaxis has been observed, and thus they trend to move upstream. These movements may occur just by swimming or walking, and in insects the compensation upstream flights of adults (Figure 1.5.14), especially females, has been pointed out and evaluated by many authors (Müller, 1982; Hubbard, 1991).

The drift and its consequences have a very important role in recolonization (Williams and Hynes, 1976), and as a consequence, in recuperating the fauna of altered stretches of river and streams.

1.5.5   RESPIRATORY MECHANISMS

Most aquatic macroinvertebrates conduct gaseous interchange through the tegument (cutaneous respiration). Many small instars and small-sized groups also maintain this respiratory mechanism. Relatively large animals, which do not develop any respiratory systems however, need to inhabit very well oxygenated flowing waters (i.e. some gill-lacking larvae of Trichoptera or Plecoptera). Some groups, such as 'bloodwoms' (Chironomidae), Notonectidae or Oligochaeta (Figure 1.5.15), have a respiratory pigment (haemoglobin) that increases the efficiency to capture oxygen. Thus, these are very abundant downstream of organically polluted sites.

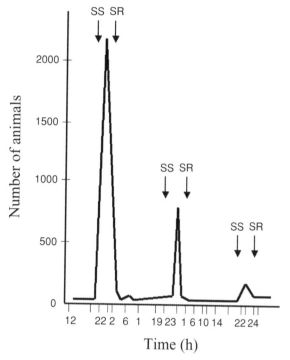

Figure 1.5.13 Drift activity of mayflies of the genus *Baetis* during artificially shortened nights (SS and SR correspond to natural sunset and sunrise, respectively), although artificial lights reduced the period of darkness to 4, 2 and 1 h, respectively (adapted from Müller, 1965 (in Allan, 1995))

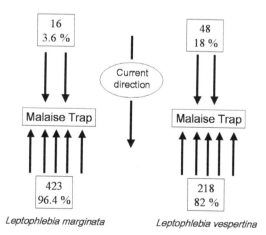

Figure 1.5.14 Compensation flights of two mayflies (adapted from Müller, 1982)

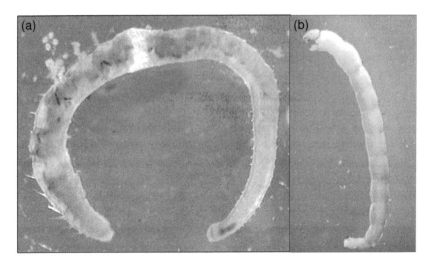

Figure 1.5.15 (a) Oligochaeta and (b) larva of Chironomidae

Most macroinvertebrates develop gills – tracheal gills in the case of insects. The animals trend to ventilate the body and/or gills by different methods (Merrit and Cummins, 1996): by undulation (i.e. Chironomidae, Trichoptera, Lepidoptera), swimming, or just positioning themselves in water-flowing areas (i.e. Plecoptera, Ephemeroptera, Zygoptera, Trichoptera, Simuliidae, Blephariceridae); beating the gills (Ephemeroptera), or by contractions of the legs which move the body up and down (Plecoptera, Zygoptera, Lestidae).

Insects have colonized aquatic habitats once they had developed a terrestrial tracheal respiratory system, a system with external spiracles. To be able to use this system in water, they have transformed the existing structures by reducing the external spiracles: totally (cutaneous respiration, or developing tracheal gills) or partially (at the end of the body). In the case that some spiracles remain, the animal may directly breathe atmospheric air. However, in a flowing habitat, reaching the surface may be quite risky. This is why many animals retain air bubbles, diminishing the frequency of displacements to the surface by using the bubbles as artificial-gases-interchange devices. However, spherical air bubbles tend to increase their content in nitrogen, so inducing the animal to go to the surface to renew the bubble. However, thinner and flat bubbles (*plastron*), such as those existing in the bug *Apheilocheirus* or in some Elmidae beetles, don't have that problem and permit the animal to dive permanently without the risk of reaching for the surface.

There are some insects, considered as 'plant breathers' (Coleoptera, Chrysomelidae, Curculionidae/Diptera, Culicidae, Ephydridae, Syrphidae), which when fixed to macrophytes can obtain oxygen directly from the plant.

In the case of oxygen decline induced by organic pollution, those animals that are not dependent on oxygen dissolved in water will be lesser affected, and thus more tolerant.

1.5.6 FEEDING MECHANISMS AND ROLE IN AQUATIC LOTIC SYSTEMS

To characterize what macroinvertebrates eat, analysis of the gut contents is normally carried out. With these methods, it is usually difficult to decide if a particular animal is a herbivore, a detritivore, or just a unspecific eater of small particles. This is an example of why it has been decided to distinguish functional food guilds on the basis of how the food is obtained, instead of what food is eaten (Cummins, 1973). Thus, different categories are distinguished (see Table 1.5.2).

Aquatic macroinvertebrates are important links within the lotic food web, playing a central role in the decomposition and processing of organic matter inputs, and in the processing of nutrients (Merrit *et al.*, 1984). According to the 'River Continuum Concept' (Vannote *et al.*, 1980), large particle shredders are common in headwaters where most energy is derived from coarse particulate organic matter (CPOM) (i.e. >1 mm, for example, leaves, twigs, wood) which enters from the adjacent watershed, but become rare in higher-order streams since organic matter is reduced to a finer particle size, and periphyton becomes more important as a food source (Figure 1.5.16). These biological processes, combined with physical abrasion, reduce the particle size of organic matter as it goes through the transport and storage process. Therefore, the particle size decreases as the stream order increases. According to this concept, downstream communities in higher-order streams depend on the inefficiency of the upstream communities when they process the CPOM.

1.5.7 MACROINVERTEBRATES AND POLLUTION

Different groups of macroinvertebrates have a different behaviour with respect to alterations, and thus lists of tolerant or non-tolerant species can be found (Hart and Fuller, 1974; Hellawell, 1986; Rosenberg and Resh, 1993). In general, Stoneflies, Trichoptera, and Ephemeroptera are considered non-tolerant, while Tubificidae or red Chironomidae are considered tolerant. In fact, the beginning of the use of macroinvertebrates to assess water quality was based on the concept of *indicator organism*, and still remains in some biomonitoring methods. However, this is a very simplistic approach. For instance, it is true that, in general, mayflies are found in good quality waters, and the number of nymph occurrences decrease in polluted waters (Figure. 1.5.17). However, some species can inhabit clearly polluted waters (Alba-Tercedor *et al.*, 1995). Thus, instead of using just the presence or absence of an indicator organism, the concept has been broadened to include *indicator communities* (Rosenberg and Resh, 1996).

According to Hellawell (1986), macroinvertebrates are suitable as bioindicators because of the following: (a) they constitute a heterogeneous assemblage of animal phyla and consequently it is probable that some members will respond to whatever stresses are placed upon them; (b) many are sedentary, which assists in detecting

Table 1.5.2 Macroinvertebrate functional groups (adapted from Merritt and Cummins, 1996, and Allan, 1995)

Functional group	Food resource[a,b]	Feeding mechanism[a]	Examples
Shredders	Living vascular hydrophyte plant tissue	Herbivores – chewers and miners of live macrophytes	Some families of Trichoptera, Plecoptera and Crustacea; some Diptera, Ephemeroptera
	Decomposing vascular plant tissue – CPOM	Detritivores – chewers of CPOM	Coleoptera and snails
	Wood	Gougers – excavate galleries in wood	Occasional taxa among Diptera, Coleoptera and Trichoptera
Collectors	Decomposing FPOM	Detritivores – filterers or suspension feeders	Net-spinning Tricho-tera, Simuliidae and some Ephemeroptera
		Detritivores – gather or deposit (sediment) feeders	Many Ephemeroptera, Chironomidae and Ceratopogonidae
Scrapers (grazers)	Periphytons – attached algae (especially diatoms)	Herbivores – grazing. Scrapers of mineral and organic surfaces	Several families of Ephemeroptera and Trichoptera; some Diptera, Lepidoptera and Coleoptera
Macrophyte piercers	Living vascular hydrophyte cell and tissue fluids or filamentous (macroscopic) algal cell fluids	Herbivores – pierce tissues or cells and suck fluids	Neuroptera and some Tricoptera (Hydroptilidae)
Predators	Living animal tissue	Engulfers – carnivores attack prey and ingest whole animals or parts. Pierces – carnivores attack prey, pierce tissues and cells, and suck fluids	Some Hirudinea, Plecoptera, Odonata, Coleoptera and Diptera

[a] CPOM, coarse particulate organic matter.
[b] FPOM, fine particulate organic matter.

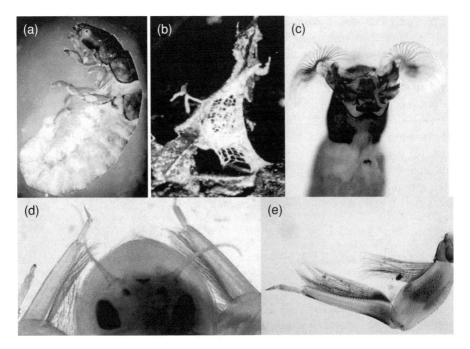

Figure 1.5.16 (Plate 1)   Examples of different mechanisms for detritivore-filters (collectors) by using silk nets (a, b) and filter setae (c–e): Trichoptera *Hydropsyche* sp. lava (a) and net (b); head of a larva of Simuliidae, showing the filtering mandibular folding fans (c); Ephemeroptera of the family Oligoneuriidae showing the long filter bristles of the internal sides of the forelegs (d, e)

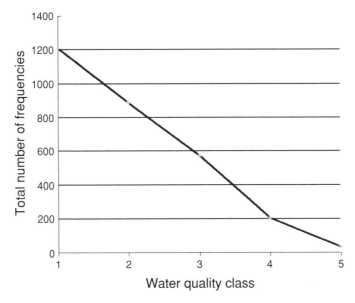

Figure 1.5.17   Frequency of captures of the Ephemeroptera species for different water qualities (from very clean, 1, to very polluted, 5) in the Guadalquivir River basin (Southern Spain)

the precise location of pollutant sources, and some have relatively long life histories (measurable in years) and this provides both a facility for examining temporal changes and also integrating the effect of prolonged exposure to intermittent discharges or variable concentrations of pollutants. This also justifies the adoption of periodic sampling. Such a procedure would be invalid for groups having short life histories; (c) qualitative sampling of benthic macroinvertebrates is relatively easy, the methodology is well developed and equipment does not need to be elaborate; (d) taxonomic keys are available for most groups although certain 'difficult" taxa exist; (e) perhaps the most useful feature of this group is that many methods of data analysis, including pollution indices and diversity indices, have been devised. Therefore the great majority of methodologies developed for biomonitoring water courses are based on benthic macroinvertebrates.

REFERENCES

Alba-Tercedor, J., Picazo-Muñoz, J. and Zamora-Muñoz, C., 1995. 'Relationships between the distribution of mayfly nymphs and water quality in the Guadalquivir River basin (Southern Spain)'. In:*Current Research on Ephemeroptera*, Corkum, L. D. and Ciborowski, J. J. H. (Eds). Canadian Scholar Press: Toronto, ON, Canada, pp. 41–54.

Allan, J. D., 1995. *Stream Ecology. Structure and Function of Running Waters*. Chapman & Hall: London, UK.

Campaioli, S., Ghetti, P. F., Minelli, A. and Ruffo, S., 1994. *Manuali per il riconoscimento dei macroinvertebrati delle acque doci italiane*, Vol. I. Agenzia Provinziale per la Protezione dell Ambiente: Provincia Autonoma di Trento, Trento, Italy.

Cummins, K. W., 1973. 'Trophic relations of aquatic insects'. *Annual Review of Entomology*, **18**, 183–206.

Dall, P. C. and Lindergaard, C. (Eds), 1995. *En oversigt over ferskvandsinvertebrater til brug ve bedommelse af forureningen i soer og vandlob*. Ferskvandsbiologisk Laboratorium, Kobenhavns Universitet.

De Pauw, N. and Vannevel, R., 1990. *Macro-invertebraten en Waterkwaliteitk Stichting* Leefmilien. State University of Ghent: Ghent, Belgium.

De Pauw, N, Ghetti, P. F., Manzini, P. and Spaggiari, R., 1992. 'Biological assessment methods for running water quality'. In: *River Water Quality. Ecological Assessment and Control*, Newman, P. J., Piavaux, P. A. and Sweeting, R. A. (Eds), EUR 14606 EN-FR. Commission of the European Communities: Brussels, Belgium.

De Pauw, N., van Damme, D. and Bij de Vaate, A., 1995. *Manual for Macroinvertebrate* Identification and Water Quality Assessment, document intended for use in the 'Integrated Programme for Implementation of the Recommended Transnational Monitoring Strategy for the Danube River Basin', a CEC PHARE/TACIS Project. State University of Ghent: Ghent, Belgium.

Fitter, R. and Manuel, R., 1986. *A Collins Field Guide to the Freshwater Life of Britain and North-West Europe*. William Collins: London, UK.

Ghetti, P. F. 1997. *Manuale di aplicazione indice biotico esteso (IBI). I macroinvertebrati nel controllo della qualità degli ambienti di acque correnti*. Agenzia Provinziale per la Protezione dell Ambiente: Provincia Autonoma di Trento, Trento, Italy.

Hart, C. W. and Fuller, S. L. H., 1974. *Pollution Ecology of Freshwater Invertebrates*. Academic Press: New York, NY, USA.

Hellawell, J. M., 1986. *Biological Indicators of Freshwater Pollution and Environmental Management*. Elsevier: London, UK.

Hubbard, M. D., 1991. 'Theoretical and practical problems involved in determination of upstream flight compensation in lotic aquatic insects'. In: *Overview and Strategies of Ephemeroptera and Plecoptera*, Alba-Tercedor, J. and Sánchez-Ortega, A. (Eds). Sandhill Crane Press: Gainesville, FL, USA, pp. 359–365.

Hynes, H. B. N., 1970. *The Ecology of Running Waters*. Liverpool University Press: Liverpool, UK.

Kolwitz, R. and Marsson, M., 1902. 'Grundzätze für die biologische Beurteilung der Wassers nach seiner Flora und Fauna'. *Prüfungsanst, Wasserversog, Abwasserreinig*, **1**, 33–72.

Merrit, R. W. and Cummins, K. W. (Eds), 1996. *An Introduction to the Aquatic Insects of North America*, 3rd Edition. Kendal/Hunt Publishing, Dubuque, Ia, USA.

Merrit, R. W., Cummins, K. W. and Burton, T. M., 1984. 'The role of aquatic insects in the processing and cycling of nutrients'. In: *The Ecology of Aquatic Insects*. Resh, V. H. and Rosenberg, D. M. (Eds). Praeger: New York, NY, USA.

Müller, K., 1982. 'The colonization cycle of freshwater insects'. *Oecologia*, **52**, 202–207.

Puig, M. A., Benito, G., Ferreras, M., Romero, J., Garcia-Aviles, J. and Soler, G., 1999. 'Els macroinvertebrats dels rius catalans'. Departament de Medi Ambient, Generalitat de Catalunya: Barcelona, Spain (in Catalan, with Spanish and English translations).

Rosenberg, D. M. and Resh, V. H. (Eds), 1993. *Freshwater Biomonitoring and Benthic Macroinvertebrates*. Chapman and Hall: New York, NY, USA.

Rosenberg, D. M. and Resh, V. H., 1996. 'Use of aquatic insects in biomonitoring'. In: *An Introduction to the Aquatic Insects of North America*, 3rd Edition, Merrit, R. W. and Cummins, K. W. (Eds). Kendall/Hunt Publishing: Dubuque, Ia, USA.

Sansoni, G., 1988. 'Atlante per il riconoscimento dei macroinvertebrati dei corsi d'acqua italiane'. Centro italiano di Biologia Ambientale: Provincia Autonoma di Trento, Trento, Italy.

Tachet, H., Richoux, P., Bournaud, M. and Usseglio-Polatera, P., 2000. *Invertébrés d'Eau Douce. Systématique, Biologie, Écologie*. CNRS Editions: Paris, France.

Vannotte, R. L., Minshall, G. W., Cummins, K. W., Sedell, J. R. and Cushing, C. E., 1980. 'The river continuum concept'. *Canadian Journal of Fisheries and Aquatic Science*, **37**, 130–137.

Williams, D. D. and Hynes, H. B. N., 1976. 'The recolonization mechanisms of stream benthos'. *Oikos*, **27**, 265–272.

1.6
Macrophytes and Algae in Running Waters

Georg Janauer and **Martin Dokulil**

1.6.1   MACROPHYTES

1.6.1.1   Ecological Relevance of Macrophytes in Rivers

Most authors consider "macrophytes" as genuinely aquatic (hydrophytes) and amphibious plants (amphiphytes) and that in most cases species determination

Biological Monitoring of Rivers   Edited by G. Ziglio, M. Siligardi and G. Flaim
© 2006 John Wiley & Sons, Ltd.

needs no microscope (Westlake, 1974; Wetzel, 2001). This covers large algae (e.g. Charophytes), bryophytes and vascular plants. In running water environments, macrophytes contribute to the oxygen budget and to the autochthonous carbon pool. Roots, stems and leaves of macrophytes provide structural elements in an otherwise unstructured water column, thus enhancing spatial diversity in the aquatic ecosystem. Microbes, invertebrates and vertebrates (Jeppesen *et al.*, 1998; Kornijow and Kairesalo, 1994) live on the surface of, and in spaces between, organs of aquatic plants. Fish prey, and fish, use aquatic plant stands as a nursery, feeding ground, or seek shelter in dense vegetation (Lillie and Budd, 1992; Ombredane *et al.*, 1995). The spatial functioning of macrophytes is influenced by the growth form of each species. The most simple scheme follows Sculthorpe (1967): emergents (helophytes), floating-leaved, free-floating (pleustophytes (Luther, 1949)) and submerged species. Submerged plants with small leaves and growing in dense stands, provide ample structure, whereas floating-leaved plants and pleustophytes provide little submerged surface, but support animals such as amphibians and water birds. As a consequence, aquatic macrophytes are one of the essential ecological components wherever they occur in running waters.

1.6.1.2 The role of river reach, catchment geology and river morphology on macrophyte development

Springs are the sources of running waters. Wherever water collects in pools or small lakes before it starts its run down the river, algae, bryophytes and even vascular plants grow. The limited nutrient load in spring ecosystems usually prevents the development of large plant mass. Few species are reported to grow in springs, e.g. macroalgae like *Nitella* species, oligotrophic *Chara* species, bryophytes like *Cratoneuron commutatum*, *Scapania undulata* and *Drepanocladus exannulatus*, and vascular plants like *Cardamine amara* or *Nasturtium officinale*.

In the upper reach, where rivers are still small and often close to their natural ecological status, two main parameters control the growth of macrophytes. The speed of water flow is often too fast and/or run-off too irregular to allow for the development of vascular plants in mountain rivers, but bryophytes (e.g. *Scapania undulata*, *Marsupella emarginata* (Lottausch *et al.*, 1980)) inhabit the hard and stable substrates found there. In the upper reach of lowland rivers, where bushes and trees grow next to the river, shading can prevent any macrophyte growth. Wherever the gradient decreases, or more light reaches the river, stands of vascular plants will develop.

In the middle and lower reaches, rivers create a system of secondary channels, meanders and oxbows. In situations where water flow is slow enough (for most vascular species < 0.8 m s^{-1}) and the width of the water body prevents heavy shading, macrophytes will become abundant. Downstream changes usually reflect an increase in species preferring higher trophic states (Kohler *et al.*, 1971; Kohler and Zeltner 1974; Holmes and Whitton, 1975, 1977a; Haslam, 1978, 1987; Janauer, 1981). Successive surveys in the same river (e.g. Fritz *et al.*, 1994; Würzbach *et al.*, 1977;

Kohler *et al.*, 2000; Veit and Kohler, 2003) clearly showed recuperation effects following, e.g. enhanced sewage treatment and phosphorus stripping in river catchments, and successful regrowth of macrophytes after severe floods (Henry *et al.*, 1994; Janauer and Wychera, 2000).

Water flow velocity is a dominant environmental parameter. In the river corridor of the Danube (original data, MIDCC database, 2003), *Potamogeton filiformis* and *Potamogeton nodosus* were associated with still water. *Myriophyllum spicatum*, *Potamogeton pusillus* and *Potamogeton natans* were indicative of water flow less than 30 cm s^{-1}. Intermediate flow (30–65cm s^{-1}) was characterized by the occurrence of *Potamogeton pectinatus*. For higher flow velocities, no statistically significant occurrence of vascular species was detected, but the absence of *Ceratophyllum demersum* and *Potamogeton perfoliatus* was significant ($p = 0.05$). In smaller rivers, *Ranunculus* species are often dominant in fast-flowing water.

In the centre of the main stem of large rivers, high flow velocities and moving sediment material prevent macrophyte growth. Near to the banks, the flow is lower, but plants rarely grow in water deeper than 1.5 m. This phenomenon is caused by shading due to either inorganic suspended material or phytoplankton growth (Westlake, 1975). However, permanent stands of plants even without roots, like *Ceratophyllum demersum*, were found in the main stem of the Danube in sheltered habitats, e.g. behind groynes (Janauer and Stetak, 2003).

Catchment geology is a prime factor in determining the occurrence of macrophyte species (Zander *et al.*, 1992; Grasmück *et al.*, 1993; Thiébaut *et al.*, 1995). Bryophytes like *Hygrohypnum ochraceum*, *Nardia compressa* and *Marsupella emarginata*, amphiphytic *Juncus* species, and hydrophytic *Myriophyllum alterniflorum*, *Potamogeton polygonifolius*, and *Ranunculus peltatus* are indicators of silicate rock areas with acidic water (Kohler and Tremp, 1996). In many cases, acidic rivers are also low in nutrient content and the occurrence of a specific species may indicate a combination of environmental factors. Bogs and mires tint streams and rivers with humic substances. The growth of submerged macrophytes is then suppressed, but species with floating leaves like water lilies, or spreading out from the bank like *Menyanthes trifoliata*, dominate the river. Typical species for rivers in regions dominated by calcareous rock are, e.g. Charophytes, *Fontinalis antipyretica*, *Cinclidotus fontinaloides*, *Callitriche obtusangula* and *Potamogeton coloratus*.

Substrate is another important factor. Rocks and hard, immobile substrates (rip-rap) are most closely associated with bryophytes, no matter what the elevation above sea level is for a certain location. Typical species on gravel are *Potamogeton perfoliatus* and *P. crispus*, but *Najas marina* is never found there. Pure sand forms the basis of many *Cyperus* stands, but is never found below *P. lucens* and *P. pectinatus*. The latter species is predominantly associated with fine inorganic sediments, together with *Phragmites australis* and *Typha angustifolia* (original data, MIDCC project). It is clear that all classification of sediment type based on field surveys is only accurate to the extent that the surveyor's eye determines the essential fraction. In many cases, small volumes between large stones or gravel contain fine sediment in which the plants actually root. Useful collections of physical and chemical habitat factors are the compilations by Haslam (1978, 1987) and the concise review by Bayerisches Landesamt (1998).

1.6.1.3 Regional distribution and abundance

The remarkable study of running water vegetation in EC states by Haslam (1987) is still a valuable source of basic information. Three main reasons for diminished aquatic plant growth were listed: highlands, where the water flow has high force, pollution and river straightening in regions with low or medium elevation above sea level throughout Europe. *Ceratophyllum demersum*, *Elodea canadensis*, *Myriophyllum spicatum*, *Nuphar lutea* and *Sparganium emersum* were found in more eutrophic habitats, and *Potamogeton crispus* in semi-eutrophic reaches. *Potamogeton pectinatus* was concentrated in polluted areas. *Callitriche hamulata* and *Myriophyllum alterniflorum* were frequent in oligotrophic and dystrophic habitats. This general picture is still valid for most regions in Europa, but as only Western European countries were included different patterns of dominance and habitat preference should be expected in other regions.

The Atlantic zone can be characterized by examples from the United Kingdom and North-Western France. In the River Tweed system (UK), the upper reach was dominated by *Hygrohypnum ochraceum*, *Scapania undulata*, *Myriophyllum alterniflorum* and *Ranunculus aquatilis* agg. In the lower reach, *Potamogeton lucens*, *P. perfoliatus*, *P. pusillus*, *P. pectinatus*, *Ranunculus fluitans* and *Zannichellia palustris* were found. Following the course of the Tweed and its tributary, Teviot, the dominance of bryophytes decreased, and that of vascular species increased with distance from the source. However, bryophytes like *Eurhynchium riparioides* or *Hygroamblystegium fluviatile*, as well as vascular plants like *Elodea canadensis*, *Potamogeton crispus* or *Ranunculus penicillatus*, occurred throughout the whole length of the river (Holmes and Whitton, 1975). The importance of bryophytes within the macrophyte assemblage is also mirrored by other rivers in this region (Holmes and Whitton, 1977b, 1977c; Haslam, 1978).

In the Bretagne (France), Haury (1996) recorded more vascular species, e.g. *Ranunculus hederaceus*, than bryophytes in the upstream parts of Kernec Brook. However, several bryophytes (e.g. *Leptodictyum riparium*) were found next to *Ranunculus pseudofluitans* in the lower reach. Studies on the relationship of species to environmental factors show the importance of bryophytes as part of macrophyte communities in Western France (Haury *et al.*, 1995; Ombredane *et al.*, 1995, Haury, 1996).

In Sweden, the rivers Kävlinge and Björka were studied by Kohler *et al.* (2000). The slightly eutrophic Kävlinge River was dominated by *Potamogeton perfoliatus* in its upper reach, while further down its course *Potamogeton pectinatus*, several *Potamogeton* hybrids and *Ceratophyllum demersum* reached high abundance, and near the mouth *Myriophyllum spicatum* and *Zostera marina* dominated the species pattern. *Potamogeton lucens*, *Nymphaea alba* and *Sagittaria sagittifolia* were among the most widespread species. As a characteristic of many northern European rivers,

46% out of 65 aquatic species were of amphiphytic character. In the eutrophic Björka River, amphiphytes dominated (56% of all species) and were most abundant throughout the whole river. With little variation, Lemnids, *Elodea canadensis* and *Nuphar lutea* were dominant among the hydrophytes. Very few truly aquatic bryophytes were found. In the humic River Bräkne (Kohler *et al.*, 1996) 57% of all species were amphiphytes, and the floating leaved species *Nuphar lutea* and *Nymphaea alba* dominated the submerged *Myriophyllum alterniflorum*. Similar to the macrophyte assemblages in England and Western France, bryophyte species were as numerous as the vascular macrophytes, and *Fontinalis antipyretica* showed medium to high abundance throughout the whole river course.

The head waters of the Portuguese River Divor were characterized by the vascular aquatic species *Callitriche stagnalis*, *Zannichellia palustris* and *Potamogeton pectinatus*. In the middle course, *Utricularia australis*, *Myriophyllum spicatum* and *Ceratophyllum demersum* occurred, but fringing herbs and helophytes became increasingly dominant. In the lower reach, *Myriophyllum aquaticum* prevailed between emergent species. Bryophytes represented only 3% of the listed species (Ferreira, 1994). Quite similar assemblages of macrophytes were found in other Iberian rivers and in drainage channels (Ferreira *et al.*, 1998a,b).

The Danube River crosses Europe from West to East over a course of over 2800 km and its catchment is highly complex. The Alps, the Carpathians and the Balkan Mountains dominate much of its run-off regime in the upper and middle part where constrained reaches and wide fluvial plains alternate, until the river reaches the lowlands of Romania, Moldavia and the Ukraine. Hydropower regimes dominate most of the river in Germany and Austria. In the head water streams Breg and Brigach, which form the Danube, several bryophytes and *Callitriche hamulata* dominate the upper parts. In the lower reach of the Brigach, *Fontinalis antipyretica* is the only bryophyte, but vascular plants occur with high abundance. In the Breg, bryophytes, as well as vascular species, are less abundant. In the Danube, bryophytes are confined to the upper reach (Germany, Austria). In the middle and lower reaches, vascular species reach high abundance in sheltered habitats, even in the main channel. Floodplain water bodies and the canals in the Danube delta rival for the maximum in species richness and abundance (Janauer, 2003b).

1.6.1.4   Macrophytes in flood plain water bodies

Fluvial corridors in Europe clearly reflect the human impact since the mid-19th Century: the larger rivers are confined to a regulated main channel and former side channels and meanders are oxbows and relict waters with reduced, or missing, permanent connection with the main stem today. Several studies (Rhone (Bornette and Amoros, 1991; Bornette *et al.*, 1994, 1998), Rhine (Buchwald *et al.*, 1995; Robach *et al.*, 1997, Fritz *et al.*, 1998; Greulich and Tremolières, 2002), Danube (Janauer, 2003a; Sarbu, 2003; Janauer and Stetak, 2003; Otahelova and Valachovic, 2003; Janauer *et al.*, 2003; Rath *et al.*, 2003)) have dealt with macrophytes in floodplain waters. In general, an increase in species richness was detected where the direct influence of the river was lower and where groundwater influence increased. In the Danube floodplain waters, species richness did not always correspond to disturbance frequency in a uniform way, but the total amount of hydrophytic vegetation clearly decreased with enhanced disturbance (Janauer, 2003a). In Hungary, a very pronounced increase in species richness was detected between the main stem of the Danube, floodplain waters, and man-made irrigation and drainage channels, where the most constant hydrological and hydraulic regime prevailed in the last category (Janauer and Stetak, 2003). In a statistical analysis (PCA, DA, sign test, $p = 0.05$) of the Danube and its secondary water bodies, the absence of *Elodea canadensis* from the main channel, of *Potamogeton perfoliatus* from large secondary channels and of *Potamogeton crispus* from small side-arms, and the presence of *Potamogeton pectinatus*, *Myriophyllum spicatum* and *Pharagmites australis* in these respective cases, were characteristic and indicative (Filzmoser and Janauer, original data). Aside from the examples presented here, many studies did not distinguish between the contribution of amphiphytes and helophytes to total species richness. Therefore, the final picture of macrophyte reaction to differences in connectivity is certainly more complex than pointed out so far. One conclusion can be drawn without any reservation: river regulation, urban development and intensive agriculture restricted wetlands, water courses and fluvial corridors to their very limits of spatial extension in almost every river catchment in Europe. The remaining floodplain water bodies are the last and most essential species refuges for aquatic vegetation in modern landscapes.

1.6.1.5   Macrophytes and pollution

Pollution can be defined as man-made impacts on water quality. Annex VIII, Indicative List of the Main Pollutants, of the Water Framework Directive (WFD, 2000), the most recent work dealing with all aspects of pollution, lists organohalogens, organophosphorus and organotin compounds, carcinogenic and mutagenic substances, persistent and bioaccumulating hydrocarbons and organic toxic substances, cyanides, metals and arsenic and their compounds, and biocides and plant protection products, suspended solids, substances causing eutrophication (nitrate, phosphate) and substances influencing the oxygen balance. Eutrophication and sewage impact are the most common threats to water quality. With regard to nitrogen compounds, the ammonium concentration is the more discriminating feature as compared to nitrate. The importance of aquatic macrophytes in supporting natural pollution reduction processes was recognized a long time ago (Liebmann, 1940). General aspects of different types of pollution and some pollution-related indices have been discussed by Haslam, (1987).

Some mosses (*Marsupella emarginata*, *Cratoneuron commutatum*) occur predominantly in oligotrophic water (Lottausch *et al.*, 1980), but Fontinalis antipyretica grows well in eutrophic conditions, too. The same is true for Charophytes: *Chara aspera* indicates predominantly oligotrophic conditions, whereas *Chara vulgaris* is most abundant in eutrophic water (Bayerisches Landesamt, 1998). Vascular plants exhibit a wider range related to plant nutrients, which differs with respect to geographical region, river type and adaptation of a species. *Isoetes* species, *Potamogeton polygonifolius* (acidic water), *Potamogeton coloratus* (carbonate-rich rivers) and *Juncus subnodulosus* indicate nutrient-poor water. On the other end of the scale, *Ceratophyllum demersum, Potamogeton crispus, Potamogeton pectinatus, Sparganium erectum* and *Sagittaria sagittifolia* become most abundant in eutrophic and eu-polytrophic conditions, but are found in oligo-meso- or mesotrophic water as well. The most extreme range is reported for *Zannichellia palustris*, which survives oligotrophic conditions as well as polytrophic water. Clear relations between macrophyte abundance and phosphate and ammonium content in running waters were summarized by Kohler (1978). The negative influence of fish farms on small acidic rivers was extensively studied in Western France (Daniel and Haury, 1995, 1996). Changes in macrophyte composition caused by eutrophication were also studied with multivariate statistical methods, e.g. by Monschau-Dudenhausen (1982) and Filzmoser *et al.* (2003). With regard to geographic regions, macrophytes may react differently to trophic levels. Three ways of assessing the trophic state of water bodies using macrophytes were developed in Europe: the "Mean Trophic Rank" (MTR in the UK, Holmes *et al.*, 1999), the "Trophic Index of Macrophytes" (TIM in Germany, Schneider *et al.*, 2001; Kohler and Schneider, 2003) and the "Indices Biologiques macrophyte en rivière" (IBMR in France, Haury *et al.*, 2002). All calculations of the indices in a survey unit include the abundance of the species present, and the focus of occurrence of those species with regard to phosphorus concentration in their environment. When applying these indices, their indicative power should be re-estimated as it may deviate from the relation between macrophytes and nutrients found in the geographic region where the indices were developed.

One consequence of eutrophication, as well as of organic pollution, can be increased turbidity, which reduces light intensity in the water. This and the deposition of silt and/or bacteria and algae on the leaves may reduce macrophyte growth to a great extent. Low and intermediate organic pollution *per se* does not seem to influence macrophytes too much, but alpha-saprobic conditions usually limit the growth of mesotrophic species.

The toxicity of detergents (Westlake, 1975; Labus, 1979) was studied in some detail. Special interest focused on bryophytes (Miller *et al.*, 1983; Say and Whitton, 1983; Wehr and Whitton, 1983; Tremp, 1991; Rath, 1995; Samecka-Cymerman and Kempers, 1998; Zhihong *et al.*, 1998) and their application for heavy-metal passive monitoring. In most cases, acidification impact parallels with heavy-metal toxicity and the same bryophyte species function as indicators. Regarding salinity tolerance, Olsen (1950) is still a valid source of information. Several limnic macrophytes like *Potamogeton natans, Sagittaria sagittifolia, Elodea canadensis, Lemna trisulca* and *Nuphar lutea* withstand oligohaline conditions (< 0.3 % salinity) and *Potamogeton pectinatus, Najas marina, Ranunculus baudotii* and *Zannichellia palustris* prevail under β-mesohaline conditions (Luther, 1951; Gessner, 1959). Most species tolerating high trophic levels survive enhanced salinity levels. Good examples are *Potamogeton pectinatus*, *Myriophyllum spicatum* and *Ceratophyllum demersum* which recently showed progressive growth in shallow lakes with high ionic concentrations, e.g. Lake Balaton (Hungary) and Neusiedler See/Lake Fertö (Austria/Hungary, Dinka *et al.*, 2004; Richter, 2004).

Other man-induced impacts on rivers are changes in flow regime and connectivity caused by river regulation or construction of hydroelectric power plants, and changes in man-induced turbidity. Such changes are readily mirrored in the composition of the aquatic vegetation. Despite their rather wide amplitude of habitat preference, macrophytes react to environmental stress and resulting changes in species composition and abundance are easily observed. For the assessment of ecological conditions in surface waters, e.g. with regard to Water Framework Directive monitoring, aquatic macrophytes are an indispensable tool in modern environmental control and research.

1.6.2 ALGAE

1.6.2.1 Algal vegetation, geology and river morphology

Algae are an essential part of river vegetation. As a polyphyletic group they are intermediate in size and generation time between smaller micro-organisms and higher plants. River algae can belong to any freshwater algal group, with Cyanobacteria (blue–green algae), Bacillariophyceae (diatoms) and Chlorophyceae (green algae) being usually most prominent. Their morphological variability ranges from unicellular, to colonial, to filamentous. Some are motile (Table 1.6.1). For taxonomic and life-cycle details, consult Ettl (1980) or Van den Hoek *et al.* (1995).

River algae grow in a multitude of heterogeneous habitats along the river corridor from springs, through brooks to large rivers. In smaller streams, algae are almost

Table 1.6.1 Morphology in the divisions of benthic algae (adapted from Stevenson, 1996)

Taxon	Unicellular[a]		Colonial[a]		Filamentous[a]		Motility by
	Mot	n-m	Mot	n-m	Mot	n-m	
Cyanobacteria (blue–green algae)		•		•	•	•	Sheaths
Rhodophyta (red algae)						•	
Chrysophyta (chrysophytes)	•	•	•	•		•	Flagella pseudopods
Xantophyta (xantophytes)						•	
Phaeophyta (phaeophytes)					•	•	Sheaths
Bacillariophyta (diatoms)	•	•		•		•	Raphe
Chlorophyta (green algae)	•	•	•	•		•	Flagella
Euglenophyta (eugleoids)	•						Flagella
Pyrrophyta (dinoflagellates)	•	•		•			Flagella
Cyryptophyta (cryptomonads)	•						Flagella

[a] Mot, motile; n-m, non-motile.

exclusively attached to a wide variety of substrates as *Periphyton* ('*Aufwuchs*'). Microscopic-sized algae are often macroscopically visible. In oligotrophic waters, they appear as coloured incrustations of less than 1 mm in thickness, while in more nutrient-rich waters, periphyton assemblages form matrices of several centremeters in thickness. These *biofilms* can be differentiated from macroscopical filamentous aggregates often more than a metre long by using the term *benthic algae* which otherwise is used synonymous with periphyton (Stevenson, 1996). On the same substratum type, however, macroalgae are in very different habitats than microalgae, because they extend farther into the water. A common criterion for habitat is the substratum type (refer to Stevenson (1996) for a detailed discussion) which is also important for the various interactions between benthic algae with their substrata, including *endolithic* algae (Burkholder, 1996). Structure and stability/instability of the substrata are crucial factors for benthic algal growth. Both factors depend on the position within the fluvial corridor and the hydrological regime affecting flood disturbance.

In more protected areas, algal aggregates can develop which are not directly attached to substrata, nor are they freely suspended in the water column. This *Metaphyton* comes in many forms and may have many origins. Usually, metaphyton are clouds of filamentous algae, like *Spirogyra, Mougeotia,* or assemblages of filamentous cyanobacteria, but can often originate from other submerged substrata as well when they become detached (Stevenson, 1996).

Larger rivers contain algae in the free-flowing water as river phytoplankton or *Potamoplankton* which prolifically reproduce in rivers, often achieving high biomass (Reynolds, 1988). River phytoplankton is usually dominated by diatoms and a variety of green algae, particularly during summer (Dokulil, 1991, 1996). Other algal groups are suppressed by the current but may become more distinct where currents are reduced. Much of the phytoplankton community, however, may derive from attached forms (*Meroplankton*) or originates from lake surface outflow, and outlets of streams and rivers. Survival is accomplished by a variety of water-retentive mechanisms, most importantly by backwater 'dead zones' (Reynolds and Descy, 1996).

Algal communities in streams and rivers can be seen as *ecotones* mediating between different habitats. Biofilms link the substratum with the overlaying water. River plankton acts as a mega-ecoton connecting the river with the floodplain (Dokulil, 2003a). Through this lateral connectivity, flood-plain waters exchange algal taxa and nutrients with river communities (e.g. Dokulil and Janauer, 1990).

1.6.2.2 Algal vegetation and environmental factors

The ability of river algae to grow and prosper, especially on substrata in streams, is the outcome of complex interactions between hydrology, water quality and biotic factors which, in turn, reflect the topography, slope, land-use and vegetation of catchments, among others. Broad-scale patterns of benthic algae in different geographic areas, and between years, reflect the geology, climate and human activity of their watersheds

(Biggs, 1996). Resources, particularly nutrients and light, are the main factors reg-ulating algal growth in streams. In addition, temperature influences metabolic rates.

In areas of negligible anthropogenic disturbance, nutrient concentrations mainly depend on the geochemical background which, in turn, is related to geology. The main differences arise from carbon availability which is linked to the pH value, and silica concentrations, a potential limiting element which is essential for cell-wall formation, especially in diatoms and Chrysophyceae. Similarly, sulfate and chloride concentrations, which are important parameters for species composition, are dependent in undisturbed streams on the geological underground (Rott *et al.*, 1999).

Light environments of the benthos are highly variable and may range from near zero to full sunlight. In addition, considerable temporal variability is observed from very short 'sun spots' to seasonal or inter-annual long-term changes. Terrestrial streamside vegetation reduces the light intensity, alters the spectral distribution and creates considerable heterogeneity. As light impinges on and penetrates into the water column, both the quantity and the spectral composition are altered. Part of the impinging radiation is reflected. The penetrating part is attenuated exponentially which, in addition, is selective because of various suspended and dissolved compo-nents in the water. The importance of light attenuation increases in running waters with stream size because of greater depths and more likely effects from phytoplank-ton and suspended particles. When light finally reaches the benthic community, it is further reduced by the matrix of the biofilm and associated inorganic particles. Ad-ditional effects come from UV irradiation penetrating considerably into clear waters or affecting algae on river banks (Hill, 1996).

Light intensity affects biomass, productivity and taxonomic composition of ben-thic algal species. When grazing pressure is low, algal biomass and productivity often correlate with the amount of stream-side vegetation. When grazing pressure is high, biomass is not correlated to light. Turbidity from inorganic particles substantially reduces the light available for photosynthesis for both plankton and benthic algae in larger streams and rivers. Differences in light response of the major taxonomic categories of algae have been suggested, although large interspecific variability is observed. Diatoms, cyanobacteria and rodophytes in general, grow better than most of the chlorophytes under low-light conditions. Motile benthic taxa, such as raphe-bearing diatoms, have a distinct advantage over non-motile taxa because they may regulate their light environment through phototaxis. The effects of light quality are much less clear than those from variation in light intensity. Most authors, however, conclude that light quantity is much more important than spectral distribution. Many benthic species are either highly resistant against high light intensities and UV irradi-ation or have protective mechanisms against the detrimental effects of UV exposure (Hill, 1996). These mechanisms include mycosporine-like amino acids, the sheath pigment scytonemin or carotenoids (reviewed by Castenholz and Garcia-Pichel, 2000).

Temperature is one of the most important environmental factors because it affects biochemical reactions. Water temperature is primarily determined by direct solar ra-diation. On a large spatial scale, temperature regimes depend on latitude, elevation, continentality and morphometry. Water temperature also varies with long-term climatic cycles which recently are increasingly influenced by anthropogenic changes in global climate. As the temperature increases, dominance of algal classes shifts from diatoms to green algae to cyanobacteria. Concomitantly, biodiversity and biomass increases to an upper limit at around 20–25°C, both decreasing at higher temperatures. Community structure and biomass of the very specialized flora on sand (*epipsammic*) are less affected by temperature than other periphyton assemblages. Temperature is not limiting in most natural communities but the degree to which primary productivity is limited by factors such as light, nutrients and grazing depends on temperature. A more complete understanding of temperature effects, especially autecological responses of (benthic) algae, is required to make accurate predictions of periphyton response in a 'warmer world' (De Nicola, 1996).

Abundance, biomass and seasonality of algal vegetation

Development of benthic algal communities in streams and rivers is governed by a complex array of factors and interactions. In addition, different sampling and analytical techniques, as well as the use of artificial substrates, which may be highly selective (e.g. Schagerl and Donabaum, 1998), make generalizations difficult. Available data, however, permit us to conclude that benthic chlorophyll-a values in streams can span four orders of magnitude and ash-free dry mass three orders in the course of a year (Biggs, 1996). Some of the variation is accounted for by the degree of enrichment, vegetation cover, development of the water-shed and depends on the type and quality of the river (Table 1.6.2). Filamentous green algal communities can have particularly high chlorophyll-concentrations (> 600 mg m^{-2}) and ash-free dry mass (> 200 g m^{-2}).

The temporal pattern of short-term benthic algal accrual is clear and generally universal (Figure 1.6.1). It starts off as a linear process by immigration/colonization, followed by exponential growth. Accumulation of biomass slows down when the loss

Table 1.6.2 Algal biomass range in rivers of different qualities (adapted from Marker and Collett, 1991)

River type	NO$_3$-N (mg l^{-1})	PO$_3$-P (mg l^{-1})	biomass range, minimum (g m^{-2})	chl-a, maximum (mg m^{-2})
Soft water, pH 4–5	0.06–0.35	< 1	26	92
Soft water, pH 6–7	0.5	1–2	40	178
Head water chalk stream	4–5	10–30	50	200
Chalk stream	4–5	30–50	15–25	150–300
Hyper–eutrophic stream	5–15	1000–3000	25–50	150–250

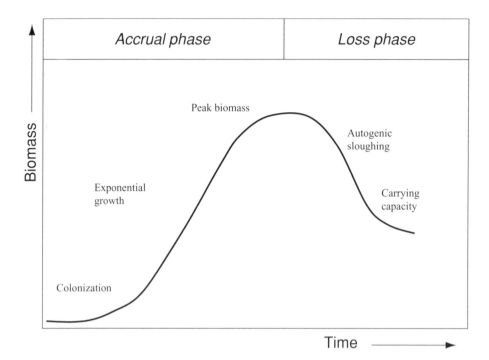

Figure 1.6.1 Idealized accumulation curve of benthic algal biomass, showing different phases (adapted from Biggs, 1996)

rates approach accrual rates, finally reaching the carrying capacity when both processes balance. Long-term or seasonal patterns of biomass can usually be classified into one of the following:

(1) more or less constant, low biomass when streams are frequently disturbed;

(2) accrual and sloughing cycles in rivers experiencing moderate or seasonal flood disturbance;

(3) seasonal cycles in community development are mediated by:

    (a) seasonality in disturbance regimes if nutrient resources are adequate;

    (b) seasonality in grazer activity when flood disturbances are rare;

    (c) seasonality in light regimes when neither (a) nor (b) are important.

Spatial patterns in the distribution of river benthic algae occur on a wide range of scales, from single sand grains to across continents. Micro-scale patterns on substrates vary according to species preference, some living in depressions or crevices while others prefer more exposed positions. Biomass is often greater on larger substrata because of higher stability. Communities on boulders are structured in zones with increasing depth (Blum, 1956). In gravel streams, algae may live as deep as 1 m within the hyporheic interstitial substrate. Assemblages on more mobile substrata are more frequently disturbed and set back to early successional stages (Peterson, 1996). Meso-scale distribution within a catchment can occur in pool, run and riffle habitats. In general, benthic algal biomass increases progressively downstream from head water to mid-catchment reaches. When river channels become wider and deeper, biomass is predicted to decrease again in lower reaches as a function of light attenuation (Vannote *et al.*, 1980). Gradients are also formed by increased nutrient loading downstream leading to severe changes in benthic algal composition. Large-scale regional (inter-catchment) patterns reflect differences in geology, flow pattern, flood regime, land-use and associated nutrient enrichment (Biggs 1996).

1.6.2.3 Ecological relevance of algal vegetation

The role of benthic algae in river and stream environments is manifold. As primary producers, algae are the basis of any food web. River algae are important components in both *macroscopic* and *microscopic* food webs (Lamberti, 1996; Bott, 1996). As *producers*, benthic algae serve as the food basis for many *consumers*, particularly for herbivores, detritivores and decomposers. Depending on the size and the reach of the river, periphyton may contribute varying amounts of *autochthonous* organic carbon to the overall energy budget. In small-order streams (order 1–2), autotrophic production may be as small as < 1 % when most organic carbon is derived from external sources (*allochthonous* carbon) but may reach as much as > 60 %. The degree of river autotrophy is related to stream size and biome type. In arid regions with limited riparian vegetation, for instance, carbon budgets are dominated by autotrophic production. In most streams, there is a delicate interplay between production and consumption. Plant–herbivore interactions are central to food web structures and energy flow. Streams may be, however, at the extreme of the environmental constancy spectrum because of their frequent disturbance. Benthic food webs therefore have fewer links and lower connectance than in other ecosystems. Such food webs may also be more easily disrupted by global environmental changes.

At the other, microscopic end, benthic algae are important because their excreted organic substances (∼ 10 % of total photosynthate) support heterotrophic organisms, particularly bacteria. Both algae and bacteria are ingested by a variety of organism groups, such as protists, which have an amazing diversity of energy acquisition, rotifers, copepods, nematodes, etc. Few, if any, of these taxa are strictly algivorous or bacterivorous. Other food, such as *detritus* fungi, and other animals, is used as well.

Benthic algae also contribute significantly to nutrient cycling in stream ecosystems directly by increasing nutrient supplies, uptake of nutrients and back-release to the river water. Through the formation of boundary zones, benthic algae alter the hydraulic characteristics of the river bottom and therefore contribute to nutrient cycling indirectly, because the probability for remineralization increases before nutrients are transported downstream. In addition, interactions between periphyton and herbivores can have positive or negative effects on nutrient cycling in streams depending on the relative importance of algal biomass (negative) or remineralization (positive).

Algae, both benthic and planktonic, positively influence the oxygen budget of streams and rivers through oxygen evolution during photosynthesis. This is especially important in rivers carrying larger pollution loads, resulting in oxygen depletion due to bacterial respiration during mineralization.

River algae as ecological indicators

Benthic algae are primarily used for ecological monitoring of streams. Phytoplankton is used in larger rivers when conditions are appropriate. The rationales, methods and techniques of bioindication and biomonitoring using algae have recently been summarized by Dokulil (2003b).

Benthic river algae possess several attributes to make them ideal organisms for water quality monitoring (Lowe and Pan, 1996):

(1) Benthic algae are primarily autotrophic and are positioned at the interface be- tween environmental and biotic components of the food web.

(2) Benthic algae are sessile and cannot avoid pollution through migration.

(3) Benthic communities are usually species-rich and therefore represent an information-rich assemblage ideal for environmental monitoring.

(4) Many benthic algae have relatively short life cycles, thus allowing rapid response to shifts in environmental conditions. Many species, however, live long enough to integrate impacts over certain periods of time.

Historically, benthic algae have been used for monitoring organic pollution of streams and rivers, by applying the concept of *saprobity* (Rott *et al.*, 1997). More recently, the emphasis has changed to direct nutritional effects using algae for trophic classifica- tion because algal growth is usually limited by a single substance (Rott *el al.*, 1999). As a consequence, structure and abundance of phytoplankton and periphyton is now included in the EC-Water Framework Directive (WFD, 2000) for water quality as- sessment and monitoring. Benthic algae can be used for *active monitoring* by using the reactions of individual species or assemblages to trace environmental changes, but may also be used in *passive monitoring* when species accumulate substances such as heavy metals from the surroundings. Summaries of algal usage for river monitoring in individual countries can be found in Whitton *et al.* (1991), Whitton and Rott (1996) and Prygiel *et al.* (1999).

Among the various algal groups in streams and rivers, diatoms are most univer- sally used in freshwaters (e.g. Dokulil *et al.*, 1997; Rott *et al.*, 2003; Poulíčková *et al.*, 2004) because many species are very sensitive indicators of changes in the surrounding environment over short (days to weeks) or even very long time periods (palaeoecology). Diatoms can be used to trace changes in trophy, organic pollution, acidification, salinity or climate (summarized in, e.g. Schönfelder, 2000). However, several different methods are in use. If amalgamated, unified and tested, diatoms could form the basis for a classification system across lakes and rivers. Diatoms can be easily identified by their shells which are highly resistant and can readily be preserved as permanent mounts. In many cases, however, identification is also possible from inspection of live cells, at least at higher taxonomic levels (Cox, 1996). Diatoms, like many other algal groups, are sensitive indicators for inorganic chemical stress (Genter, 1996), toxic organic substances (Hoaglund *et al.*, 1996) or the effects of acidification (Planas, 1996). Benthic algal biomass and production may not be drastically altered in response to organic toxicants. At the community level, however, shifts in species composition and structure are common. A variety of organic toxicants, particularly herbicides, produce dramatic impacts to benthic algal communities at concentrations already below the $\mu g \ l^{-1}$ range (Dokulil, 2003b; Fent, 2003). During acidification of streams and rivers, carbon may become limiting, metals, especially aluminium, become toxic, nitrogen becomes more available, and microbial–algal interactions are altered. As a result, species composition will change and affect herbivore structure. In contrast to most algal groups which are sensitive to toxicants, cyanobacteria are potential producers of cyanotoxins which can be detrimental for fish, insects and mammals. In some instances, human health might be affected as well (Dow and Swoboda, 2000). All of these effects may become more pronounced and deleterious through food-web disrupture when global environmental changes impact on fresh waters.

ACKNOWLEDGEMENTS

Norbert Exler is thanked for extracting information from, and Peter Filzmoser is much acknowledged for statistical analyses of, the MIDCC Danube River database. Michaela Bohenzky assisted in referencing.

REFERENCES

Bayerisches Landesamt, 1998. *Trophiekartierung von Aufwuchs- und makrophyten-dominierten Fließgewässern*. Bayerisches Landesamt für Wasserwirtschaft: München, Germany.

Biggs, B. J. F., 1996. 'Patterns in benthic algae of streams'. In: *Algal Ecology. Freshwater Benthic Ecosystems*, Stevenson, R. J., Bothwell, M. L. and Lowe, R. L. (Eds). Academic Press: San Diego, CA, USA, pp. 31–58.

Blum, J. L., 1956. 'The ecology of river algae'. *Biol. Rev.*, **2**, 291–341.

Bornette, G. and Amoros, C., 1991. 'Aquatic vegetation and hydrology of a braided river floodplain'. *J. Veg. Sci.*, **2**, 497–512.

Bornette, G., Amoros, C., Castella, C. and Beffy, J. L., 1994. 'Succession and fluctuation in the aquatic vegetation of two former Rhone River channels'. *Vegetation*, **110**, 171–184.

Bornette, G., Amoros, C. and Lamouroux, N., 1998. 'Aquatic plant diversity in riverine wetlands: the role of connectivity'. *Freshwater Biol.*, **39**, 267–283.

Bott, T. L., 1996. 'Algae in microscopic food webs'. In: *Algal Ecology. Freshwater Benthic Ecosystems*, Stevenson, R. J., Bothwell, M. L. and Lowe, R. L. (Eds). Academic Press: San Diego, CA, USA, pp. 574–608.

Buchwald, R., Carbiener, R. and Tremolières, M., 1995. 'Synsystematic division and syndynamics of the Potamogeton coloratus community in flowing waters of Southern Central Europe'. *Acta Bot. Gallica*, **142**, 659–666.

Burkholder, J. M., 1996. 'Interactions of benthic algae with their substrata'. In: *Algal Ecology. Freshwater Benthic Ecosystems*, Stevenson, R. J., Bothwell, M. L. and Lowe, R. L. (Eds). Academic Press: San Diego, CA, USA, pp. 253–298.

Castenholz, R. W. and Garcia-Pichel, F., 2000. 'Cyanobacterial responses to UV-radiation'. In: *The Ecology of Cyanobacteria. Their Diversity in Time and Space*, Whitton, B. A. and Potts, M. (Eds). Kluwer Academic Publishers: Dordrecht, The Netherlands, pp. 591–611.

Cox, E. J., 1996. *Identification of Freshwater Diatoms from Live Material*. Chapman & Hall, London, UK.

Daniel, H. and Haury, J., 1995. 'Effects of fish farm pollution on phytoceonoses in an acidic river (the River Scorff, South Brittany, France)'. *Acta Bot. Gallica*, **142**, 639–650.

Daniel, H. and Haury, J., 1996. 'Les macrophytes aquatiques: une métrique de l'environnement en rivière'. *Cybium*, **20**, 3–129.

De Nicola, D. M., 1996. 'Periphyton responses to temperature at different ecological levels'. In: *Algal Ecology. Freshwater Benthic Ecosystems*, Stevenson, R. J., Bothwell, M. L. and Lowe, R. L. (Eds). Academic Press: San Diego, CA, USA, pp. 150–183.

Dinka, M., Agoston-Szabo, E., Berczik, A. and Kutrucz, G., 2004. 'Influence of water level fluctuation on the spatial dynamic of the water chemistry at Lake Fertö/Neusiedler See'. *Limnologica* **34**, 48–56.

Dokulil, M., 1991. 'Review on recent activities, measurements and techniques concerning phytoplankton algae of large rivers in Austria. In: *Use of Algae for Monitoring Rivers*, Whitton, B. A., Rott, E. and Friedrich, G. (Eds). E. Rott: University of innsbruck, Innsbruck, Austria, pp. 53–57.

Dokulil, M. T., 1996. 'Evaluation of eutrophication potentials in rivers: The River Danube example, a review'. In: *Use of Algae for Monitoring Rivers II*, Whitton, B. A, and Rott, E. (Eds). E. Rott: University of Innsbruck, Innsbruck, Austria, pp. 173–178.

Dokulil, M. T., 2003a. 'Horizontale und vertikale Interaktionen im Mega- und Mikroökoton einer Flussaue: Phytoplankton und Phytobenthos'. In: *Ökotone Donau-March. Veröffentlichungen des Österreichischen MaB-Programms 19*, Janauer, G. A. and Hary, N. (Eds). Österreichische Akademie der Wissenschaften, Universitätsverlag Wagner: Innsbruck, Austria, pp. 113–155.

Dokulil, M. T., 2003b. 'Algae as ecological bio-indicators'. In: *Bioindicators and Biomonitors*, Markert, B. A., Breure, A. M. and Zechmeister, H. G. (Eds). Elsevier, Amsterdam, The Netherlands, pp. 258–327.

Dokulil, M. T. and Janauer, G. A., 1990. 'Nutrient input and trophic status of the 'Neue Donau', a high-water control system along the river Danube in Vienna, Austria'. *Water Sci. Technol.*, **22**, 137–144.

Dokulil, M. T., Schmidt, R. and Kofler, S., 1997. 'Benthic diatom assemblages as indicators of water quality in an urban flood-water impoundment, Neue Donau, Vienna, Austria'. *Nova Hedwig.*, **65**, 273–283.

Dow, C. S. and Swoboda, U. K., 2000. 'Cyanotoxins'. In: *The Ecology of Cyanobacteria. Their Diversity in Space and Time*, Whitton, B. A. and Potts, M. (Eds). Kluwer Academic Publishers: Dordrecht, The Netherlands, pp. 613–632.

Ettl, H., 1980. *Grundgriss der allgemeinen Algologie*. G. Fischer Verlag, Stuttgart, Germany.

Fent, K., 2003. *Ökotoxikologie*, 2nd Edition. Thieme Verlag, Stuttgart, Germany.

Ferreira, M. T., 1994. 'Aquatic and marginal vegetation of the River Divor and its relation to land use'. *Verh. Int. Verein. Limnol.*, **25**, 2309–2315.

Ferreira, M. T., Catarino, L. and Moreira, I., 1998a. 'Aquatic weed assemblages in an Iberian drainage channel system and related environmental factors'. *Weed Res.*, **38**, 291–300.

Ferreira, M. T., Godinho, F. N. and Cortes, R. M. 1998b. 'Macrophytes in a southern Iberian river'. *Verh. Int. Verein. Limnol.*, **26**, 1835–1841.

Filzmoser, P., Janauer, G. A. and Exler, N., 2003. 'A statistical method for finding indicators of water quality'. *TemaNord*, **547**, 19–23.

Fritz, R., Zeltner, H. G. and Kohler, A., 1994. 'Flora und Vegetation der Brenz und der Hürbe (Ostalb) – Ihre Entwicklung von 1987 bis 1993'. *Hohenheimer Umwelttagung*, **26**, 233–238.

Fritz, R., Tremp, H. and Kohler, A., 1998. *Klassifizierung und Bewertung der südbadischen Rheinseitengewässer*.

Genter, R. B., 1996. 'Ecotoxicology of inorganic chemical stress to algae'. In: *Algal Ecology. Freshwater Benthic Ecosystems*, Stevenson, R. J., Bothwell, M. L. and Lowe, R. L. (Eds). Academic Press: San Diego, CA, USA, pp. 408–468.

Gessner, F., 1959. *Hydrobotanik. Die physiologieschen Grundlagen der Pflanzenverbreitung im Wasser. II. Stoffhaushalt*. VEB Deutscher Verlag der Wissenschaften: Berlin, Germany.

Grasmück, N., Haury, J., Lèglize, L. and Muller, S., 1993. 'Analyse de la végétation aquatique fixèe des cours d'eau lorrains en relation avec les paramètres d'environnement'. *Ann. Limnol.*, **29**, 223–237.

Greulich, S. and Tremolières, M., 2002. 'Distribution and expansion of Elodea nuttallii in the Alsatian Upper Rhine floodplain (France)'. In: *Gestion des Plants Aquatiques*, Dutartre, A. and Montel, M. H. N. (Eds). EWRS, CEMAGREF, Conseil Général des Landes, INRA, ENSAR.

Haslam, S. M., 1978. River Plants. *The Macrophytic Vegetation of Watercourses*. Cambridge University Press: London, UK.

Haslam, S. M., 1987. *River Plants of Western Europe*. Cambridge University Press: Cambridge, London, UK.

Haury, J., 1996. 'Assessing functional typology involving water quality, physical features and macrophytes in a Normandy river'. *Hydrobiologia*, **340**, 43–49.

Haury, J., Balignière, J.-L., Cassou, A.-I. and Maisse, G., 1995. 'Analysis of spatial and temporal organisation in a salmonid brook in relation to physical factors and macrophytic vegetation'. *Hydrobiologia*, **300/301**, 269–277.

Haury, J., Peltre, M.-C., Tremolières, M., Barbe, J., Thiebaut, G., Bernez, I., Daniel, H., Chatenet, P., Muller, S., Dutarte, A., Laplace-Treyture, C., Cazaubon, A. and Lambert-Servien, E., 2002. 'A method involving macrophytes to assess water trophy and organic pollution: the mMacrophyte Biological Index of Rivers (IBMR). – Application to different types of rivers and pollutions'. In: *Gestion des Plants Aquatiques*, Dutartre, A. and Montel, M. H. N. (Eds). EWRS, CEMAGREF, Conseil Général des Landes, INRA, ENSAR.

Henry, C. P., Bornette, G. and Amoros, C., 1994. 'Differential effects of floods on the aquatic vegetation of braided channels of the Rhone River'. *J. North Am. Benthol. Soc.*, **13**, 439–467.

Hill, W., 1996. 'Effects of light'. In: *Algal Ecology. Freshwater Benthic Ecosystems*, Academic Press: Stevenson, R. J., Bothwell, M. L. and Lowe, R. L. (Eds). San Diego, CA, USA, pp. 121–149.

Hoaglund, K. D., Carder, J. P. and Spawn, R. L., 1996. 'Effects of organic toxic substances'. *Algal Ecology. Freshwater Benthic Ecosystems*, Stevenson, R. J., Bothwell, M. L. and Lowe, R. L. (Eds). Academic Press: San Diego, CA, USA, pp. 469–496.

Holmes, N. T. H. and Whitton, B. A., 1975. 'Submerged bryophytes and angiosperms of the River Tweed and its tributaries'. *Trans. Bot. Soc. Edinburgh*, **42**, 383–395.

Holmes, N. T. H. and Whitton, B. A., 1977a. 'Macrophytes of the River Wear: 1966–1976'. *Naturalist*, **102**, 53–73.

Holmes, N. T. H. and Whitton, B. A., 1977b. 'The macrophytic vegetation of the River Tees in 1975: observed and predicted changes'. *Freshwater Biol.*, **7**, 43–60.

Holmes, N. T. H. and Whitton, B. A., 1977c. 'Macrophytic vegetation of the River Swale, Yorkshire'. *Freshwater Biol.*, **7**, 545–558.

Holmes, N. T. H., Newman, J. R., Chadd, S., Rouen, K. J., Saint, L. and Dawson, F. H., 1999. *Mean Trophic Rank: A User's Manual*. Environmental Agency: Bristol, UK.

Janauer, G. A., 1981. 'Die Zonierung submerser Wasserpflanzen und ihre Beziehung zur Gewässerbelastung am Beispiel der Fischa (Nierderösterreich)'. *Verh. Zool. Bot. Ges. Österreich*, **120**, 73–98.

Janauer, G. A., 2003a. 'Makrophyten der Augewässer'. In: *Ökotone Donau – March. Österr. MaB-Programm, Österr*, Janauer, G. A. and Hary, N. (Eds). Akad.Wissenschaften: Vienna, Austria, pp. 156–200.

Janauer, G. A., 2003b. 'Overview and final remarks'. *Arch. Hydrobiol., Suppl. 135, Large Rivers*, **14**, 217–229.

Janauer, G. A. and Stetak, D. 2003. 'Macrophytes of the Hungarian Lower Danube Valley (1498–1468 river-km)'. *Arch. Hydrobiol., Suppl. 147, Large Rivers*, **14**, 167–180.

Janauer, G. A. and Wychera, U., 2000. 'Biodiversity, succession and the functional role of macrophytes in the New Danube (Vienna, Austria)'. *Arch. Hydrobiol., Suppl. 135, Large Rivers*, **12**, 61–74.

Janauer, G. A., Vukov, D. and Igic, R., 2003. 'Aquatic macrophytes of the Danube River near Novi Sad (Yugoslavia, river-km 1255–1260)'. *Arch. Hydrobiol., Suppl. 147, Large Rivers*, **14**, 195–203.

Jeppesen, E., Sondergaard, Ma., Sondergaard Mo. and Christoffersen, K. (Eds), 1998. *The Structuring Role of Submerged Macrophytes in Lakes*, Ecological Studies, Vol. 131. Springer-Verlag: New York, NY, USA.

Kohler, A., 1978. 'Wasserpflanzen als Bioindikatoren'. *Beih. Veröff. Natuschutz Landschaftspflege Bad.-Württ.*, **11**, 259–281.

Kohler, A. and Schneider, S., 2003. 'Macrophytes as bioindicators'. *Arch. Hydrobiol. Suppl. 147, Large Rivers*, **14**, 17–31.

Kohler, A. and Tremp, H., 1996. 'Möglichkeiten zur Beurteilung des Säuregrades und der Versauerungsgefährdung von Fließgewässern mit Hilfe submerser Makrophyten'. *Verh. Ges. Ökologie*, **25**, 195–203.

Kohler, A. and Zeltner, G., 1974. 'Verbreitung und Ökologie von Makrophyten in Weichwasserflüssen des Oberpfälzer Waldes (Naab, Pfreimd und Schwarzach)'. *Hoppea*, **33**, 171–232.

Kohler, A., Vollrath, H. and Beisl, E., 1971. 'Zur Verbreitung, Vergesellschaftung und Ökologie der Gefäß-Makrophyten im Fließgewässersystem Moosach (Münchener Ebene)'. *Arch. Hydrobiol.*, **69**, 333–365.

Kohler, A., Sipos, V. and Björk, S., 1996. 'Makrophyten-Vegetation und Standorte im humosen Bräkne-Fluss (Südschweden)'. *Bot. Jahrb. Syst.*, **118**, 451–503.

Kohler, A., Sipos, V., Sonntag, E., Penksza, K., Pozzi, D., Veit, U. and Björk, S., 2000. 'Makrophyten-Verbreitung und Standortqualität im eutrophen Björka-Kävlinge-Fluss (skane, Sübschweden)'. *Limnologica*, **30**, 281–298.

Kornijow, R. and Kairesalo, T., 1994. '*Elodea canadensis* sustains rich environment for macroinvertebrates'. *Verh. Int. Verein. Limnol.*, **25**, 4098–4111.

Labus, B. C., 1979. *Der Einfluss des Waschrohstoffes Marlon A auf das Wachstum und die Photosynthese verschiedener submerser Makrophyten*, PhD Thesis. Hohenheim, Germany.

Lamberti, G. A., 1996. 'The role of periphyton in benthic food webs'. In: *Algal Ecology. Freshwater Benthic Ecosystems*, Stevenson, R. J., Bothwell, M. L. and Lowe, R. L. (Eds). Academic Press: San Diego, CA, USA, pp. 533–573.

Liebmann, H., 1940. 'Über den Einfluss der Verkrautung auf den Selbstreinigungsvorgang in der Saale unterhalb Hof'. *Wasser* **14**, 92–102.

Lillie, R. and Budd, J., 1992. 'Habitat architecture of *Myriophyllum spicatum* L. as an index to habitat quality for fish and macroinvertebrates'. *J. Freshwater Ecol.*, **7**, 113–125.

Lottausch, W., Buchloh, G. and Kohler, A., 1980. 'Vegetationskundliche Untersuchungen in kryptogamenreichen Gebirgsbächen'. *Verh. Ges. Ökologie*, **8**, 351–356.

Lowe, R. L. and Pan, Y., 1996. 'Benthic algal communities as biological monitors'. In: *Algal Ecology. Freshwater Benthic Ecosystems*, Stevenson, R. J., Bothwell, M. L. and Lowe, R. L. (Eds). Academic Press, San Diego, CA, USA, pp. 705–740.

Luther, H., 1949. 'Vorschlag zu einer ökologischen Grundeinteilung der Hydrophyten'. *Acta Bot. Fenn.*, **44**, 1–15.

Luther, H., 1951. 'Verbreitung und Ökologie der Höheren Wasserpflanzen im Brackwasser der Ekenäs-Gegend in Südfinnland. II. Spezieller Teil'. *Acta Bot. Fenn.*, **49**, 1–354.

Marker, A. F. H. and Collett, G. D., 1991. 'Biomass, pigment and species composition'. In: *Use of Algae for Monitoring Rivers*, Whitton, B. A., Rott, E. and Friedrich, G. (Eds). E. Rott: University of Innsbruck, Innsbruck, Austria, pp. 21–24.

Miller, G. E., Wile, I. and Hitchin, G. G., 1983. 'Pattern of accumulation of selected metals in members of the soft-water macrophyte flora of central Ontario lakes'. *Aquat. Bot.*, **15**, 53–64.

Monschau-Dudenhausen, K., 1982. 'Wasserpflanzen als Belastungsindikatoren in Fließgewässern dargestellt am Beispiel der Schwarzwaldflüsse Nagold und Alb- Beih'. *Naturschutz Landespflege Bd.-Württ.*, **28**, 1–118.

Olsen, S., 1950. 'Aquatic plants and hydrospheric factors. I. Aquatic plants in SW-Jutland, Denmark'. *Svensk Bot. Tidskrift*, **44**, 1–34.

Ombredane, D., Haury, J. and Chapon, P. M., 1995. 'Heterogeneity and typology of fish habitat in the main stream of a Breton coastal river (Elorn-Finistère, France)'. *Hydrobiologia*, **300/301**, 259–268.

Otahelova, H. and Valachovic, M., 2003. 'Distribution of macrophytes in different water-bodies influenced by the Gabcikovo hydropower station (Slovakia) – present status'. *Arch. Hydrobiol., Suppl. 147, Large Rivers*, **14**, 97–115.

Poulíčková A., Duchoslav M. and Dokulil, M. T., 2004. 'Littoral diatom assemblages as bioindicators for lake trophy: A case study from alpine and pre-alpine lakes in Austria'. *Eur. J. Phycol.*, **39**, 143–152.

Peterson, Ch. G., 1996. 'Response of benthic algal communities to natural physical disturbance'. In: *Algal Ecology. Freshwater Benthic Ecosystems*, Stevenson, R. J., Bothwell, M. L. and Lowe, R. L. (Eds). Academic Press: San Diego, CA, USA, pp. 375–401.

Planas, D., 1996. 'Acidification effects'. In: *Algal Ecology. Freshwater Benthic Ecosystems*, Stevenson, R. J., Bothwell, M. L. and Lowe, R. L. (Eds). Academic Press: San Diego, CA, USA.

Prygiel, J., Whitton, B. A. and Bukowska, J. (Eds), 1999. *Use of Algae for Monitoring Rivers III*. Agence de L'Eau: Douai, France.

Rath, B., 1995. 'Potamogeton pectinatus-Bestände als Bioindikatoren der Schwermetallbelastung im Hauptarm der Donau'. *Opusc. Zool. Budapest*, **27**, 167–174.

Rath, B., Janauer, G. A., Pall, K. and Berczik, A., 2003. 'The auqatic macrophyte vegetation in the Old Danube/Hungarian bank, and other water bodies of the Szigetköz wetlands'. *Arch. Hydrobiol., Suppl. 147, Large Rivers*, **14**, 129–142.

Reynolds, C. S., 1988. 'Potamoplankton: Paradigms, paradoxes and prognoses'. In: *Algae and the Aquatic Environment*, Round, F. E. (Ed.). Biopress Ltd: Bristol, UK, pp. 285–311.

Reynolds, C. S. and Descy, J. P., 1996. 'The production, biomass and structure of phytoplankton in large rivers'. *Arch. Hydrobiol., Suppl., Large Rivers*, **10**, 161–187.

Richter, M., 2004. *Die Makrophyten des Neusiedler Sees*, PhD Thesis. Univ. Vienna: Vienna, Austria.

Robach, F., Eglin, I. and Tremolières, M., 1997. 'Species richness of aquatic macrophytes in former channels connected to a river: a comparison between two fluvial hydrosystems differing in their regime and regulation'. *Global Ecol. Biogeog. Lett.*, **6**, 267–274.

Rott, E., Hofmann, G., Pall, K., Pfister, P. and Pipp, E., 1997. *Indikationslisten für Aufwuchsalgen. Teil 1: Saprobielle Indikation*. Wasserwirtschaftskataster, BMLF: Vienna, Austria.

Rott, E., Pfister, P., van Dam, H., Pipp, E., Pall, K., Binder, N. and Ortler, K., 1999. *Indikationslisten für Aufwuchsalgen. Teil 2: Trophieindikation und autökologische Anmerkungen*. Wasserwirtschaftskataster, BMLF: Vienna, Austria.

Rott, E., Pipp, E. and Pfister, P., 2003. 'Diatom methods developed for river quality assessment in Austria and a cross-check against numerical trophic indication methods used in Europe'. *Arch., Hydrobiol., Suppl., Algological Studies*, **110**, 91–115.

Samecka-Cymerman, A. and Kempers, A. J., 1998. 'Bioindication of gold by aquatic bryophytes'. *Acta Hydrochim. Hydrobiol.*, **26**, 90–94.

Sarbu, A., 2003. 'Inventory of aquatic plants in the Danube Delta: a pilot study in Romania'. *Arch. Hydrobiol., Suppl., Large Rivers*, **14**, 205–216.

Say, P. J. and Whitton, B. A., 1983. 'Accumulation of heavy metals by aquatic mosses. 1. Fontinalis antipyretica Hedw'. *Hydrobiologica*, **100**, 245–260.

Schagerl, M. and Donabaum, K., 1998. 'Epilithic algal communities on natural and artificial substrata in the River Danube near Vienna (Austria)'. *Arch. Hydrobiol., Suppl. 113, Large Rivers*, **11**, 153–165.

Schneider, S., Schranz, C. and Melzer, A., 2001. 'Indicating the trophic state of running waters by submersed macrophytes and epilithic Diatoms'. *Limnologica*, **30**, 1–8.

Schönfelder, I., 2000. 'Indikation der Gewässerbeschaffenheit durch Diatomeen'. In: *Handbuch Angewandte Limnologie*, Steinberg, Ch., Calmano, W., Klapper, H. and Wilken, R.-D. (Eds). Ecomed: Landsberg, Germany, Bd. 3, Teil viii–7.2, pp. 1–61.

Sculthorpe, C. D., 1967. *The Biology of Aquatic Vascular Plants*. Arnold: London, UK.

Stevenson, R. J., 1996. 'An introduction to algal ecology in freshwater benthic habitats'. In: *Algal Ecology. Freshwater Benthic Ecosystems*, Stevenson, R. J., Bothwell, M. L. and Lowe, R. L. (Eds). Academic Press: San Diego, CA, USA, pp. 3–30.

Thiébaut, G., Guérold, F and Muller, S., 1995. 'Impact de l'acidification des eaux sur les macrophytes aquatiques dans les eaux faiblement minéralisées des Vosges de Nord. Premiers résultats'. *Acta Bot. Gallica*, **142**, 617–626.

Tremp, H., 1991. 'Passives Monitoring mit Wassermoosen zur Überwachung der Versauerungsdynamik in pufferschwachen Fließgewässern'. *Verh. Ges. Ökologie*, **20**, 529–535.

Van den Hoek, C., Mann, D. G. and Jahns, H. M., 1995. *Algae. An Introduction to Phycology*. Cambridge University Press, Cambridge, UK.

Vannote, R. L., Minshall, G. W., Cummins, K. W., Sedell, J. R. and Cushing, C. E., 1980. 'The River Continuum Concept'. *Can. J. Fish. Aquat. Sci.*, **37**, 130–137.

Veit, U. and Kohler, A., 2003. 'Long-term study of the macrophyteic vegetation in the running water of the Friedberger Au (near Augsburg, Germany)'. *Arch. Hydrobiol., Suppl. 147, Large Rivers*, **14**, 65–86.

Wehr, J. D. and Whitton, B. A., 1983. 'Accumulation of heavy metals by aquatic mosses. 2. Rhynchostegium riparioides'. *Hydrobiologia*, **100**, 261–284.

Westlake, D. F., 1974. 'Macrophytes'. In: *A manual on Methods for Measuring Primary Production in Aquatic Environments*, Vollenweider, R. A. (Ed.), IBP Handbook Vol. 12. Blackwell Scientific Publications, Oxford, UK, pp. 25–32.

Westlake, D. F., 1975. 'Macrophytes'. In: *River Ecology*, Whitton, B. A. (Ed.), Studies in Ecology, Vol. 2. Blackwell Scientific Publications, Oxford, UK.

Wetzel, R. G., 2001. Limnology. *Lake and River Ecosystems*, 3rd Edition. Academic Press; San Diego, CA, USA.

WFD, 2000. 'Directive 2000/60/EC of the European Parliament and of the Council establishing a framework for community action in the field of water policy'. *Official J. Eur. Communities*, **L327**, Brussels, Belgium.

Whitton, B. A., Rott, E. and Friedrich F. (Eds), 1991. *Use of Algae for Monitoring Rivers*, Vol. I. E. Rott: University of Innsbruck, Innsbruck, Austria.

Whitton, B. A. and Rott, E. (Eds), 1996. *Use of Algae for Monitoring Rivers*, Vol. II. E. Rott: University of Innsbruck, Innsbruck, Austria.

Würzbach, R., Zeltner, G.-H. and Kohler, A., 1977. 'Die Makrophyten-Vegetation des Fließgewässersystems der Moosach'. *Ber. Inst. Landschafts-Pflanzenökologie Hohenheim*, **4**, 7–241.

Zander, B., Wohlfahrt, U. and Wiegleb, G., 1992. *Typisierung und Bewertung der Fließgewässervegetation der Bundesrepublik Deutschland*. Bundesministerium für Umwelt, Naturschutz und Reaktorsicherheit: Berlin, Germany.

Zhihong, Y., Baker, A. J. M. and Wong, M. H., 1998. 'Zinc, lead and cadmium accumulation and tolerance in Typha latifolia as affected by iron on the root surface'. *Aquat. Bot.*, **61**, 55–67.

Section 2
Biological Monitoring of Rivers

2.1

River Monitoring and Assessment Methods Based on Macroinvertebrates

Niels De Pauw, Wim Gabriels and **Peter L. M. Goethals**

Biological Monitoring of Rivers   Edited by G. Ziglio, M. Siligardi and G. Flaim
© 2006 John Wiley & Sons, Ltd.

2.1.1   IMPORTANCE OF BIOLOGICAL MONITORING AND ASSESSMENT

Biological monitoring and assessment methods are important tools to support decision making within river management. The biotic component of an aquatic ecosystem may indeed be considered as an 'integrating-information-yielding unit' for assessment of its quality. Biological communities also integrate the effects of mixed types of stress and in certain cases already respond before analytical detection allows (De Pauw and Hawkes, 1993).

Among the biological communities, the macroinvertebrates are by far the most frequently used group for bioindication in standard water management (Woodiwiss, 1980; Helawell, 1986; De Pauw *et al.*, 1992; Rosenberg and Resh, 1993; Metcalfe-Smith, 1994; Hering *et al.*, 2004). The term 'macroinvertebrates', however, does not correspond to a taxonomical concept but to an artificial delimitation of part of the groups of invertebrate animals (Alba-Tercedor, 2006: Chapter 1.5 in this book). In general, in running waters, one considers macroinvertebrates as those organisms large enough to be caught with a net or retained on a sieve with a mesh size of 250 to 1000 μm, and thus can be seen with the naked eye. In fact, most of them are larger than 1 mm (e.g. Cummins, 1975; Sladecek, 1973; De Pauw and Vanhooren, 1983; Rosenberg and Resh, 1993; Ghetti, 1997; Tachet *et al.*, 2002).

The majority of aquatic macroinvertebrates have a benthic life and inhabit the bottom substrates (sediments, debris, logs, macrophytes, filamentous algae, etc.) and for this reason, in the literature about biological water quality assessment methods one often refers to them as benthic macroinvertebrates or macrozoobenthos (Rosenberg and Resh, 1993). Other representatives of the macroinvertebrates, however, also serving as bioindicators, are pelagic and freely swimming in the water column, or pleustonic and associated with the water surface (Tachet *et al.*, 2002).

2.1.2   ADVANTAGES OF MACROINVERTEBRATES

The reasons for macroinvertebrates being so popular in bioassessment are numerous (e.g. Hawkes, 1979; Sladecek, 1973; Helawell, 1986; Metcalfe, 1989; Rosenberg and Resh, 1993; Hering *et al.*, 2004). Macroinvertebrates are ubiquitous and abundant throughout the whole river system in the crenal and rhitral as well as the potamal part (Illies, 1961). They play an essential role in the functioning of the river continuum food web (e.g. Vannote *et al.*, 1980; Giller and Malmqvist, 1998).

Since macroinvertebrates are a heterogeneous collection of evolutionary diverse taxa, this means that at least some will react to specific changes in the aquatic environment, natural as well as imposed. They are not merely affected by different types of physical-chemical pollution (e.g. organic enrichment, eutrophication, acidification), but as well by physical changes and anthropogenic manipulation of the aquatic habitat (e.g. canalization, impoundment, river regulation) (Figure 2.1.1).

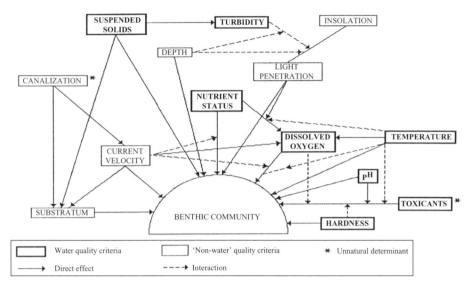

Figure 2.1.1 Water quality and 'non-water quality' determinants of benthic communities in rivers (adapted from Hawkes, 1979, and De Pauw and Hawkes, 1993)

Macroinvertebrates can thus be used for the assessment of the water as well as the habitat quality (Armitage *et al.*, 1983) and enable a holistic assessment of streams.

Macroinvertebrates have, furthermore, the advantage to be relatively easy to collect and identify, and to be confined for most part of their life to one locality on the river bed and are therefore indicative of the changing water qualities. As such, they act as continuous monitors of the water flowing over them as opposed to chemical samples of the water taken at one time. Having long life spans, macroinvertebrates integrate environmental conditions over longer periods (weeks, months, years) and thus sampling may be less frequent (De Pauw and Hawkes, 1993; Giller and Malmqvist, 1998; Tachet *et al.*, 2002).

2.1.3 LIMITATIONS OF MACROINVERTEBRATES

Using macroinvertebrates as monitors of river (water) quality, however, has also its limitations. Quantitative sampling, for example, is difficult because of their non-random distribution in the river bed. Because of the seasonality of the life cycles of some invertebrates, e.g. insects, they may not be found at some times of the year (e.g. Linke *et al.*, 1999; D'heygere *et al.*, 2002; Tachet *et al.*, 2002). An appreciation of this seasonality enables this to be taken into account in interpreting the data. As shown in Figure 2.1.1, factors other than water quality are also important determinants of benthic communities. Of these, the related factors of current velocity and nature of the substratum are overriding ones determining the nature of the community, especially in relation to invertebrates. Since these factors differ along the river in different zones, different communities become established at different sites with the same water quality (Giller and Malmqvist, 1998). Therefore, in practice where possible, sampling sites having similar benthic conditions are selected or a typology is developed consisting of distinct river types with adapted sampling and assessment systems (e.g. Hering *et al.*, 2004). Some assessment systems, e.g. RIVPACS (Wright *et al.*, 1993), even predict the reference communities on the basis of a set of local river features as a basis for the assessment.

A last limitation of macroinvertebrates is their restricted geographic distribution, the incidence and frequency of occurrence of some species being different in rivers throughout the region. Furthermore, because of their geographic distribution, species at the edge of their natural distribution range are theoretically more sensitive to additional stress – pollution than those at the centre of their distribution. It would therefore not be possible to have a universal system of biological assessment based on the response of the same species/taxa (Sandin *et al.*, 2000).

2.1.4 ELEMENTS OF BIOLOGICAL MONITORING AND ASSESSMENT METHODS BASED ON MACROINVERTEBRATES

The main elements of biological monitoring and assessment methods are summarized in Figure 2.1.2. Monitoring includes the sampling and sample analysis, that is, the collection of information, while assessment on the other hand is the interpretation of the data (Chapman, 1992). The assessment involves the numerical evaluation and

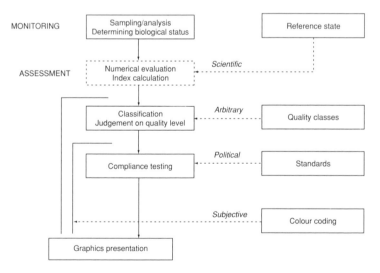

Figure 2.1.2   Elements of biological monitoring and assessment methods (adapted from Knoben *et al.*, 1995)

index calculation, the classification of the indices into quality classes, the testing of compliance with standards, and finally the graphical presentation. These elements will be briefly commented upon in relation to the existing methods and approaches based on macroinvertebrates. Not all monitoring and assessment methods, however, apply all the elements presented.

2.1.4.1 Sampling and sample analysis of macroinvertebrates

The devices with which macroinvertebrates can be collected are quite diverse. Commonly used in river monitoring programmes are different types of nets (e.g. handnet, surber net, dredges), grabs (e.g. Van Veen, Ponar), core samplers, artificial substrates or colonization samplers (e.g. standard Aufwuchs sampler, bag sampler, Hester–Dendy), drift and emergence samplers. Overviews of these instruments and convenient handling can be found in Schwoerbel (1970), Mason (1981), Rosenberg and Resh (1982), APHA (1998) and Ghetti (1997). Important for the nets are the mesh size (between 250 μm and 1 mm) (e.g. ISO 7828, 1985) for the grabs and cores, the volume and sediment surface sampled (e.g. ISO 8265, 1988) for the artificial substrates, the material, volume and colonization (exposure) time (e.g. Rosenberg and Resh, 1982; ISO 9391, 1993; De Pauw *et al.*, 1994).

The selected sampling method is dependent on the physical characteristics of the aquatic environment (depth, current velocity, sediment structure), the objective of the assessment method (qualitative, semi-quantitative, compartments of the ecosystem to be sampled river bed, sediment, riffles, pools, water column, banks, aquatic vegetation) and taxonomic groups considered (the whole benthic community of macroinvertebrates or specific groups, such as oligochaetes, chironomids, molluscs, gammarids and isopods).

For an overall assessment of running waters, all habitats, or specific habitats are often being sampled for periods roughly proportional to their extent on the site (e.g. Woodiwiss, 1980; De Pauw and Vanhooren, 1983; Ghetti, 1997). Each site is indeed composed of a mosaic of biotopes (e.g. Fontoura and De Pauw, 1994; Tachet *et al.*, 2002; Hering *et al.*, 2004), i.e. areas where the environmental conditions are uniform and clearly defined and which need to be examined. Different sampling tools may be used for exploring different habitats. Samples may be completed by hand picking of macroinvertebrates on hard substrates (stones, debris, plants) along the banks. Most sampling methods for assessments with macroinvertebrates are qualitative or semi-quantitative in approach which means that the sampling effort is often linked to examining the site during a fixed period of time (depending on the size of the river, 3 to 5 minutes) and within a certain stretch (for example, 10 to 20 m) (e.g. De Pauw and Vanhooren, 1983).

In shallow rivers, it is common practice to apply the kick sampling technique with a handnet (e.g. Woodiwiss, 1980).

For taking samples of macroinvertebrates in deeper waters, artificial substrates (colonization samplers) or grab samplers are recommended (ISO 9391, 1993).

Commercial (e.g. Multiplate or Hester–Dendy samplers), as well as self-made artificial substrates, can be used. Important is the placement of the substrates, not in static areas but in the main flow of the river. To prevent drifting, substrates may need to be weighted. One should, however, prevent samplers to be sinking in the mud, that is, otherwise leading to a disturbed colonization by macroinvertebrates. Three to four weeks of colonization is the recommendable practice for obtaining a satisfactory assessment (e.g. De Pauw *et al.*, 1994; Ghetti, 1997; APHA, 1998). Samples of the bottom and sediments can be directly collected by means of grabs.

In large and deep rivers, a combination of different sampling methods may have to be applied to get a complete picture of the macroinvertebrate communities present (e.g. handnet along the shallow river banks, examination of stones along the banks, dredging net and/or grabs or artificial substrates in the deeper parts of the river). Precise protocols however have not been described so far.

2.1.4.2   Processing of samples

Samples have to be labelled carefully and immediately registered. The macroinvertebrate samples can be fixed *in situ* or in the laboratory with formalin, or preferably, the less toxic 70 % denatured alcohol, or specific preservatives (e.g. APHA, 1998; De Pauw and Vannevel, 1991; Ghetti, 1997). Live examination may, however, be preferred for identification and sorting out which should be done *in situ* or soon after capture to avoid predation (Bervoets *et al.*, 1989). After collection, samples are sorted with 250, 500, 750 or 1000 μm sieves to facilitate separation of the different groups of macroinvertebrates, often using a pile of sieves (for example 500 μm, 1, 2, 5 and 10 mm). Sieving can be done *in situ* (with river water) or in the laboratory (with running tap water). Sieve contents are placed in white trays and sorted, usually according to different groups, by means of tweezers and/or pipettes. The organisms are transferred into small closed glass or plastic containers and conserved in (denaturated) alcohol (70 %) (e.g. AFNOR, 1992; De Pauw and Vannevel, 1991; NBN, 1984; Ghetti, 1997; APHA, 1998; Tachet *et al.*, 2002). Depending on the assessment method, organisms are quantified (eventually as abundance classes, e.g. saprobic indices) or not (e.g. BBI (De Pauw and Vanhooren, 1983), IBGN (AFNOR, 1992), DSFI (Skriver *et al*, 2001)).

2.1.4.3   Identification

Macroinvertebrate communities may include a large number of taxonomic groups belonging to worms, molluscs, crustaceans and insects, with arthropods usually represent the great majority of taxa. The identification level chosen is often the result of a practical trade-off between taxonomic precision and time constraints and financial resources (Guerold, 2000; Adriaenssens *et al.*, 2004c; Gabriels *et al.*, 2004). Essential for the successful application of assessment methods based on macroinvertebrates is the availability of suited keys of identification up to the required level and this in the local language (e.g. De Pauw and Vannevel, 1991; Roldan, 1992; Schmedtje and Kohmann, 1992; Campaioli *et al.*, 1994; Puig, 1999; Tachet *et al.*, 2002). In addition, for the interpretation of the data, precise information on the ecology of the different macroinvertebrate species is urgently needed (e.g. Tachet *et al.*, 2002; Dedecker *et al.*, 2004).

2.1.5 BIOASSESSMENT SYSTEMS OF RIVER (WATER) QUALITY

The history of bioassessment of rivers is a good hundred years old, taking a definite start in Europe in 1902 with the development of the saprobic system introduced by Kolkwitz and Marsson and in the USA in 1913 with the development of a river water quality classification system by Forbes and Richardson (Richardson, 1928) (cf. reviews by Hynes, 1971; Sladecek, 1973; Persoone and De Pauw, 1979; Woodiwiss, 1980; Metcalfe, 1989; De Pauw *et al.*, 1992; Rosenberg and Resh, 1993; Sandin *et al.*, 2000; Hering *et al.*, 2004). Although the main focus in the beginning was on microorganisms (among others, plankton), macroinvertebrates as bioindicators rapidly gained in importance (cf. Bartch and Ingram, 1966; Mackenthun, 1969; Sladecek, 1973; Rosenberg and Resh, 1993; Hering *et al.*, 2004). The earlier systems were purely descriptive or qualitative and mainly based on the presence or absence of indicator species, in the first place, related to discharges of domestic sewage, i.e. organic pollution. Since the early 1950s, however, biologists have felt the need to convey their complex biological data in a numerical form such as indices or scores (e.g. Beck, 1954; Knöpp, 1954; Pantle and Buck, 1955). Today, over 100 different biotic indices have been described (De Pauw *et al.*, 1992; Ghetti and Ravera, 1994; Hering *et al.*, 2004). Yet, many ecologists remain sceptical regarding the possibility and advisability of expressing complex biological communities in terms of a single numerical value. Nevertheless, the 'pseudo-accuracy' of a biotic index is apparently more acceptable to the non-biologist and administrator than biological survey data expertly interpreted. To ensure biological information is made more comprehensible and therefore more acceptable in decision-making, the use of indices is probably justifiable, although by using them information is inevitably lost. Having reduced the original data to a 'number' there is a danger that it can then be more readily misused (Seegert, 2000).

2.1.6 DIFFERENT ASSESSMENT APPROACHES BASED ON MACROINVERTEBRATES

Analysis of the macroinvertebrate communities in rivers can theoretically be structural, functional, taxonomical and non-taxonomical in approach (Matthews *et al.*, 1982). Most of the actually used bioassessment systems are, however, structural and taxonomical, which means relying, for example, on the presence or absence of particular taxa, the sensitivity of particular taxa, the taxa richness, taxa abundance and taxa diversity. All of this information can be converted into numerical values, including indices and scores. Most assessment methods are based on the analysis of species assemblages or populations of particular taxonomic groups of benthic macroinvertebrates (e.g. oligochaetes, chironomids). Assessment methods based on organism-level indicators (biochemical, physiological, morphological deformities, behavioural responses and life-history responses) are not considered here.

Reviewing the common assessment methods in Europe based on structural–taxonomical analysis, Metcalfe (1989) distinguishes three major approaches to assess the response of macroinvertebrate communities to pollution, namely, the saprobic, biotic and diversity approaches. In recent years, however, several new approaches were also developed. Since the 1980s, for example, the use of multimetric assessment systems, such as the Index of Biotic Integrity, initiated in the USA, became more and more popular (Karr and Chu, 1999). Another approach was the introduction in the UK of RIVPACS, which led to methods in which the comparison with reference conditions became central, a principle which was later adopted in the assessment proposal of the European Water Framework Directive (WFD) (EU, 2000). A final approach, although existing for some time already and for which a growing interest exists, is the use of multivariate analysis to distinguish among different river typologies and communities and which can be considered as a type of similarity indices. In the next sections, a brief overview of these major approaches is presented.

2.1.6.1 Saprobic approach

The saprobic approach was the first river assessment system to be developed, already at the beginning of the 20th Century by Kolkwitz and Marsson (1902), and later on expanded by, among others, Zelinka and Marvan, 1961, Liebmann (1962) and Sladecek (1973). The objective is to provide a water quality classification based on the pollution tolerance of the indicator species present (= Response A in Table 2.1.1). Every species has a specific dependency of organic substances and, thus, of the dissolved oxygen content: this tolerance is expressed as a saprobic indicator value.

Table 2.1.1 Biocoenotic responses of indicator value induced by pollution discharges (adapted from De Pauw and Hawkes, 1993)

Response class	Species versus community response	Response description
A	Species	The appearance or disappearance of individual species
B	Community	A reduction in numbers of species/taxa present, i.e. a reduction in diversity
C	Community	A change in the population of individual species
D	Community	A change in the proportional species composition of the community

The advantage is a quick classification of the investigated community by means of a saprobic index, which can be made on a universal scale (e.g. DEV, 1988–1991). A major problem is the identification of the organisms up to species level. The saprobic index calculation also requires the assessment of abundances. The indicator system furthermore implies more knowledge than actually exists: pollution tolerances are highly subjective and based on ecological observations and are rarely confirmed by experimental studies.

2.1.6.2   Diversity approach

The diversity approach uses three components of community structure: richness (B), evenness (D) and abundance (C) (Washington, 1984). Diversity indices developed from information theory methods (Shannon and Weaver, 1949) have been used by Patten (1962), Wilhm and Dorris (1966), and others. The objective aims at evaluating the community structure with respect to occurrence of species. The diversity indices relate the number of observed species (richness) to the number of individuals (abundance). The principle is that disturbance of the water ecosystem or communities under stress leads to a reduction in diversity. The advantages of diversity indices lies in the fact that they are easy to use and calculate, applicable to all kinds of watercourses, have no geographical limitations and are best used for comparative purposes. Having no clear endpoint or reference level is, however, the main problem; the diversity in natural undisturbed waters can indeed vary considerably, moreover, all species have equal weight. This is probably the reason why not one country in Europe has been adopting a diversity index as a national standard for biological water quality assessment (see De Pauw *et al.*, 1992; Ghetti and Ravera, 1994; Nixon, 2003).

2.1.6.3   Biotic approach

The biotic approach, on the other hand, incorporates desirable features of the saprobic and diversity approaches by combining a quantitative measure of species diversity (B) with qualitative information on the ecological sensitivities of individual taxa (A) into a single numerical expression (cf. Table 2.1.1). Woodiwiss (1980) rightly distinguishes between biotic indices and biotic scores, which although using the same responses A + B, do so in quite different ways. In the biotic index approach, the index is directly taken from a table in which the taxa richness is combined with the presence of the most sensitive taxon (e.g. the Trent Biotic Index, Woodiwiss, 1964). In the biotic score system, on the other hand, a score is allocated to each taxon. The score for the site is then derived by summing the individual scores. The biotic score may also include a measure of abundance of the organisms (e.g. Chandler, 1970). The objective of biotic indices or scores is to assess the biological water quality of running waters, in most cases based on macroinvertebrates, and to measure various types of environmental stress, organic waters, acid waters, etc.

The principle is that macroinvertebrate groups disappear as pollution increases and that the number of taxonomic groups is reduced as pollution increases. MacKenthun (1969) identified the following stepwise disappearance of macroinvertebrates subsequent to increasing pollution: stoneflies (Plecoptera), mayflies (Ephemeroptera), caddisflies (Trichoptera), scuds (Amphipoda), aquatic sowbugs (Isopoda), midges (Diptera) and bristle worms (Oligochaeta).

The advantages are that only qualitative sampling is required and that identification is mostly at family or genus level and that there is no need to count abundances per taxon (e.g. biotic indices like BBI, IBGN, EBI, ASPT and BMWP). The problems, on the other hand, are how to determine representative reference communities to which the investigated stations can be compared to. In addition, should an optimal biological assessment be achieved through regional adaptations?

2.1.6.4 Multimetric approach

The first 'true', explicitly called, multimetric systems were developed in the USA by Karr (1981) for assessments with fish. Recently, similar systems have also being designed for benthic macroinvertebrate communities (USEPA, 1996; Barbour *et al.*, 1992; Karr and Chu, 1999; Hering *et al.*, 2004). In multimetric systems, several metrics representing different characteristics of the macroinvertebrate community are summed up into one index value or score (e.g. Barbour and Yoder, 2000) which is an expression of the overall quality. It is expected that working with more descriptors will result in an index being representative for a specific aquatic environment (e.g. the Acidification Index developed by Johnson (1998)). Multimetric systems may include structure metrics, community balance metrics, tolerance metrics, feeding group metrics and others (e.g. USEPA, 1996). Within the context of the implementation of the EU Water Framework Directive (WFD), the European project AQEM has been proposing a strategy and methodology for the establishment of multimetric assessment systems for different streams in Europe based on macroinvertebrates (Hering *et al.*, 2004). Most of the multimetric systems do not aim to separate the impact of different stressors (Lorenz *et al.*, 2004). However, it has been recommended that the developed multimetric systems should be 'stressor-specific' (e.g. for organic pollution, acidification, morphological degradation) to ease the cause allocation under conditions of deterioration. Examples of such stressor-specific systems can be found in Brabec *et al.* (2004) and Buffagni *et al.* (2004). For the assessments of the sediment quality in rivers, a TRIAD approach has been developed, combining physico-chemical, ecotoxicological and biological information based on macroinvertebrates (De Cooman *et al.*, 1999; De Pauw and Heylen, 2001).

2.1.6.5 Ecological quality ratio approach

The assessment value obtained with any index system can be compared with a reference status to be reached, by calculating the proportion between both values. This is called the Ecological Quality Ratio (EQR), according to the EU Water Framework Directive (EU, 2000). The reference can be based on real samplings, expert knowledge, historical data or predictive models, or a combination of these. An example of an EQR is the Environmental Quality Index (EQI) based on the 'River Invertebrate Prediction and Classification System' (RIVPACS) developed in the UK (Armitage *et al.*, 1983; Wright *et al.*, 1993, 2000). The principle of RIVPACS is that on the basis of the physical-chemical features of the river it is possible to predict which macroinvertebrate taxa should be present under these conditions. The predicted reference conditions can then be compared with the observed macroinvertebrate communities. The RIVPACS EQI can be calculated with different metrics or indices, for example, the BMWP, the ASPT or the number of taxa (NOT) (Sweeting *et al.*, 1992). Based on RIVPACS, other similar models have been developed in Australia (AUSRIVAS, 'Australian River Assessment Scheme', e.g. Davies (2000) and Smith *et al.* (1999)) and Canada (BEAST, 'Benthic Assessment of Sediment', Reynoldson *et al.*, 2000).

2.1.6.6   Multivariate approach

Several multivariate techniques have been applied in water quality assessment using macroinvertebrates (Norris and Georges, 1993). The basis for the multivariate approach is the similarity index (Sandin *et al.*, 2001). The most commonly used similarity index is the Jaccard index (Jaccard (1908) and Washington (1984)). This index expresses the percentage of species shared between two sites. Other examples are the percentage similarity index (Whittaker, 1952), Bray–Curtis dissimilarity index (Bray and Curtis, 1957), Sorensen index (Sorensen, 1948) and the Euclidean or ecological distance (Williams, 1971). All of these indices give an indication of how much a biological community at each sampled site is similar to the median of all reference communities and are not resulting in an assessment class as such.

Multivariate techniques are, since the 1990s, also commonly applied for the development of multimetric systems. The selection of the metrics is based on how complementary or explanatory these are. The complementary of score systems is necessary to guarantee that correlated metrics do not dominate the overall assessment, while the explanatory aspects are interesting for obtaining insight in the causes of deterioration. Since the new millennium, also a shift in use from multivariate statistical (classification, ordination, regression, clustering, etc., based on data distribution functions) to soft computing (based on heuristic search methods, e.g., artificial neural networks, inductive logic programming, etc.) techniques has started. Major examples of assessment systems using multivariate approaches are RIVPACS and AusRivAS (Davies, 2000).

2.1.6.7   Examples of assessment approaches

Examples of indices of the different assessment approaches based on macroinvertebrates are given in Table 2.1.2.

Table 2.1.2   Examples of commonly applied biological assessment methods based on macroinvertebrates

Approch/method	Country	Reference
Saprobic approach		
Saprobic Index (S)	Austria	Moog, 1995
German Saprobic Index (S)	Germany	DEV, 1988–1991
Biotic approach		
Belgian Biotic Index (BBI)	Flanders, Belgium	De Pauw and Vanhooren, 1983; NBN, 1984
Bulgarian Biotic Index (BGBI)	Bulgaria	Uzunov *et al.*, 1998
Indice Biotique Global Normalisé (IBGN)	France Belgium, Wallonia	AFNOR, 1992 Vanden Bossche and Josens, 2003
Danish Stream Fauna Index (DSFI)	Denmark	Skriver *et al.*, 2001
Indice Biotico Esteso (IBE)	Italy	Ghetti, 1997
BMWP, ASPT	UK	Armitage *et al.*, 1983
IBMWP	Spain	Alba-Tercedor and Sanchez-Ortega, 1988
Family Biotic Index (FBI)	USA	Hilsenhoff, 1988
IBPAMP	Argentina	Rodriguez Capitulo *et al.*, 2001
NEPBIOS	Nepal	Sharma and Moog, 2001
South African Score System (SASS)	South Africa	Chutter, 1972
Acid Class	Germany	Braukmann, 2001
Diversity approach		
Diversity index (H′)	Various	Shannon and Weaver, 1949
Sequential Comparison Index (SCI)	USA	Cairns *et al.*, 1968
Multimetric approach		
Index of Biotic Integrity (IBI)	USA	Barbour *et al.*, 1992
Acidification Index	Sweden	Johnson, 1998
EBEOSWA	Netherlands	STOWA, 1992
Environmental Quality Ratio (EQR) approach		
Environmental Quality Index (EQI)	UK	Sweeting *et al.*, 1992
RIVPACS	UK	Wright *et al.*, 2000
AUSRIVAS	Australia Indonesia	Smith *et al.*, 1999 Sudaryanti *et al.*, 2001
SWEPACS	Sweden	Sandin, 2001
Other approaches		
Index of Trophic Completeness (ITC)	Russia, Netherlands	Pavluk *et al.*, 2000
Gammarus/Asellus Index	UK	MacNeil *et al.*, 2002

Presently, the most commonly applied indices in Europe are based on the saprobic and biotic approach. According to Nixon (2003), eleven countries (mainly Central and Eastern Europe) are assessing river water quality by means of the saprobic system, while another eleven are using one or another biotic index. The saprobic system would produce comparable results, whereas the biotic indices used by one country may not necessarily be comparable with that used in another. Recently, however, as an incentive of the European Water Frame Directive, also (stressor-specific) multimetric systems, originating from the USA, are now being developed and introduced. In contrast with the saprobic and biotic indices, which are solely based on a community structure analysis, the multimetric assessment systems may also include functional and non-taxonomic characteristics. In addition, diversity indices which are nowhere used as a national standard are now being included as a separate metric in these multimetric systems (e.g. Hering *et al.*, 2004). For the assessments based on macroinvertebrates, also more and more use is being made of multivariate analysis which has the advantage to clearly link the biological communities to the river typology. Other characteristics which have received attention during the last decade in assessments are the macroinvertebrate community structure related to the feeding strategy, migration or habitat use (e.g. Index of Trophic Completeness (ITC), Pavluk *et al.*, 2000) and the use of key or target species (in how far does the species or taxa composition correspond with the expected composition of a particular type of surface water?) (e.g. Lorenz *et al.*, 2004).

2.1.7 CLASSIFICATION AND COMPLIANCE TESTING

Index values or scores are usually divided into river (water) quality classes. Depending on the assessment methods used, these classes are an expression of a general (ecological) quality or refer more specifically to oxygen conditions linked with organic pollution, eutrophication linked with enrichment of organic and inorganic nutrients, or acidification linked with pH. Quality classes are, however, a subjective and arbitrary element, often merely based on a consensus among (a select group of) ecologists and water managers. This has resulted in different sets of quality classes used in European countries (Newman, 1988; Sandin *et al.*, 2001).

Presently, the WFD (EU, 2000) distinguishes five classes of ecological quality which are the result of the assessment of several biological communities, including the macroinvertebrates, together with the physical-chemical and hydromorphological quality. The worst quality of the biological communities is determining the final (ecological) quality. The quality of the communities is determined by the ecological quality ratio (EQR), which is the proportion between the observed versus the (predicted) reference condition. This value varies between 0 and 1, with zero meaning a bad status, and a value near 1 meaning a very good or reference status with no, or very slight, deviation of the undisturbed status.

An often essential but critical point in the assessment is the establishment of reference conditions. These should be based on scientific grounds (see also Section

2.1.6.5). According to the EU (2000), these reference conditions have to be described for the different types of rivers in different ecoregions. In several cases, at least in Europe, reference conditions are not always available anymore and have to be searched for in other similar river basins located in the same ecoregion.

Compliance, finally, is a political decision. Depending on the quality objectives, a goal can be set, based on a biotic index, for example. In Flanders, a BBI value of 7 on a scale of 0 to 10 is considered as a basic water quality standard (De Pauw *et al.*, 1992; De Pauw and Hawkes, 1993). In other cases, no clear biological baseline is set (e.g. saprobic index, BMWP, IBGN). The interpretation of the index values is mostly left to the water managers. The WFD (EU 2000), however, is clearly setting a goal, aiming at least at a good ecological status of all water bodies. This means that only a small deviation from the reference condition is allowed for. The status is, however, based not only on the analysis of several biological communities, including the macroinvertebrates, but as well on the physical-chemical and hydromorphological features. The lowest quality index of the different biological communities is in the first place determining the ecological status of the river.

2.1.8   GRAPHICAL REPRESENTATION AND MAPPING

Important in the overall strategy of water management, and for the creation of public awareness, is the mapping and visualization of the river quality by means of a colour code. Colours provide a simple but strong signal (e.g. De Pauw *et al.*, 1992; Knoben *et al.*, 1995). Rather than adopting a standardized biological assessment method, the European Centre for Normalization (CEN) has been opting for a harmonization of the colour code which is an expression of the water quality (ISO 8689-1, 2000; ISO 8689-2, 2000). For the presentation of biotic indices based on macroinvertebrates, usually five colours are used: blue (very good quality, unpolluted), green (good quality, slightly polluted), yellow (moderately polluted, critical situation), orange (bad quality, heavily polluted) and red (very bad quality, very heavily polluted). Often these colours correspond to the quality classes. Sometimes, black is also included to express an extremely bad quality with no macroinvertebrates present at all, except Syrphidae or Eristalinae (Rat-tailed maggot larvae). The river quality can be mapped by means of full lines representing the quality of the watercourse, or dots representing the quality of the sampled site.

2.1.9   CHALLENGES AND FUTURE DEVELOPMENTS

Even after a century of endeavour, as can be seen from this overview, worldwide the interest in the further development of biological methods for monitoring and assessment of rivers is still growing and in full evolution. Whereas in many developing countries a serious start is now being given to introducing and developing biological methods for river assessment, in numerous developed countries, on the other hand, the existing methods applied over many years are now in the process of being optimized and internationally standardized. After gaining experience with several approaches, an important evolution seems to be a common shift towards the application and development of multimetric indices based on score systems related to reference conditions (ecological quality ratios) (Hering *et al.*, 2004). In addition to expert knowledge, also multivariate data analysis and modelling techniques start playing a more and more crucial role in the development, evaluation and optimization of these indices.

These data analysis and modelling techniques are also becoming more and more important to gain insight in the biogeographic distribution of the macroinvertebrates and their habitat preferences (Adriaenssens *et al.* 2004a,b, Dedecker *et al.* 2004 and D'heygere *et al.* 2003.) The assessment methods could thus be developed in a more reliable way by gathering information on the river conditions the organisms prefer and consequently on their vulnerability, habitat specificity and synecology. These insights would also deliver valuable information for decision support in river restoration management (Goethals and De Pauw, 2001), and also on cause detection of river deterioration (cf. the actual development of stressor-specific metrics). Actually, the set of tools useful for this purpose is growing increasingly (e.g. Lorenz *et al.*, 2004).

In addition, improvements are necessary within the whole information chain from data collection (including sampling) towards assessment incentives (Gabriels *et al.*, 2005). Therefore, an integrated uncertainty analysis of this chain (data quality control and assurance), from the perspective of the information needs of the river managers, is necessary to guarantee that the assessment and data collection is based on methods with the needed precision and accuracy (Karr and Chu, 1999). Decision makers should thus also be aware of the level of uncertainty of the used methods and the impact of this uncertainty on the reliability of their planning and restoration actions.

In particular, one has to be aware of the specificity of the ecological indices used (Seegert, 2000). Many water systems do not only show large natural differences, but also have multiple uses and related stresses (Verdonschot and Nijboer, 2002). As a consequence, the importance (and related value) of the water system conditions can differ a lot between regions (leading to different scores merely on the basis of its uses and values). Therefore, most regions opt for a local assessment system, leading to the use of particular monitoring and assessment methods, as has been the case, for instance, in the different member states of Europe up until now. An important phase in the ecological assessment will be the international intercalibration of the different biological approaches and methods, not only in Europe in the context of the European Water Framework directive, but also worldwide. To ease and allow for the development of international databases on river ecology, the exchange of data collection and handling methods of, for example, macroinvertebrates but also of physical, chemical and hydromorphological river characteristics, will be of major importance.

REFERENCES

Adriaenssens, V., De Baets, B., Goethals, P. L. M. and De Pauw, N. 2004a. 'Fuzzy rule-based models for decision support in ecosystem management'. *Sci. Total Environ.*, **319**, 1–12.

Adriaenssens, V., Goethals, P., Charles, J. and De Pauw, N. 2004b. 'Application of Bayesian Belief Networks for the prediction of macroinvertebrate taxa in rivers'. *Ann. Limnol.*, **40**, 181–191.

Adriaenssens, V., Simons, F., Nguyen, L. T. H., Goddeeris, B., Goethals, P. L. M. and De Pauw, N., 2004c. 'Potential of bio-indication of Chironomid communities for assessment of running water quality in Flanders (Belgium)'. *Belg. J. Zool.*, **134**, 15–24.

AFNOR, 1992. *Essai des Eaux: Determination de l'Indice Biologique Global* Normalise (IBGN), Norme Française, NF T90–350. AFNOR, Paris, France.

Alba-Tercedor, J., 2006. 'Aquatic macroinvertebrates'. In: *Biological Monitoring of Rivers: Applications and Perspectives*, Ziglio, G., Siligardi, M. and Flaim, G. (Eds). Wiley, Chichester, UK, pp. 71–87.

Alba-Tercedor, J. and Sanchez-Ortega, A., 1988. 'Un metodo rapido y simple para evoluar la qualidad biologica de las aquas corrientes basado en el de Helawell (1978)'. *Limnetica*, **4**, 51–56.

APHA, 1998. Standard Methods for the Examination of Water and Wastewater, 20th Edition. American Public Health Association: Washington, DC, USA.

Armitage, P. D., Moss, D., Wright, J. F. and Furse, M. T., 1983. 'The performance of a new biological water quality score system based on macroinvertebrates over a wide range of unpolluted running-water sites'. *Water Res.*, **17**, 333–347.

Barbour, M. T. and Yoder, C. O., 2000. 'The multimetric approach to bioassessment, as used in the United States of America'. In: *Assessing the Biological Quality of Freshwaters – RIVPACS and Other Techniques*, Wright, J. F., Sutcliffe, W. and Furse, M. T. (Eds). Freshwater Biological Association: Ambleside, UK, pp. 281–292.

Barbour, M. T., Gerritsen, J., Snyder, B. D. and Stribling, J. B., 1992. *Rapid Bioassessment Protocols for Use in Wadeable Streams and Rivers. Periphyton, Benthic Macroinvertebrates and Fish*. Office of Water, USEPA: Washington, DC, USA.

Bartch, A. F. and Ingram, W. M., 1966. 'Biological analysis of water pollution in North America'. *Verh. Int. Ver. Limnol.*, **16**, 786–800.

Beck, W. M., 1954. 'Studies in stream pollution biology. A simplified ecological classification of organisms'. *Q. J. Fla. Ac. Sci.*, **17**, 211–277.

Bervoets, L., Bruylants, B., Marquet, P., Vandelannoote, A. and Verheyen, R., 1989. 'A proposal for modification of the Belgian biotic index method'. *Hydrobiologia*, **179**, 223– 228.

Brabec, K., Zahradkova, S., Nemejcova, D., Paril, P., Kokes, J. and Jarkovsky, J. 2004. 'Assessment of organic pollution effect considering differences between lotic and lentic stream habitats'. *Hydrobiologia*, **516**, 331–346.

Braukmann, U., 2001. 'Stream acidification in South Germany – chemical and biological assessment methods and trends'. *Aquat. Ecol.*, **35**, 207–232.

Bray, J. R. and Curtis, J. T., 1957. 'An ordination of the upland forest communities of Southern Wisconsin'. *Ecol. Monogr.*, **27**, 325–349.

Buffagni, A., Erba, S., Cazolla, M. and Kemp, J. L., 2004. 'The AQEM multimetric system for the southern Italian Apennines: assessing the impact of water quality and habitat degradation on pool macroinvertebrates in Mediterranean rivers'. *Hydrobiologia*, **516**, 313–329.

Cairns, J., Albaugh, D. W., Busey, F. and Chaney, M. D., 1968. 'The Sequential Comparison Index. A simplified method to estimate relative differences in biological diversity in stream pollution studies'. *J. Water Poll.Control Fed.*, **40**, 1607–1613.

Campaioli, S., Ghetti, P. F., Minelli, A. and Ruffo, S., 1994. *Manuale per il Riconscimento dei Macroinvertebrati delle Acque Dolci Italiane*, Vol. I. Provincia Autonoma di Trento: Trento, Italy.

Chandler, J. R., 1970. 'A biological approach to water quality management'. *Water Poll. Control*, **69**, 415–422.

Chapman, D., 1992. Water Quality Assessments. *A Guide to the Use of Biota, Sediments and Water in Environmental Monitoring*. Chapman & Hall: London, UK.

Chutter, F. M., 1972. 'An empirical biotic index of the quality of water in South African streams and rivers'. *Water Res.*, **6**, 19–30.

Cummins, K. W., 1975. 'Macroinvertebrates'. In: *River Ecology*, Whitton, B. A. (Ed.). Blackwell, London, UK, pp. 170–198.

Davies, P. E., 2000. 'Development of a national river bioassessment system (AUSRIVAS) in Australia'. In: *Assessing the Biological Quality of Freshwaters – RIVPACS and Other Techniques*, Wright, J. F., Sutcliffe, W. and Furse, M. T. (Eds). Freshwater Biological Association: Ambleside, UK, pp. 113–124.

De Cooman, W., Florus, M., Vangheluwe, M., Janssen, C., Heylen, S., De Pauw, N., Rillaerts, E., Meire, P. and Verheyen, R., 1999. 'Sediment characterisation of rivers in Flanders. The Triad approach'. In: *Characterisation and Treatment of Sediments*, CATS 4 Proceedings, De Schutter, G. (Ed.). PIH: Antwerp, Belgium, pp. 351–363.

Dedecker, A. P., Goethals, P. L. M., Gabriels, W. and De Pauw, N., 2004. 'Optimization of Artificial Neural Network (ANN) model design for prediction of macroinvertebrates in the Zwalm river basin (Flanders, Belgium)'. *Ecol. Modelling*, **174**, 161–173.

De Pauw, N. and Hawkes, H. A., 1993. 'Biological monitoring of river water quality'. In: *River Water Quality Monitoring and Control*, Walley, W. J. and Judd, S. (Eds). Aston University: Birmingham, UK, pp. 87–111.

De Pauw, N. and Heylen, S., 2001. 'Biotic index for sediment quality assessment of watercourses in Flanders, Belgium'. *Aquat. Ecol.*, **35**, 121–133.

De Pauw, N. and Vanhooren, G., 1983. 'Method for biological quality assessment of watercourses in Belgium'. *Hydrobiologia*, **100**, 153–168.

De Pauw, N. and Vannevel, R., 1991. *Macroinvertebrates and Water Quality*. Dossier Stichting Leefmilieu 11: Antwerp, Belgium (in Dutch).

De Pauw, N., Ghetti, P. F., Manzini, P. and Spaggiari, P., 1992. 'Biological assessment methods for running water'. In: *River Water Quality – Assessment and Control*, Newman, P., Piavaux, A. and Sweeting, R. (Eds), EUR 14606 EN-FR, 1992-III. Commission of the European Communities: Brussels, Belgium, pp. 217–248.

De Pauw, N., Lambert, V., Van Kenhove, A. and Bij de Vaate, A., 1994. 'Performance of two artificial substrate samplers for macroinvertebrates in biological monitoring of large and deep rivers and canals in Belgium and The Netherlands'. *Environ. Monitoring Assessment*, **30**, 25–47.

DEV, 1988–1991. *Methoden der Biologisch-Ökologischen Gewässeruntersuchung, Gruppe M: Fliessende Gewässer*, Teil 1 und 2. Deutsche Einheitsverfahren zur Wasser, Abwasser- und Schlammuntersuchung (DEV): Weinheim, Germany.

D'heygere, T., Goethals, P. and De Pauw, N., 2002. 'Optimisation of the monitoring strategy of macroinvertebrate communities in the river Dender, in relation to the EU Water Framework Directive'. *Sci. World J.*, **2**, 607–617.

D'heygere, T., Goethals, P. L. M. and De Pauw, N., 2003. 'Use of genetic algorithms to select input variables in decision tree models for the prediction of benthic macroinvertebrates'. *Ecol. Modelling*, **160**, 291–300.

EU, 2000. 'Directive 2000/60/EC of the European Parliament and of the Council of 23 October 2000 establishing a framework for Community action in the field of water policy'. *Off. J. Eur. Communities*, **L327**, 1–72.

Fontoura, A. P. and De Pauw, N., 1994. 'Microhabitat preference of stream macrobenthos and its significance in water quality assessment'. *Verh. Int. Verein. Limnol.*, **25**, 1936–1940.

Gabriels, W., Goethals, P. and De Pauw, N., 2005. 'Implications of taxonomic modifications and alien species on biological water quality assessment as exemplified by the Belgian Biotic Index method'. *Hydrobiologia*, **542**, 137–150.

Ghetti, P. F., 1997. *Manuale di Applicazione Indice Biotico Esteso (IBE)*. Provincia Autonoma di Trento: Trento, Italy.

Ghetti, P. F. and Ravera, O., 1994. 'European perspective on biological monitoring. 46'. In: *Biological Monitoring of Aquatic Systems*. Loeb, L. and Spacie, A. (Eds). Lewis Publishers, Boca Raton, FL, USA.

Giller, P. S. and Malmqvist, B., 1998. *The Biology of Streams and Rivers*, Oxford University Press, Oxford, UK.

Goethals, P. and De Pauw, N., 2001. 'Development of a concept for integrated river assessment in Flanders, Belgium'. *J. Limnol.*, **60**, 7–16.

Guerold, F., 2000. 'Influence of taxonomic determination level on several community indices'. *Water Res.*, **34**, 487–492.

Hawkes, H. A., 1979. 'Invertebrates as indicators of river water quality. 2'. In: *Biological Indicators of Water Quality*, James, A. and Evison, L. (Eds). Wiley: Chichester, UK, pp. 1–45.

Helawell, J. M., 1986. *Biological Indicators of Freshwater Pollution and Environmental Management*. Elsevier Applied Science Publishers, London, UK.

Hering, D., Moog, O., Sandin, L. and Verdonschot, P., 2004. 'Overview and application of the AQEM assessment system'. *Hydrobiologia*, **516**, 1–20.

Hilsenhoff, W. L., 1988. 'Rapid field assessment of organic pollution with a family-level biotic index'. *J. North Am. Benthol. Soc.*, **7**, 65–68.

Hynes, H. B. N., 1971. *The Biology of Polluted Waters*. University of Toronto Press: Toronto, ON, Canada.

Illies, J., 1961. 'Versuch einer allgemeinen biozönotische Gliedering der Fliesgewässer'. *Int. Rev. Ges. Hydrobiol.*, **46**, 205–213.

ISO 7828, 1985. *Water Quality – Methods for Biological Sampling – Guidance on Hand Net Sampling of Benthic Macroinvertebrates*, EN 27828, 1985. International Standards Organization: Geneva, Switzerland.

ISO 8265, 1988. *Water Quality – Methods of Biological Sampling – Guidance on the Design and Use of Quantitative Samplers for Benthic Macroinvertebrates on Stony Substrata in Shallow Waters*, EN 28265,1988. International Standards Organization: Geneva, Switzerland.

ISO 9391, 1993. *Water Quality – Sampling in Deep Waters for Macroinvertebrates – Guidance on the use of Colonization, Qualitative and Quantitative Samplers*, EN ISO 9391, 1993. International Standards Organization: Geneva, Switzerland.

ISO 5667-3, 1995. *Water Quality – Sampling – Part 3: Guidance on the Preservation and Handling of Samples*, EN ISO 5667-3, 1995. International Standards Organization: Geneva, Switzerland.

ISO 8689-1, 2000. *Biological Classification of Rivers, PART I: Guidance on the Interpretation of Biological Quality Data from Surveys of Benthic Macroinvertebrates in Running Waters*, EN ISO 8689-1, 2000. International Standards Organization: Geneva, Switzerland.

ISO 8689-2, 2000. *Biological Classification of Rivers, PART II: Guidance on the Presentation of Biological Quality Data from Surveys of Benthic Macroinvertebrates in Running Waters*, EN ISO 8689-2, 2000. International Standards Organization: Geneva, Switzerland.

Jaccard, P., 1908. 'Nouvelles reserches sur la distribution florale'. *Bull. Soc. Vaud. Sci. Nat.*, **XLIV**, 223–269.

Johnson, R. K., 1998. 'Classification of Swedish lakes and rivers using benthic macroinvertbrates'. In: *Bakgrundsrapport 2 till Bedömningsgrunder för Sjöar och Vattendrag – Biologiska*

Parametrar, Report 4921, Wiederholm, T. (Ed.). Swedish Environmental Protection Agency: Stockolm, Sweden.

Karr, J. R., 1981. 'Assessment of biotic integrity using fish communities'. *Fisheries*, **6**, 21–27.

Karr, J. R. and Chu, E. W., 1999. *Restoring Life in Running Waters. Better Biological Monitoring*. Island Press: Washington, DC, USA.

Knoben, R. A. E., Roos, C. and van Oirschot, M. C. M., 1995. *Biological Assessment Methods for Watercourses*, UN/ECE Task Force on Monitoring and Assessment, Vol. 3. RIZA: Lelystad, The Netherlands.

Knöpp, H., 1954. 'Ein neuer Weg zur Darstellung biologischer Vorfluteruntersuchungen, erläutert an einem Gütelangschnitt des Maines'. *Wasserwirtschaft*, **45**, 9–15.

Kolkwitz, R. and Marsson, M., 1902. 'Grundsatze für die biologische Beurteiung des Wassers nach zeiner Flora und Dauna'. *Prüfungsanst.Wasserversorg. Abwasserrein.*, **1**, 33–72.

Lafont, M., 1984. 'Oligochaete communities as biological descriptors of pollution in the fine sediment of rivers'. Hydrobiologia, **115**, 127–129.

Liebmann, H., 1962. *Handbuch der Frischwasser- und Abwasserbiologie*, Vol.1., 2nd Edition. R. Oldenburg: München, Germany.

Linke, S., Bailey, R. C. and Schwindt, J., 1999. 'Temporal variability of stream bioassessments using benthic macroinvertebrates'. *Freshwater Biol.*, **42**, 575–584.

Lorenz, A., Hering, D., Feld, C. K. and Rolauffs, P., 2004. 'A new method for assessing the impact of hydromorphological degradation on the macroinvertebrate fauna of five German stream types'. *Hydrobiologia*, **516**, 107–127.

Mackenthun, K. M., 1969. *The Practice of Water Pollution Biology*. FWPCA: *Washington*, DC, USA.

MacNeil, C., Dick, J. T. A., Bigsby, E., Elwood, R. W., Montgomery, W. I., Gibbis, C. N. and Kelly, D. W., 2002. 'The validity of the Gammarus: Asellus ratio as an index of organic pollution: abiotic and biotic influences'. *Water Res.*, **36**, 75–84.

Mason, C. F., 1981. *Biology of Freshwater Pollution*. Longman: London, UK.

Matthews, R. A., Buikema, A. L., Cairns, J. and Rogers, J. H., 1982. 'Biological monitoring Part IIA – Receiving system functional methods, relationship and indices'. *Water Res.*, **6**, 129–139.

Metcalfe, J. L., 1989. 'Biological water quality assessment of running water based on macroinvertebrate communities: history and present status in Europe'. *Environt. Pollut.*, **60**, 101–139.

Metcalfe-Smith, J. L., 1994. 'Biological water quality assessment of rivers: use of macroinvertebrate communities'. In: *The Rivers Handbook*, Vol. 2. Calow, P. and Petts, G. E. (Eds). Blackwell Scientific Publications: Oxford, UK, pp. 144–170.

Moog, O., 1995. *Fauna Aquatica Austriaca – a Comprehensive Species Inventory of Austrian Aquatic Organisms with Ecological Data*, Ist Edition. Wasserwirtschafskataster, Bundesministeriumf für Land- und Forstwirtschaft: Vienna, Austria.

NBN, 1984. Belgian Standard T92-402. Biological quality of watercourses. Determination of the Belgian Biotic Index based on aquatic macroinvertebrates Belgian Institute for Normalisation, Brussels, Belgium.

Newman, P. J., 1988. *Classification of Surface Water Quality. Review of the Schemes Used in EC Member States*. Heinemann Professional Publishers: Oxford, UK.

Nixon, S., 2003. 'An overview of the biological assessment of surface water quality in Europe'. In: *Biological Evaluation and Monitoring of the Quality of Surface Waters*, Symoens, J. J. and Wouters, K. (Eds). National Committee of Biological Sciences and SCOPE National Committee: Brussels, Belgium, pp. 9–15.

Norris, R. H. and Georges, A., 1993. 'Analysis and interpretation of benthic macroinvertebrate surveys'. In: *Freshwater Biomonitoring and Benthic Macroinvertebrates*, Rosenberg, D. M. and Resh, V. H. (Eds). Chapman & Hall: New York, NY, USA, pp. 243–286.

Pantle, R. and Buck, H., 1955. 'Die biologische Uberwachung der Gewasser und die Darstellung der Ergebnisse'. *Gas-u Wasserfach*, **96**, 604.

Patten, B. C., 1962. 'Species diversity in net phytoplankton of Raritan Bay'. *J. Marine Res.*, **20**, 57–75.

Pavluk, T. I., Bij de Vaate, A. and Leslie, H. A., 2000. 'Development of an Index of Trophic Completeness for benthic macroinvertebrate communities in flowing waters'. *Hydrobiologia*, **427**, 135–141.

Persoone, G. and De Pauw, N., 1979. 'Systems of biological indicators for water quality assessment'. In: Ravera O (ed.), Biological Aspects of Freshwater Pollution, Ravera, O. (Ed.) Pergamon Press: Oxford, UK, pp. 39–75.

Puig, M. A., 1999. *Els Macroinvertebrats dels Rius Catalans. Guia Illustrada.* Generalitat de Catalunya, Departament de Medi Ambient: Barcelona, Spain.

Reynoldson, T. B., Day, K. E. and Pascoe, T., 2000. 'The development of the BEAST: a predictive approach for assessing sediment quality in the North American Great Lakes'. In: *Assessing the Biological Quality of Freshwaters – RIVPACS and Other Techniques*, Wright, J. F., Sutcliffe, W. and Furse, M. T. (Eds). Freshwater Biological Association: Ambleside, UK, pp. 165–180.

Richardson, R. E., 1928. 'The bottom faunaof the middle Illinois River 1913–1925: its distribution, abundance, valuation and index value in the study of stream pollution'. *Bull. Illinois State Nat. Hist. Survey*, **17**, 387–475.

Rodriguez Capitulo, A., Tangorra, M. and Ocon, C., 2001. 'Use of benthic macroinvertebrates to assess the biological status of Pampean streams in Argentina'. *Aquat. Ecol.*, **35**, 109–119.

Roldan, G., 1992. *Guia Para el Estudio de los Macroinvertebrados Acuaticos de Antioquia, Colombia.* Universidad de Antioquia, Bogota, Colombia.

Rosenberg, D. M. and Resh V. H., 1982. 'The use of artificial substrates in the study of freshwater benthic macroinvertebrates'. In: *Artificial Substrates*, Cairns, J. (Ed.) Ann Arbor Science: Ann Arbor, MI, USA, pp. 175–235.

Rosenberg, D. M. and Resh, V. H., 1993. *Freshwater Biomonitoring and Benthic Macroinvertebrates.* Chapman & Hall, New York, NY, USA.

Sandin, L., 2001. 'SWEPACS: a Swedish running water prediction and classification system using benthic macroinvertebrates'. In: *Classification of Ecological Status of Lakes and Rivers*, Bäck, S. and Karttunen, K. (Eds.). TemaNord Environment 2001, pp. 44–46.

Sandin, L., Sommerhäuser, M., Stubauer, I., Hering, D. and Johnson, R., 2000. *Stream Assessment Methods, Stream Typology Approaches and Outlines of a European Stream Typology.* AQEM – EU Project EVK1-CT1999-00027.

Sandin, L., Hering, D., Buffagni, A., Lorenz, A., Moog, O., Rolauffs, P. and Stubauer, I., 2001. *Experiences with Different Stream Assessment Methods and Outlines of an Integrated Method for Assessing Streams Using Benthic Macroinvertebrates.*, AQEM–EU Project EVK1-CT1999-00027.

Schmedtje, U. and Kohmann, F., 1992. *Bestimmungsschlüssel fur die Saprobier-DIN-Arten (Macroorganismen).* Informationsberichte Heft 2/88. Bayerisches Landesamt für Wasserwirtschaft: Münich, Germany.

Schwoerbel, J., 1970. *Methods of Hydrobiology. Freshwater Biology.* Pergamon Press: London, UK.

Seegert, G., 2000. 'The development, use, and misuse of biocriteria with an emphasis on the index of biotic integrity'. *Environ. Sci. Policy*, **3**, 51–58.

Shannon, C. E. and Weaver, W., 1949. *The Mathematical Theory of Communication.* University of Illinois Press: Urbana, IL, USA.

Sharma, S. and Moog, O., 2001. *Introducing the NEPBIOS method of Surface Water Quality Monitoring.* Aquatic Ecology Centre, Kathmandu University: Kathmandu, Nepal.

Skriver, J., Friberg, N. and Kirkegaard, J., 2001. 'Biological assessment of watercourse quality in Denmark: Introduction of the Danish Stream Fauna Index (DSFI) as the official biomonitoring method'. *Verh. Int. Verein. Limnol.*, **27**, 1822–1830.

Sladecek, V., 1973. 'The reality of three British biotic indices'. *Water Res.*, **7**, 995–1002.

Smith, M. J., Kay, W. R., Edward, D. H. D., Papas, P. J., Richardson, K. S. J., Simpson, J. C., Pinder, A. M., Cale, D. J., Horwitz, P. H. J., Davis, J. A., Yung, F. H., Norris, R. H. and Halse, S. A., 1999. AusRivAS: using macroinvertebrates to assess ecological condition of rivers in Western Australia'. *Freshwater Biol.*, **41**, 269–282.

Sorensen, T., 1948. 'A method of establishing groups of equal amplitude in plant sociology based on similarity of species content and its application to analysis of the vegetation on Danish commons'. *Biologiske Skifter. Det Konglige Danske Videnskabernes Selskab*, **5**, 1–34.

STOWA, 1992. *Ecologische Beoordeling en Beheer van Oppervlaktewater: Beoordelingsstseem voor Stromende Wateren op Basis van Macrofauna*, STOWA Rapport 92-8. STOWA: Utrecht, The Netherlands, pp. 1–86.

Sudaryanti, S., Trihadiningrum, Y., Hart, B. T., Davies, P. E., Humphrey, C., Norris, R., Simpson, J. and Thurtell, L., 2001. 'Assessment of the biological health of the Brantas River, East Java, Indonesia using the Australian River Assessment System (AUSRIVAS) methodology'. *Aquat. Ecol.*, **35**, 135–146.

Sweeting, R., Lowson, D., Hale, P. and Wright, J., 1992. '1990 Biological assessment of rivers in the UK'. In: *River Water Quality – Assessment and Control*, Newman, P., Piavaux, A. and Sweeting, R. (Eds), EUR 14606 EN-FR, 1992-III. Commission of the European Communities: Brussels, Belgium, pp. 319–326.

Tachet, H., Richoux, P., Bournaud, M. and Usseglio-Polatera, P., 2002. *Invertébrés d'Eau Douce. Systématique, Biologie, Écologie.* CNRS Editions: Paris, France.

USEPA, 1996. Biological Criteria: Technical Guidance for Streams and Small Rivers, EPA-822-B96-001. Office of Water, U.S. Environmental Protection Agency: Washington, DC, USA.

Uzunov, Y., Penev, L., Kovachev, S. and Baev, P., 1998. 'Bulgarian Biotic Index (BGBI) – an express method for bioassessment of the quality of running waters'. *Comptes Rendus Acad. Bulg. Sci.* **51**, 117–120.

Vanden, Bossche, J. P. and Josens, G., 2003. 'Macrozoobenthos biodiversity and biological quality monitoring of watercourses in Wallonia (Belgium)'. In: *Biological Evaluation and Monitoring of the Quality of Surface Waters*, Symoens, J. J. and Wouters, K. (Eds). National Committee of Biological Sciences and SCOPE National Committee: Brussels, Belgium, pp. 77–79.

Vannote, R. L., Minshall, G. W., Cummins, K. W., Sedell, J. R. and Cushing, C. E., 1980. 'The river continuum concept'. *Can. J. Fisheries Aquat. Sci.*, **37**: 130–137.

Verdonschot, P. F. M. and Nijboer, R. C. 2002. 'Towards a decision support system for stream restoration in the Netherlands: an overview of restoration projects and future needs'. *Hydrobiologia*, **478**, 131–148.

Washington, H. G., 1984. 'Diversity, biotic and similarity indices: a review with special relevance to aquatic ecosystems'. *Water Res.*, **18**, 653–694.

Whittaker, R. H. 1952. 'A study of summer foliage insect communities in the Great Smoky Mountains'. *Ecol. Monogr.*, **22**, 1–44.

Wilhm, J. L. and Dorris, T. C., 1966. 'Species diversity of benthic macroinvertebrates in a stream receiving domestic and oil refinery effluents'. *Am. Midl. Nature*, **76**, 427–449.

Williams, W. T., 1971. 'Principles of clustering'. *Rev. Ecol. Syst.*, **2**, 303–326.

Woodiwiss, F. S., 1964. 'The biological system of stream classification used by the Trent River Board'. *Chem. Ind.*, **14**, 443–447.

Woodiwiss, F. S., 1980. *Biological Monitoring of Surface Water Quality. Summary Report*, ENV/787/80-EN. Commission of the European Communities: Brussels Belgium.

Wright, J. F., Furse, M. T. and Armitage, P. D., 1993. 'RIVPACS – a technique for evaluating the biological quality of rivers in the UK'. *Eur. Water Quality Control*, **3**, 15–25.
Wright, J. F., Sutcliffe, W. and Furse, M. T. 2000. *Assessing the Biological Quality of Freshwaters – RIVPACS and Other Techniques*. Freshwater Biological Association: Ambleside, UK.
Zelinka, M. and Marvan, P., 1961. 'Zur präzisierung der biologische Klassifikation der reinheit fliessender gewässer'. *Arch. Hydrobiol.*, **57**, 389–407.

2.2

Monitoring Methods
Based on Fish

Michele Scardi, **Lorenzo Tancioni** and **Stefano Cataudella**

2.2.1   INTRODUCTION

Fish have been used as biological indicators of environmental quality in many different aquatic ecosystems (Fausch *et al.*, 1990; Whitfield, 1996) and their role as indicators has been explicitly mentioned in the legislation about water resources and aquatic environments, not only in the European Union and USA, but also in other countries (European Commission, 1992; European Union, 2000; Kurtz *et al.*, 2001). In all cases, these legislations focus on the need for an improvement in environmental quality, as well as for the assessment of ecological integrity as key tools for conservation, restoration and management activities.

During the last twenty years, biological monitoring has gained a broader consensus with respect to chemical monitoring, because the latter often misses relevant anthropogenic perturbations (e.g. channelization of river stretches) that may induce severe habitat degradation with little or no impact on water quality. Moreover, biotic responses usually integrate over time and space the effect of such perturbations, thus reducing the sampling error, which is usually smaller for biological attributes than for chemical ones. In fact, physical and chemical attributes of water are unsuccessful as surrogates for measuring biotic integrity (Karr and Dudley, 1981). This opinion is also supported by Oberdorff and Hughes (1992), who used data about the fish assemblage in the Seine River catchment to assess water quality. They found that comparisons between a biotic index and an independent water quality index (based on water chemistry) indicated that the former was a more sensitive and robust measure of water quality.

The usage of fish species as indicators in aquatic ecosystems is based on the assumption that fish species and fish assemblages are sensitive indicators, which are able to detect subtle environmental changes (Karr, 1981; Shamsudin, 1988). Obviously, there are both advantages and disadvantages in using fish fauna as a biotic indicator, which depend on its particular ecological characteristics and on the scale of its interactions with the environment.

The advantages of biotic indicators of water quality based on fish fauna are numerous. For instance, fish assemblages are present in all aquatic ecosystems and their structure not only depends on biotic, physical and chemical constraints, but also on the hydromorphological continuity of rivers and streams.

Most fish species are easy to identify and specimens can be sampled and then immediately released after identification and biometric measurements, while the biology of fish species is usually well know, in many cases even at the physiological level.

From an ecological point of view, fish assemblages often include species that belong to different trophic levels, as well as to different trophic guilds. Changes in the structure of fish populations and assemblages integrate responses to environmental disturbances over long time intervals, while, depending on the characteristics of each species, they integrate those responses over different spatial scales: sedentary species respond to local disturbances, whereas mobile species respond to diffuse stressors. Moreover, a useful characteristic of fish as indicators is that many species are well suited for ecotoxicological tests, as they can be easily kept in captivity. Finally, there is a broad public awareness of the iconic role of fish fauna in aquatic ecosystems and this awareness, in conjunction with the economic value of many species, makes it easier to enforce monitoring activity based on fish or to take into account fish fauna in economic analyses.

Obviously, there are also disadvantages in the usage of fish as biotic indicators of environmental quality. In fact, unbiased sampling may be difficult or impossible to attain in some cases (e.g. large rivers). Fishing gear may be more or less selective, depending on environmental characteristics, as well as on fish species and size, while sampling may be biased by the mobility of fish according to seasonal, daily, nicte-meral or occasional patterns. As for their ecological response to perturbations, some fish species may tolerate pollutants that are dangerous to other biota and they can also actively avoid anthropogenic disturbance. The advantages in the usage of fish

Table 2.2.1 Effects of disturbances on fish fauna at different levels of biotic organization

Level	Undisturbed fish fauna	Disturbed fish fauna
Cells	Normal cell functionality, stable lysosomes, genetic integrity	Ongoing detoxifying activity, genetic damages
Individuals	No morphological anomalies, low parasitism, normal behaviour, normal condition factor	Fin damages and other lesions or anomalies, tumours, abnormal behaviour, impaired condition factor
Populations	Self-sustaining populations, adequate larval recruitment, normal demographic structure, all the age classes are present, predictable spatial distribution	Insufficient or null larval recruitment, low number of juveniles and sub-adults, altered spatial distribution
Assemblages	Normal to high diversity, many guilds are present, complex biotic interactions, expected seasonal cycles	Low diversity, some guilds are not present, loss of sensitive species, reduced biotic interactions, altered seasonal cycles

as indicators, however, outweigh the disadvantages, especially in the case where environmental quality is not to be assessed on a very small spatial scale. The expected effects of disturbances on fish fauna are summarized in Table 2.2.1, with respect to different levels of biotic organization.

Independently of the level of biological organization, as well as of the spatial and temporal scale, the evaluation of the ecological status of an aquatic ecosystem based on biotic indicators (fish, as well as other organisms) is only possible if a suitable metric (or a set of metrics) is defined, which is appropriate for measuring deviation of the observed conditions from their expected status.

This goal can be achieved in different ways, but two main options are available: (1) measuring the deviation of a set of functional and structural features of fish assemblages from those that are expected for pristine fish assemblages; (2) measuring the distance or similarity between observed fish assemblages and properly selected reference fish assemblages. In the latter case, the reference can be obtained from a number of sources, such as previous observations, historical records, biogeographical inferences, mathematical models, empirical knowledge, etc.

In both the above-mentioned cases, the whole fish assemblage is usually taken into account. This does not imply that single species cannot be used as indicators. On the contrary, single species can be used as biomonitoring tools, for instance, in ecotoxicological tests or by screening DELT (i.e. deformities, eroded fins, lesions, and tumours) anomalies. In fact, at the lowest end of the biotic organization scale, the effects of ecological disturbances on fish fauna can be detected by analysing specific molecular, genetic, metabolic or morphological responses, i.e. by taking into account appropriate *biomarkers*.

Although the use of biomarkers is gaining momentum (for a thorough review, see van der Oost *et al.*, 2003), it should only be considered in fish-based monitoring in case it is really needed and supported by a specific ecological rationale (e.g. in order to trace the bioaccumulation or biomagnification of pollutants that are not as concentrated in abiotic matrices as to directly affect community structure and functioning). Most fish species belong to the upper trophic levels and therefore they are often involved into such processes. Moreover, fish are complex and sensitive organisms that are well suited for setting up laboratory experiments, but they can also be sampled in a non-destructive way (e.g. by collecting blood samples).

As a matter of fact, the occurrence of a fish species is always a direct evidence for ecological conditions that are at least compatible with its survival, although it does not imply the existence of a self-sustaining, healthy population. On the other hand, the interpretation of biomarkers that are not directly related to short-term physiological responses is not as straightforward because of the mobility of most fish species. Moreover, biomarker-based approaches are typically species-specific and therefore difficult to generalize. Therefore, in this chapter biomarkers and fish physiological responses will be only taken into account as attributes of the fish assemblage in assemblage-oriented evaluation procedures, which, in contrast, will be discussed in detail.

2.2.2   MULTIMETRIC INDICES

Most of the monitoring methods based on fish fauna rely upon biotic indices and, in particular, upon multimetric indices, which combine different indicators (i.e. metrics) into a single score. Each candidate metric is separately tested and calibrated in order to adequately scale its values to obtain a unitless score, which can be easily aggregated to other scores into a multimetric index.

Different approaches have been used to arrange and to analyse biotic data, but they all start out with a list of the organisms that were collected and identified. In the past, the distribution of a few indicator species was used to assess watershed health, but the assessment procedure was often a little more complicated than just recording the occurrence of these indicator species. More recently, other sources of ecological information, such as population structure, presence of anomalies or diseases, etc., have been also taken into account and combined into multimetric indices.

Such multimetric indices were first called *biotic indices* because they scored the pollution tolerance of many different species, which were regarded as biological indicators. While biotic indices were expanding in use, other indices, such as those based on species diversity, grew in popularity and were used for many years. Recently, multimetric indices, such as the Index of Biotic Integrity (Karr, 1981), have become a standard.

Four main points should be taken into account in developing an effective multi-metric biotic index:

(1) Classifying homogeneous biotopes within or across ecoregions (e.g., large or small streams; high- or low-gradient streams, etc.).

(2) Selecting those metrics that reflect the most relevant and reliable responses to the effects of disturbances.

(3) Defining sampling designs and protocols that allow us to accurately measure the selected metrics in the field.

(4) Analysing collected data to relate this information to the watershed health as straightforwardly as possible (e.g. by means of simple linear functions).

Obviously, even the best index, if not adequately tested and widely accepted, is practically useless. Therefore, it is of paramount importance to rely on the widest base of testers and collaborators when developing a multimetric index.

Even when a multimetric index has been correctly developed (see, for instance, Hughes *et al.*, 1998), its ability to measure environmental perturbations depends on the right combination of metrics that are taken into account. Of course, there are no general rules that can help in selecting proper metrics, but usually the latter are chosen on the basis of their presumed general ecological relevance or because of their responsiveness to specific perturbations (e.g. as shown by ecotoxicological tests). In any case, the selection of a metric should be also supported by adequate statistical evidences, and only metrics that are linearly or monotonically related to specific stressors or to general environmental conditions should be taken into account. The selection of metrics, however, is often based on literature references, on expert judgement (especially when existing indices are adapted to different regional conditions) or on *a priori* belief. A typical example of potentially biased metrics is the species richness, which is considered as a key metric in most multimetric indices. The intermediate disturbance hypothesis (Connell, 1978) and many field evidences, in fact, suggest that species richness is not monotonically related to environmental quality, as the highest values are usually observed in moderately perturbed sites rather than in pristine ones. Thus, the overall efficiency of a multimetric index can be impaired by a poor or biased selection of the underlying metrics.

2.2.3 INDEX OF BIOTIC INTEGRITY

The Index of Biotic Integrity (IBI) gained considerable popularity as a multimetric method for the assessment of the integrity of fish assemblage during the last two decades, not only in the USA (Karr and Dudley, 1981; Karr *et al.*, 1986; Plafkin *et al.*, 1989; Fausch *et al.*, 1990), but also in other countries (Hughes and Oberdorff, 1999). The original version of the IBI includes twelve assemblage attributes (see Table 2.2.2) that are compared to values expected for an unperturbed stream of the same size in the same ecoregion (Plafkin *et al.*, 1989). The assemblage attributes can be grouped into three main categories, i.e. species richness and assemblage composition, trophic composition and fish abundance and condition.

These attributes are scored according to the uneven integers 1, 3, and 5, which stand, respectively, for conditions that deviate strongly, moderately or slightly from

Table 2.2.2   Metrics of the Index of Biotic Integrity as originally developed (adapted from Karr, J. R., Fausch, F. D., Angermeir, P. L., Yant, P. R. and Schlosser, I. J., 1986, *Assessing Biological Integrity in Running Waters: A Method and its Rationale*, Special Publication 5, Illinois Natural History Survey, Chicago, IL, USA)

Category	Metric
Species richness and assemblage composition	1. Total number of fish species
	2. Number and identity of darter species
	3. Number and identity of sunfish species
	4. Number and identity of sucker species
	5. Number and identity of intolerant species
	6. Proportion of individuals as green sunfish (tolerant species)
Trophic composition	7. Proportion of individuals as omnivores
	8. Proportion of individuals as insectivorous cyprinids (minnows)
	9. Proportion of individuals as top carnivores
Fish abundance and condition	10. Number of individuals in sample
	11. Proportion of individuals as hybrids
	12. Proportion of individuals with disease, tumours, fin damage or skeletal anomalies

situations at reference sites (Fausch *et al.*, 1990). The overall score, obtained by summing up those of the twelve assemblage attributes, is then categorized into discrete classes according to expert judgment, in order to provide an integrated scoring system for ecological integrity. A description of the compositional attributes of the fish assemblage is usually provided for each quality class. Usually, IBI scores are presented directly, or may be expressed as a percentage of the maximum (Lyons *et al.*, 1995).

The original version of the IBI has been modified more or less substantially in order to preserve its rationale independently of the ecoregion in which it is applied. Therefore, it is probably more correct to think about the IBI as a flexible conceptual framework that can be easily adapted on a regional scale.

Simon and Lyons (1995) pointed out that the IBI is not an index based on community analysis, but rather a procedure involving several hierarchical biotic levels and based on a sample of the assemblage. Since its first application, the IBI has been criticized by some Authors (e.g. by Suter, 1993), although strong counter arguments have been also presented by others (e.g. Simon and Lyons, 1995; Karr and Chu, 1997; Hughes *et al.*, 1998).

Although the ecological principles on which the IBI is based are sound, its application in the original form, as well as in other adapted versions, may present problems when considered for use in some ecological regions.

In fact, the IBI relies on biotic attributes that require detailed historical and ecological information which is often not available. A particular problem is to be tackled when dealing with attributes involving proportions of fish species or functional groups, as little reference (e.g. pre-impact) information on them is usually available. Moreover, available data are often biased from a quantitative point of view. Lyons *et al.* (1995), for instance, developed a preliminary IBI for streams in west central Mexico but expressed concerns on the scarcity of fish community data.

Another potential drawback of the IBI is its lack of sensitivity with respect to some disturbances. For instance, in the USA it was found that the IBI did not indicate a degradation of biotic integrity in prairie streams following intensive testing of armored vehicles. Fishes were naturally adapted to droughts and flash floods and their presence, as well as the structure of the assemblage, depended on their rate of colonization rather than on habitat changes (Bramblett and Fausch, 1991). A similar situation can be expected in other ecoregions in which high variability in environmental conditions is observed, especially when associated with variability and unpredictability of rainfall and runoff within seasons and between years. In other words, a naturally high disturbance regime to which fish are adapted may affect the ability of the IBI to evaluate ecological quality, especially because there are several anthropogenic changes that may actually mimic these natural disturbance regimes.

A major problem in the adaptation of the IBI to European and other ecoregions is the number of fish species, which may be significantly less than in North American rivers. This problem affects all of the European ecoregions, but it is much more evident in Southern European rivers and streams, where only a handful of species is found. In such situations, it is not possible to adapt the original IBI scheme and deeper changes are needed in order to optimally exploit the relevant information conveyed by a simplified fish assemblage structure. The presence of juveniles, or even the age structure of fish populations, for instance, can be assumed as an example of an alternate metric in species-poor situations.

Notwithstanding these problems, the original IBI has been successfully adapted to many ecoregions. Several examples of localized IBIs exist not only for the United States and European countries (see for instance, Oberdorff and Hughes, 1992), but also in many other countries, like, for instance, Canada, Mexico, Australia, South Africa, Guinea, Namibia, Cameroon, Korea, etc. (see, respectively, Steedman, 1988; Lyons *et al.*, 1995; Kleynhans, 1999; Harris, 1995; Hugueny *et al.*, 1996; Hay *et al.*, 1996; Kamdem Toham and Teugels, 1999; An *et al.*, 2002).

2.2.4   OTHER BIOTIC INDICES

IBIs are not the only methods for evaluating environmental quality on the basis of fish assemblages. In fact, several other approaches have been proposed, and some of them rely upon different rationales. However, it is very difficult to trace the exact boundary between IBI-inspired methods and other procedures, because the underlying ecological concepts are obviously related.

MuLFA, for instance, is a method proposed by Schmutz *et al.* (2000) for the Austrian rivers and streams, which is certainly different from IBI. In fact, it takes into account multiple level of organization of fish fauna and river-type-specific assessment criteria, namely, presence of river-type-specific species, presence of self-sustaining populations, shifts in fish region, number of missing guilds, alterations in guild composition, changes in biomass and density, and changes in population age structure.

The final assessment of ecological integrity is obtained as a weighted average of the scores assigned to each criterion. An advantage of the MuLFA approach relative to other biotic indices is that the weighting scheme is not the same for all of the river types, thus allowing us to optimize the assessment of ecological integrity on a functional and regional basis. Another advantage of this simple and flexible method is that it encompasses different temporal and spatial scales, thus allowing the detection of a wide range of environmental effects of human alterations.

Other biotic indices are more closely related to the original IBI, although they focus on more sophisticated – and, probably, objective – methods for selecting or weighting an optimal set of metrics. An example of such an index is the one proposed for the Walloon part of the Meuse river basin (Belgium) by Kestemont *et al.* (2000). This index was originally developed on the basis of twelve metrics by adapting the IBI rationale to the local ecological conditions. Then, a multivariate statistical procedure based on Principal Component Analysis (PCA) was used to identify the most relevant metrics with respect to local disturbances. In particular, a subset of metrics was selected on the basis of their PCA factor loadings, thus allowing the downscaling of the index to a simplified version which preserved most of the efficiency of the original version in assessing the ecological integrity of local rivers and streams.

Several indices include metrics based on population structure data, but the most straightforward implementation of this concept is probably the one that Badino *et al.* (1992) proposed for Italian rivers. Their Ichthyologic Index (II) is computed by multiplying species richness by a linear combination of two factors, obtained from tables provided by the authors, which accounts for population structure (presence of juveniles, sub-adults and adults) and for abundance of fish fauna relative to species richness. In this way, the II takes into account the two main components of fish assemblage diversity, namely the species richness to fish abundance ratio and the average demographic complexity of local fish populations. The II score can be also discretized in order to express the estimate of ecological integrity according to a scale of five quality classes.

2.2.5   COMPARISON WITH REFERENCE CONDITIONS

Most problems with the application of IBI and related indices arise in situations in which the fish assemblage structure is too simple and therefore it cannot convey enough information about the biotic response to environmental perturbations. Moreover, the growing number of different implementation of IBI-inspired multimetric indices, although necessary for a better adaptation to ecoregional conditions, is certainly narrowing the number of users of each local index, which often cannot be validated on the basis of a large number of independent applications.

Therefore, measuring distance or similarity between the observed fish assemblage and a reference one seems a more objective approach to the assessment of ecological integrity than computing biotic indices. Of course, biotic indices may provide useful insights into ecosystem quality and they are certainly adequate in a number of practical applications, but they are always based on metrics that are selected on a subjective basis. It is obvious that a biotic index can be unreliable if wrong or scarcely relevant metrics are used, but it is still inherently subjective even when ecologically sound and sensitive metrics are carefully selected. For instance, in many cases closely related metrics are simultaneously used in the same index, thus increasing the relative influence of a single underlying ecological criterion. The most obvious example of such a lack of independence between metrics is provided by the number of species and the number of individuals in a given sample, which are both very common in biotic indices. Even though in low diversity assemblages it is possible to observe many individuals belonging to a very limited number of species, it is not possible to record many species in case only a few individuals have been observed. Moreover, in ecologically homogeneous sites the assemblage diversity is not likely to vary too much, and the number of species and number of individuals are more closely dependent on each other (and both of them depend on the sampling techniques). Thus, taking simultaneously into account both of these metrics increases the actual weight of species richness in the multimetric score.

Comparing observed and reference species richness is the simplest way to evaluate environmental quality using fish assemblages. This method has been applied in many cases and the observed to expected number of species ratio (O/E) has been regarded as an ecological indicator. According to this approach, the ratio should be larger than one in unperturbed conditions and smaller than one in perturbed conditions. The rationale supporting this method is absolutely straightforward and it has been applied in many different cases, such as, for instance, in the RIVPACS approach based on benthic macroinvertebrates (Wright *et al.*, 1989). However, it can be misleading. In fact, it is certainly true that strong disturbances induce a decrease in species richness because of the exclusion of non-tolerant species, but it is also true that moderate perturbations may actually favour an increase in species richness. As already mentioned about the selection of metrics in multimetric indices, the intermediate disturbance hypothesis (Connell, 1978) provides a theoretical background for this empirical evidence. Therefore, species richness should not be regarded as a criterion for evaluating environmental quality because in many cases it is not monotonically related to disturbance.

A problem with multimetric indices that is sometimes overlooked is that the latter cannot be readily applied in the case of "information-poor" situations, such as those often found in Africa or in other developing countries. Therefore, alternate solutions should be considered instead, such as those based on community analysis, in order to assess ecological quality by comparing fish assemblage composition in a set of different sites. An application of this approach was presented by Ramm (1988), who developed a community degradation index (CDI) based on the principles of the Jaccard similarity index (Jaccard, 1900, 1901, 1908), which involves the number of species that are found both in the site to be evaluated and in a reference site, excluding other ecological aspects (i.e. trophic specialization, habitat specialization and intolerance) and considering abundance information unreliable.

Methods based on the comparison of the observed fish assemblage to reference conditions are not affected by problems related to the selection of a suitable set of metrics. However, a critical step (and not a minor one) is the selection of an appropriate measure for distance or similarity between observed and reference assemblages. In fact, this measure must be accurately selected according to the quality and to the nature of the available information, which may be more or less affected by errors. For instance, information may be inaccurate from the quantitative point of view, or it may be not completely reliable as far as absence of species (that is usually an implicit assumption if no specimens are collected) is concerned. In particular, when species-absence data are reliable, they contribute to the overall information about the structure of a fish assemblage, whereas in many cases information about absence is not completely dependable, and species that are reported as absent may be actually present. In the first case, the meaning of absence and presence data is opposite, but equivalent, and symmetrical similarity coefficient can be used (e.g. simple matching coefficient). On the contrary, in the second case only the meaning of species presence is unequivocal, and absence records should be ignored. In this case, asymmetrical similarity coefficients are more appropriate (e.g. Jaccard's similarity).

In case quantitative differences are relevant, Euclidean distance or 'city-block' (Manhattan) metrics are usually adequate, whereas other distance coefficients are more appropriate in the case where relative abundances are to be compared. The complement of the Whittaker's association index (Whittaker, 1952) is a suitable choice for comparing abundance data normalized relative to sample total abundance, and it is particularly effective if the species lists to be compared are quantitatively heterogeneous. Finally, in case the role species play is independent of their abundance, the Canberra metric (Lance and Williams, 1966) assigns to each species the same weight. The Bray–Curtis distance (Bray and Curtis, 1957), which is frequently used in community studies, varies within the [0,1] interval, thus allowing us to compare different distance values very easily. Providing further details on similarity and distance coefficients is beyond the scope of this present chapter, but a complete presentation, including many other coefficients, as well as relevant information about their usage, can be found in Legendre and Legendre (1998).

In order to measure the distance from the reference conditions, the ecological meaning of "reference", as well as the associated fish assemblages, have to be clearly defined. The most obvious solution is to select a set of pristine sites and to assume their fish assemblages as the reference, thus considering as disturbed those sites in which, given environmental conditions similar to those of a reference site, the fish assemblage is different.

However, sampling fish assemblages at pristine sites is not the only way to define reference conditions, and in many cases it is not even possible because of the lack of really pristine sites. In such cases, reference conditions may be defined on the basis of other sources of information, such as historical records or species-distribution models.

2.2.6 MODELING FISH ASSEMBLAGES

While defining reference conditions according to historical records is a straightforward approach and data availability is the only real constraint, modeling species distribution involves a more complex procedure.

In fact, when information about species composition of fish assemblages is not available, neither from past studies nor from other records, the only viable solution for obtaining information about reference assemblages is based on mathematical models, which can predict the presence (or the abundance) of fish species, given an adequate amount of ecological information. In particular, the physical structure of rivers and streams can be often described by a relatively small number of morphodynamic attributes that, in turn, directly affect the distribution of fish species or which are related to other relevant attributes that are more difficult to quantify (e.g. discharge).

The best suited mathematical models for this task are based on an empirical approach, i.e. on the direct extraction of information from existing data sets. Statistical models, like those based on multiple linear or logistic regression, are a typical example of this category of models. An example of this approach was presented by Oberdorff *et al.* (2001), who developed a set of probabilistic models based on logistic regression that are aimed at predicting the occurrence of 34 fish species in French rivers and streams. The models were based on eight predictive variables (including gradient, elevation, July and January mean daily maximum air temperature, stream width, mean depth and distance from headwater sources) and on the average explained about 60 % of the total variation in species richness. This result, according to the authors, was comparable to those of more sophisticated techniques, like, for instance, Artificial Neural Networks (ANNs), and the models were then applied for defining a fish-based index aimed at a nation-wide application in French rivers.

However, a clear trend in the literature about species distribution modeling shows that empirical models based on ANNs have become more and more popular during the last decade, proving to be the best tools for these applications. In fact, although several different modeling techniques may provide similar results in terms of overall accuracy, ANNs produced the greatest number of statistically significant models as far as fish fauna composition is concerned (Olden and Jackson, 2002).

Even though the idea of modeling fish fauna composition on the basis of environmental variables is not new (e.g. Faush *et al.*, 1988), only recently have ANNs been applied to this problem. ANNs have been used to predict fish species richness (e.g. Guegan *et al.*, 1998), as well as density and biomass of single fish populations (Baran *et al.*, 1996; Lek *et al.*, 1996a,b; Mastrorillo *et al.*, 1997a,b) and ecological characteristics of fish assemblages (Aguilar Ibarra *et al.*, 2003). As far as fish assemblages composition at river basin scale is considered, only a few models have been developed so far (e.g. Boët and Fhus, 2000; Olden and Jackson, 2001; Joy and Death, 2005; Scardi *et al.*, 2004, 2005). A complete description of the way ANNs work is obviously beyond the scope of this chapter and readers who are interested in the technical details will find more information and an introduction to ecological applications of ANNs in Lek and Guégan (1999) or in Fielding (1999).

An example of the potentialities of ANN models in this context was presented by Scardi *et al.* (2004, 2005) in a recent study on Italian fish fauna, in which very reliable predictions about fish fauna composition were obtained on the basis of an ANN model on a regional scale. In particular, 20 environmental attributes (see Table 2.2.3) were used to predict the presence or absence of 32 fish species in North-Eastern Italian streams and rivers.

The ANN model was then able to correctly predict presence or absence of fish species in the majority of the cases included in an independent data set, which was only used for testing purposes. The percentage of Correctly Classified Instances (CCIs), i.e. the percentage of correct predictions about species presence or absence, ranged from 79 to 99 %, with 91.6 % as the average value for the whole fish assemblage. Evaluating a species distribution model on the basis of CCIs, however, may be misleading, because very rare and very frequent species tend to be correctly

Table 2.2.3   Environmental attributes used as predictive variables by an Artificial Neural Network model for predicting fish assemblage composition (adapted from Scardi *et al.*, 2004, 2005)

---

  1 Elevation (m)
  2 Mean depth (m)
  3 Runs (surface) (%)
  4 Pools (surface) (%)
  5 Riffles (surface) (%)
  6 Mean width (m)
  7 Boulders (surface) (%)
  8 Rocks and pebbles (surface) (%)
  9 Gravel (surface) (%)
 10 Sand (surface) (%)
 11 Silt and clay (surface) (%)
 12 Stream velocity (score, 0–5)
 13 Vegetation covering (surface) (%)
 14 Shade (%)
 15 Anthropogenic disturbance (score, 0–4)
 16 pH
 17 Conductivity (μS cm^{-1})
 18 Gradient (%)
 19 Catchment's area surface (km^2)
 20 Distance from source (km)

---

predicted in most cases, even by models that return constant outputs. It is obvious, for instance, that a species that occurs in 5 out of 100 records will be predicted with a 95 % CCI accuracy by a model that always returns an 'absence' prediction. Therefore, a more accurate assessment of the ANN model performance was based on alternate validation strategies, including both K statistics (Cohen, 1960) and the Mantel test (Mantel, 1967). In the first case, model predictions were significantly different from those of a random model for 27 out of 32 species and the model only failed with very rare species ($\leq 3\%$ occurrence), i.e. when not enough information was available in data for describing species ecological properties and therefore for *training* (i.e., in ANN jargon, for calibrating) a model. In the second case, two Rogers and Tanimoto similarity matrices (Rogers and Tanimoto, 1960) were computed between 67 sites, based, respectively, on observed and predicted species composition. The resulting standardized Mantel statistics, which is a measure of the overall correlation between the two matrices, was highly significant ($R = 0.84$, $p = 1.0$, 10^5 permutations). This result provided a very clear evidence for the ability of the ANN model to consistently reproduce fish assemblage composition over a broad range of environmental conditions.

Further developments of ANN modeling of fish assemblage composition include quantitative predictions (species abundance or biomass) and, at least in the case of the most abundant species, predictions about the population structure (ranging from presence or absence of juveniles to frequency of age classes). It is very important to stress the fact that the feasibility of these advances in fish assemblage modeling depends only on the amount of available field data, as no methodological issues hinder such developments.

The need for large data bases as prerequisite for modeling is obvious, but it can be less strict in practice than in theory. In fact, modeling presence or absence of 32 fish species, as in the case that was just presented, implies 2^{32} (i.e. about 4 billion) different assemblage compositions. In practice, however, the number of combinations that really occur is much smaller because of the effects of biotic constraints and interactions (e.g. about a hundred in the North-Eastern Italian fish fauna). Therefore, the number of records that are needed to calibrate complex models (and ANNs in particular) is not as large as expected on a theoretical basis and practical applications are actually feasible.

Fish assemblage composition models are obviously very useful tools for monitoring activities. In fact, the evaluation of environmental quality may be directly based on the comparison between observed and modeled fish assemblage composition, using an appropriate similarity or distance coefficient for defining the deviation from the expected structure of the assemblage. The selection of an appropriate coefficient is not a trivial task, of course, and it must be carefully based on the characteristics and limits of both the model and the field data to be analysed. For instance, in large rivers the absence of a species may be due to sampling problems rather than to unfavourable environmental conditions, whereas many models systematically fail in predicting the presence of rare species. In all such cases, an asymmetrical coefficient (e.g. Jaccard's similarity) is certainly more appropriate than a symmetrical one (e.g. the Rogers and Tanimoto similarity), as the latter relies on the assumption that information about fish assemblage composition is complete. Moreover, there are differences among similarity coefficients (both symmetrical and asymmetrical) which depend on the relative weight that is assigned to concordances and discordances in species composition. Selecting different coefficients implies changes in 'contrast' (using a pictorial analogy) between 'lights' and 'shades' in the environmental quality picture. Obviously, it is up to the modeler to select the optimal 'contrast' settings, i.e. the best suited similarity or distance coefficient.

Another key issue in applying species distribution models to environmental quality assessment is the selection of an appropriate data set for model calibration (i.e. training, in the case of ANNs) and validation. In fact, it is obvious that a model aimed at predicting undisturbed community structure as a reference for environmental quality assessment should be only based on information about pristine sites. However, given the more or less severe environmental alterations that affect many regions, records from really pristine sites may be very difficult to find, even taking into account a whole ecoregion and both contemporary and historical data. In such cases, a different strategy should be considered, which includes in calibration and validation data sets not only pristine sites, but a broader range of environmental conditions. The predicted fish assemblage, in this case, cannot be regarded as an absolute reference, as it reflects the average conditions within a heterogeneous set of records. Nevertheless, predictions about the fish assemblage structure in hypothetical pristine conditions can still be obtained from simulations performed by properly 'tuning' the model inputs.

Finally, predicting fish assemblage structure is not only relevant to the definition of reference conditions aimed at the assessment of environmental quality (Olden and Jackson, 2002). In fact, it is an important achievement in the light of conservation and management strategies, and it can also help in optimizing sampling design for further research (Jackson and Harvey, 1997). Other applications include prediction and evaluation of habitat alteration due to changes in land use (Oberdorff *et al.*, 2001), assessment of the potential risk of invasion and spread of exotic species (Peterson and Vieglais, 2001), optimization of strategies for species reintroduction (Evans and Oliver, 1995) and simulation of changes in fish assemblage induced by environmental restoration (Scardi *et al.*, 2004, 2005).

2.2.7 CONCLUSIONS AND PERSPECTIVES

The assessment of ecological integrity based on biotic indices is becoming a standard practice, and fish-based methods certainly play a major role in this field. Their main advantages over methods based on other organisms are two. The first one is the minimal taxonomical knowledge that is needed to support a fish-based study. In fact, the number of fish species is always much smaller than that of other types of aquatic organisms, such as, for instance, benthic macroinvertebrates or benthic diatoms, while the average size of specimens is much larger, so that it is easy to identify species at a glance. The second advantage is related to the public awareness of the ecological role of fish fauna, which is also associated with a clear understanding of its social and economical value. In fact, while other organisms are only relevant in technical or scientific contexts, the role of fish fauna is well known and broadly accepted, even from people who do not have such background.

Model-based methods are rapidly emerging, as they are the only tools that allow us to bridge the gap between a limited amount of field data and an increasing demand for ecological assessment based on comparisons with reference conditions. In some cases, models may also play a role in reconstructing pristine assemblage structures that have disappeared as a consequence of diffuse anthropic disturbance. Although statistical models provided very good results in predicting fish species distribution on the basis of environmental variables (e.g. Oberdorff *et al.*, 2001), Machine Learning techniques and, in particular, Artificial Neural Networks usually outperform conventional models (Olden and Jackson, 2002).

In the future, when more and more field data will become available, the role of biotic indices will probably become less important, as well as that of model-based assessment methods, while real data and real time series will be eventually used to set reference conditions for further evaluations.

However, model-based evaluation procedures will still play a major role in river reaches that are deeply modified by anthropic impacts. It is obvious that in these cases no reference data about pristine conditions will ever be available. Therefore, the only way to assess ecological integrity will be based on the deviation from the expected fish assemblage, i.e. from the potential fish fauna that would be present if all of the anthropic disturbances were removed.

Independently of the assessment method, however, it is very important to bear in mind that no entirely objective procedures exist. Multimetric biotic indices may be subjective in the selection and in the scaling of metrics, but also comparing observed fish assemblages to reference ones is not as objective as it may seem. In fact, the deviation from reference conditions cannot be univocally defined, as it depends on the choice of a similarity or distance coefficient, which in turn is inherently subjective. As a general rule of thumb, the shorter the path from data to ecological assessment, then the better the method.

In this framework, it is certainly useful to stress the role played by sampling procedures, which is usually not critical in streams and small wadeable rivers, whereas it is a major one in larger rivers. In fact, an exhaustive census of the whole fish assemblage can only be obtained by means of proper electrofishing, which is very common in sampling activities focused on small rivers and streams, while it usually fails in larger rivers because of the physical complexity of the aquatic environment. In such cases, only a combination of sampling techniques (e.g. gill nets, fyke nets and seines) may provide reliable information about the fish assemblage structure, and only from a qualitative point of view, as abundance data and occurrence of rare species are often affected by severe errors. Moreover, reliable quantitative data can be obtained only for abundant species, provided that sampling activities are not occasional and that fishing gear is properly standardized and operated.

Finally, a caveat is needed about multimetric indices, biomarkers and other approaches not involving data and methods that closely and accurately represent the ecological complexity of aquatic ecosystems. These tools may play a role and are certainly useful in many cases, especially when specific disturbances are to be monitored. However, our ability to understand complex ecological processes relies on the amount of field data we collect and on the work of skilled ecologists, who are able to identify species, analyse relevant information, model biotic and abiotic relationships, and understand relevant results. Therefore, oversimplifying the interpretation of ecological data or inferring complex ecological properties on the basis of simple chemical analyses, although sometimes attractive from the point of view of water policy and management, is not a strategy that is likely to pay off in the long run. On the contrary, the growing demand for assessment of ecological integrity, e.g. according to the EU Water Framework Directive (European Union, 2000), must be regarded as a unique opportunity for raising a new generation of 'real' aquatic ecologists.

REFERENCES

Aguilar Ibarra, A., Gevrey, M., Park, Y. -S., Lim, P. and Lek S., 2003. 'Modelling the factors that influence fish guilds composition using a back-propagation network: assessment of metrics for indices of biotic integrity'. *Ecol. Model.*, **160**, 281–290.

An, K. -G., Park, S. -S. and Shin, J. -Y., 2002. 'An evaluation of a river health using the index of biological integrity along with relations to chemical and habitat conditions'. *Environ. Int.*, **28**, 411–420.

Badino, G., Forneris, G., Lodi E. and Ostracoli, G., 1992. 'Ichthyological Index, a new standard method for the river biological water quality assessment'. In: *River Water Quality. Ecological Assessment and Control.* Commission of the European Communities: Brussels, Belgium, pp. 729–730.

Baran, P., Lek, S., Delacoste, M. and Belaud, A., 1996. 'Stochastic models that predict trout population density or biomass on a mesohabitat scale'. *Hydrobiologia*, **337**, 1–9.

Boet, P. and Fuhs, T., 2000. 'Predicting presence of fish species in the Seine River basin using artificial neuronal networks'. In: *Artificial Neuronal Networks: Application to Ecology and Evolution*, Environmental Science, Lek, S. and Gueguan, J.-F. (Eds). Springer-Verlag: Berlin, Germany, pp. 187–201.

Bramblett, R. G. and Fausch, K. D., 1991. 'Variable fish communities and the index of biotic integrity in a Western Great Plains river'. *Trans. Am. Fish. Soc.*, **120**, 752–769.

Bray, R. J. and Curtis, J. T., 1957. 'An ordination of the upland forest communities of southern Wisconsin'. *Ecol. Monogr.*, **27**, 325–349.

Cohen, J., 1960. 'A coefficient of agreement of nominal scales'. *Edu. Psychol. Measurement*, **20**, 37–46.

Connell, J. H., 1978. 'Diversity in tropical rainforests and coral reefs'. *Science*, **199**, 1302–1310.

European Commission, 1992. *Council Directive 92/43/ECC of 21st May 1992 on the Conservation of Natural Habitats and of Wild Fauna and Flora*, Official Journal, **L206**. European Commission: Brussels, Belgium.

European Union, 2000. *Parliament and Council Directive 2000/60/EC of 23rd October 2000. Establishing a Framework for Community Action in the Field of Water Policy.* Official Journal, **PE-CONS 3639/1/00 REV 1**. European Union: Brussels, Belgium.

Evans, D. O. and Oliver, C. H. 1995. 'Introduction of lake trout to inland lakes of Ontario, Canada: Factors contributing to successful colonization'. *J. Great Lakes Res.*, **21** (*Suppl. 1*), 30–53.

Faush, K. D., Hawkes, C. L. and Parsons, M. G., 1988. *Models that Predict the Standing Crop of Stream Fish from Habitat Variables: 1950–1985*, General Technical Report PNW-GTR-213. US Department of Agriculture, Forest Service, Pacific North Reaserch Station: Portland, OR, USA.

Fausch, K. D., Lyons, J., Karr, J. R. and Angermeier, P. L., 1990. 'Fish communities as indicators of environmental degradation'. *Am. Fish. Soc. Symp.*, **8**, 123–144.

Fielding, A. H., 1999. 'An introduction to machine learning methods'. In: *Machine Learning Methods for Ecological Applications*, Fielding, A. H. (Ed.). Kluwer Academic Publishers: Boston, MA, USA, pp. 1–35.

Guegan, J. -F., Lek, S. and Oberdorff, T., 1998. 'Energy availability and habitat heterogeneity predict global riverine fish diversity'. *Nature*, **391**, 382–384.

Harris, J. H., 1995. 'The use of fish in ecological assessments'. *Aus. J. Ecol.*, **20**, 65–80.

Hay, C. J., Van Zyl, B. J. and Steyn, G. J., 1996. 'A quantitative assessment of the biotic integrity of the Okavango River, Namibia, based on fish'. *Water SA*, **22**, 263–284.

Hughes, R. M. and Oberdorff, T., 1999. 'Application of IBI concepts and metrics to waters outside the United States and Canada'. In: *Assessing the Sustainability and Biological Integrity of Water Resources using Fish Communities*, Simon, T. P. (Ed.). CRC Press: Boca Raton, FL, USA, pp. 79–93.

Hughes, R. M., Kaufmann, P. R., Herlihy, A. T., Kincaid, T. M., Reynolds, L. and Larsen, D. P., 1998. 'A process for developing and evaluating indices of fish assemblage integrity'. *Can. J. Fish. Aquat. Sci.*, **55**, 1618–1631.

Hugueny, B., Camara, S., Samoura, B. and Magassouba, M., 1996. 'Applying an index of biotic integrity based on fish assemblages in a West African river'. *Hydrobiologia*, **331**, 71–78.

Jaccard, P., 1900. 'Contribution au problème de l'immigration post-glaciaire de la flore alpine'. *Bull. Soc. Vaudoise Sci. Nat.*, **36**, 87–130.

Jaccard, P., 1901. 'Etude comparative de la distribution florale dans une portion des Alpes et du Jura'. *Bull. Soc. Vaudoise Sci. Nat.*, **37**, 547–579.

Jaccard, P., 1908. 'Nouvelles recherches sur la distribution florale'. *Bull. Soc. Vaudoise Sci. Nat.*, **44**, 223–270.

Jackson, D. A. and Harvey, H. H., 1997. 'Qualitative and quantitative sampling of lake fish communities'. *Can. J. Fish. Aquat. Sci.*, **54**, 2807–2813.

Joy, M. K. and Death, R. G., 2005. 'Modelling of freshwater fish and macrocrustacean assemblages for biological assessment in New Zealand'. In: *Modelling Community Structure in Freshwater Ecosystems*, Lek, S., Scardi, M., Verdonschot, P. F. M., Descy, J. -P. and Park, Y. -S. (Eds). Springer-Verlag: Berlin, Germany, pp. 76–89.

Kamdem Toham, A. and Teugels, G. G., 1999. 'First data on an Index of Biotic Integrity (IBI) based on fish assemblages for the assessment of the impact of deforestation in a tropical West African river system'. *Hydrobiologia*, **397**, 29–38.

Karr, J. R., 1981. 'Assessment of biotic integrity using fish communities'. *Fisheries*, **6**, 21–27.

Karr, J. R. and Chu, E. W., 1997. *Biological Monitoring and Assessment: Using Multimetric Indexes Effectively*, EPA 235-R97-0001. US Environmental Protection Agency: University of Washington, Seattle, WA, USA.

Karr, J. R. and Dudley, D. R., 1981. 'Ecological perspective on water quality goals'. *Environ. Management.*, **5**, 55–68.

Karr, J. R., Fausch, F. D., Angermeier, P. L., Yant, P. R. and Schlosser, I. J., 1986. *Assessing Biological Integrity in Running Waters: A Method and Its Rationale*, Special Publication 5. Illinois Natural History Survey: Chicago, IL, USA.

Kestemont, P., Didier, J., Depiereux, E. and Micha, J. C., 2000. 'Selecting ichtyological metrics to assess river basin ecological quality'. *Arch. Hydrobiol.*, **121**, 321–348.

Kleynhans, C. J., 1999. 'The development of a fish index to assess the biological integrity of South African rivers'. *Water SA*, **25**, 265–278.

Kurtz, J. A., Jackson L. E. and Fisher, W. S., 2001. 'Strategies for evaluating indicators based on guidelines from the Environmental Protection Agency's Office of Research and Development'. *Ecol. Indicators*, **1**, 49–60.

Lance, G. N. and Williams, W. T., 1966. 'Computer programs for classification'. In: *Proceedings of the ANCCAC Conference*, May, 1966, Canberra, Australia, Paper 12/3.

Legendre, P. and Legendre, L., 1998. *Numerical Ecology*, 2nd Edition (in English). Elsevier Amsterdam, The Netherlands.

Lek, S. and Guégan, J. F., 1999. 'Artificial neural networks as a tool in ecological modelling, an introduction'. *Ecol. Modelling*, **120**, 65–73.

Lek, S., Belaud, A., Baran, P., Dimopoulos, I. and Delacoste, M., 1996a. 'Role of some environmental variables in trout abundance models using neural networks'. *Aquat. Living Res.*, **9**, 23–29.

Lek, S., Delacoste, M., Baran, P., Dimopoulos, I., Lauga, J. and Aulagnier, S., 1996b. 'Application of neural networks to modelling nonlinear relationships in ecology'. *Ecol. Modelling*, **90**, 39–52.

Lyons, J., Navarro-Pérez, S., Cochran, P. A., Santana, E. and Guzmán-Arroyo, M., 1995. 'Index of biotic integrity based on fish assemblages for the conservation of streams and rivers in West-Central Mexico'. *Conserv. Biol.*, **9**, 569–584.

Mantel, N., 1967. 'The detection of desease clustering and a generalized regression approach'. *Cancer Res.*, **27**, 209–220.

Mastrorillo, S., Lek, S. and Dauba, F., 1997a. 'Predicting the abundance of minnow *Phoxinus phoxinus* (Cyprinidae) in the River Ariege (France) using artificial neural networks'. *Aquat. Living Res.*, **10**, 169–176.

Mastrorillo, S., Lek, S., Dauba, F., and Belaud, A., 1997b. 'The use of artificial neural networks to predict the presence of small-bodied fish in a river'. *Freshwater Biol.*, **38**, 237–246.

Oberdorff, T. and Hughes, R. M., 1992. 'Modification of an index of biotic integrity based on fish assemblages to characterize rivers of the Seine Basin, France'. *Hydrobioly*, **228**, 117–130.

Oberdorff, T., Pont, D., Hugueny, B. and Chessel, D., 2001. 'A probabilistic model characterizing fish assemblages of French rivers: a framework for environmental assessment'. *Freshwater Biol.*, **46**, 399–415.

Olden, J. D. and Jackson, D. A., 2001. 'Fish–habitat relationships in lakes: gaining predictive and explanatory insight by using artificial neural networks'. *Trans. North Am. Fish. Soc.*, **130**, 878–897.

Olden, J. D. and Jackson, D. A., 2002. 'A comparison of statistical approaches for modelling fish species distributions'. *Freshwater Biol.*, **47**, 1976–1995.

Peterson, A. T. and Vieglais, D. A., 2001. 'Predicting species invasions using ecological niche modeling: new approaches from bioinformatics attack a pressing problem'. *Bioscience*, **51**, 363–371.

Plafkin, L. P., Barbour, M. T., Porter, K. D., Gross, S. K. and Hughes, R. M., 1989. *Rapid Bioassessment Protocols for Use in Streams and Rivers: Benthic Macroinvertebrates and Fish*. US Environmental Protection Agency: Washington, DC, USA.

Ramm, A. E., 1988. 'The community degradation index: A new method for assessing the deterioration of aquatic habitats'. *Water Res.*, **22**: 293–301.

Rogers, D. J. and Tanimoto, T. T., 1960. 'A computer program for classifying plants'. *Science*, **132**, 1115–1118.

Scardi, M., Cataudella, S., Ciccotti, E., Di Dato, P., Maio, G., Marconato, E., Salviati, S., Tancioni, L., Turin, P. and Zanetti, M., 2004. 'Previsione della composizione della fauna ittica mediante reti neurali artificiali'. *Biol. Ambient.*, **18**, 1–8.

Scardi, M., Cataudella, S., Ciccotti, E., Di Dato, P., Maio, G., Marconato, E., Salviati, S., Tancioni, L., Turin, P. and Zanetti, M., 2005. 'Optimisation of artificial neural networks for predicting fish assemblages in rivers'. In: *Modelling Community Structure in Freshwater Ecosystems*, Lek, S., Scardi, M., Verdonschot, P. F. M., Descy, J. -P. and Park, Y. -S. (Eds). Springer-Verlag: Berlin, Germany, pp. 114–129.

Schmutz, S., Kaufmann, M., Vogel, B., Jungwirth, M. and S. Muhar, 2000. 'A multi-level concept for fish-based, river-type-specific assessment of ecological integrity'. *Hydrobiologia*, **422/423**, 279–289.

Shamsudin, P., 1988. 'Water quality and mass fish mortality at Mengabang lagoon, Trengganu, Malaysia'. *Malay. Nat. J.*, **41**, 515–527.

Simon, T. P. and Lyons, J., 1995. 'Application of the index of biotic integrity to evaluate water resource integrity in freshwater ecosystems'. In: *Biological Assessment and Criteria: Tools for Water Resource Planning and Decision Making*, Davis, W. S. and Simon, T. P. (Eds). Lewis Publishers: Boca Raton, FL, USA, pp. 245–262.

Steedman, R. J., 1988. 'Modification and assessment of an index of biotic integrity to quantify stream quality in southern Ontario'. *Can. J. Fish. Aquat. Sci.*, **45**, 492–501.

Suter, G. W., 1993. 'A critique of ecosystem health concepts and indexes'. *Environ. Toxicol. Chem.*, **12**, 1533–1539.

van der Oost, R., Beyer, J. and Vermeulen, N. P. E., 2003. 'Fish bioaccumulation and biomarkers in environmental risk assessment: a review'. *Environ. Toxicol. Pharmacol.*, **13**, 57–149.

Whitfield, A. K., 1996. 'Fishes and the environmental status of South African estuaries'. *Fish. Management Ecol.*, **3**, 45–57.

Whittaker, R. H., 1952. 'A study of summer foliage insect communities in the Great Smoky Mountains. *Ecol. Monogr.*, **22**, 1–44.

Wright, J. F., Armitage, P. D. and Furse, M. T., 1989. 'Prediction of invertebrate communities using stream measurements'. *Regulated Rivers Res. Management*, **4**, 147–155.

2.3
Monitoring Methods Based on Algae and Macrophytes

Jean Prygiel and **Jacques Haury**

2.3.1   INTRODUCTION

Algae and macrophytes are widely used for river quality assessments (Prygiel *et al.*, 1999; Whitton and Kelly 1995; Rott *et al.*, 2003) and have regained interest in Europe thanks to the reinforcement of the legislation with the Urban Waste-water Directive (EC, 1991), and more recently with the Water Framework Directive

Biological Monitoring of Rivers   Edited by G. Ziglio, M. Siligardi and G. Flaim
© 2006 John Wiley & Sons, Ltd.

(WFD) (EC, 2000) which indicates both macrophytes and phytobenthos, but also phytoplankton as relevant water quality elements. In this review, mainly biocenotic methods will be discussed, including those using the aquatic flora for micropollution monitoring but excluding those using the bioaccumulation capacities of the aquatic flora and (eco)toxicity tests. Besides diatoms, desmids (Coesel, 1977) and cyanobacteria (Doutelero *et al.*, 2004) are sometimes used for monitoring rivers as well as filamentous algae which are rarely considered alone (Dell'Uomo, 1991; Pirsoo *et al.*, 1999, in Prygiel *et al.*, 1999). Rather, they are included in periphyton studies (Rott *et al.*, 2003) or in macrophytes studies. Macrophytes are here considered as a heterogeneous floristic group according the CEN standard (CEN, 2003). In some cases, fungi and heterotrophic bacteria are also considered as, for example, in the French IBMR Standard (AFNOR, 2003). In the monitoring programmes, the accuracy of determination, as well as the level (species, genus, hybrids for *Potamogeton* or *Ranunculus*), differ following the methods used and the authors. Kohler and Janauer (1995) don't take into account neither filamentous algae (except stoneworts), nor bryophytes, while Dawson *et al.* (1999a) or Haury *et al.* (1996, in Haury *et al.*, 2001, 2002) take into account most groups.

2.3.2   PHYTOPLANKTON

Phytoplankton is an important component of rivers, especially those of third order and above (Reynolds and Descy, 1996). However, qualitative and quantitative phytoplankton networks are very few and mostly limited to large international rivers (http://www.midcc.at/) (Friedrich *et al.*, 1998; Ibelings *et al.*, 1998). This can be attributed to the high inherent spatiotemporal variability which requires large numbers of samples and make surveillance expensive, but also to the lack of references (Köhler and Descy, 2003) and the complex determinism of phytoplankton assemblages (Wehr and Descy, 1998). This complexity led to the elaboration of models for river management (Everbecq *et al.*, 2001). National networks are then often limited to chlorophyll *a* content, which do not allow distinguishing the different phytoplankton groups. High performance liquid chromatography (HPLC) can be used to characterize phytoplankton community structure for the phytoplankton functional groups on the basis of separation and quantification of a variety of photopigments (Wright *et al.*, 1991). Statistical procedures, such as 'Chemtax' (Mackey *et al.*, 1996), can then be applied to partition the total pool of chlorophyll (Chl) *a* into the Chl *a* contributed by each functional group. This methodology has been successfully applied to freshwater environments by Descy *et al.* (2000) who note, however, that validation of pigment data is still necessary. Accurate regional studies imply validation of HPLC analysis by microscopy observations (Antosegui *et al.*, 2001). Use of biological traits like size, nitrogen fixation or shape have also been proposed to characterize functional aspects of phytoplankton (Weithoff, 2003). According to Kohler and Descy (2003), no actual assessment methods fulfil the requirements of the EU WFD. However, monthly Chl *a* monitoring using ISO standards is necessary for water bodies for which the retention time exceeds six days, with enumeration and identification of species when the Chl *a* content is above 20 μg l^{-1}.

2.3.3 PHYTOBENTHOS

Undoubtly, diatoms are the most widely used phytobenthic algae (Stevenson and Pan, 2003) and their advantages have been largely described (McCormick and Cairns, 1994; Reid *et al.*, 1995). Except for Patrick *et al.* (1954) in the USA, who developed a methodology based on the structure of diatom assemblages, the first methods largely originated from Europe in the framework of the saprobic system and were first devoted to organic content monitoring and then for salinity, eutrophication, acidification and global water quality (Kelly, 1998a; Prygiel *et al.*, 1999; Rott *et al.*, 2003). Recent methods concern habitat disturbance (Kutka and Richards, 1996) and metal contamination monitoring (Hirst *et al.*, 2002; Nunes *et al.*, 2003), some of them based on the occurrence of deformities (Fore and Grafe, 2002). Most techniques require species identification but some elaborated for routine surveillance only require genus identification (Wu, 1999) or both genus and species identification (Kelly, 1998b; Prygiel *et al.*, 1996, in Prygiel *et al.*, 1999). Others used lumped taxa (AFNOR, 2000; Hürlimann and Niederhauser, 2002). Epilithon assemblages are commonly used, sometimes with centric diatoms excluded because of their occurrence in epilithon only for low discharges (Kelly, 1998b). Epipelon is also used (Gomez and Licursi, 2001), as well as tychoplankton (Almeida and Gil, 2001). Artificial substrates can be used too as they reproduce diatom communities quite well (Cattaneo and Amireault, 1992; Lane *et al.*, 2003). Numerous studies describing the relationships between European diatoms indices and chemistry have been carried out, thanks to specific programmes (Lecointe *et al.*, in Ector *et al.*, 1999) and largely confirmed the validity of diatom indices for monitoring rivers in Europe (Dokulil *et al.*, 1997; Eloranta and Soininen, 2002; Kwandrans *et al.*, 1998; Montesanto *et al.*, 1999) but also in other continents (Fawzi *et al.*, 2002; Rott *et al.*, 1998; Sgro and Johansen, 1998). Indices based on the structure of diatom assemblages also exist, such as diversity, eveness, dominance, similarity, and also rank the distributions of size–morphological groups (Maksimov *et al.*, 1997). These kinds of indices are largely used, especially the Shannon–Weaver index and are much debated (Van de Vijver and Beyens, 1998). Basis ones, such as species number, are often the most effective (Economou-Amilli, 1980). Canonical Correspondence Analysis is often used in conjunction with weighting average regression as a basis for comparison and distribution of diatom assemblages (Sabater *et al.*, 1988) and allows the calculation of various water quality parameters by inverse regression, such as pH (Pan *et al.*, 1996, in Winter and Duthie, 2000), nutrients (Winter and Duthie, 2000), conductivity (Gomez and Licursi, 2001), ionic strength (Potapova and Charles, 2003) or salinity (Leland *et al.*, 2001). Recently, diatom multimetric indices (Fore and Grafe, 2002) and periphyton multimetric indices (Hill *et al.*, 2003) have been proposed in the USA in order to assess the biological integrity of rivers. These integrity indices are based on the definition of hydroecoregions and on deviations from reference conditions. This supposes a good knowledge of environmental gradients before using diatom metrics in order that diatom-environmental conditions are not masked by natural gradients (Potapova and Charles, 2002). Many of the large-scale indirect factors (geology, land use, etc.) constrain the expression of small-scale direct factors such as velocity and nutrients (Munn *et al.*, 2002). In contrast to biological integrity, the assessment of ecosystem health led to the elaboration of models to predict diatom communities at genus level which would be expected at a given site on the basis of natural environmental features (Chessman *et al.*, 1999).

2.3.4   MACROPHYTES

Different methods involving macrophytes have been developed in relation to the aims of monitoring (biodiversity, habitat or water quality assessment), the physical characteristics of surveyed media (i.e. large rivers versus small streams) and the cost of the surveys (CEN, 2003). Macrophytes can be used for general monitoring plans for many purposes (see, for example, reviews in Amoros *et al.*, 2000; Haslam and Wolseley, 1987; Haury *et al*, 2001; Whitton, 1979), including mapping, estimation of cover and biomass, biodiversity and conservation biology, physical assessment, water trophy, toxicants, radionuclides, thermal stress, etc. Although many studies have been made on the possibility of using bryophytes (Empain *et al.*, 1980) as bioaccumulators for heavy metals, and even for radionuclides, our purpose is to present available field methods and indices, which have been tested not only by the authors but also by field surveyors.

2.3.4.1   Field methods based on species ecophysiology and morphology

Macrophytes are quite rarely used as sentinel organisms to monitor the presence of toxic pollutants, unless some cases of visible toxicity are recorded, as Giovanni and Haury (1995) observed with a *Phalaris arundinacea* L. population where high concentrations in atrazine were measured in yellow stands. Very few examples of field toxicity are given in the literature, either for pesticides, heavy metals or acidification and no general grid is presently available to use symptoms as toxicant-specific alert tools. From another point of view, the establishment of relationships between plant morphology and available nutrients is regarded as a useful element to assess the level of trophy (Ali *et al.*, 1999; Chatenet *et al.*, 2002, in Dutartre and Montel, 2002). Tissue enrichment in phosphorus (Garbey *et al.*, 2004) or pigments for bryophytes (Lopez *et al.*, 1997) are possible methods to verify how healthy and well fed are the macrophytes. These approaches could be used to strengthen the methods based on community surveys.

2.3.4.2   Field studies involving communities

Biotypologies to assess reference phytocenoses

Many studies dealing with river macrophyte distribution, and biotypes, have been carried out since the beginning of the 20th Century, either by using phytosociology (Kohler, 1975; Wiegleb, 1983; Carbiener *et al.*, 1990, in Amoros *et al.*, 2000; Muller, 1990, in Haury *et al.*, 2001; Haury *et al.*, 1995; Pot, 1996; Dawson *et al.*, 1999a; Kohler and Schneider, 2003) or stretch surveys giving species assemblages (i.e. Holmes and Whitton, 1977; Grasmück *et al.*, 1995; Ferreira and Moreira, 1999; Vanderpoorten *et al.*, 2000; Riis, 2000). Most of these studies have been carried out for monitoring purposes, either for conservation or water quality assessment. Now models are being developed to predict the community composition from both physical and chemical variables (Barendregt and Bio, 2003).

2.3.4.3   Physical habitat monitoring and vegetation structure

The relationships between physical habitats and macrophytes is an increasing challenge, because, with mapping, macrophytes can also easily be used to monitor physical habitat. The different cover scaling takes into account physical constraints and most biotypologies are based both on physical and chemical features. The Mass Index (Janauer *et al.*, 1993, in Janauer *et al.*, 2003) gives a new approach of species macrophyte development which is very useful for mapping.

Analysis of structure (Den Hartog and Van Den Velde, 1988) becomes compulsory due to pheno-plasticity and selection of life-history traits (Willby *et al.*, 2000) by physical habitat constraints (Bornette *et al.*, 1995). Such an ecomorphological approach provides linkage between the different schools dealing with macrophytes (phytosociology, conservation biology, functional ecology, etc.), as shown in Pot (1996), Dawson and Szoszkiewicz (1999) or Dawson *et al.* (1999b).

2.3.4.4   Water quality assessment

Floristic indices are mostly used in this case. For all of these, using local and/or regional studies, databases and literature reviews, the authors have proposed species scores assessing 'water quality' (Harding, 1987, in Haury *et al.*, 2001), trophy level (Newbold and Holmes, 1987, in Haury *et al.*, 2001; Haury *et al.*, 1996, in Haury *et al.*, 2001; Dawson *et al.*, 1999a; Schneider, 2000, in Kohler and Schneider, 2003;

Haury *et al.*, 2002), organic pollution (Husak *et al.*, 1989) and acidification (Tremp and Kohler, 1995; Thiébaut and Muller, 1999). Amoros *et al.* (2000) have given a general view of these approaches and synthetic tables on species ecology. In a second step, some proposals include cover (IBMR, TIM, MTR) and ecological tolerance mainly to eutrophication (Newbold and Holmes, 1987, in Haury *et al.*, 2001 (IBMR, TIM).

The Bryophytic Index of metal toxiphoby (Empain *et al.*, 1980) assesses the changes in bryophyte community due to heavy metal contamination. It has been applied as a separate index in Belgium and Northern France, but, more recently, bioaccumulation studies have been preferred. The three main floristic indices which are most widely used are as follows.

The *Mean Trophic Rank* (MTR) index (Dawson *et al.*, 1999a; Holmes *et al.*, 1999) is regularly applied in the UK by various Environmental Agencies (Jarvie *et al.*, 2002) and was also tested in Poland (Szoszkiewicz *et al.*, 2002). In the UK, their results are discussed versus ten types of phytocenoses (simplified from Holmes *et al.*, 1998). Nevertheless, after having tested this index versus plant traits, Ali *et al.* (1999) concluded that the most powerful prediction could be made with the latter. The MTR index is also being tested within the STAR project (and in France, in comparison with the IBMR index – Peltre, University of Metz, personal communication).

The *Biological Macrophyte Index for Rivers* (IBMR) (Haury *et al.*, 2002; AFNOR, 2003) has been developed for many rivers in France. Field tests have been performed by independant companies in four Water Agencies, in Mediterranean areas, as well as in mountain regions.

The *Trophic Index for Macrophytes* (TIM) (Schneider, 2000, in Kohler and Schneider, 2003) takes into account the sensivity of stoneworts and higher plants to sediment and water phosphorus. It has been built for some rivers and then applied to many sites in Germany and in the River Danube catchment.

2.3.4.5   Community indices

A general framework on communities has been built up, partially with phytosociology tools (Carbiener *et al.*, 1990, in Robach *et al.*, 1996; Thiébaut and Muller, 1999; Robach *et al.*, 1996; Haan-Archipof *et al.*, 2002, in Dutartre and Montel, 2002; Kohler and Schneider, 2003), where the quantitative indices are due to general zonation from oligotrophy to hypertrophy, with some (up to six) trophy types.

Only one community index, assessing that loss of habitat diversity and eutrophication and/or pollution lead to many changes in vegetation, stands. *Damage Rating* (Haslam, 1982, in Haslam and Wolseley, 1987) can be considered as pointing out the difference between expected and observed phytocenoses. Unfortunately, at present, precise knowledge on reference phytocenoses to improve the Damage Rating is still lacking, although a general framework to assess such differences has been given. Other methods, including both structural aspects and a diagnosis on trophy, have been proposed, such as the *Plant Community Description method* (Harding, 1987, in Haury *et al.*, 2001), or with consideration of rarity (De Lange and Van Zon, 1974, in Haury *et al.*, 2001), but these do not seem to be in current use.

Sampling design and mapping

Three types of sampling design for river monitoring have been used (and reviewed in Haury *et al.*, 2001), i.e. upstream–downstream – a pin-point pollution, a stratified plan and extensive inventories with or without belt sub-sampling. The latter leads to mapping results.

The most developed example of mapping and river macrophyte monitoring is the Danube survey (Janauer *et al.*, 2003), following the proposed method by Kohler and Janauer (1995): more than 2300 km have been surveyed by using the same method. In such an approach, reaches are contiguous but as homogeneity is expected, their length is very variable. Two levels are taken into account: the species distribution and the community development, using previous biotypologies. A trophic diagnosis within four zones is given when considering the community composition. With all other methods, the challenge is to extend a diagnosis obtained in a station to the entire stretch. The possibility to survey from the bank and point out the discontinuities leading to a map (Haury and Muller, 1991 for example) gives a real advantage to macrophytes.

2.3.5 CONCLUSION: FUTURE ORIENTATIONS

2.3.5.1 WFD and its implications for algae and macrophytes

The improvement of the European framework directive requires a complete revision of bioindicative methods to compare assessments to reference conditions identified for each large ecoregion and intercalibration exercises to allow data and results comparability (Rott *et al.*, 2003).

Some CEN standards have been produced or are being produced for macrophytes, diatoms, periphytic algae and phytoplankton in CEN/TC230/WG2. As biology will have a key role in assessing the ecological status and will largely influence decision-making and investment programmes, assurance quality and quality control become a major point in monitoring ecological status (Alverson *et al.*, 2003; Kelly, 2001; Prygiel *et al.*, 2002; Rott *et al.*, 2003) and a guidance for assuring the quality of biological and ecological assessments in the aquatic environment is also in progress by the CEN.

2.3.5.2 Perspectives for diatoms

For the time being, current methods will be used with special attention to reference conditions (Rott *et al.*, 2003). IBD is a standardized diatom index (AFNOR,

2000) and is largely used in France (Prygiel, 2002). A study to identify reference values, as well as limits for very good/good and good/moderate status, for each ecoregion occurring in France has been undertaken by the Cemagref (Coste *et al.*, 2004) using the PAEQANN approach (Predicting Aquatic Ecosystems Quality Using Artificial Networks (Tison *et al.*, 2005)). A total of 836 valid inventories from French monitoring networks have been analysed by using neuronal networks (Kohonen, 1995) and characteristics species have been identified according to Dufrêne and Legendre (1997). Five diatom assemblages, linked to ecoregional features such as climate, soils, vegetation, etc., have been identified (Figure 2.3.1). Reference sites have then been selected both from taxonomic characteristics and from anthropogenic disturbances. IBD reference values and 'very good'/'good' status limits correspond, respectively, to the median of the IBD values and to the 25 % percentile of these sites. A very first proposal for reference values and ecological status limits for IBD have been set, as well as ecological quality ratios (Table 2.3.1). Note that present networks

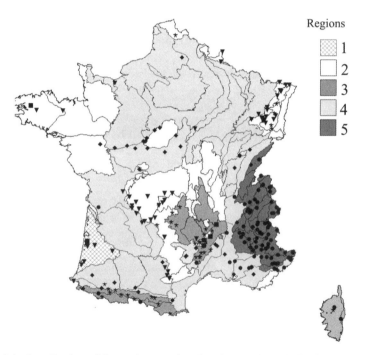

Figure 2.3.1 Localization of first reference sites for diatoms through the five French natural diatom regions. See text for further explanation: Region 1, acid rivers (i.e. Landes); Region 2, siliceous river substrates (i.e. Brittany); Region 3, mountain rivers with low mineralization (i.e. Pyrenean rivers); Region 4, calcareous plain rivers (i.e. Northern France); Region 5, mountain calcareous rivers (i.e. Alps). Thin black limits show the borders of French 'natural level 2 hydroecoregions', as described by Wasson *et al.*, 2002. Reproduced by permission of J. G. Wasson from Coste, M., Tison, J. and Delmas, F., 2004, 'Flores diatomiques des cours d'eau: propositions de valeurs limites du "bon état" pour l'IPS et l'IBD', Document de travail – 10/02/2004, Cemagref, Bordeaux, France

Table 2.3.1 First proposals for IBD (AFNOR, 2000) reference values and 'very good' and 'good' status limits. See text for further explanation: EQR, Ecological Quality ratio: Region 1, acid rivers (i.e. Landes); Region 2, siliceous river substrates (i.e. Brittany); Region 3, mountain rivers with low mineralization (i.e. Pyrenean rivers); Region 4, calcareous plain rivers (i.e. Northern France); Region 5, mountain calcareous rivers (i.e. Alps). Reproduced by permission of J. G. Wasson from Coste, M., Tison, J. and Delmas, F., 2004, flores diatomiques des cours d'eau: propositions de valeurs limites du "bon état" pour l'IPS et l'IBD', Document de Travail – 10/02/2004, Cemagref, Bordeaux, France

Diatom region	Reference value	IBD first limits			
		'Very good' status limit	EQR	'Good' status limit	EQR
1	19	18	0.93	15	0.71
2	16	14	0.87	11	0.67
3	18	16	0.85	14	0.69
4	16	13	0.8	11	0.67
5	19	17	0.86	14	0.64

mainly aim to assess impacts. Thus, reference sites are not numerous enough and are irregularly distributed through out the French natural hydroecoregions identified by Wasson *et al.* (2002) (Figure 2.3.1). These first proposals will be updated when results from diatom reference networks become available.

2.3.5.3 Perspectives for macrophytes

To obtain a better complementarity between European policies dealing with freshwaters, the field surveyors should develop not only a water quality approach, but also have a conservation view, because most of the macrophyte vegetations are concerned by the Habitat Directive, either as '*Ranunculus* communities' (including bryophytes) or 'Mediterranean rivers'.

In many European countries, namely the Nordic ones (Baattrup-Pedersen *et al.*, 2001), data are just not available in most rivers: surveys need to be carried out as soon as possible. Field and conceptual intercalibrations between methods and indices should be conducted by specialists in each approach. Long-term studies (such as Whitton *et al.*, 1998) show that macrophytes are pertinent tools to monitor ecological processes.

2.3.6 GENERAL CONCLUSIONS

Proposals have also been made in France to take into account macrophytes and algae data (Agences de l'eau, 2003). Floristic indicators among macrophytes (IBMR index) or algae (IBD index and chlorophyll *a*) are first selected according to their ecological representation along the river continuum. The more relevant the index, then the more weight it has. Similar choices are made for macroinvertebrates and fish. In a second step, a global biological quality is derived from information on flora and fauna. Fish are considered as the most important indicator, then macroinvertebrates and finally flora. This approach has been proven to be more relevant than the mean between biological indicators and the more downgrading.

Most European methods have aimed at monitoring specific features of water quality, such as organic content or trophic status, and macrophytes and algae are often used simultaneously (Jarvie *et al.*, 2002; Triest *et al.*, 2001), as well as macrophytes and invertebrates (Thiébaut *et al.*, 2002). Attempts to develop methods closer to the directive requirements with a shift from disturbance indices approaches, towards a more ecological dimension approach, are very few (Bukhtiyarova *et al.*, 1996, in Prygiel *et al.*, 1999), but will lead to a more comprehensive diagnosis on ecosystem quality and stability. In many cases, a gap between research results and monitoring programs remains, due to excessive complexity or cost (time or analytical requirements) of the methods.

REFERENCES

AFNOR, 2000. 'Détermination de l'indice biologique diatomées (IBD)', NF T90-354, June, 2000. AFNOR: Saint-Denis La Plaine, France.

AFNOR, 2003. 'Détermination de l'indice biologique macrophytique en rivière (IBMR)', NF T90-395, October, 2003. AFNOR: Saint-Denis La Plaine, France.

Agence de l'eau, 2003. 'Utilisation des indices biologiques pour évaluer l'état écologique des cours d'eau au sens de la Directive Cadre sur l'Eau'. Document final provisoire, April, 2003. Agence de l'Eau Rhône-Méditerranée-Corse, Lyon, France.

Ali, M. M., Murphy, K. J. and Abernethy, V. J., 1999. 'Macrophyte functional variables versus species assemblages as predictors of trophic status in flowing waters'. *Hydrobiologia*, **415**, 131–138.

Almeida, S. F. P. and Gil, M. C. P., 2001. 'Ecology of freshwater diatoms from the central region of Portugal'. *Cryptogamie Algologie*, **22**, 109–126.

Alverson, A. J., Manoylov, K. M. and Stevenson, R. J., 2003. 'Laboratory sources of error for algal community attributes during sample preparation and counting'. *Journal of Applied Phycology*, **15**, 357–369.

Amoros, C., Bornette, G. and Henry, C. P., 2000. 'A vegetation-based method for the ecological diagnosis of riverine wetlands'. *Environmental Management*, **25**, 211–227.

Antosegui, A., Trigueros, J. M. and Orive, E., 2001. 'The Use of Pigment Signatures to Assess Phytoplankton Assemblage Structure in Estuarine Waters'. *Estuarine Coastal and Shelf Science*, **52**, 689–703.

Baattrup-Pedersen, A., Andersson, B., Bandrud, T. E., Karttumen, K., Tiis, T. and Toivinen, H., 2001. '7-Macrophytes'. In: *Biological Monitoring in Nordic Rivers and Lakes*, Skriver, J. (Ed.). TemaNord: Copenhagen, pp. 53–60.

Barendregt, A. and Bio, A. M. F., 2003. 'Relevant variables to predict macrophyte communities in running waters'. *Ecological Modelling*, **160**, 205–217.

Bornette, G., Henry, C., Barrat, M. H. and Amoros, C., 1995. 'Theoretical habitat templets, species traits, and species richness: aquatic macrophytes in the Upper Rhône River and its floodplain'. *Freshwater Biology*, **31**, 487–505.

Cattaneo, A. and Amireault, M. C., 1992. 'How artificial are artificial substrates for periphyton?'. *Journal of the North American Benthological Society*, **11**, 244–256.

CEN, 2003. *European Standard EN 14184: 2003 E: Water quality – Guidance standard for the surveying of aquatic macrophytes in running waters*. European Committee for Standardization: Brussels, Belgium.

Chessman, B., Growns I. O., Currey J. and Plunkett-Cole, N., 1999. 'Predicting diatom communities at the genus level for the rapid biological assessment of rivers'. *Freshwater Biology*, **41**, 317–331.

Coesel, P. F. M., 1977. 'On the ecology of Desmids and the suitability of these algae in monitoring the aquatic environment'. *Hydrobiology Bulletin*, **11**, 20–21.

Coste, M., Tison, J. and Delmas, F., 2004. 'Flores diatomiques des cours d'eau: propositions de valeurs limites du 'bon état' pour l'IPS et l'IBD', Document de travail – 10/02/2004. Cemagref: Bordeaux, France.

Dawson, F. H. and Szoszkiewicz, K., 1999. 'Relationships of some ecological factors with the associations of vegetation in British rivers'. *Hydrobiologia*, **415**, 117–122.

Dawson, F. H., Newman, J. R., Gravelle, M. J., Rouen, K. J. and Henville, P., 1999a. *Assessment of the Trophic Status of Rivers Using Macrophytes – Evaluation of the Mean Trophic Rank*, Research and Development, Technical Report E39. Environment Agency: Bristol, UK.

Dawson, F. H., Raven, P. J. and Gravelle, M. J., 1999b. 'Distribution of the morphological groups of aquatic plants for rivers in the UK'. *Hydrobiologia*, **415**, 123–130.

Dell'Uomo, A., 1991. 'Use of benthic macroalgae for monitoring rivers in Italy'. In: *Use of Algae for Monitoring Rivers*, Düsseldorf, Germany, Whitton, B. A., Rott, E. and Friedrich, G. (Eds). Institut für Botanik, Universität Innsbruck: Innsbruck, Austria, pp. 129–137.

Den Hartog, C. and Van der Velde, G., 1988. 'Structural aspects of aquatic plant communities'. In: *Handbook of Vegetation Science*, Vol.15/1, *Vegetation of Inland Waters*, Symoens, J. J. (Ed.). Kluwer Academic Publisher: Dordrecht, The Netherlands, pp. 113–153.

Descy, J.-P., Higgins, H. W., Mackey, D. J., Hurley, J. P. and Frost, T. M., 2000. 'Pigment ratios and phytoplankton assessment in northern Wisconsin lakes'. *Journal of Phycology*, **36**, 274–286.

Dokulil, M. T., Schmidt, R. and Kofler, S., 1997. 'Benthic diatom assemblages as indicators of water quality in an urban flood-water inpoundment, Neue Donau, Vienna, Austria'. *Nova Hedwigia*, **65**, 273–283.

Doutelero, I., Perona, E. and Mateo, P., 2004. 'Use of cyanobacteria to assess water quality in running waters'. *Environmental Pollution*, **127**, 377–384.

Dufrêne, M. and Legendre, P., 1997. 'Species assemblages and indicator species: the need for a flexible asymmetrical approach'. *Ecological Monographs*, **67**, 345–366.

Dutartre, A. and Montel, M. H. N. (Eds), 2002. *Gestion des Plantes Aquatiques, Proceedings of the 11th International Symposium on Aquatic Weeds – EWRS*, 3–7 September, 2002, Moliets et Maâ (40), Cemagref, Conseil Général des Landes, INRA, ENSAR: Bordeaux, France, pp. 448.

Economou-Amilli, A., 1980. 'Marine Diatom from Greece. I. Diatoms from the Saronikos Gulf'. *Nova Hedwigia*, **32**, 63–104.

Ector, L., Loncin, A. and Hoffmann, L., 1999. 'Compte rendu du 17èmc colloque de l'Association des diatomistes de langue française'. *Cryptogamie – Algologie*, **20**, 105–148.

Eloranta, P. and Soininen, J., 2002. 'Ecological status of some Finnish rivers evaluated using benthic diatom communities'. *Journal of Applied Phycology*, **14**, 1–7.

Empain, A., Lambinon, J., Mouvet, C. and Kirchamann, R., 1980. 'Utilisation des bryophytes aquatiques et subaquatiques comme indicateurs biologiques de la qualité des eaux courantes'. In: *La Pollution des Eaux Continentales – Incidences sur les Biocénoses Aquatiques*, 2nd Edition, Pesson, P. (Ed.). Gauthier-Villars: Paris, pp. 195–223.

EC, 1991. 'Council directive of 21 May 1991 concerning urban waste water treatment (91/271/EEC)'. *Official Journal of the European Communities*, **L135**, 40–52.

EC, 2000. 'Directive 2000/60/EC of the European Parliament and of the council of 23 October 2000 establishing a framework for Community action in the field of water policy'. *Official Journal of the European Communities*, **L327/1**.

Everbecq, E., Gosselain, V., Viroux, L. and Descy, J. P., 2001. 'Potamon: A dynamic model for predicting phytoplankton composition and biomass in lowland rivers'. *Water Research*, **35**, 901–912.

Fawzi, B., Loudiki, M., Oubraim, S., Sabour, B. and Chlaida, M., 2002. 'Impact of Wastewater Effluent on the Diatom Assemblages Structure of a Brackish Small Stream: Oued Hassar (Morocco)'. *Limnologica*, **32**, 54–65.

Ferreira, M. T. and Moreira, I. S., 1999. 'River plants from an Iberian basin and environmental factors influencing their distribution'. *Hydrobiologia*, **415**, 101–107.

Fore, L. S. and Grafe, C., 2002. 'Using diatoms to assess the biological condition of large rivers in Idaho (USA)'. *Freshwater Biology*, **47**, 2015–2037.

Friedrich, G., Gerhardt, V., Bodemer, U. and Pohlmann, M., 1998. 'Phytoplankton Composition and Chlorophyll Concentration in Freshwaters: Comparison of Delayed Fluorescence Excitation Spectroscopy, Extractive Spectrophotometric Method, and UTERMÖHL-Method'. *Limnologica*, **28**, 323–328.

Garbey, C., Murphy, K. J., Thiébaut, G. and Muller, S., 2004. 'Variation in P-content in aquatic plant tissues offers an efficient tool for determining plant growth strategies along a resource gradient'. *Freshwater Biology*, **49**, 346–356.

Giovanni, R. and Haury, J., 1995. 'Pesticides et milieu aquatique'. In: *Colloque Qualité des Eaux et Produits Phytosanitaires: du Diagnostic à l'Action. Bilan de 5 Années d'Études et Propositions de la CORPEP en Bretagne*, 27 November 1995, Bretagne Eau Pure, Rennes, France, pp. 57–70.

Gomez, N. and Licursi, M., 2001. 'The Pampean Diatom Index (IDP) for assessment of rivers and streams in Argentina'. *Aquatic Ecology*, **35**, 173–181.

Grasmuck, N., Haury, J., Leglize, L. and Muller, S., 1995. 'Assessment of the bioindicator capacity of aquatic macrophytes using multivariate analysis'. *Hydrobiologia*, **300–301**, 115–122.

Haslam, S. M. and Wolseley, P. A., 1987. *River Plants of Western Europe. The Macrophytic Vegetation of Watercourses of the European Economic Community*. Cambridge University Press: Cambridge, UK.

Haury, J. and Muller, S., 1991. 'Variations écologiques et chorologiques de la végétation macrophytique des rivières acides du Massif Armoricain et des Vosges du Nord (France)'. *Revue des Sciences de l'Eau*, **4**, 463–482.

Haury, J., Thiébaut, G. and Muller, S., 1995. 'Les associations rhéophiles des rivières acides du Massif armoricain, de Lozère et des Vosges du Nord, dans un contexte Ouest-Européen'. 37th International Colloquim of the International Association of Vegetation Science, 'Large Area Vegetation Surveys', Bailleul, 1994. *Colloquia Phytosociology*, **23**, 145–168.

Haury, J., Peltre, M.-C., Muller, S., Thiébaut, G., Trémolières, M., Demars, B., Barbe, J., Dutartre, A., Guerlesquin, M. and Lambert, E., 2001. *Les Macrophytes Aquatiques Bioindicateurs des Systèmes Lotiques – Intérêts et Limites des Indices Macrophytiques. Synthèse Bibliographique des Principales Approches Européennes*, Etudes sur l'Eau en France, No. 87. Ministère de l'Ecologie et du Développement Durable: Paris, France (www.eaufrance.tm.fr).

Haury, J., Peltre, M.-C., Trémolieres, M., Barbe, J., Thiebaut, G., Bernez, I., Daniel, H., Chatenet, P., Muller, S., Dutartre, A., Laplace-Treyture, C., Cazaubon, A. and Lambert-Servien, E., 2002. 'A method involving macrophytes to assess water trophy and organic pollution: the Macrophyte Biological Index for Rivers (I.B.M.R.) – Application to different types of rivers and pollutions'. In: *Gestion des Plantes Aquatiques, Proceedings of the 11th International Symposium on Aquatic Weeds – EWRS*, Dutartre, A. and Montel, M. H. N. (Eds), 3–7 September, 2002, Moliets et Maâ (40), Cemagref, Conseil Général des Landes, INRA, ENSAR): Bordeaux. France, pp. 247–250.

Hill, B. H., Herlihy, A. T., Kaufmann, P. R., Decelles, S. J. and Vanderborgh, M. A., 2003. 'Assessment of streams of the eastern United States using a periphyton index of biotic integrity'. *Ecological Indicators*, **2**, 325–338.

Hirst, H., Jüttner, I. and Ormerod, S. J., 2002. 'Comparing the responses of diatoms and macroinvertebrates to metals in upland streams of Wales and Cornwall'. *Freshwater Biology*, **47**, 1752–1765.

Holmes, N. T. H. and Whitton, B. A., 1977. 'The macrophytic vegetation of the river Tees: observed and predicted changes'. *Freshwater Biology*, **7**, 43–60.

Holmes, N. T. H., Boon, P. J. and Rowell, T. A., 1998. 'A revised classification system for British rivers based on their aquatic plant communities'. *Aquatic Conservation: Marine Freshwater Ecosystems*, **8**, 555–578.

Holmes, N. T. H., Newman, J. R., Chadd, J. R., Rouen, K. J., Saint, L. and Dawson, F. H. 1999. *Mean Trophic Rank: A User's Manual*, Research and Development, Technical Report E38. Environment Agency: Bristol, UK.

Hürlimann, J. and Niederhauser, P., 2002. *Méthodes d'Étude et d'Appréciation de l'État de Santé des Cours d'Eau: Diatomées – Niveau R (Région)*. Office Fédéral de l'Environnement, des Forêts et du Paysage (OFEFP): Berne, Switzerland.

Husák, S., Sládecek, V. and Sládecková, A., 1989. 'Freshwater macrophytes as indicators of organic pollution'. *Acta Hydrochimica Hydrobiology*, **17**, 693–697.

Ibelings, B., Admiraal, W., Bijerk, R., Ietswaart, T. and Prins, H., 1998. 'Monitoring of algae in Dutch rivers: does it meet its goals?'. *Journal of Applied Phycology*, **10**, 171–181.

Janauer, G., Hale, P. and Sweeting, R. (Eds), 2003. 'Macrophyte inventory of the River Danube: a pilot study'. *Archiv für Hydrobiologie, Supplement*, **147**(1–2).

Jarvie, H. P., Lycett, E., Neal, C. and Love, A., 2002. 'Patterns in nutrient concentrations and biological quality indices across the upper Thames river basin, UK'. *Science of the Total Environment*, **282–283**, 263–294.

Kelly, M. G., 1998a. 'Use of community-based indices to monitor eutrophication in European rivers'. *Environmental Conservation*, **25**, 22–29.

Kelly, M. G., 1998b. 'Use of the trophic diatom index to monitor eutrophication in rivers'. *Water Research*, **32**, 236–242.

Kelly, M. G., 2001. 'Use of similarity measures for quality control of benthic diatom samples'. *Water Research*, **35**, 2784–2788.

Kohler, A., 1975. 'Submerse Makrophyten und ihre Gesellschaften als Indikatoren der Gewässerbelastung'. *Beiträge Naturkundliche Forschung in Südwest Deutschland*, **34**, 149–159.

Köhler, J. and Descy, J. P., 2003. 'Main results of a workshop on phytoplankton in European rivers. *FBA News*, (22), 7.

Kohler, A. and Janauer, G. A., 1995. 'Zur Methodik der Untersuchung von aquatischen Makrophyten in Fließgewässern'. In: *Handbuch Angewandte Limnologie*, VIII, 1.1.3, Steinberg, Ch., Bernhardt, H. and Klapper, H. (Eds). Ecomed-Verlag: Landsberg, Germany.

Kohler, A. and Schneider, S., 2003. 'Macrophytes as bioindicators'. *Archiv für Hydrobiologie, Supplement Large Rivers*, **147**(1–2), 17–31.

Kohonen, T., 1995. *Self-Organizing Maps*, 2nd Edition, Springer Series in Information Sciences, Vol. 30. Springer-Verlag: Berlin, Germany.

Kutka, F. J. and Richards, C., 1996. 'Relating diatom assemblage structure to stream habitat quality'. *Journal of the North American Benthological Society*, **15**, 469–480.

Kwandrans, J., Eloranta, P., Kawecka, B. and Wojtan, K., 1998. 'Use of benthic diatom communities to evaluate water quality in rivers of southern Poland'. *Journal of Applied Phycology*, **10**, 193–201.

Lane, C. M., Taffs, K. H. and Corfield, J. L., 2003. 'A comparison of diatom community structure on natural and artificial substrata'. *Hydrobiologia*, **493**, 65–79.

Leland, H. V., Brown, L. R. and Mueller, D. K., 2001. 'Distribution of algae in the San Joaquin River, California, in relation to nutrient supply, salinity and other environmental factors'. *Freshwater Biology*, **46**, 1139–1167.

Lopez, J., Retuerto, R. and Carballeira, A., 1997. 'D665/D665a Index vs frequencies as indicators of bryophyte response to physicochemical gradients'. *Ecology*, **78**, 261–271.

Mackey, M. D., Mackey, D. J., Higgins, H. W. and Wright, S. W., 1996. 'CHEMTAX – a program for estimating class aboundances from chemical markers – application to HPLC measurements of phytoplankton'. *Marine Ecology Progress Series*, **144**, 265–283.

Maksimov, V. N., Bulgakov, N. G. and Dzhabrueva, L. V., 1997. 'Rank Distributions of Size–Morphological Groups of the Algae in Periphyton and Their Relationship to the Level of Water Body Pollution'. *Biology Bulletin*, **24**, 575–582.

McCormick, P. V. and Cairns, J. J., 1994. 'Algae as indicators of environmental change'. *Journal of Applied Phycology*, **6**, 509–526.

Montesanto, B., Ziller, S. and Coste, M., 1999. 'Diatomées épilithiques et qualité biologique des ruisseaux de mont Stratonikon, Chalkidiki (Grèce)'. *Cryptogamie – Algologie*, **20**, 235–251.

Munn, M. D., Black, R. W. and Gruber, S. J., 2002. 'Response of benthic algae to environmental gradients in an agriculturally dominated landscape'. *Journal of the North American Benthological Society*, **21**, 221–237.

Nunes, M. L., Ferreira Da Silva, E. and De Almeida, S. F. P., 2003. 'Assessment of water quality in the Caima and Mau river basins (Portugal) using geochemical and biological indices'. *Water, Air, and Soil Pollution*, **149**, 227–250.

Patrick, R., Hohn, M. H. and Wallace, J. H., 1954. 'A new method for determining the pattern of the diatom flora'. *Notulae Naturae*, **259**, 2–12.

Pot, R., 1996. 'Monitoring watercourse vegetation, a synecological approach to dynamic gradients'. *Hydrobiologia*, **340**, 59–65.

Potapova, M. G. and Charles, D. F., 2002. 'Benthic diatoms in USA rivers: distributions along spatial and environmental gradients'. *Journal of Biogeography*, **29**, 167–187.

Potapova, M. and Charles, D. F., 2003. 'Distribution of benthic diatoms in US rivers in relation to conductivity and ionic composition'. *Freshwater Biology*, **48**, 1311–1328.

Prygiel, J., 2002. 'Management of the diatom monitoring networks in France'. *Journal of Applied Phycology*, **14**, 19–26.

Prygiel, J., Bukowska, J. and Whitton, B. A., 1999. *Use of Algae for Monitoring Rivers, III*. Agence de l'Eau Artois-Picardie: Douai, France.

Prygiel, J., Carpentier, P., Almeida, S., Coste, M., Druart, J-C., Ector, L., Guillard, D., Honoré, M-A., Iserentant, R., Ledeganck, P., Lalanne-Cassou, C., Lesniak, C., Mercier, I., Moncaut, P., Nazart, M., Nouchet, N., Peres, F., Peeters, V., Rimet, F., Rumeau, A., Sabater, S., Straub, F., Torrisi, M., Tudesque, L., Van de Vijver, B., Vidal, H., Vizinet, J. and Zydek, N., 2002. 'Determination of the biological diatom index (IBD NF T 90-354): results of an intercomparison exercise'. *Journal of Applied Phycology*, **14**, 27–39.

Reid, M. A., Tibby, J. C., Penny, D. and Gell, P. A., 1995. 'The use of diatoms to assess past and present water quality'. *Australian Journal of Ecology*, **20**, 57–64.

Reynolds, C. S. and Descy, J. P., 1996. 'The production, biomass and structure of phytoplankton in large rivers'. *Archiv für Hydrobiologie, Supplement*, **113**(1–4), 161–187.

Riis, T., 2000. *Distribution and Abundance of Macrophytes in Danish Streams*, PhD Thesis. University of Copenhagen: Copenhagen, Denmark.

Robach, F., Thiébaut, G., Trémolières, M. and Muller, S., 1996. 'A reference system for continental running waters: plant communities as bioindicators of increasing eutrophication in alkaline and acidic water in north-east France'. *Hydrobiologia*, **340**, 67–76.

Rott, E., Duthie, H. C. and Pipp, E., 1998. 'Monitoring organic pollution and eutrophication in the Grand River, Ontario, by means of diatoms'. *Canadian Journal of Fisheries and Aquatic Sciences*, **55**, 1443–1453.

Rott, E., Pipp, E. and Pfister P., 2003. 'Diatom methods developed for river quality assessment in Austria and a cross-check against numerical trophic indication methods used in Europe'. *Algological Studies*, **110**, 91–115.

Sabater, S., Sabater, F. and Armengol, J., 1988. 'Relationships between diatom assemblages and physico-chemical variables in the river Ter (NE Spain)'. *Internationale Revue dergesamten Hydrobiologie*, **73**, 171–179.

Sgro, G. V. and Johansen, J. R., 1998. *Algal Periphyton Bioassessment Methods For Lake Erie Estuaries*, Vol. 1, *Metric Development*. Lake Erie Office, Toledo, OH, USA.

Stevenson, R. J. and Pan, Y., 2003. 'Assessing environmental conditions in rivers and streams with diatoms'. In: *The Diatoms: Applications for the Environmental and Earth Sciences*, Smol, J. P. (Ed.). Cambridge University Press, Cambridge, UK, pp. 11–40.

Szoszkiewicz, K., Karolewicz, K., Lawniczak, A. and Dawson, F. H., 2002. 'An assessment of the MTR aquatic plant bioindication system for determining the trophic status of Polish rivers'. *Polish Journal of Environmental Studies*, **11**, 421–427.

Thiébaut, G. and Muller, S., 1999. 'A macrophyte communities sequence as an indicator of eutrophication and acidification levels in weakly mineralised streams in North-Eastern France'. *Hydrobiologia*, **410**, 17–24.

Thiébaut, G., Guerold, F. and Muller, S., 2002. 'Are trophic and diversity indices based on macrophyte communities pertinent tools to monitor water quality?'. *Water Research*, **36**, 3602–3610.

Tison, J., Giraudel, J. L., Park, Y.-S., Coste, M. and Delmas, F., 2005. 'Classification of stream diatom communities using a self-organizing map'. In: *Modelling Community Structure in Freshwater Ecosystems*, Vol. 5.5, Lek, S., Scardi, M., Verdonschot, P. F. M., Descy, J.-P. and Park, Y. S. (Eds.). Springer-Verlag: Berlin, Germany, pp. 304–316.

Tremp, H. and Kohler, A., 1995. 'The usefullness of macrophyte monitoring systems, examplified on eutrophication and acidification of running water'. *Acta Botanica Gallica*, **142**, 541–550.

Triest, L., Kaur, P., Heylen, S. and De Pauw, N., 2001. 'Comparative monitoring of diatoms, macroinvertebrates and macrophytes in the Woluwe River (Brussels, Belgium)'. *Aquatic Ecology*, **35**, 183–194.

Van de Vijver, B. and Beyens, L., 1998. 'Diatoms and water quality in the kleine Nete, a Belgian lowland stream'. *Limnologica*, **28**, 145–152.

Vanderpoorten, A., Thiebaut, G., Tremolieres, M. and Muller, S., 2000. 'A model for assessing water mineralization and trophic level by using aquatic bryophyte assemblages in Eastern France'. *International Association of Theoretical and Applied Limnology*, **27**, 807–810.

Wasson, J. G., Chandesris, A., Pella, H. and Blanc, L., 2002. 'Définition des hydro-écorégions françaises métropolitaines', Rapport final, BEA/LHQ, Ministère de l'Ecologie et du Développement Durable, June 2002. Cemagref: Lyon, France.

Wehr, J. D. and Descy, J.-P., 1998. 'Use of phytoplankton in large river management'. *Journal of Phycology*, **34**, 741–749.

Weithoff, G., 2003. 'The concepts of "plant functional types" and "functional diversity" in lake phytoplankton – a new understanding of phytoplankton ecology?'. *Freshwater Biology*, **48**, 1669–1675.

Whitton, B. A., 1979. 'Plants as indicators of river water quality'. In: *Biological Indicators of Water Quality*, James, A. and Evison, L. (Eds). Wiley: Chichester, UK, pp. 5.1–5.34.

Whitton, B. A. and Kelly, M. G., 1995. 'Use of algae and other plants for monitoring rivers'. *Australian Journal of Ecology*, **20**, 45–56.

Whitton, B. A., Boulton, P. N. G., Clegg, E. M., Gemmel, J. J., Graham, G. G., Gustar, R. and
Moorhouse, T. P., 1998. 'Long-term changes in macrophytes of British rivers; 1. River Wear'.
Science of the Total Environment, **210/211**, 411–426.

Wiegleb, G., 1983. 'A phytosociological study of the macrophytic vegetation of running waters in
Western Lower Saxony (Federal Republic of Germany)'. *Aquatic Botany*, **17**, 251–274.

Willby, N. J., Abernethy, V. J. and Demars B. O. L., 2000. 'Attribute-based classification of
European hydrophytes and its relationship to habitat utilization'. *Freshwater Biology*, **43**, 43–
74.

Winter, J. G. and Duthie, H. C., 2000. 'Epilithic diatoms as indicators of stream total N and total
P concentration'. *Journal of the North American Benthological Society*, **19**, 32–49.

Wright, S. W., Jeffreys, S. W., Mantoura, R. F. C., Llewellyn, C. A., Bjornland, T., Repeta, D.,
and Welschmeyer, N., 1991. 'Improved HPLC method for the analysis of chlorophylls and
carotenoids from marine phytoplankton'. *Marine Ecology Progress Series*, **77**, 183–196.

Wu, J.-T., 1999. 'A generic index of diatom assemblages as bioindicator of pollution in the Keeung
River of Taiwan'. *Hydrobiologia*, **397**, 79–87.

2.4

Organization of Biological Monitoring in the European Union

Laura Mancini

2.4.1   INTRODUCTION

Both scientific community and political institutions in Europe are more and more involved in the conservation and restoration of aquatic environments. Biological monitoring is a key element in environmental defense programmes and regulations of many European countries and in the near future, the implementation of the Water Framework Directive (WFD) 2000/60 including CIS (Common Implementation Strategies for WFD) will increase the relevance of biological monitoring with respect to other environmental monitoring methods (EU, 2000).

The WFD is a fundamental tool guiding all actions aimed at improving aquatic ecosystems health. This indicates strategies for monitoring activities, sets quality targets and tries to push toward an integrated approach to ecosystem management. From a wider point of view, the WFD marks a shift from an anthropocentric vision to a holistic and ecosystem-centred approach. In this vision, man can be seen as one element of the system and the main user of its services (Mancini and Zapponi, 2002).

The new focus on biological indicators is a substantial step forward with respect to previous environmental directives. These previous directives make extensive use of measurements of physical, chemical and microbiological parameters, whereas the WFD concentrates on the assessment of the effects of pollution on biotic communities. In this revolutionary approach, plants and animals are considered to be privileged indicators of water bodies' health for they are able to integrate pressures coming from both biotic and abiotic components of the ecosystem (EU, 1999; Nixon, 2002; CIS, 2003a) As a general rule, animal and vegetal populations are the best indicators of impacts on the environment. This is due to several reasons. First, one of the main limits of many environmental indicators lies in the difficulty in 'catching' the strong temporal and spatial variability of environmental phenomena. The use of living organisms can partly overcome this problem. They can keep memories of previous natural or anthropogenic stressing events that could go unnoticed by physio-chemical analysis. Secondly, biological parameters can usually detect synergies and antagonisms between substances and phenomena and are able to give early warning of modifications within populations and communities; in other terms, organisms respond more to habitat modifications than to single factors and are able to express cumulative effects. Furthermore, ecological and physiological differences among organisms living in a water body can provide detection tools for a wide range of phenomena. Finally, biological indicators are highly communicative and useful for a participatory approach to environmental management (Premazzi and Chiaudiani, 1992; De Zwart, 1995; Boon *et al.*, 2000).

Biological water assessment methods have been developed in the past to measure the response of key species or higher taxa to pollution. Results are expressed in various indices. Many of them share common features such as the relative abundance of each species or group and their known pollution range 'reliability' as indicators. Furthermore, methods always refer to a reference condition, which is assumed as the condition of no human impact on the water body (Begon *et al.*, 1996). Different elements of river ecosystems can be used as indicators of freshwater biological quality: benthic invertebrates, macrophytes, benthic algae, fish and phytoplankton. Benthic fauna is already widely used as an ecological indicator of rivers' 'health conditions' in many European countries and following the WFD, the biological assessment carried out on benthos is part of the overall environmental quality assessment defining 'ecological quality'.

The achievement of a 'good' ecological status for surface and groundwater is the primary goal of EU Member States in the implementation of the Water Framework Directive. The WFD defines five classes of environmental quality which range from 'high' to 'bad'. The assessment of the environmental status of surface waters, particularly, investigates the alteration of water and sediment chemistry, aquatic biota and hydromorphology.

The first action that had to be taken after the WFD became effective was the description of each river basin district, to be completed within four years. This description had to define the main features of different types of water bodies and had to identify point and non-point pollution sources, flow regulation structures and morphological alterations. Land-use models had to be provided and risks to fail the good ecological status had to be analysed.

The definition of reference conditions for different types of water bodies must be done with respect to hydromorphological, physio-chemical and biological parameters. These conditions must refer to a good status. For which biological conditions concerned, European States will set up a reference network entirely composed by sites having a good quality.

After the phase of description, the WFD enjoins that each member state should ensure the establishment of programmes for the monitoring of water status, after six yerars, at the latest, from the date of entry into force of this Directive, in order to establish a coherent and comprehensive overview of water status within each river-basin district. For surface waters, such programmes shall cover the ecological and chemical status and ecological potential, the volume and level or rate of flow (Article 8).

Annex V of the WFD gives detailed indications about monitoring activities, including monitoring network design, frequency of monitoring, quality elements to be measured, and normative definitions of ecological status classifications. Member States shall establish surveillance monitoring programmes and operational monitoring programmes. Surveillance monitoring programmes are aimed at assessing impacts, long-term changes in natural conditions and long-term changes resulting from widespread anthropogenic activity. Operational monitoring shall be undertaken in order to establish the status of those bodies identified as being at risk of failing to meet their environmental objectives, and assess any changes in the status of such bodies resulting from the programmes of measures. An additional investigative monitoring shall be carried out where the reason for any exceedances is unknown, where surveillance monitoring indicates that the objectives for a body of water are not likely to be achieved or to ascertain the magnitude and impacts of accidental pollution (Annex V, 1.3).

The major relevance of biological elements in the definition of ecological status is clearly expressed in Annex V. For riverine ecosystems, the biological elements to be monitored are the composition and abundance of aquatic flora, benthic invertebrate and fish fauna. In the case of fish fauna, age structure is also to be studied.

In addition to these strictly biological elements, relevance is given also to hydromorphological, chemical and physico-chemical elements supporting the biological ones. The hydromorphological elements are hydrological regime, quantity and dynamics of water flow, connection to groundwater bodies, river continuity, morphological conditions, river depth and width variation, structure and substrate of the river bed and structure of the riparian zone. The chemical and physico-chemical elements are as follow: thermal conditions, oxygenation conditions, salinity, acidification status, nutrient conditions and pollutants.

Even though biological monitoring in the European Union encompasses a wide range of well-established techniques and methods, the most important challenge for both the scientific community and the environmental officers will be the comparability of methods and results. In other words, the crucial point in the WFD implementation is the equivalence of the statements about the ecological status that result from different assessment methods used throughout the EU member states.

To this aim, for any water body in the EU the results of the biological monitoring will be given as 'Ecological Quality Ratios' (EQRs) to ensure comparability. This ratio represents the gap from the 'reference condition' of a given type of water body. Working Group 2.5 Intercalibration (Water Framework Directive (WFD), Common Implementation Strategy) stated: 'Establishing comparable boundaries between good and moderate quality is particularly important in order to have an equal level of ambition in achieving "good status" of surface waters in different Member States'. The value for the boundary between the classes of high and good status, and the value for the boundary between good and moderate status shall be established through a intercalibration exercise carried out on a network of sites encompassing all water body types and ecoregions (CIS, 2003a,b, 2004a,b; Chave, 2001; EU, 2001).

2.4.2 BIOLOGICAL MONITORING IN THE EUROPEAN UNION

Nearly all countries in the European Union have a national monitoring programme for rivers, generally based on physical, chemical and microbiological indicators of quality. Some countries add biological monitoring to other indicators. Biological indicators can be used either independently or in combination with other indicators to measure the ecological status of water bodies.

River basins across Europe are extremely variable in size and structure. Even though their response to a wide range of pressures has been long studied, monitoring the impacts on biological communities is still a complex task.

Among all existing methods, the use of macroinvertebrates to monitor the ecological status of rivers is the most commonly used, widely known and robust. In order of diffusion, macroinvertebrates are followed by diatoms which are mainly used for describing the impacts of eutrophication. Macrophytes and fish occupy the next step of the scale (Table 2.4.1). Macrophytes are mainly used to study the effects of morphological modification and seem to be very promising in the monitoring of artificial canals where macroinvertebrates might fail. Fish are characterized by a long life cycle which allows them to detect cumulative effects due to hormones or heavy metals.

Table 2.4.1 Biological quality elements used in river monitoring for countries in the EU (adapted from Nixon, 2002)

Diatoms	Macrophytes	Macroinvertebrates	Fish
Austria	Austria	Austria	Austria
Belgium	Belgium	Belgium	Belgium
France	France	Denmark	France
Germany	Sweden	Finland	Ireland
Ireland	The Netherlands	France	Norway
UK	Luxembourg	Germany	UK
Italy		Ireland	
Denmark		Italy	
Sweden		Luxembourg	
The Netherlands		The Netherlands	
Luxembourg		Portugal	
		Norway	
		Spain	
		Sweden	
		UK	

Table 2.4.2 summarizes the national monitoring programmes for river waters and is based on work undertaken by the European Topic Centre on Inland Waters for the European Environment Agency during 1995 and 1996. There may, therefore, have been some changes since then. For the working group 2A Ecostat of CIS, countries were asked to complete a questionnaire on their use of classification schemes for rivers and lakes. The premise is that if countries classify rivers and lakes on the basis of specific quality elements, then they must also monitor for these elements. The reverse may not be true, however, as countries may monitor for quality elements but not subsequently classify (European Environmental Agency, 1996a,b).

Table 2.4.2 indicates the institution responsible for monitoring, the starting date of monitoring, the frequency and the geographical coverage, for each European country.

First, we find that the approaches to typologies are very diverse (Mancini and Spaggiari, 2000). Besides the classification systems which use single abiotic parameters (aquatic geochemistry in Greece) or are based on abiotic factors and functional elements (e.g. France), some typologies have already been built up on abiotic factors and biocoenoses, mainly macroinvertebrates (e.g. Netherlands) (CIS, 2003b; Kristensen and Bogestrand, 1996).

Most of the elements used in monitoring need to be adapted to indications coming from the WFD. A fundamental element of the WFD is that all of the biological components, from phytoplankton to fish, must be addressed in national monitoring programmes. Substantial revision of existing methods will come from the ongoing intercalibration exercise. The European trend in monitoring practices is a progressive inclusion of more and more elements, from chemicals to hydromorphology, all of

Table 2.4.2 Biological assessment monitoring programmes for river waters (from Commission of the European Communities, 1992 (adapted from Chave, 2001 and CIS, 2003a,b), Iversen *et al.*, 2000 and Boon *et al.*, 2000)

Country	WHO monitor	Monitoring since	Frequency	Geographical coverage
Austria	Federal Ministry of Agriculture and Forest	1968	6 per year	National
Belgium (Flanders)	Flemish Environmental Agency	1989	1 per year	Flanders region/National coverage
Belgium (Wallon region)	Division de la Police de l'Environnement	1980	1 per year	Walloon region/national coverage
Cyprus	No monitoring	—	—	—
Denmark	Danish Environmental Protection Agency; National Environmental Research Institute, Ministry of Environment and Energy	1989	1–2 per year	261 national sampling sites, with a further 10 000 local sites/national coverage
Finland	—	—	—	Regional coverage
France	Réseau National de Bassin			National basin network, 1082 monitoring sites/national coverage
Greece	No monitoring	—	—	—
Germany	Joint Water Commission of the Federal States	1976	1 per 5 years	All main flowing waters/national coverage
Ireland	Environmental Protection Agency	1971	1 per year	3000 sites in 1200 rivers/national coverage
Italy	Agenzie Regionali e Provinciali per la Protezione e l'Ambiente (ARPA e APPA)	1999	4 per year or 2 per year (depending on region)	All main flowing waters/national coverage
Luxembourg	Administration des Eaux et Forêts	1972	Heavily polluted, 1 per year; others, every 3–5 years	Main rivers/national coverage
Malta	No monitoring	—	—	—

Table 2.4.2 (*Continued*)

Country	WHO monitor	Monitoring since	Frequency	Geographical coverage
The Netherlands	Institute for Inland Water Management and Waste Water Treatment		1–13 per year	Regional coverage
Portugal	—	—	—	—
Spain	Ministerio de Obras Públicas y Urbanismo, Centro de Estudios y Experimentación de Obras Públicas	1980	4 per year	847 national sampling points/national coverage
Sweden	The Swedish University of Agricultural Science, Department of Environmental Assessments	1993	1 per year	35 streams/national coverage and regional coverage
United Kingdom	National River Authority, England and Wales; The Scottish Office Environment Department, River Purification Boards, Scotland; Department of Environment, Northern Ireland	Early 1970s	2–3 per year, every 5 years	8266 national sampling points/national coverage

which must be related to the biological elements of the ecosystem. The WFD substantially modifies the concept of monitoring and stresses the point that monitoring is only important as long as it supports restoration processes and the achievement of good ecological status (Mancini, 2003).

2.4.2.1 Benthic invertebrates

Macroinvertebrates are the most commonly used elements to assess the effects of organic pollution of rivers for the biological classification of rivers in Europe.

The distribution, abundance and productivity of benthic organisms are influenced by several ecological processes: the historical events that have allowed or prevented a species from reaching a habitat; the physiological limitations of the species at all stages of the life cycle; the availability of energy resources; the ability of the species to tolerate competition, predation, parasitism and natural mortality (Wetzel, 2001). The benthic animals of running water are extremely diverse (Oligochaeta, Leeches, Isopods, Amphipods, Molluscs, Aquatic insects). The basic principle behind the study of macroinvertebrates is that some are more sensitive to pollution than others. Macroinvertebrates 'in stream' are defined as invertebrates retained by a 0.5 mm sieve or net. Macroinvertebrates from the river bottom (macro-zoobenthos) are the most used pool of these organisms, as they can be placed at an intermediate level in the food chain, representing the link between higher (e.g. fish) and lower (e.g. micro-decomposers) levels; they are rather easy to collect and to identify, and most of their species are quite well studied, so much so that it is possible to know both the autoecological and sinecological aspects of their natural history. Some taxa are very sensitive to pollutants, resulting in good indicators of water quality, while others are extremely resistant (Boyle and Fraleigh, 2003). Due to their fundamental role in energy and organic nutrient fluxes, macroinvertebrate communities are a key component of freshwater ecosystems (Petersen *et al.*, 1995; Ormerod *et al.*, 1993).

Macroinvertebrates are large and readily visible invertebrate animals which colonize the substrata of all rivers. The main constituents of this group are young aquatic stages of insects. Within this bottom dwelling community, the sensitivity and tolerance to various types of pollution vary considerably between taxa. Some species are, for example, very sensitive to the decrease in dissolved oxygen and will only be found in areas where oxygen levels are consistently high. A characteristic feature of polluted environments is a reduction in the overall community diversity and an increase in the density of tolerant species. Therefore, the composition of a macroinvertebrate community in any point of a river reflects the average water quality at that particular point. For this reason, macroinvertebrates are widely used for the assessment of river quality (Sandin and Hering, 2004).

In European countries, there is a long tradition for the assessment of river quality on the basis of macroinvertebrate communities. However, these assessments have primarily been made by local organizations responsible for managing and monitoring specific rivers. At present, all countries have developed these activities into national surveys of the biological quality (Table 2.4.2). These national surveys of main rivers are generally based on the measurements made by local authorities in accordance with harmonized and standardized procedures (e.g. sampling methods, criteria for site selection, classification schemes, etc.). Some countries use macroinvertebrate indicators to define the ecological status of water bodies, whereas others use them in combination with additional physico-chemical or microbiological indicators.

Several indices based on macroinvertebrate sampling are used worldwide to assess biological water quality (Metcalfe, 1989; Resh *et al.*, 1996; Lammert and Allan, 1999). The assessment techniques applied in Europe have been summarized by Woodiwiss (1964) and Nixon *et al.*, (1996). The first biological assessment method was the *Saprobic System*, which focused on species presence in relation to organic pollution (Kolkwitz and Marsson, 1902, 1908, 1909; Liebman, 1962). This was quantified by Pantle and Buck (1955) and Zelinka and Marvan (1961).

Biological indices and indicators elaborated during the last few decades in Europe may be gathered within three main categories (Newman, 1988; Metcalfe, 1989; De Pauw *et al.*, 1992), as follows:

- *Saprobic Indices*, based on the identification of species which are indicators of different trophic levels of the ecosystem (Sládecek, 1973).

- *Diversity Indices*, based on the sensitivity of species diversity to disturbances and stress. Among these indices, the Shannon–Weaver formula (Shannon and Weaver, 1949) is the most frequently used; this is based on the number of species and the abundance of each species. Other important methods were developed by Washington (1984), Hellawell (1989) and Boyle *et al.* (1990).

- *Biotic Indices*, mainly based on the structure of community, the abundance of each taxon and the tolerance to pollution of some of these (Woodiwiss, 1964; Tuffery and Verneaux, 1968; De Pauw and Vanhoren, 1983; AFNOR, 1992; Ghetti, 1997). Some biotic indices add a 'score', which is a function of each taxon sensitivity (Armitage *et al.*, 1983, Alba-Tercedor and Sanchez-Ortega, 1988). Wright *et al.* (1984) used key physical, chemical and geographical variables to predict the macroinvertebrate community of a given site and compare it with the community recorded. The River Invertebrate Prediction and Classification System (RIVPACS) was developed following this approach.

Important overviews on macroinvertebrate methods have been given by De Pauw *et al.* (1992) and Metcalfe (1989), and in technical papers of European projects, such as PAEQANN and AQUEM (Sandin *et al.*, 2000; Verdonschot and Dohet, 1999).

Table 2.4.3 gives an overview of the assessment methods for river waters, based on benthic macroinvertebrates most frequently applied in the EU member states. The WFD indicates that monitoring activities should investigate composition, abundance and diversity of benthic invertebrates communities, together with the presence of sensitive taxa. Monitoring based on benthic invertebrates is the most important and widespread biological monitoring approach carried out in Europe and is currently the most common tool used for ecological classification. Classification systems are in place and can be adapted to incorporate requirements of the WFD. Standards for sampling methodology are also in place.

The typical sampling frequency is one or two times a year, with samplings usually carried out in summer and winter. Habitats sampled are riffles, pools, edges and macrophytes; sampling is normally easy but some difficulties may arise in deep or fast flowing rivers. Identification is relatively straightforward to genera level, but usually requires expert knowledge to species level. This phase may be more difficult if specimens are damaged during sampling or preservation. The results can be influenced by sampling period, since the benthic invertebrates community has

Table 2.4.3 Current use of benthic macroinvertebrates in biological monitoring or classification in the EU ('Guidance of monitoring') assessment programmes for river waters (adapted from Chave, 2001 and CIS, 2003a,b, and from Iversen et al., 2000 and Nixon et al., 1996)

Country	Method used	Reference	Method	Taxa used	Sampling	Analysis	Identification	Range	Used in monitoring or classification
Austria	Saprobic Water Quality Assessment, Austria (ONORM M6232)	Moog, 1995; Moog et al., 1999; Osterreichisches Normungsinstitut, 1997	Biologic/ saprobic	Macroinvertebrates	Qualitative	Quantitive	Order family genus species	—	Yes
Belgium (Flanders)	Belgium Biotic Index (BBI)	De Pauw and Vanhooren, 1983; De Pauw et al., 1992	Biologic	Macroinvertebrates	Qualitative	Qualitative	Order family genus	0–10	Yes
Belgium (Wallon region)	Belgium Biotic Index (BBI)	De Pauw and Vanhooren, 1983; De Pauw et al., 1992	Biologic	Macroinvertebrates (macrophytes, phytoplankton)	Qualitative	Qualitative	Order family genus	0–10	Yes
Cyprus	No index	—	—	—	—	—	—	—	No
Denmark	Danish Strema Fauna Index (DSFI)	Skriver et al., 2001	Biologic	Macroinvertebrates (phytobenthos)	Qualitative	Qualitative	Family species genus	1–4	No
Finland	Various indexes, e.g. BMWP River Oligochaeta–Chironomidae Index (ROCI Index)	Paasavirta, 1990	Biologic	Macroinvertebrates	—	—	—	—	Yes

Country	Index	Reference	Type	Organisms	Qualitative/ quantitive	Qualitative/ Qualitative	Taxonomic level	Range	
France	Indice Biologique de la Qualità Generale Fance (IBGN)	AFNOR, 1985	Biologic	Macroinvertebrates (invertebrates, fish, macrophytes, phytoplankton)	Qualitative/ quantitive	Qualitative	Family	0–20	Yes
Greece	Belgium Biotic Index (BBI)	De Pauw and Vanhooren, 1983; De Pauw et al., 1992	—	—	—	—	—	—	No
Germany	Saprobic system (BEOL)/ Saprobien Index (DIN 38410)	DEV, 1992	Saprobic	Macroinvertebrates, microflora, microfauna	Qualitative	Quantitive	Species	0–100/ 1–4	Yes
Ireland	Quality Rating System (Q-rating) Average Score Per Taxon BMWP/ASPT	De Pauw and Vanhooren, 1983; De Pauw et al., 1992; Armitage et al., 1983; Chesters, 1980; Wright et al., 1984	Biologic	Macroinvertebrates macrophytes, algae, Siltation	Qualitative	Qualitative	Family species genus	0–5	Yes
Italy	Indice Biotico Esteso (IBE)	Ghetti, 1997	Biologic	Macroinvertebrates	Qualitative	Qualitative	Order family genus	0–14	Yes
Luxemburg	Index Biotic (IB)	—	Biologic	Macroinvertebrates (plankton, macrophytes occasionally)	Qualitative	Qualitative	Order family	0–10	Yes
Malta	No index	—	—	—	—	—	—	—	No

(*Continued*)

Table 2.4.3 (*Continued*)

Country	Method used	Reference	Method	Taxa used	Sampling	Analysis	Identification	Range	Used in monitoring or classification
The Netherlands	Quality Index (K135) AMOEBAEKOE-BEOSWA	Gardeniers and Tolkamp, 1976; Ten Brink et al., 1991; STOWA, 1992; Peeters et al., 1994	Biologic	Macroinvertebrates (fish, birds, phytoplankton, zooplankton, macrophytes)	Qualitative	Qualitative	Family species genus	100–500	Yes
Portugal	Belgium Biotic Index (BBI)	De Pauw and Vanhooren, 1983; De Pauw et al., 1992	Biologic	Macroinvertebrates	Qualitative	Qualitative	Order family genus	0–10	Yes
Norway	No index	—	—	Macroinvertebrates	—	—	—	—	Yes
Spain	Biological Monitoring Working Party Modified (BMWP)	Alba-Tercedor and Sanchez-Ortega, 1988	Biologic	Macroinvertebrates	Qualitative	Qualitative	Family	0–>150	Yes
Sweden	Acidification Index Average Score per Taxon (BMWP/ASPT)	Henrikson and Medin, 1986; Johnson, 1998; Armitage et al.,1983; Chesters, 1980; Wright et al., 1984	—	Macroinvertebrates (periphyton)	—	—	—	—	Yes
United Kingdom	Average Score per Taxon (BMWP/ASPTBMWP) Score	Armitage et al.,1983; Chesters, 1980; Wright et al., 1984	Biologic	Macroinvertebrates	Qualitative	Qualitative	Family	0–>150/0–10	Yes

a high seasonal variability and is influenced by climatic events (e.g. rainfall and flooding).

The methods still need adaptation to meet the requirements of the WFD and are affected by a few drawbacks, such as high substrate-related spatial variability, high temporal variability due to hatching of insects, variation of water flow and presence of exotic species in some European rivers (CIS, 2003b; Di Dato *et al.*, 2004).

2.4.2.2 Fish

Fishes permanently live in aquatic ecosystems and they are the longest living freshwater aquatic organisms. They constitute a highly visible component of aquatic communities with a consequently high public profile. Only 301 species of fish have been identified in European rivers and 44 of these are not native species but were introduced into Europe (Noble and Cowx, 2002). Freshwater fishes belong to two zoological groups, namely super-class Agnatha and super-order Teleostei (Cowx, 2002).

Fish communities can be useful biological indicators since they tend to be sensitive to both poor water quality and poor habitat quality, with effects manifesting themselves as shifts in parameters such as species composition, population size, age structure and growth rates (Hughes and Gammon, 1987).

In water quality terms, parameters such as dissolved oxygen, unionized ammonia and heavy metals have a strong influence on fish communities. Many other physico-chemical parameters can be relevant. The pH can affect the toxicity of metals (e.g. aluminium dissociates to its unionized, more toxic form at lower pH values). The percentage of unionized ammonia increases with pH, and also with temperature; however, it decreases with increasing salinity.

Tolerance to decline in water quality varies between species, with salmonids being particularly vulnerable to reduced levels of dissolved oxygen and elevated ammonia concentrations. Apart from physico-chemical features, poor quality physical habitat can have as great an influence as poor water quality in many instances, particularly in lowland rivers (Nixon *et al.*, 1996). Fish are good witnesses to the history of the river and they are able reveal long-term variations. Thus, parameters of fish communities can provide a precise and detailed analysis of the water body quality (Lafaille *et al.*, 1999; Odderdoff *et al.*, 1998, 2001).

There is a wide range of fish indices types. Biological indices to monitor water quality were originally developed using concepts such as 'indicator species' or 'diversity indices'. More recently, the shift has been towards a typology approach (Huet, 1959; Verneaux, 1976a,b), an integrated approach (Washington, 1984) and a community-based approach (Fausch *et al.*, 1990) which can be also used to assess impacts from human activities (Karr, 1981, 1991). Odderdoff *et al.* (2001) developed for France a probabilistic model based on species occurrence to define a fish-based index applicable at national level. In the UK, Milner *et al.* (1998) developed the HABSCORE system for assessing the suitability of habitats for salmonid communities. Less frequently, fish have been used in river classification systems for their behavioural characteristics (e.g. mobility, seasonal upstream or downstream migration and avoidance to pollution).

Generally speaking, fish are useful indicators of long-term impacts as they have long life cycles, they are can be found in several trophic levels and are relatively simple to identify. The CIS also recommends fish as one of the key elements for monitoring habitat and morphological changes (CIS, 2003b).

The WFD indicates that monitoring activities should investigate composition and abundance, sensitive species diversity and age structure of fish communities.

The typical sampling frequency for fish monitoring of rivers is once a year, but samplings period can vary significantly. All types of habitat are sampled and specialized sampling equipment is often required. Identification is normally simple to species, except some cyprinids which require expert knowledge. Important advantages of these methods are the existence of river classification systems and the possibility of adapting existing classification systems to incorporate the requirements of the WFD.

Major drawbacks of the methods are the high seasonal variability in community structure and abundance, high interannual variation due to age structure, horizontal and vertical distribution patterns and high mobility.

Generally, fish are not used for routine monitoring in the European Union; however, existing classification systems based on fish are used in Belgium, France, Netherlands and the UK, and occasionally in Austria, Ireland and Norway. Unfortunately, there is no standard for sampling methodologies yet and fish are still far less relevant than benthic invertebrates for ecological classification in Europe.

Monitoring of vegetal components is essential for a correct ecosystem health assessment. Photosynthetic organisms are particularly relevant in riverine environments and the importance of their monitoring has been widely recognized in European Countries. In the WFD, phytoplankton, macrophytes and benthic algae are indicated as key elements for biological quality assessment of rivers. Benthic diatoms and macrophytes can give early warning of organic pollution that benthic invertebrates could underestimate (Kelly and Whitton, 1995; Robach *et al.*, 1996). An important advantage of the methods using vegetal species is that many riverine photosynthetic organisms are present all across Europe (Bielli *et al.*, 1999). The most important monitoring techniques studying aquatic flora in Europe are concerned with benthic algae, mainly diatoms, and macrophytes.

2.4.2.3   Benthic algae

The freshwater phytoplankton is composed of algae belonging to almost every major taxonomic group, with the addition of cyanobacteria. Many of these groups have different physiological features and vary in response to physical and chemical parameters. The photosynthetic cyanobacteria, formerly known as blue–green algae, constitute a major component of the phytoplankton. The other components are green algae, yellow–green algae, golden-brown algae, cryptomonads, dinoflgellats, euglenoids, diatoms, and brown and red algae. For monitoring activities, diatoms are the most important group of algae, even though most species are sessile and associated with littoral substrate. The primary characteristic of diatoms, also called Bacillarophyce, is the silicified cell wall. They are commonly divided into centric diatoms (*Centrales*) which have radial symmetry, and pennate diatoms *(Pennales)* which exhibit essentially bilateral symmetry. They are mainly unicellular organisms, their size ranging from a few microns to more than 500 microns. Each cell is composed of a siliceous box called frustule (Rumeau and Coste, 1988).

Diatoms are a good environmental indicator and are one of the key groups of oganisms recommended by the WFD for the identification of ecological quality gradients in rivers. They have a worldwide diffusion, live in all running waters and represent an important element in aquatic ecosystem. The taxonomic diversity of diatoms constitutes a valuable tool when monitoring general water quality and trophic status (Prygiel and Coste, 1995) or more specific phenomena such as eutrophication and acidification.

For several years, diatoms have been used as bioindicators in different countries of Europe for evaluating the quality status of rivers (Ector, 1999). Table 2.4.4 shows the diatoms methods used in European countries.

Diatom indices are very promising tools for ecosystem health assessment of rivers in Europe. Diatoms are easy to sample in shallow waters, widely diffused and respond quickly to changes in environmental conditions and to anthropogenic pressures. Diatoms communities have a high seasonal variability and are influenced by climatic events, light and nutrient availability. The habitats sampled are benthic or artificial substrates and typical sampling frequencies are from two to four times a year.

Only a few countries presently use diatom indices in their routine monitoring activities. Most of these countries are currently carrying out experimentation on diatoms, while some use them for studying particular situations. There are many reasons for the limited diffusion of diatoms indices. Their measurement requires a specific expertise for species identification (CIS, 2003b), they are difficult to sample in deep rivers, have a high seasonal variability and high substrate-related spatial variability, many species can only be identified using an electron microscope (Prygiel *et al.*, 1997) and the number of frustules to be identified is generally high (some hundreds) (CEN, 2003, CEN, 2004). Some advancement might come from new methods based on identification at genus level or from a better standardization of existing methods at international level. CEN sampling methodologies are under development.

2.4.2.4 Macrophytes

The term 'macrophytes' encompasses many vegetal species having a macroscopic size that live by or within surface water bodies. Macrophytes are mainly represented by angiosperms (*Fanerogamae* erbacee, but also some *Pteridophydae* and numerous

Table 2.4.4 Diatoms methods used in European countries (adapted from Ector, 1999, and from Nixon *et al.*, 1996)

Country	Method used	Reference	Used in monitoring or classification
Austria	Zelinka–Marvan Index	Zelinka and Marvan,1961	Yes
Belgium	Descy Index (DES), Fabri and Leclercq method, Leclercq and Maquet Index (ILM), CEE Index	Descy, 1979; Fabri and Leclercq, 1984; Leclercq and Maquet, 1987; Descy and Coste, 1991	Yes
Cyprus	No index	—	No
Denmark	No index	—	Yes
Finland	Specific Polluosensibility Index (IPS), Generic Diatom Index (GDI), Trophic Diatom Index (TDI), Diatom Index of Artois Picardie (IDAP)	CEMAGREF, 1982; Rumeau and Coste, 1988; Coste and Ayphassoro, 1991; Schiefele and Kohmann, 1993; Prygiel *et al.*, 1996;	Yes – only experimental
France	Specific Polluosensibility Index (IPS), CEE Index, General Diatom Index (IDG), Diatom Index of Artois Picardie (IDAP), Biological Diatom Index (IBD), Standardized Diatom Index (IBD)	CEMAGREF, 1982; Descy and Coste, 1991; Round, 1993; Prygiel *et al.*, 1996; Lenoir and Coste, 1996; Prygiel and Coste, 1995, 2000; AFNOR, 2000;	Yes, since 1999
Greece	Specific Polluosensibility Index (IPS)	CEMAGREF, 1982	No – only experimental
Germany	Differentiating species system, Trophic Diatom Index (TDI)	Lange-Bertalot, 1979; Schiefele and Kohmann, 1993	Yes
Ireland	—	—	Yes
Italy	Eutrophication Pollution Index Diatoms (EPI-D)	Dell'Uomo, 1996	No – only experimental
Luxembourg	Specific Polluosensibility Index (IPS), CEE Index	CEMAGREF, 1982; Descy and Coste, 1991	Yes, since 1994
Malta	No index	—	No
The Netherlands	—	—	Yes
Norway	—	—	Yes
Portugal	Specific Polluosensibility Index (IPS), Sladek Index (SLA), CEE Index	CEMAGREF, 1982; Sládecek, 1986; Descy and Coste, 1991	No – only experimental

Table 2.4.4   (*Continued*)

Country	Method used	Reference	Used in monitoring or classification
Spain	Specific Polluosensibility Index (IPS), CEE Index	CEMAGREF,1982; Descy and Coste, 1991	Yes – in a region; in others, only experimental
Sweden	No index	—	Yes
United Kingdom	Trophic Diatom Index (TDI), Diatom Quality Index (DQI)	Harding and Kelly, 1999	Yes, only in special monitoring, e.g. investigative monitoring

Bryophytae and Mosses) and macroscopic algae (Bielli *et al*, 1999; Siligardi *et al.*, 2000; Azzollini *et al.*, 2003).

Many studies have addressed the problem of linking composition and structure of macrophytes populations to ecosystems health status (Butcher, 1933; Holmes and Whitton, 1977; Wiegleb, 1981; Carpenter and Lodge, 1986; Haslam, 1987, 1997; Leglize *et al.*, 1990; Carbiener *et al.*, 1995; Tremp and Kohler, 1995; Botineau and Ghestem, 1995; Grasmuck *et al.,* 1993; Haury *et al.*, 1996, Bielli *et al.*, 1999; Minciardi *et al.*, 2003). Generally speaking, macrophytes are good environmental indicators and their applicability to rivers and lakes ranges from moderate to high. The WFD places them among key biological quality elements for both natural and artificial water bodies (CIS, 2003a).

The most important factors regulating the distribution and biomass of angiosperm macrophytes in undisturbed streams are river depth, riverbed type, transparency and morphology, flow regime, light, sediment, water quality and velocity. As a consequence, the species composition changes markedly as a function of the stream size (Nixon *et al.*, 1996).

Macrophytes are abundant in suitable habitat but fast-flowing streams can be a limiting factor. They are characterized by high seasonal variability in community structure and abundance.

Macrophytes are mainly used to study river dynamics, including hydropower effects. They can also be used to detect eutrophication but there is no good evidence that angiosperm biomass increases with increased nutrient concentration (Thyssen *et al.*, 1990) and it is generally believed that low concentrations of nutrients may subtly change species composition (Haslam, 1997). Macrophytes are good indicators of changes in flow downstream of reservoirs and for linear physical alterations such as flood works, as well as for the assessment of regulated lakes because they are sensitive to water level fluctuation.

Macrophytes indices account for composition, abundance and presence of sensitive taxa. Alteration in the populations are used to assess overall alteration of water bodies status (Caffrey, 1987; Dennison *et al.*, 1993; Peltre and Leglize, 1992; Haury and Peltre, 1993; Haury *et al.*, 1996; Kelly and Whitton, 1995).

The most diffused type of indices is the 'score indices'. In these methods, a number of indicator taxa/species are given a specific sensitivity/tolerance index. Different 'score indices' vary, depending on the considered taxa and sensitivity/tolerance indices. All of the most important macrophytes indices used in European counties belong to the category of 'score indices'. Among these are 'Plant Score' (Harding, 1981, 1996), 'Trophic Index' (Newbold and Holmes, 1987) and the group of 'Macrophytes Indices', developed by the 'Groupement d'Intérêt Scientifique' (GIS) (Holmes and Whitton, 1977; Haury *et al.*, 1996).

The typical sampling frequency is one or two times a year and samplings usually occur from mid to late summer. The habitats sampled are littoral and deposition areas (e.g. pools) and sampling is normally simple due to a fixed position and a general proximity to the banks.

The main advantages of the macrophytes are that they are easy to sample and identify, except for some genera (e.g. potamogeton), and that they have a low interannual variability; the main disadvantages are due to the lack of standardization and information for comparison. Furthermore, the number of species strongly depends on the stream. As a consequence, there is no commonly accepted European classification of streams based on macrophytes.

Regarding Mediterranean ecoregion and some other European areas, the limiting factors are often the insufficient development of aquatic macrophytes populations and the small number of indicator organisms considered by many methods with respect to the great environmental diversity of riverine ecosystems. Macrophytes indices largely used in many European countries, such as the UK and Ireland, can only be applied in the Mediterranean area in small rivers having laminar flow (Minciardi *et al.*, 2003).

Macrophytes indices are widely used only in the UK, Ireland, France and Austria (Haury and Peltre, 1993; Kelly and Whitton, 1995; Haury *et al.*, 1996; Haslam, 1997). Less extensive use is made in Belgium, Sweden, The Netherlands and Luxembourg.

2.4.3 BIOLOGICAL MONITORING OF RIVERS IN THE WATER FRAMEWORK DIRECTIVE

This Directive is the result of a process of more than five years of discussions and negotiations between a wide range of experts, stakeholders and policy makers. This process has stressed the widespread agreement on key principles of modern water management that form today the foundation of the Water Framework Directive (EU, 2000).

What are the key actions that Member States need to take?

• To identify the individual river basins lying within their national territory and assign them to individual River Basin Districts (RBDs) and identify competent authorities by 2003 (Article 3, Article 24).

- To characterize river basin districts in terms of pressures, impacts and economics of water uses, including a register of protected areas lying within the river basin district by 2004 (Article 5, Article 6, Annex II, Annex III).

- To carry out intercalibration of the surface water ecological quality status assessment systems by 2006 (Annex V).

- To make operational the monitoring networks by 2006 (Article 8).

- Based on sound monitoring and the analysis of the characteristics of the river basin, to identify, by 2009, a programme of measures for achieving the environmental objectives of the Water Framework Directive cost-effectively (Article 11, Annex III).

- To produce and publish River Basin Management Plans (RBMPs) for each RBDm, including the designation of heavily modified water bodies by 2009 (Article 13, Article 4.3).

- To implement water-pricing policies that enhance the sustainability of water resources by 2010 (Article 9).

- To make the measures of the programme operational by 2012 (Article 11).

- To implement the programmes of measures and achieve the environmental objectives by 2015 (Article 4).

Biological monitoring of rivers is a key element of the directive. This provides tools for the characterization of surface water body types, for the establishment of reference conditions and for the assessment of surface water ecological status and potential. Article 8 states that Member States shall ensure the establishment of programmes for the monitoring of water status in order to establish a coherent and comprehensive overview of water status within each river basin district.

The most relevant step forward, with respect to previous environmental regulations, is that biological elements are given the highest priority in the classification of ecological status. Chemical, physico-chemical and hydromorphological elements are considered important, as far as they support the biological elements.

The WFD defines different types of monitoring, depending on their purposes. Member States are asked to set up programmes for surveillance monitoring, operational monitoring and investigative monitoring.

Surveillance monitoring programmes should provide information for supplementing and validating the impact assessment procedures, for the efficient and effective design of future monitoring programmes, the assessment of long-term changes in natural conditions and change resulting from widespread anthropogenic activity. Surveillance monitoring shall be carried out on sufficient surface water bodies to provide an assessment of the overall surface water status within each catchment or sub-catchment within the river basin district. In selecting these bodies, the Member States shall ensure that monitoring is carried out at points where the rate of water flow is significant within the river basin district as a whole, the volume of water present is significant, including large lakes and reservoirs, and where significant bodies of water cross a Member State boundary. Surveillance monitoring shall be carried out for each monitoring site for a period of one year during the period covered by a river basin management plan for parameters indicative of all biological, hydromorphological and physico-chemical quality elements, for priority-list pollutants which are discharged into the river basin or sub-basin and for other pollutants discharged in significant quantities in the river basin. The results of such monitoring shall be reviewed and used, in combination with the impact assessment procedure, to determine the requirements for monitoring programmes in the current and subsequent river basin management plans.

Operational monitoring shall be undertaken in order to establish the status of those bodies identified as being at risk of failing to meet their environmental objectives. It shall also assess any changes in the status of such bodies resulting from the programmes of measures. Operational monitoring shall also be undertaken for those bodies of water into which priority list substances are discharged. The frequency of the monitoring can be reduced where an impact is found not to be significant or the relevant pressure is removed. For bodies at risk from significant point sources, diffuse sources or hydromorphological pressures, sufficient monitoring points within each body will be chosen in order to assess the magnitude and impact of such sources.

Investigative monitoring shall be carried out where the reason for any exceedance is unknown, where surveillance monitoring indicates that the objectives for a body of water are not likely to be achieved and operational monitoring has not already been established, in order to ascertain the causes of a water body or water bodies failing to achieve the environmental objectives. Investigative monitoring should be set up to ascertain the magnitude and impacts of accidental pollution, and to inform the establishment of a programme of measures for the achievement of the environmental objectives and specific measures necessary to remedy the effects of accidental pollution.

For each type of monitoring, the WFD also defines the frequency of monitoring. It also sets standards for monitoring of biological quality elements, such as Macroinvertebrates, Macrophytes, Fish and Diatom sampling. All methods shall conform to the national or international standards, in order to ensure the provision of data of an equivalent scientific quality and comparability.

The WFD, in Annex V.1.3.5, deals with the additional monitoring requirements for biological monitoring of protected areas (EU, 2000; CIS, 2003b).

Protected Areas include bodies of surface water and groundwater used for the abstraction of drinking water, habitat and species protection areas, bathing waters, vulnerable and sensitive zones (European Commision, 1992).

The WFD stresses that protected areas are the starting point for monitoring plans and must achieve higher levels of biological quality before the rest of the watershed, because the main roles of protected areas are the maintenance and improvement of the ecological conditions of targeted biological features (Poiani *et al.*, 1998; Groves *et al.* 2002; Ormerod, 2003). As a matter of fact, protected areas favour the preservation of biodiversity, preventing the loss of species, and have an important ecological role in the landscape due to their function as a biological corridor and a source of faunistic recolonization (Simberloff and Abele, 1982; Soule', 1991; Pressey *et al*. 1993; Prendergast *et al*., 1993). In addition, the protection of a drainage basin, or of a part of it, might increase the functionality of the whole catchment. For example, sustainable management of the buffer strips, such as the riparian zone and the floodplain, could improve the functionality of the system, keeping pollutants from reaching the water bodies through run-off and drainage (Fennessy and Cronk, 1997; Wallace *et al*., 1997). Due to the worldwide decline in the biological quality and functioning of lotic ecosystems (Loh *et al*., 1998; Saunders *et al*., 2002), the interest in better planning of protected areas aimed at freshwater wildlife conservation is increasing (Allan and Flecker, 1993; Ward, 1998; Lowe, 2002). Restoration and sustainable management of protected areas to recuperate the river's natural functions and to reduce sources of pollution are essential for improving biological quality of running waters (Osborne and Kovacic, 1993; Friberg *et al*., 1994; Muotka and Laasonen, 2002). Moreover a important aspect is connected with the protected area size (Mancini *et al*., 2005)

The ecological status of surface waters can be monitored using many indicators or set of indicators. The diversity and differences of biological indicators in Europe depend on many historical, scientific, technical and environmental factors. Indicators have often been developed to describe the status of specific water bodies types, located in particular ecoregions.

Biological characteristics among ecoregions can vary significantly and have usually led to the selection of different taxa for the definition of biological indices. In addition to this, within each ecoregion, river features can be very different depending on catchment area, altitude, geomorphology, geology and flow regime (Illies, 1978). These parameters, defining a water body type, must be taken into account when comparing results of different indicators.

Such considerations make the comparability of different biological methods very challenging.

On the other hand, the WFD should be implemented in a coherent and harmonious manner in order to hit the final target of 'good' ecological status for all European water bodies by 2015.

In order to ensure comparability of different monitoring systems, ANNEX V of the WFD addresses the problems of classification and presentation of ecological status and intercalibration of results.

For the purposes of classification of ecological status, the results of the systems operated by each Member State shall be expressed as ecological quality ratios. These ratios shall represent the relationship between the values of the biological parameters observed for a given body of surface water and the values for these parameters in the reference conditions applicable to that body. The ratio shall be expressed as a numerical value between 'zero' and 'one', with high ecological status represented by values close to 'one' and bad ecological status by values close to 'zero'. Each Member State shall divide the ecological quality ratio scale for their monitoring system for each surface water category into five classes ranging from 'high' to 'bad' ecological status, by assigning a numerical value to each of the boundaries between the classes. The correct identification of boundaries between classes is essential for the ultimate comparability of methods. The value for the boundary between the classes of 'high' and 'good' status, and the value for the boundary between 'good' and 'moderate' status shall be established through an intercalibration exercise. The Joint Research Centre (Institute of Environment and Sustainability) has the responsibility of the leadership and animation of the working group dedicated to prepare guidance for the Intercalibration exercise.

Within the WFD, intercalibration deals with the comparability of biological monitoring results, involving a 'one-off' intercalibration exercise between countries. Intercalibration of individual parameters is difficult because the different Member States may employ different methods for a given biological quality element. Therefore, the biological quality elements should be the level for intercalibration. Biological quality elements will be monitored at those sites included in the intercalibration network, consisting of sites from a range of surface water body types and ecoregions. Whenever possible, monitoring methods of the different Member States sharing the same natural water body should undertake measurements simultaneously, in order to permit a real comparison of the assessment of good status. The main purpose of the intercalibration exercise is to define the boundaries between 'high' and 'good' and between 'good' and 'moderate' status. The achievement of 'good' status is one of the major Environmental Objectives of the Directive and hence its level will determine how many water bodies require measures to be applied to achieve 'good' status. The definition of this boundary is thus a crucial aspect of the implementation of the Directive.

Errors inevitably occur, both in the process of sampling and in the analysis of water samples. Therefore, quality assurance (QA) procedures should be implemented for each monitoring institution, as well as in data collection centres. The aim of an appropriate quality assurance procedure is to quantify and control the errors. Quality assurance procedures may take the form of standardization of sampling and analytical methods, replicate analyses, ionic balance checks on samples and laboratory accreditation schemes. QA measures should encompass all operational facets of a monitoring programme, including field sampling and sample receipt, sample storage and preservation, and laboratory analysis.

2.4.4 CONCLUSIONS

Biological monitoring is a key element in environmental defence programmes and regulations of many European countries and, in the near future, the implementation of the Water Framework Directive.

The analysis of the status of biological monitoring across European countries has shown a strong heterogeneity, both in methodologies and in the implementation level. Benthic macroinvertebrates are the only exception to this rule, since they are used by almost all countries in their monitoring activities, at times together with other indicators. As far as macroinvertebrates are concerned, the major effort should be spent in the harmonization of the methods used across Europe. Fish are not used for routine monitoring in the European Union and unfortunately there is no standard for sampling methodologies yet. Macrophytes have yet to undergo a large development phase throughout all of Europe. Diatoms have been used as bioindicators in different countries of Europe and seem to be a very promising tool for the ecosystem health assessment of rivers.

The proper implementation of the Water Framework Directive asks for two major actions by European Countries: first, the development and standardization of monitoring methodologies and secondly, the application of these methodologies in the current monitoring activities, which requires the training of all operators involved. The Common Implementation Strategy for the WFD supports this activity at the continental level; each country must take charge of the national follow-up.

REFERENCES

AFNOR, 1985. 'Essais des eaux. Détermination de l'indice biologique global (IBG)', AFNOR T. 90–350. AFNOR: Saint-Denis La Plaine CEDEX, France.

AFONOR, 1992. 'Essai des auux: détermination de l'indice biologique global normalisé (IBGN)'. Norme francaise NFT 90–350. AFNOR: Saint-Denis La Place CEDEX, France.

AFNOR, 2000. 'Determination de l'Indice Biologique Diatomeés (IBD)', Norme francaise NFT 90–354. AFNOR: Saint-Denis La Place CEDEX, France.

Alba-Tercedor, J. and Sanchez-Ortega, A., 1988. 'Un metodo rapido y simple para evoluar la calidad biologica de las aguas corrientes basado en el de Hellawell (1978)'. *Limnética*, **4**, 51–56.

Allan, J. D. and Flecker, A. S., 1993. 'Biodiversity conservation in running waters'. *BioScience*, **43**, 32–43.

Armitage, P. D., Moss, D., Wright, J. F. and Furse, M. T., 1983. 'The performance of a new biological water quality scores system based on macroinvertebrates over a wide range of unpolluted running-water sites'. *Water Research*, **17**, 333–347.

Azzollini, R., Betta, G. and Minciardi, M. R., 2003. 'Uso di macrofite acquatiche per il biomonitoraggio delle acque dei canali irrigui: prima applicazioni in un'area del vercellese'. Bollettino del Museo Regionale di Storia Naturale del Piemonte: Torino/talia, pp. 269–292.

Begon, M., Harper, J. L. and Townsend, C. R., 1996. *Ecology: Individuals, Populations and Communities*, 3rd Edition. Blackwell Science Ltd: Cambridge, MA, USA.

Bielli, E., Buffagni, A., Cotta Ramusino, M., Crosa, G., Galli, P., Guzzi, L., Guzzella, L., Minciardi, M. R., Spaggiari, R. and Zoppini, A., 1999. 'Linee guida per la classificazione biologica delle acque correnti superficiali', Manuale UNICHIM 191. Associazione perl 'unificazione nel settore del' industria chimica: Milano, Italia.

Boon, P. J., Davies, B. R. and Petts, G. E. (Eds), 2000. *Global Prespectives on River Conservation: Science, Policy and Practice*. Wiley: Chichester, UK.

Botineau, M. and Ghestem, A., 1995. 'Caractérisation des communautés des macrophytes aquatiques (plantes vasculaires, bryophytes, lichen) en Limousin. Leurs relations avec la qualité de l'eau'. *Acta Botanica Gallica*, **142**, 585–594.

Boyle, T. P. and Fraleigh, Jr, H. D., 2003. 'Natural and anthropogenic factors affecting the structure of the benthic macroinvertebrate community in an effluent-dominated reach of the Santa Cruz River, AZ'. *Ecological Indicators*, **3**, 93–117.

Boyle, T. P., Gmillie, G. M. Anderson, J. C. and Beeson, D. R., 1990. 'A sensitivity analysis of nine diversity and seven similarity indices'. *Journal of the Water Pollution Control Federation*, **62**, 749–762.

Butcher, R. W., 1933. 'Studies on the ecology of rivers – I: on the distribution of macrophytic vegetation in the rivers of Britain'. *Journal of Ecology*, **21**, 58–91.

Caffrey, J. M., 1987. 'Macrophytes as biological indicators of organic pollution in Irish rivers'. In: *Biological Indicators of Pollution, Dublin, 1986*, Richardson, D. H. S. (Ed.). Royal Irish Academy: Dublin, Ireland, pp. 77–87.

Carbiener, R., Trémolières, M. and Muller, S., 1995. 'Végétation des eaux courantes et qualité des eaux: une thèse, des dèbats, des perspectives'. *Acta Botanica Gallica*, **142**, 489–531.

Carpenter, S. R. and Lodge, D. M., 1986. 'Effects of submerged macrophytes on ecosystem processes'. *Aquatic Botany*, **26**, 341–370.

CEMAGREF, 1982. 'Etude des méthodes biologiques quantitatives d'appréciation de la qualité des eaux', Rapport QE, AFB. CEMAGREF: Rhône-Méditerranée-Corse, Lyon, France.

CEN, 2003. Water Quality Guidance Standard for the Routine Sampling and Pre-treatment of Benthic Diatoms from Rivers. EN 13946. CEN: Brussels, Luxembourg.

CEN 2004. Water Quality Guidance Standard for the Identification and Enumeration of Benthic Diatoms Samples from Rivers and their Interpretation. EN 14407. CEN: Brussels, Luxembourg.

Chave, P., 2001. *The EU Water Framework Directive – An Introduction*. IWA Publishing: London, UK.

Chesters, R. K., 1980. 'The 1978 National Testing Exercise', Biological Monitoring Working Party', Technical Memorandum 19. Department of the Environment: London, UK.

CIS, 2003a. 'Overall Approach to the Classification of Ecological Status and Ecological Potential – Common Implementation Strategy', Working Group 2A, Ecological Status. ECOSTAT: Luxembourg.

CIS, 2003b. 'Guidance on Monitoring for the Water Framework Directive. Water Framework Directive – Common Implementation Strategy', Working Group 2.7, Guidance on Monitoring. ECOSTAT: Luxembourg.

CIS, 2004a. 'Guidance on the Intercalibration Process' (draft), Working Group 2.5, Intercalibration. ECOSTAT: Luxembourg.

CIS, 2004b. 'Guidance on a Protocol for Intercalibration of the Surface Water Ecological Quality Assessment Systems in the EU' (draft), Working Group 2.5, Intercalibration. ECOSTAT: Luxembourg.

Commission of the European Communities, 1992. River Water Quality. *Ecological Assessment and Control*, Newman, P. J., Piavaux, M. A. and Sweeting, R. A. (Eds), 1992, EUR 14606 EN-FR. Commission of the European Communities: Brussels, Belgium.

Coste, M. and Ayphassoro, H., 1991. 'Etude de la qualité des eaux du bassin Artois-Oicardie a L'aide des communautés de diatomées bentiques. Application des indices diatomiques au réseau', Rapport Convention d'Etude No. 90 X 3300, 19 June, 1990. Agence de l'Eau Artoisi-Picardie, CHEMAGREF: Bordeaux, France.

Cowx, I. G., 2002. *Factors Influencing Coarse Fish Populations in Rivers: A Literature Review*, Research and Development Publication 18. Environment Agency, Bristol, UK.

Dell'Uomo, A., 1996. 'Assessment of water quality of an Apennine river as a pilot study for diatom monitoring of water courses'. In: *Use of Algae for Monitoring Rivers, II*, Whitton, B. A., Rott, Rott, E., and Friedrich, G. (Eds), Institut für Botanik, Universität Innsbruck: Innsbruck, Austria, pp. 65–72.

Dennison, W. C., Orth, R. J., Moore, K. A., Steveneson, J. C., Carter, V., Kollar, S., Bergstom, P. W. and Batiuk, R. A., 1993. 'Assessing water quality with submerged aquatic vegetation'. *BioScience*, **43**, 86–94.

De Pauw, N. and Vanhooren, G., 1983. 'Method for biological quality assessment of watercourses in Belgium'. *Hydrobiologia*, **100**, 153–168.

De Pauw, N., Ghetti, P. F., Manzini, D. P. and Spaggiari, R., 1992. 'Biological assessment methods for running water'. In: *River Water Quality. Ecological Assessment and Control*, Newman, P. J., Piavaux, M. A. and Sweeting, R. A. (Eds), EUR 14606 EN-FR. Commission of the European Communities: Brussels, Belgium, pp. 217–248.

Descy, J. P., 1979. 'A new approach to water quality estimation using diatoms'. *Nova Hedwigia*, **64**, 305–323.

Descy, J. P. and Coste, M., 1991. 'A test of methods for assessing water quality based on diatoms'. *Verhandlungen internationale Vereinigung Limnologie*, **24**, 2112–2116.

DEV, 1992. 'Biologisch-ökologische Gewässergüteuntersuchung: Bestimmung des Saprobienindex (M2)'. In: *Deutsche Einheitsverfahren zur Wasser-, Abwasser- und Schlammuntersuchung*. VCH: Weinheim, Germany, pp. 1–13.

De Zwart, D., 1995. *Monitoring Water Quality in the Future, Vol. 3, Biomonitoring*. National Institute of Public Health and Environmental Protection (RIVM), Bilthoven, The Netherlands.

Di Dato, P., Mancini, L., Tancioni, L. and Scardi, M. 2004. 'A neural network approach to the prediction of bentic macroinvertebrate fauna composition in rivers'. In: *Modelling Community Structure in Freshwater Ecosystems*, Lek, S., Scardi, M., Vardonscot, P. and Jorgensen, S. (Eds). Springer-Verlag, Berlin, Germany, pp. 147–157.

Ector, L., 1999. 'Diatom ecology and use for river quality assessment'. Deliverable 3. 'Predicting Aquatic Ecosystem Quality using Artificial Neural Networks PAEQANN', Technical Paper, EVK-1999-00125. Available at: http://aquaeco.UPS-TLSE.FR.

EU, 1999. 'Amended proposal for Council Directive establishing a framework for Community action in the field of water policy', DGI, ENV 68 PRO-COOP 46. European Union: Brussels, Belgium.

EU, 2000. 'Directive 2000/60/EC of the European Parliament and the Council of 23 October 2000 establishing a framework for Community action in the field of water policy', *Official Journal of the European Communities*, **L327**, 1–72.

EU, 2001. 'Common Strategy on Implementation of Water Framework Directive'. Strategic Document, 2 May 2001. European Union: Brussels, Belgium.

European Commission, 1992. 'EC Council Directive 92/43/EEC of 21 May 1992 on the conservation of natural habitats and wild fauna and flora'. European Commission: Brussels, Belgium.

European Environmental Agency, 1996a. 'European Freshwater Monitoring Network Design', Topic Report No. 10/1996. European Environmental Agency: Copenhagen, Denmark.

European Environmental Agency, 1996b. 'Requirements for water monitoring', Topic Report No. 1/1996. European Environmental Agency: Copenhagen, Denmark.

Fabri, R. and Leclercq, L., 1984. *Etude Écologique des Rivières du Nord du Massif Ardennais (Belgiques) Flore et Végétation de Diatomées et Physico-Chimie des Eaux. Contexte Mèsologique, Mèthodes, Analyses Physico-Chimiques, Synthése Taxonomique, Écologique et Floristique, Iconographie, Bibliographie*. University of Liége, Station Science: Hautes-Fagnes, Robertville, Belgium.

Fausch, K. D., Lyons, J., Kar, J. R. and Angermeier, P. L., 1990. 'Fish communities as indicators of environmental degradation'. *American Fisheries Society Symposium*, **8**, 123–144.

Fennessy, M. S. and Cronk, J. K., 1997. 'The effectiveness and restoration potential of Riparian Ecotones for the management of nonpoint source pollution, particularly nitrate'. *Critical Reviews in Environmental Science and Technology*, **27**, 285–317.

Friberg, N., Kronvang, B., Svendsen, L. and Hansen, H. O., 1994. 'Restoration of a channelized reach of the River Gelsa°, Denmark: effects on the macroinvertebrate community'. *Aquatic Conservation: Marine and Freshwater Ecosystems*, **4**, 289–296.

Gardeniers, J. J. P. and Tolkamp, H. H., 1976. 'Hydrobiologische kartering, waadering en schade de beekfauna in Achterhoekse beken'. In: *Modelonderzoek 71–74*, Nes, Th. (Ed). Comm. Best. Waterhuist. GLD., Arnhem, The Netherland, pp. 26–29.

Ghetti, P. F., 1997. *Manuale di Applicazione Indice Biotico Esteso (I.B.E.). I Macroinvertebrati nel Controllo Della Qualità Degli Ambienti di Acque Correnti.* Provincia Autonoma di Trento: Trento, Italy.

Grasmuck, N., Haury, J., Leglize, L. and Muller, S., 1993. 'Analyse de la végétation aquatique fixée des cours d'eau lorrains en relation avec les paramètres d'environnement'. *Annals de Limnologie*, **29**, 223–237.

Groves, C. R., Jensen, D. B., Valutis, L. L., Redford, K. H., Shaffer, M. L. and Scott, J. M., 2002. 'Planning for biodiversity conservation: putting conservation science into practice'. *BioScience*, **52**, 499–512.

Harding, J. P. C., 1981. *Macrophytes as Monitors of River Quality in the Southern NWWA Area*, Ref. TS-BS-81-2. Rivers Division, North West Water Authority: UK.

Harding, J. P. C., 1996. 'Use of algae for monitoring rivers in the United Kingdom. Recent developments'. In: *Use of Algae for Monitoring Rivers*, Whitton B. A. and Rott, E. (Eds). Institüt fur Botanik, University of Innsbruck, Innsbruck, Austria, pp. II: 125–133.

Harding, J. P. C. and Kelly, M. G., 1999. 'Recent developments in algal-based monitoring in the United Kingdom'. In: *Use of Algae for Monitoring Rivers III*, Prygiel, B. A., Whitton, B. A. and Bukowska, J. (Eds). Agence de l'Eau Artois Picerdie, France, pp. 26–34.

Haslam, S. M., 1987. *River Plants of Western Europe – The Macrophytic Vegetation of Watercourses of the European Economic Community.* Cambridge University Press: Cambridge, UK.

Haslam, S. M., 1997. *The River Scene.* Cambridge University Press: Cambridge, UK.

Haury, J. and Peltre, M. C., 1993. 'Intérêts et limites des "indices macrophytes" pour qualifier la mésologie et la physicochimie des cours d'eau: examples armoricains, picards et lorrains'. *Annals de Limnologie*, **29**, 239–253.

Haury, J., Peltre, M. C., Muller, S., Tremolieres, M., Barbe, J., Dutartre, A. and Guerlesquin, M., 1996. 'Des indices macrophytes pour estimer la qualite des cours d'eau francais: premières proposition'. *Écologie*, **3**, 233–244.

Hellawell, J. M., 1989. *Biological Indicators of Freshwater Pollution and Environmental Management.* Elsevier Applied Science: London.

Henrikson, L. and Medin, M., 1986. 'Biologisk bedömning av försurningspåverkan på Lelångens tillflöden och grundområden 1986. Aquaekologerna', Rapport till länsstyrelsen i Älvsborgs län.

Holmes, N. T. H. and Whitton, B. A., 1977. 'The macrophytic vegetation of the River Tees in 1975: observed and predicted changes'. *Freshwater Biology*, **7**, 43–60.

Huet, M., 1959. 'Profiles and biology of Western European as related to fish management'. *Transactions of the American Fisheries Society*, **88**, 155–163.

Hughes, R. M. and Gammon, J. R., 1987. 'Longitudinal changes in fish assemblages and water quality in the Willamette River, Oregon'. *Transactions of the American Fisheries Society*, **116**, 196–209.

Illies, J. (Ed.), 1978. *Limnofauna Europea.* Gustav Fischer Verlag: Stuttgart, Germany. Swets & Zeitlinger, Amsterdam: The Netherlands.

Iversen, T. M., Madsen, B. L. and Bogestrand, J., 2000. 'River conservation in the Europe Community , including Scandinavia'. In: *Global Perspectives on River Conservation: Science, Policy and Practice*, Boon, P. J., Davies, B. R. and Petts, G. E. (Eds). Willey: Chichester, UK, pp. 79–103.

Johnson, R. K., 1998. 'Classification of Swedish lakes and rivers using benthic macroinvertebrates'. In: *Bakgrundsrapport 2 till Bedömningsgrunder för Sjöar och Vattendrag – Biologiska Parametrar*, Wiederholm, T. (Ed.), Swedish Environmental Protection Agency

Report 4921. Swedish Environmental Protection Agency: Stockholm, Sweden, pp. 85–166.

Karr, J. R., 1981. 'Assessment of biological integrity using fish communities'. *Fisheries*, **6**, 21–27.

Karr, J. R., 1991. 'Biological integrity : a long-neglected aspect of water resource management'. *Ecological Applications*, **1**, 66–84.

Kelly, M. G. and Whitton, B. A., 1995. Workshop: 'Plants for Monitoring Rivers', Workshop, Durham, UK, 26–27 September, 1994. National Rivers Authority: Durham, UK.

Kolkwitz, R. and Marsson, L., 1902. 'Grundsätze fur die biologische Beurteilung des Wassers nach seiner Flora und Fauna'. *Prüfungsanst. Wasserversorg. Abwasserrein.*, **1**, 32–73.

Kolkwitz, R. and Marsson, M., 1908. 'Ökologie der pflanz-lichen Saprobien'. *Ber. Dtsch. Bot Ges.*, **26**, 505–519.

Kolkwitz, R. and Marsson, M., 1909. 'Ökologie der tierischen Saprobien'. *Int. Rev. Ges. Hydrobiol.*, **2**, 126–152.

Kristensen, P. and Bogestrand, J., 1996. *Surface Water Quality Monitoring*. European Topic Center on Inland Waters. National Environmental Research Institute: Copenhagen, Denmark. (European Environmental Agency, 1996c. 'Surface water quality monitoring', Topic Report No. 2/1996).

Lafaille, P., Lek, S. and Oberdorff, T., 1999. 'Fish ecology, and use for river monitoring in Europe'. Deliverable 3. 'Predicting Aquatic Ecosystem Quality using Artificial Neural Networks (PAEQANN)'. Technical Paper, EVK-1999-00125. Available at: http://aquaeco.UPS-TLSE. FR.

Lammert, M. and Allan, J. D., 1999. 'Assessing biotic integrity of streams: effects of scale in measuring the influence of land use-cover and habitat structure on fish and macroinvertebrates'. *Environmental Management*, **23**, 257–270.

Lange-Bertalot, H., 1979. 'Pollution tolerance of diatoms as a criterion for water quality estimation'. *Nova Hedwigia*, **64**, 285–304.

Leclercq, L. and Maquet, B., 1987. 'Deux nouveaux indices chimique et diatomique de qualité de l'eau courante. Application au Samson et à ses affluents 8 bassin de la Meuse Belge). Comparaison avec d'autres indices chimiques, biocénotiques et diatomiques'. *Inst. Royal Sci. Nat. Belgique. Doc. Trav.*, **38**, 1–113.

Leglize, L., Peltre, M. C., Decloux, J. P., Duval, T., Paris, P. and Zumstein, J. F. 1990. *Caractérisation des Milieux Aquatiques D'Eaux Courantes et Végétation Fixée*. In: Journées internationales d'études sur la lutte contre les mauvaises herbes. 14th Conférence du COLUMA, Versailles, France, 23–24 January, 1990, pp. 237–245.

Lenoir, A. and Coste, M., 1996. 'Development of a practical diatom index of overall water quality applicable to the French National Water Board Network'. In: *Use of Algae for Monitoring Rivers, II*, Whitton, B. A., Rott, E. and Friedrich, G. (Eds), Institut für Botanik, Universität Innsbruck: Innsbruck, Austria, pp. 29–45.

Liebman, H., 1962. *Handbuch der Frischwasser und Abwasserbiologie*, Band I. R. Oldenburg: Munich, Germany.

Loh, J., Randers, J., MacGillivray, J., Kapos, V., Jenkins, M., Groombridge, B. and Cox, N., 1998. *Living Planet Report*. World Wide Fund for Nature: Gland, France.

Lowe, W. H., 2002. 'Landscape-scale spatial population dynamics in human-impacted stream systems'. *Environmental Management*, **30**, 225–233.

Mancini, L., 2003. 'Bioindicators: need for new developments following the application of the decree 152/99 and the taking in of the Water Framework Directive 2000/60/CE'. Atti della 7th Conferenza Nazionale delle Agenzie Ambientali, Milano, Italy.

Mancini, L. and Spaggiari, R., 2000. 'Gli indici biotici nei paesi dell'Unione Europea. Elementi comuni e differenze tra quattro indici biologici: IBE, BBI, BMWP', RIVPACS'. *Biol. Amb.*, **14**, 77–80.

Mancini, L. and Zapponi, G. A., 2002. "Health and ecotoxicology." *Annali dell 'Institute Superiore di 'Sanits*, **38**, 109–155.

Mancini, L., Formichetti, P., Anselmo, A., Tancioni, L., Marchini, S. and Sorace, A., 2005. 'Biological quality of running waters in protected areas: the influence of size and land use'. *Biodiversity and Conservation*, **14**, 351–364.

Metcalfe, J. L., 1989. 'Biological water quality assessment of running waters based on macroinvertebrate communities: history and present status in Europe'. *Environmental Pollution*, **60**, 101–139.

Milner, N. J., Broad, K. and Wyatt, R. J., 1998. 'HABSCORE – applications and future development of related habitat models'. *Aquatic Conservation: Marine and Freshwater Ecosystems*, **8**, 633–644.

Minciardi, M. R., Rossi, G. L., Azzolini, R. and Betta, G., 2003. 'Linee guida per i biomonitoraggio di corsi d'acqua in ambiente alpino'. ENEA: Provincia di Torino, Torino, Italy.

Moog, O. (Ed.), 1995. Fauna Aquatica Austriaca – *A Comprehensive Species Inventory of Austrian Aquatic Organisms with Ecological Notes*, 1st Edition. Bundesministerium für Land- und Forstwirtschaft, Wasserwirtschaftskataster: Vienna, Austria.

Moog, O., Chovanec, A., Hinteregger, J. and Romeret, A., 1999. *Richtlinie zur Bestimmung der Saprobiologischen Gewassergutebeurteilung von Fliebgewasser (Guidelines for the Saprobiological Water Qualty Assessment in Austria)* Bundesministerium fur Land -und Forstwirtschaft, Wasserwirtschaftskataster: Vienna, Austria.

Muotka, T. and Laasonen, P., 2002. 'Ecosystem recovery in restored headwater streams: the role of streams in the agricultural landscape'. *Freshwater Biology*, **27**, 295–306.

Newbold, C. and Holmes, N. T. H., 1987. 'Nature conservation: water quality criteria and plants as water quality monitors'. *Water Pollution Control*, **86**, 345–364.

Newman, P. J., 1988. *Classification of Surface Water Quality*. Review of schemes used in the EC Member States. Heinemann: Oxford, UK.

Nixon, S., 2002. 'Towards a common understanding of the monitoring requirements under the Water Framework Directive', Working Draft, Version 3, 25 January, 2002. European Topic Centre on Water: Brussels, Belgium.

Nixon, S. C., Mainstone, C. P., Iversen, T. M., Ktristensen, P., Jeppensen, E., Friberg, N., Papathanassiou, E., Jensen, A. and Pedersen, F., 1996. 'The harmonised monitoring and classification of ecological quality of surface water in the European Union', WRC Report No. CO 4150. Water Research Corporation: Medmenton, UK.

Noble, R. and Cowx, I., 2002. 'Development of a river – Tyoe classification system (D1) – Compilation and harmonisation of fish species classification (2)', Final Report, FAME, EVK1-CT-2001-00094. (fame.boku.ac.al).

Odderdoff, T., Hugueny, B., Compin, A. and Belkssam, D., 1998. 'Non-interactive fish communities in the coastal streams of North-Western France'. *Journal of Animal Ecology*, **67**, 472–484.

Odderdoff, T., Pont, D., Hugueny, B. and Chessel, D., 2001. 'A probabilist model characterizing fish assemblage of French rivers: a framework for environmental assessment'. *Freshwater Biology*, **46**, 399–415.

Ormerod, S. J., 2003. 'Restoration in applied ecology', Editor's Introduction. *Journal of Applied Ecology*, **40**, 44–50.

Ormerod, S. J., Rundle, S. D., Lloyd, E. C. and Douglas, A. A., 1993. 'The influence of riparian management on the habitat structure and macroinvertebrate communities of upland streams draining plantation forests'. *Journal of Applied Ecology*, **30**, 13–24.

Osborne, L. L. and Kovacic, D. A., 1993. 'Riparian vegetated buffer strips in water-quality restoration and stream management'. *Freshwater Biology*, **29**, 243–258.

Österreichisches Nurmungsinstitut, 1997. 'Guidelines for the eological study and assessment of rivers', ÖNORM M 6232. Österreichisces Normungsinstitut: Vienna, Austria.

Paasavirta, L., 1990. *The Macrozoobenthos Studies in the Upper Part of the Vanajavesi Catchment Area in the Years of 1985 and 1988, with a Comparison to Earlier Data*. Ass. War. Poll. Control (the Kokemaenjoki River) Publication, Vol. **225**, pp. 1–24.

Pantle, R. and Buck, H., 1955. 'Die biologische Überwachung der Gewässer und die Darstellung der Ergebnisse'. *Gas Wasserfach*, **96**, 604.

Peeters, E. T. H. M., Gardeniers, J. J. P. and Tolkamp, H. T., 1994. 'New methods to assess the ecological status of surface waters in the Netherlands. Part 1: Running waters'. *Verhandlungen internationale Vereinigung Limnologie*, **25**, 1914–1916.

Peltre, M. C. and Leglize, L., 1992. *Essais d'Application d'un Protocole Hiérarchisé pour l'Étude des Peuplements Végétaux Aquatiques en Eau Courante*. In: Journées internationales d'ètudes sur la lutte contre les muvaises herbes. 15th Conférence du COLUMA, Versailles, France, 2–4 December, 1992.

Petersen, Jr, R. C., Gíslason, G. M. and Vought, L. B.-M., 1995. *Rivers of the Nordic Countries. River and Stream Ecosystems – Ecosystems of the World*, Vol. 22 (Cushing, C. E., Cummins, K. W. and Minshall, G. W. (Eds)) Elsevier: Amsterdam, The Netherlands.

Poiani, K. A., Baumgartner, J. V., Buttrick, S. C., Green, S. L., Hopkins, E. and Ivey, G. D., 1998. 'A scale-independent site conservation planning framework in The Nature Conservancy'. *Landscape and Urban Planning*, **43**, 143–156.

Premazzi, G. and Chiaudiani, G., 1992. 'Ecological quality of surface waters'. JRC Commission of the European Communities, EUR 1453 EN. European Communities: Brussels, Belgium.

Prendergast, J. R., Quinn, R. M., Lawton, J. H., Eversham, B. C. and Gibbons, D. W., 1993. 'Rare species, the coincidence of diversity hotspots and conservation strategies'. *Nature*, **365**, 335–337.

Pressey, R. L., Humphries, C. J., Margules, C. R., VaneWright, R. I. and Williams, P. H., 1993. 'Beyond opportunism: key principles for systematic reserve selection'. *Trends in Ecology and Evolution*, **8**, 124–128.

Prygiel, J. and Coste, M., 1995. 'Les diatomées et le diagnostic de la qualité des eaux courants continentales: les principales méthodes indicielles'. *Vie Milieu*, **45**, 179–186.

Prygiel, J. and Coste, M., 2000. 'Guide méthodologique pour la mise en ouvre de l'Indice Biologique Diatomées', NF T 90–354. Agencie de l'Eau Artois Picardie: DOUAI CEDEX, France.

Prygiel, J., Leveque, L. and Iserentant, R., 1996. 'Un nouvel indices diatomique pratique pour l'évaluation de la des eaux en réseau de sorveillance'. *Reviews in Science*, **9**, 97–113.

Prygiel, J., Coste, M. and Bukowska, J., 1997. 'Review of the major diatom-based techniques for the quality assessment of continental surface waters – A state of the art in Europe', IIIrd European Symposium 'Use of Algae for Monitoring Rivers', Douai, France, September 29–October 1, 1997.

Resh, V. H., Meyers, M. J. and Hannaford, M. J., 1996. 'Macroinvertebrates as indicators of environmental quality'. In: *Methods in Stream Ecology*, Hauer, F. R. and Lamberti, G. A. (Eds). Academic Press: San Diego, CA, USA, pp. 647–698.

Robach, F., Thiébault, G., Trémolières, M. and Muller, S., 1996. 'A reference system for continental running waters: plant communities as bioindicators of increasing eutrophication in alkaline and acid waters in North-East France'. *Hydrobiologia*, **340**, 67–76.

Round, F. E., 1993. *A Review and Methods for the Use of Epilithic Diatoms for Detecting and Monitoring Changes in River Water Quality. Methods for the Examination of Waters and Associated Materials*. HM Stationary Office: London, UK.

Rumeau, A. and Coste, M., 1988. 'Inititatio à la systématique des Diatomées d'eau douce'. *Bull. Fr. Peche Pisc.*, **309**, 1–69.

Sandin, L. and Hering, D., 2004. 'Comparing macroinvertebrate indices to detect organic pollution across Europe: a contribution to the EC Water Framework Directive Intercalibration'. *Hydrobiologia*, **616**, 55–68.

Sandin, L., Sommerhäuser, M., Stubauer, I., Hering, D. and Johnson, R., 2000. 'The Development and Testing of an Integrated Assessment System for the Ecological Quality of Streams and Rivers throughout Europe using Benthic Macroinvertebrates' Deliverable 1, 31/8/00, entitled: 'Stream assessment methods, stream typology approaches and outlines of a European stream typology a project under the 5th Framework Programme Energy, Environment and Sustainable Development, Key Action 1, Sustainable Management and Quality of Water' (www.aqem.de).

Saunders, D. L., Meeuwig, J. J. and Vincent, A. C. J., 2002. 'Freshwater protected areas: strategies for conservation'. *Conservation Biology*, **16**, 30–41.

Schiefele, S. and Kohmann, F., 1993. *Bioindikation der Trophie in Fliessgewassern*, Forschungsbericht No. 1001 01 504. Umweltforschungsplan des Bundesministers fur Umwelt, Naturschutz und Reaktorsicherheit: Berlin, Germany.

Shannon, C. E. and Weaver, W., 1949. *The Mathematical Theory of Communication*. University of Illinois Press, Urbana, IL, USA.

Siligardi, M., Cappelletti, C., Chierici, M., Ciutti, F., Egaddi, F., Maiolini, B., Mancini, L., Monauni, K., Minciardi, M. R., Rossi, G. L., Sansoni, G., Spaggiari, R. and Zanetti, M., 2000. *Indice di Funzionalità Fluviale IFF*, Manuale di applicazione. ANPA: Rome, Italy.

Simberloff, D. S. and Abele, L. G., 1982. 'Refuge design and island biogeographic theory'. *American Naturalist*, **120**, 41–50.

Skriver, J., Friberg, N. and Kirkegaard, J., 2001. 'Biological assessment of watercourse quality in Denmark: Introduction of the Danish Stream Fauna Index (DSFI) as the official biomonitoring method'. *Verhandlungen internationale Vereinigung Limnologie*, **27**, 9–36.

Sládecek, V., 1973. 'Systems of water quality from the biological point of view'. *Archiv für Hydrobiolie*, Beiheft **7**, 1–217.

Sládecek, V., 1986. Diatoms as indicators of organic pollution. *Acta Hydrochimica et Hydrobiologie*, **14**, 555–566.

Soule', M., 1991. 'Conservation: tactics for a constant crisis'. *Science*, **253**, 744–750.

STOWA, 1992. 'Ecologische Beoordeling en beheer van oppervlaktewater: Beoordelingssysteem voor stromende wateren op basis van macrofauna'. STOWA Report 92-8. STOWA: Utrecht, The Netherlands, pp. 92–98.

Ten Brink, B. J. E., Hosper, S. H. and Colin, F., 1991. 'A quantitative method for description and assessment of ecosystems: the AMOEBA-approach', Deliverable 1, 31/8/00, Contract No. EVK1-CT1999–00027 43. *Marine Pollution Bulletin*, **23**, 265–270.

Thyssen, N., Erlandsen, M., Krovang, B. and Svendsen, L. M., 1990. 'Vandlebsmodeller – biologisk struktur og stofomsaetning', Npo-forskining fra Miljostyrelsen, No. C 10. Danmarks Miljøundersøgelser, Afdeling for Ferskvandsøkologi: Copenhagen, Denmark (in Danish).

Tremp, H. and Kohler, A., 1995. 'The usefulness of a macrophyte monitoring system, exemplified on eutrophication and acidification of running waters'. *Acta Botanica Gallica*, **142**, 541–550.

Tuffery, G. and Verneaux, J., 1968. 'Méthode de détermination de la qualité biologique des eaux courantes. Exploitation codifiée des inventaires de la faune du fond'. Centre National d'Etudes Techniques et de Recherches Technologiques pour L'Agriculture, les Forêts et L'Équipment Rural (CERAFER), Section Pêche et Pisciculture, Ministère de l'Agriculture: Paris, France.

Verdonschot, P. and Dohet, A., 1999. 'Typology, assessment systems and prediction techniques based on macroinvertebrates', Deliverable 3, 'Predicting Aquatic Ecosystem Quality using Artificial Neural Networks (PAEQANN)', Technical paper EVK-1999–00125. Available at: http://AQUAECO.UPS-TLSE.FR.

Verneaux, J., 1976a. 'Biotypologie de l'écosystème "eau corante" la structure typologique'. *Comptes Rendu, Acadamie de Science, Paris*, **282D**, 1663–1666.

Verneaux, J., 1976b. 'Biotypologie de l'écosystème "eau corante" la structure typologique'. *Comptes Rendu, Acadamie de Science, Paris*, **283D**, 1791–1793.

Wallace, J. B., Eggert, S. L., Meyer, J. L. and Webster, J. R., 1997. 'Multiple trophic levels of a forest stream linked to terrestrial litter inputs'. *Science*, **277**, 102–104.

Ward, J. V., 1998. 'Riverine landscapes: biodiversity patterns, disturbance regimes and aquatic conservation'. *Biological Conservation*, **83**, 269–278.

Washington, H. G., 1984. 'Diversity, biotic and similarity indices. A review with special relevance to aquatic ecosystems'. *Water Research*, **18**, 653–694.

Wetzel, R. G., 2001. *Limnology – Lake and River Ecosystems*. Elsevier Science, Academic Press: San Diego, CA, USA.

Wiegleb, G., 1981. 'Récherches Métodologiques sur les Groupments Végétaux des Eaux Courantes'. In: *Proceedings of the 10th Colloquium of the Phyto Society: Végétations Aquatiques*, pp. 69–83.

Woodiwiss, F. S., 1964. 'The biological system of stream classification used by the Trent River Board'. *Chemistry and Industry*, **14**, 443–447.

Wright, J. F., Moss, D., Armitage, P. D. and Furse, M. T., 1984. 'A preliminary classification of running water sites in Great Britain based on macroinvertebrate species and the prediction of community type using environmental data'. *Freshwater Biology*, **14**, 221–256.

Zelinka, M. and Marvan, P., 1961. 'Zur Prazisierung der biologischen Klassifikation des Reinheit fliessernder Gewasser'. *Archiv Für Hydrobiologie*, **57**, 389–407.

2.5

Biomonitoring in North American Rivers: A Comparison of Methods Used for Benthic Macroinvertebrates in Canada and the United States

James L. Carter, Vincent H. Resh, David M. Rosenberg and
Trefor B. Reynoldson

Biological Monitoring of Rivers   Edited by G. Ziglio, M. Siligardi and G. Flaim
© 2006 John Wiley & Sons, Ltd.

2.5.1   INTRODUCTION

Biomonitoring is an essential part of water quality assessment programs in North America. The implementation of biomonitoring in the United States (US) has been more rapid than predicted. Only 26 years ago, Cairns (1979, p.18) noted that: 'Although biological monitoring was recommended by the United States Congress as an effective means to obtain an integrated assessment of the effects of pollutant stress on aquatic biota. . . . .implementation has been almost entirely ignored'. However, today biomonitoring is an integral part of environmental assessments in almost every state and represents a substantial component of several federal water-quality programs in the US. It is also an important component of many Canadian provincial, territorial and national water-quality programs; however, efforts in Mexico are only just beginning.

Even though biomonitoring is extremely widespread in North America, only a few US states have specific numeric criteria codified into law, however, numerous states have narrative criteria (written description without specific numeric goals). Even with the existence of national guidelines (Barbour *et al.*, 1999), substantial differences exist in currently used field and laboratory methods (Carter and Resh, 2001). These differences inhibit identifying of generalized biotic responses to ecosystem impairment by reducing the comparability among studies (Palmer *et al.*, 2004) and severely limit the development of numeric biocriteria (USEPA, 1993) for the analysis of water quality nationally.

Benthic macroinvertebrates are the group of organisms most commonly used in biomonitoring lotic systems. For example, in the United States, the number of programs (state, native American tribes and territories) based on these organisms has increased from 39 in 1989 to 44 in 1999 to 56 in 2001. All states, with the exception of Hawaii, monitor using macroinvertebrates, while Hawaii's macroinvertebrate monitoring program is under development (USEPA, 2002). In many states, fish and algae are also used for biomonitoring. Currently about two-thirds of US state programs monitor fish and about one-third also monitor using algae. A variety of taxonomic groups are also used for monitoring in Canada but there, as in the US, macroinvertebrates are the most popular group of organisms used for biomonitoring in lotic systems (Reynoldson *et al.*, 1999).

In the US, biomonitoring programs have been initiated in different regions of the country at different times; some eastern and midwestern states have long-running programs, while those of the far west tend to be of more recent origin. There has also been a temporal shift in the total number and kinds of biomonitoring studies done in the US. Far fewer studies were done in the 1970s than are done today. These earlier studies tended to emphasize quantitative approaches and were based on experimental designs similar to research-oriented studies conducted at that time. The book by Green (1979) on experimental design was a major influence on the development of these quantitative, point-source-type studies. However, in the 1980s, there was a shift in basic study design to larger-scale studies for evaluating non-point-source impacts. These larger-scale studies were no longer based on fixed-area samples, on-site sample replication and inferential statistical analysis, as were earlier point-source studies. The term 'rapid bioassessment' was applied to these latter types of assessments (Barbour and Gerritsen, 2006; Chapter 3.5 in this text) and the work of Plafkin *et al.* (1989) was the seminal influence that initiated the widespread use of the **R**apid **B**ioassessment **P**rotocols (RBPs) in the US. Currently, a range of approaches based on the multimetric approach (Karr, 1981; Barbour *et al.*, 1999 – an outgrowth of Plafkin *et al.*, 1989) and others based on multivariate approaches (Hawkins *et al.*, 2000) are used within a variety of US state and federal programs.

Almost every Canadian province and territory has biomonitoring programs that use macroinvertebrates (Reynoldson *et al.*, 1999) in evaluations. The largest, most consistently applied program in Canada is the **E**nvironmental **E**ffects **M**onitoring (EEM) program that evaluates the environmental effects of both the pulp and paper industry and the metal mining industry (see below). This nationally standardized program was designed to evaluate point-source impacts (Environment Canada, 2003), is based on study designs similar to those presented by Green (1979), and is similar to point-source studies in the US. Large-scale, regional lotic monitoring, however, is still developing within Canada's provinces and territories (Reynoldson *et al.*, 1999). In addition, trans-boundary programs between Canada and the US are currently under development (Nancy Glozier, Environment Canada, Saskatoon, SK – personal communication).

Biomonitoring is just beginning in Mexico. To date, most biomonitoring in Mexico using macroinvertebrates has been done by foreign investigators (e.g., Weigel *et al.*, 2002). However, data on fish distributions are being collected by various Mexican university researchers and will be amenable to developing a number of biomonitoring approaches, such as the multimetric Index of Biotic Integrity (IBI) (Karr, 1981; John Lyons, Wisconsin DNR, Monona, WI – personal communication). Several studies are being proposed for the future to supplement existing chemical and microbiological investigations as well. In addition, Mumme and Pineda (2002), in an analysis of issues associated with binational water use (principally quantity) along the US–Mexico border, indicate that water use for ecological purposes is beginning to be considered in this region.

In this article, we provide the basic legislative rationale for biomonitoring lotic systems in Canada and the US, and describe the field and laboratory methods used in North America for evaluating water quality using benthic macroinvertebrates. We have the following objectives: (1) to compare and contrast the specific field and laboratory techniques used for biomonitoring by Canadian provinces and territories and US states, and (2) to compare and contrast North America and European Union (EU) biomonitoring field and laboratory methods that use macroinvertebrates. Because programs in Mexico are just beginning, we must restrict evaluations to the two northern countries and leave inclusion of our southern neighbor to a future review when sufficient data are available.

2.5.2 LEGISLATIVE MANDATES

In Canada, biomonitoring is enabled under three Federal Government acts: the 1867 Fisheries Act, the 1970 Clean Water Act, and the 1999 Canadian Environmental Protection Act (http://canada.justice.gc.ca/FTP/EN/Regs/Chap/F/F-14/index.html). However, only the Fisheries Act, which has been amended repeatedly since inception, is used in a practical sense. Sections 36 and 37 of the Fisheries Act are the pertinent parts: Section 36 prohibits the deposition of substances deleterious to fish or fish habitat, and Section 37 requires that such activities, and plans for their mitigation, be reported 'to the Minister'. Legal responsibility for the Fisheries Act rests with Fisheries and Oceans Canada, but administrative responsibility for Sections 36 to 42, which are related to the pollution control provisions of the Act, were assigned to Environment Canada in 1978 (http://www.ec.gc.ca/ele-ale/policies/c_and_e_fisheries_act/main_e.asp?print=1).

The strongest part of the Fisheries Act *vis-à-vis* biomonitoring started with the Pulp and Paper Effluent Regulations, which were developed pursuant to Section 36(4)(b) of the Fisheries Act and have been in effect since 1971. These regulations were designed primarily to control pollution from pulp and paper mills to protect fish and fish habitat. The Regulations were amended in 1992, requiring more stringent limits for some effluent variables and requiring Environment Effects Monitoring (EEM) pursuant to Section 37(1) of the Fisheries Act, and were amended again in 2004 to streamline monitoring and reporting requirements. 'The combination of concerns about inconsistent monitoring, the potential for unseen effects, and the potential for 'modern' effluents to be associated with effects of concern, led to the development of a requirement for environmental effects monitoring to be included in the revised pulp and paper effluent regulations' (Munkittrick *et al.*, 2002, p. 3; see also Walker *et al.*, 2002). In 2002, the Metal Mining Effluent Regulations were added to the EEM program to assess the effectiveness of regulations in mitigating effluents from the metal mining industry (Dumaresq *et al.*, 2002). Plans exist to expand the EEM program to other areas such as aquaculture and municipal waste water. The EEM program also addresses Canadian Environmental Protection Act regulations prescribing limits for dioxins and furans (Pulp and Paper Aquatic Environmental Effects Monitoring Requirements, Annex 1 to EEM/1997/1, Environment Canada, 10/15/97).

According to Canada's 1867 Constitution Act, Canadian provinces have the primary responsibility for management of natural resources, including water, with the exception of Federal lands (e.g. National Parks), the reserves of Canada's First Nations and in the Canadian territories (http://laws.justice.gc.ca/en/const/c1867_e.html#pre). (A brief overview of jurisdiction in relation to water can be found at http://www.ec.gc.ca/water/en/policy/coop/e_juris.htm.) Thus, Canadian provinces (and territories) may also do biomonitoring (Reynoldson *et al.*, 1999). Provincial acts enabling these activities vary from province to province, as does the strength of the actual biomonitoring programs (see below). For example, biomonitoring in the province of British Columbia (BC) has been done under the BC Environmental Management Act and the BC Waste Management Act (to be merged into the BC Environmental Management Act) (I. Sharpe, Water, Land and Air Protection, Environmental Protection Division, BC Government, Smithers, BC – personal communication). In Manitoba, the Environment Act, the Water Protection Act and the Manitoba Water Quality Standards, Objectives and Guidelines provide either a mandate or at least discretionary powers to do biomonitoring (D. Williamson, Department of Water Stewardship, Manitoba Government, Winnipeg, Manitoba – personal communication). Federal and provincial agreements may also exist (e.g. the Canada and Ontario Agreement for Great Lakes Biomonitoring).

An attempt by Reynoldson *et al.* (1999) to start a Canadian national biomonitoring program (**C**anadian **A**quatic **Bi**omonitoring **N**etwork (CABIN)) having a similar scope to already extant national programs in the UK (Wright, 2000) and Australia (Davies, 2000) has, thus far, been unsuccessful. However, the CABIN approach is being used in a regional context, most notably in BC on the Fraser River catchment, in nearshore waters of the Great Lakes and in northern Ontario by an industry–government consortium as part of the EEM program. The Canadian program, like the ones in the UK and Australia, uses the reference condition approach (Reynoldson *et al.*, 1999; see also Reynoldson *et al.*, 2000 and Rosenberg *et al.*, 2000).

In the US, the development of biological monitoring programs is an outgrowth of the Mandates covered in the Clean Water Act (CWA) of 1972 and its amendments. A key statement in the CWA is: 'The objective of this Act is to restore and maintain the chemical, physical, and biological integrity of the Nation's waters' (PL92-500, 101. (a)). The term *biological integrity* has become a catch phrase of biomonitoring, especially with the development of the Index of Biological Integrity by Karr (1981). As in Canada, the primary responsibility for resource management in the US lies with the states. In regards to state and federal responsibilities and the CWA specifically, section 101. (b) states:

> *It is the policy of the Congress to recognize, preserve and protect the primary respon-*
> *sibilities and rights of States to prevent, reduce and eliminate pollution, to plan the*
> *development and use (including restoration, preservation and enhancement) of land*
> *and water resources, and to consult with the [USEPA] Administrator in the exercise of*
> *his authority under this Act. . . . .It is further the policy of the Congress to support and*
> *aid research relating to the prevention, reduction, and elimination of pollution, and to*
> *provide Federal technical services and financial aid to State and interstate agencies and*
> *municipalities in connection with the prevention, reduction and elimination of pollution.*

The CWA contains many key provisions that relate to the implementation of biological monitoring programs in the US. For example, Section 305(b) established a process whereby information on the quality of waters is compiled in a biennial National Water Quality Inventory Report to the US Congress. Biological monitoring provides data for specific reporting requirements including: determination of the status of water resources (i.e. are designated, beneficial and aquatic life uses being met?), evaluation of causes of water resource degradation, descriptions of activities to assess and restore waterways, determination of the effectiveness of pollution control programs, and measurement of the success of water management plans (Barbour *et al.*, 1999).

Other legislative amendments that influence biomonitoring include CWA Section 303(d), which instituted the Total Maximum Daily Load (TMDL) approach that limits the total amount of a pollutant to a system. TMDLs provide a quantitative approach for allocating inputs among pollutant sources. Biological monitoring has been widely used to estimate which water bodies need the establishment of TMDLs. In the first few decades of the CWA, most emphasis was placed on point-source pollution. CWA Section 319 established a national program to assess and control non-point-source pollution as well. Karr (1991) has noted that biomonitoring is the most effective method for examining this type of contamination. Finally, Section 303(c) of the CWA is a collaborative effort in which the US Environmental Protection Agency (USEPA) develops regulations and policies for individual states, which in turn have the primary responsibility for setting and enforcing water quality standards. The above legislative actions are described in detail in Barbour *et al.* (1999, Table 2.1).

It appears that Mexico does not have a detailed legislative arrangement that serves as the basis for biomonitoring programs (John Lyons, Doris Vidal, Southern California Coastal Water Research Project, Westminster, CA – personal communications).

2.5.3   ANALYSIS APPROACH

One step toward developing a uniformly applied set of methods for collecting and processing benthic macroinvertebrates for general biomonitoring is documenting and then comparing and contrasting current commonly used methods. In this review, Canada and the US are compared first, and then the similarities and differences between North American and EU biomonitoring programs are discussed. Finally, suggestions are presented for establishing a minimum-effort sample, along with how changes could affect ongoing programs.

The questionnaire, which is available in its entirety in the Appendix of Carter and Resh (2001), is the basis for comparing programs in Canada and the US. The questionnaire was originally sent to representatives of state agencies in the US; it was subsequently sent to government and/or university representatives developing or conducting biological assessments for each Canadian province or territory. More than one person was usually contacted and multiple responses were tallied per province/territory when appropriate. Several participants were employees of Environment Canada (EC), which provided a more national perspective.

The questionnaire consisted of four parts (Carter and Resh, 2001). The first part addressed procedures associated with sample collection. Questions ranged from what device was used for collecting macroinvertebrates, to specification of the length of stream that defined the sampled reach. The second part addressed the processing of samples in the field. Questions ranged from whether the sample was sorted only by eye in the field, to the sample-preservation techniques used. The third part addressed laboratory procedures. Questions ranged from whether and how samples were subsampled, to the level of taxonomic identification used. The fourth part of the questionnaire evaluated data quality. Questions ranged from whether Quality Assurance/Quality Control procedures were used, to whether and how reference collections were maintained.

The data used for the analyses represent the responses from 48 states and the District of Columbia in the US, and 12 provinces and territories in Canada. Analyses were limited to US state and Canadian provincial and territorial biomonitoring programs because all jurisdictions have similar water quality information needs. US state programs were contacted based on the list of participants in Davis *et al.* (1996) and Canadian programs were contacted based on a list provided by two of the present authors (DMR and TBR). Each respondent was requested to submit one questionnaire for each unique combination of field-collecting method, and field and/or laboratory processing method. Results were based on approximately 116 responses. Percentages were calculated based on the number of responses to each specific question. Some questions eliminated the need to answer other questions, and so not all questions were answered on each questionnaire. In addition, some respondents did not answer all of the possible questions. Therefore, the total number of responses varied among questions. Percentages in most tables and figures were based on the 'within-country' percentage because of the large disparity between the number of respondents between Canada and the US. Statistical analyses and percentages were calculated using STATISTICA (1999, STATISTICA for Windows, StatSoft Inc., Tulsa, OK, USA).

2.5.4   RESULTS

2.5.4.1   Sample collection

Many procedures have been developed for collecting benthic macroinvertebrates for biomonitoring studies (Merritt and Cummins, 1996). Some studies have evaluated certain aspects of these procedures (Williams and Feltmate, 1992), while other studies have evaluated the comparability of collections made using different protocols (Gurtz and Muir, 1994). However, limited research has addressed the specific effects of various procedures on detecting impacts to the benthos.

Type of sampler used

The sampling device selected for collecting macroinvertebrates may influence the number and type of organisms collected. Although a variety of samplers is used by both Canadian and US programs (Figure 2.5.1), most use a kick-type sampler. In

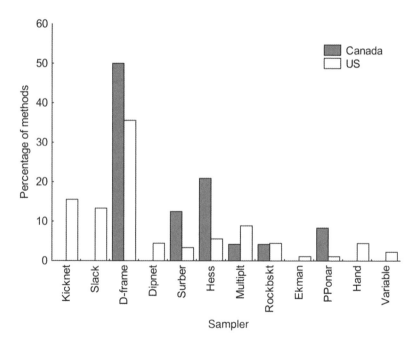

Figure 2.5.1   Types of sampler used in the field. Percentages are based on the total number of responses within each country: Multiplt, multiplate artificial substrate (e.g. Hester–Dendy); Rockbskt, rock basket artificial substrate; PPonar, petit Ponar dredge; Hand, visual searching and collecting by hand; Variable, responses that listed more than one method per response

Canada, the D-frame sampler (~ 0.3 m wide) is used by 50 % of the programs. In the US, a D-frame is used by only 35.6 % of the programs; however, similar types of kick-type samplers, such as the 0.5 m wide Slack-sampler and the 1 m wide seine, bring the total use of this semi-quantitative sampling method to 64.3 %. Carter and Resh (2001) found that the D-frame net was the only sampler used uniformly among all 10 EPA regions in the US. Similarly, all but one Canadian province reported using the D-frame net.

Surber and Hess samplers are used far more frequently in Canada (33.3 %) than in the US (8.8 %). The more frequent use of fixed-area samplers in Canada is likely a function of the more quantitative design of the spatially limited EEM programs compared to the generally larger-scale RBP-type programs in the US. The use of other sampling devices such as artificial substrates (Hester–Dendy samplers or rock baskets), and grabs and dredges, is relative rare; however, they make up the principal sampling device in some US state programs (e.g. Ohio, Maine).

Mesh size

The size of the mesh used on a sampling device in lotic studies can influence the number of and types of organisms collected more than many other methodological factors (Slack *et al.*, 1991). The tables in Resh (1979) provide general references about this bias. Mesh sizes reported ranged almost over an order of magnitude from 150 to 1200 µm. Because of this extreme variability, mesh sizes were grouped to the nearest 100 µm (Figure 2.5.2). Many (35 %) of Canadian programs use a mesh size that is < 250 µm, which is extremely small by US standards. However, most (39 %) Canadian programs use a mesh size in the 500 to 600 µm range. US programs are far more uniform in the mesh size used on sampling devices and most (> 80 %) programs use a mesh size in the 500 to 600 µm range.

Compositing of samples

Historically, samples acquired for lotic studies consisted of collections made from a well-defined area (e.g. ~ 0.9 m^2 when using a Surber sampler) sampled to a depth of ~ 10 cm, with all invertebrates collected being sorted and identified. These individual collections were often considered replicates and used for estimating error; however, relative to most study designs these individual samples were actually 'pseudo-replicates' (Hurlbert, 1984). In the last few decades, a more qualitative (often termed 'semi-quantitative') approach has been instituted and many programs now represent the fauna at a site by forming a composite sample, which consists of combining a number of individual non-contiguous collections. The individual collections used to form the composite are often collected with known effort (area or time). The advantages in collecting composited samples (= composites) include

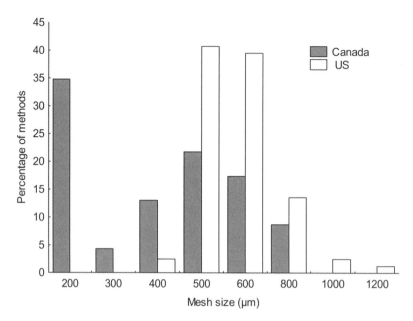

Figure 2.5.2 Mesh sizes for field collecting macroinvertebrates. Responses were categorized to the closest 100 μm, and percentages are based on the total number of responses within each country

a reduction in the variability of organism densities often found in spatially small benthic samples (Hornig and Pollard, 1978) and a more thorough representation of the species present within a reach.

Significant differences exist between Canada and the US in the use of composites. In Canada, 50 % of the programs report collecting composites, whereas in the US almost 70 % of the programs collect composites. Differences in basic study design between the two countries, with Canada often using a more traditional sampling design (i.e. with 'within-site' replication) and the US more often using the more qualitative RBPs, likely leads to these differences.

Area sampled

Whether or not composites are formed greatly influences the total area of a sample, as does the per unit area sampled by the sampler. The mean area sampled by Canadian programs is 0.78 m^2, whereas the mean area sampled by US programs is 1.7 m^2. Although there is considerable overlap in the per sample area, all but one Canadian program samples a total area $\leq$ 1 m^2. Conversely, 20 programs in the US collect macroinvertebrates from an area that is $\geq$ 1 m^2. The recommendation in Plafkin *et al.* (1989) to use the 1 m wide seine as a sampling device probably influences the total area sampled in many RBP-based assessments in the US.

Length of stream and habitats sampled

As stated above, composites are often collected to 'average-out' the effects of habitat heterogeneity on estimating the species composition at a site. Composites may be formed by combining multiple collections from a single habitat type (e.g. within or among riffles) or multiple collections from a variety of habitat types (riffles, pools, margins, etc.). Composites may be collected over a relatively short section of stream, such as a single riffle, or over a very long section of stream. The length over which the collection is made can be defined based on the geomorphology of the stream, such as a single pool-riffle sequence, multiple channel widths, etc., or simply as a fixed length of stream.

Both the length of stream and the method used to determine reach length varied among programs (Figure 2.5.3), although there was relatively little difference between Canada and the US in this regard. Canadian and US programs most often defined the sample reach as either a single pool-riffle sequence (27 % Canada, cf. 37 % US) or a fixed length of stream (36 %, cf. 32 %). Sampling over multiple pool-riffle sequences and/or multiple channel widths was also used by both countries.

There was even greater similarity between the two countries in the habitats sampled. By far the most frequently sampled habitat was fast water, which is represented

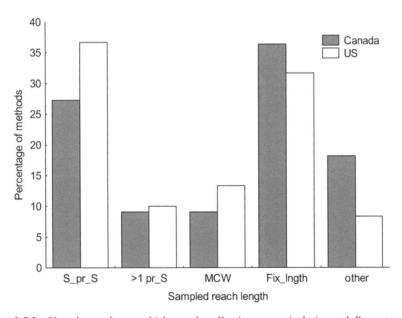

Figure 2.5.3 How the reach over which sample collecting occurs is designated. Percentages are based on the total number of responses within each country: S_pr_S, single pool-riffle sequence; >1 pr_S, more than 1 pool-riffle sequence; MCW, multiples of the channel width; Fix_lngth, fixed length of stream

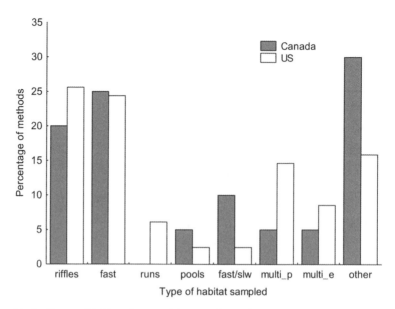

Figure 2.5.4 Types of habitat from which macroinvertebrates are collected. Percentages are based on the total number of responses within each country: fast/slw, fast and slow water; multi_p, multiple habitats sampled based on the proportion of each type of habitat present in the reach; multi_e, multiple habitats sampled based on equal effort per type of habitat present in the reach

by riffles and runs (Figure 2.5.4). The least frequently sampled was pool habitat. The sampling of multiple habitats, whether in proportion to the presences of each habitat type or with equal effort, is done by just less than 30 % of the programs.

Placement of the sampling device

In earlier quantitative lotic studies, the method most often used to locate a sampling device was by using random numbers. The intent was to develop an unbiased estimate of the variable of interest – often density. However, recent arguments suggest that the goal of sampling should be to acquire the most representative sample possible for a designated stream section (Underwood, 1997, p. 138, Section 3.11). Our results show that collecting randomly located samples is rarely done by modern assessment programs. Only about 5 % of programs in Canada and the US still use this technique (Figure 2.5.5).

However, other differences exist between the two countries in the method chosen to place the sampling device. Almost 50 % of Canadian programs place the sampling device in a systematic fashion, whereas only 10 % of US programs use this method. Conversely, 70 % of US programs place the sampling device using expert opinion but only 35 % of Canadian programs rely on expert opinion to locate the sampler.

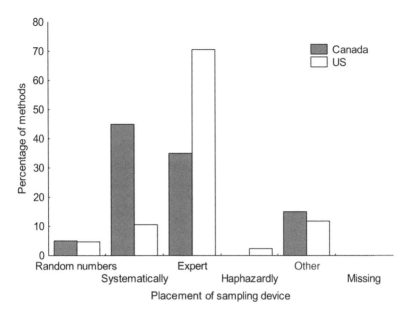

Figure 2.5.5   Methods used for the placement of the sampling device in the stream; percentages
are based on the total number of responses within each country

Replication

In traditional point-source study designs, replicate samples are often collected at
both the reference site and putatively impaired site. Although this is a common
study design, these replicates are often considered 'pseudo-replicates' (Hurlbert,
1984; Underwood, 1997). On-site replication is far less common in large-scale studies, because the site itself is likely considered a sample unit. Although the RBPs
suggest collecting replicates (duplicates) from at least 10 % of the sites sampled,
these are collected more for quality control/quality assurance checks (Barbour *et al.*,
1999) than for an estimate of pure error. Canadian programs collect replicates far
more frequently than US programs (Table 2.5.1). Approximately 75 % of Canadian
programs collect replicates at 100 % of the sites. In the US, 50 % of the programs
collect replicates at only 10 % of the sites.

2.5.4.2   Field processing

Benthic samples consist of living (e.g. macroinvertebrates) and non-living (e.g. mineral and non-living organic particles) materials. The separation of these, if done
inaccurately, may lead to bias in the results obtained and in interpretations about
the status of a site. Various types of procedures are used for pre-processing samples
in the field. To our knowledge, no studies have addressed the effects of the various
procedures used.

Table 2.5.1 The number and percentage of programs in Canada and the US that (1) collect replicates, (2) the number of replicates collected per site, and (3) the percentage of sites at which replicates are collected

Aspect	Canada		US		Total[a]	
	No.	Percentage	No.	Percentage	No.	Percentage
Replication						
Yes	*13*	65.0	*46*	56.1	*59*	50.9
No	*7*	35.0	*36*	43.9	*43*	37.1
Total	*20*	100.0	*82*	100.0	*14*$_M$	12.1$_M$
Number of replicates per site						
1	*0*	0	*1*	2.1	*1*	0.9
2	*0*	0	*31*	66.0	*31*	26.7
3	*6*	46.2	*14*	29.8	*20*	17.2
4	*1*	7.7	*0*	0	*1*	0.9
5	*6*	46.2	*1*	2.1	*7*	6.0
Total	*13*	100.0	*47*	100.0	*56*$_M$	48.3$_M$
Percentage of sites at which replicates are collected						
2.5	*0*	0	*3*	6.5	*3*	2.6
5	*0*	0	*4*	8.7	*4*	3.4
8	*0*	0	*1*	2.2	*1*	0.9
10	*3*	25.0	*21*	45.7	*24*	20.7
20	*0*	0	*1*	2.2	*1*	0.9
55	*0*	0	*1*	2.2	*1*	0.9
100	*9*	75.0	*15*	32.6	*24*	20.7
Total	*12*	100.0	*46*	100.0	*58*$_M$	50.0$_M$

[a] M, missing.

Field sorting

One of the most time-consuming tasks associated with using macroinvertebrates in lotic assessments is sorting animals from the organic and inorganic matrix collected (Resh *et al.*, 1985). Some programs only sort organisms in the field, and early RBPs advocated field sorting for low-intensity studies, particularly for reconnaissance-type studies (Plafkin *et al.*, 1989). On average, approximately 20 % of the programs in Canada and the US sort organisms in the field (Table 2.5.2). Clearly, field sorting is less thorough than laboratory sorting and is dependent on the knowledge and skill of the sorter as well as other factors such as weather conditions, etc.

Preliminary field processing

Many programs collect large, composited samples and most samples are sorted in the laboratory, and so well over 50 % of the programs remove extraneous organic and inorganic debris while still in the field. Canadian programs, in general, field-process slightly less frequently than US programs, most likely because the total sampled

Table 2.5.2 The number and percentage of programs in Canada and the US that use various techniques for processing benthic macroinvertebrate collections while still in the field

Aspect	Canada		US		Total	
	No.	Percentage[a]	No.	Percentage[a]	No.[a,b]	Percentage[a,b]
Only sort by eye in the field						
Yes	*4*	17.4	*19*	22.9	*23*	19.8
No	*19*	82.6	*64*	77.1	*83*	71.6
		100.0$_T$		100.0$_T$	*10*$_M$	8.6$_M$
					116$_T$	100.0$_T$
Sample 'cleaning'						
None	*7*	30.4	*18*	27.7	*25*	21.6
Elutriate and clean of debris	*3*	13.0	*13*	20.0	*16*	13.8
Clean of debris	*9*	39.1	*32*	49.2	*41*	35.3
Other	*4*	17.4	*2*	3.1	*6*	5.2
		100.0$_T$		100.0$_T$	*28*$_M$	24.1$_M$
					116$_T$	100.0$_T$
Field Sieving						
Yes	*7*	33.3	*38*	56.7	*45*	38.8
No	*14*	66.7	*29*	43.3	*43*	37.1
		100.0$_T$		100.0$_T$	*28*$_M$	24.1$_M$
						100.0$_T$
Field subsampling						
Yes	*0*	0.0	*7*	10.6	*7*	6.0
No	*20*	100.0$_T$	*58*	87.9	*78*	67.2
No response			*1*	1.5	*1*	0.9
					30$_M$	25.9$_M$
					116$_T$	100.0$_T$

[a] T, total.
[b] M, missing.

area is generally less than half that sampled by US programs and because Canadian programs frequently collect discrete, small ($\sim$ 0.1 m^2) samples, which contain much less debris than larger samples.

US programs tend to sieve (wash) samples about twice as often in the field, again likely because a larger, more debris-filled sample is collected. In addition, although no Canadian programs subsample samples in the field, about 10 % of the US programs subsample large samples prior to returning to the laboratory (Table 2.5.2).

Sample preservation

The choice of preservative has traditionally been a function of the type(s) of measurements made (e.g. counts of individual taxa, measurements of body lengths, biomass estimates) and the taxa of interest. However, preservatives must be applicable to a variety of taxa, and choices are often dictated by agency health and safety officers, shipping restrictions and the cost of waste disposal in large programs. Canadian programs most often (74 %) use formalin, whereas in the US it is used by only 18 % of the programs. The preservative most often used by US state programs (64 %) is ethyl alcohol in concentrations ranging from 70 to 95 % with most programs using either 95 % or 70 %. Some (21 %) of Canadian programs also use ethyl alcohol in the field to preserve samples. However, even though different preservatives are used, virtually all programs rely on taxa identification and enumeration for their assessments.

2.5.4.3 Laboratory processing

Much has been written (Barbour *et al.*, 1996; Courtemanch, 1996; Vinson and Hawkins, 1996; Cao *et al.*, 1998, 2002) about the effects of various laboratory procedures on the evaluation of benthic macroinvertebrates. However, very little research has been done to specifically address the effects of most of the individual procedures used in the laboratory for processing benthic samples (but see Moulton *et al.*, 2000). This aspect of sample processing can be quite costly and can significantly affect the comparability of data among programs. Therefore, an understanding of the effects of these procedures should be an important aspect of biomonitoring research.

Subsampling

The density (individuals per unit area or volume) of macroinvertebrate assemblages can be extremely variable. Densities can range from 0 to $\sim 100\,000$ individuals m^{-2} (JLC, unpublished National Water-Quality Assessment program data). Therefore, large, composited samples often require subsampling to reduce the effort and cost necessary for their analysis. Canadian programs tend to composite less than US programs and the overall area sampled tends to be smaller; consequently, only about 67 % of Canadian programs subsample. US programs generally collect samples that are over twice the size of Canadian samples, and therefore, most (96 %) of US programs subsample. In addition, RBPs, which are commonly used by US programs, advocate subsampling by sorting a fixed count (a pre-determined number of individuals) of organisms from each sample (see below).

  All Canadian programs and 95 % of US programs that subsample in the laboratory, subsample randomly. However, many US programs also sort large–rare (see below) organisms from the sample in the laboratory. By far the most commonly used subsampling device is a gridded tray or screen. These devices range from simple acrylic frames with gridded screen bottoms (Moulton *et al.*, 2000), to Caton subsamplers (Caton, 1991), to the Marchant box (Marchant, 1989). To our knowledge, only the efficacy of the Marchant box has been tested (Marchant, 1989). Other less

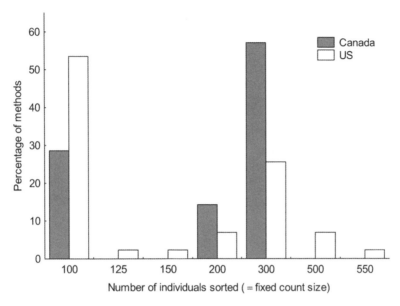

Figure 2.5.6 The target values listed for the fixed-count type of subsampling; percentages are based on the total number of responses within each country

commonly used devices include the Imhoff cone (Wrona *et al.*, 1982) and the Folsom plankton splitter.

Subsamples can be formed by using a number of techniques. Relatively few programs report subsampling by sorting a fixed percentage of the total sample. However, when this method is used, programs generally split (50 %) or quarter (25 %) the original sample. One program reported taking a variable percentage of the total sample, which was dependent on the perceived number of individuals in the subsample.

By far the most often-used subsampling technique is to sort a fixed-count of organisms. Therefore, the percentage of the original sample that is sorted is determined by the number of individual organisms it contains. This method is advocated by the RBPs (Barbour *et al.*, 1999). A number of similarities exist between Canada and the US. Most Canadian programs that subsample, sort a fixed-count of 300 organisms; however, the next most commonly sorted fixed-count size is 100 organisms (Figure 2.5.6). In contrast, most (23) US programs sort 100 organisms, with 300 organisms the next most frequently chosen fixed-count size. US programs have greater variability in the number of organisms sorted than do Canadian programs. In the US, fixed-count sizes sorted by different programs range from 100 to > 500 organisms.

Large–rare taxa

The separation of larger, more obvious organisms from benthic samples was first described by Cuffney *et al.* (1993). However, that procedure involved sorting and placing the few larger, often rare, individuals in a separate container while still in the field to better preserve their characteristics for later identification in the laboratory and/or to keep these larger organisms from destroying smaller, more fragile organisms when the sample was transported to the laboratory. Larger organisms are often designated as 'large–rare' organisms. More recently, sorting large–rare organisms in the laboratory has been advocated (Moulton *et al.*, 2000). The rationale lies in the observation that when large benthic samples are subsampled in the laboratory, there are inevitably organisms that do not occur in the subsample because of their rarity. Some researchers surmise that these organisms, particularly when large, are ecologically important and have diagnostic meaning; consequently, they should be represented in the sample even if they are not part of a random subsample. Clearly, small organisms may also have diagnostic value; however, consistently sorting them from a large sample is difficult.

There is no standard or universally accepted method for sorting large–rare organisms. Some programs sort large–rare organisms before the sample is subsampled and some programs sort large–rare organisms from the remainder of the sample after the subsample is removed (Figure 2.5.7). In general, many Canadian programs and one US program sieve samples in the laboratory by using a nest of sieves. Larger, often less abundant organisms are separated from smaller, very abundant organisms using 1000–2000 µm mesh sieves. Generally, all of the larger organisms are sorted and identified; the portion of the sample that passes these larger sieves is then subsampled

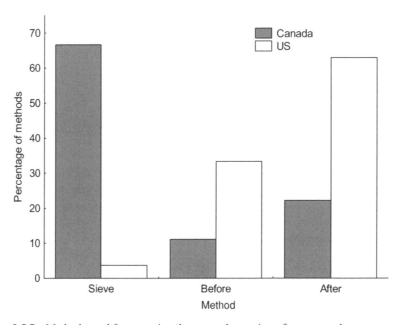

Figure 2.5.7 Methods used for removing 'large–rare' organisms from a sample; percentages are based on the total number of responses within each country

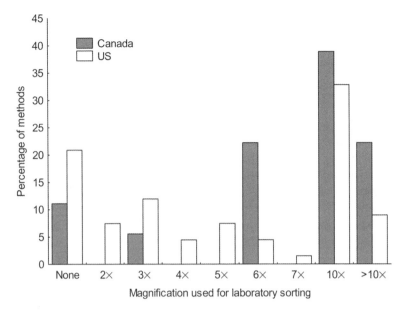

Figure 2.5.8 Magnifications used in the laboratory for sorting macroinvertebrates from a sample or subsample; percentages are based on the total number of responses within each country

using standard procedures. In contrast, most US programs, which collect very large samples, tend to subsample first and then sort large–rare organisms from the remainder of the sample.

Magnification used for sorting

The magnification at which a sample is sorted can have dramatic effects on the number of organisms acquired. The magnification most frequently used for sorting by Canadian and US programs is 10× (Figure 2.5.8). Only about 17 % of Canadian programs sort at a magnification ≤ 5×, whereas over half (52 %) of US programs sort at these lower magnifications.

Taxonomic level generally used

The level of taxonomy needed in biomonitoring has been widely debated (e.g. Bailey *et al.*, 2001; Lenat and Resh, 2001). In general, Canadian programs identify organisms to a higher taxonomic level (i.e. to a less resolved level such as family) than US programs. Approximately 35 % of the Canadian programs limit their identifications to the family level or higher, whereas only 17 % of US programs limit identifications to those levels. However, both countries report that organisms are most often identified to genus.

2.5.4.4   Data quality

Biomonitoring programs can potentially last for decades. Therefore, the maintenance of data quality is critical for evaluating long-term changes in the benthos and confidently detecting differences among locations.

Quality assurance/quality control (QA/QC)

Both Canadian and US programs report having some form of QA/QC for sample collecting and processing (94 and 97 %, respectively). In addition, 76 % of Canadian programs and 97 % of US programs have written 'standard operation procedures'. However, fewer Canadian programs (56 %) compared to 70 % of US programs have specific numerical criteria that must be met associated with their QA/QC plans.

Sample collecting and processing

Major differences exist between Canada and the US in the personnel that both collect and process the samples collected for biomonitoring. Over 90 % of US programs report that the samples are collected by the individual in charge of the program, whereas in only 61 % of the Canadian programs are samples collected solely by the person in charge. In only 5.5 % of Canadian and 1.3 % of US programs is sample collection done by contractors alone.

In contrast to sample collection, sample processing is frequently done only by contractors that are independent of the agency. Canadian programs report that 39 % of their samples are processed solely by contractors but only 21 % of US programs use only contractors for sample processing. Possibly because Canadian programs more often have their samples processed by contractors, 82 % report regular use of taxonomic specialists, whereas only 71 % of US programs use them.

Sample retention and reference collections

A variety of responses were received regarding how long samples are retained. Almost 50 % of programs in both countries report keeping samples indefinitely. The two next longest periods that samples are retained are 1 (Canadian = 7 % and US = 8 %) and 5 years (Canadian = 11 % and US = 10 %).

Almost two-thirds of programs in each country maintain reference collections. The location at which collections are maintained can be classified into three groups: in house, at the contractor's facility or in a museum. Canadian programs keep 50 % in house, 25 % at the contractor's facility, and 25 % in either local (university or provincial) or national museums. In contrast, 88 % of US programs keep reference collections in house, possibly because more programs in the US process their own samples. There, only 3 % are kept by the contractor and 10 % in museums.

2.5.4.5   Extent of monitoring programs

The number of samples collected differed widely between Canada and the US and within each country as well. In separating the number of samples into a minimum collected per year and a maximum collected per year per program, Canadian programs collected approximately one half (86) the number of samples on an annual basis as did US programs (178). The minimum number of samples collected per year was only 10, while the maximum reported by the US was 3600.

2.5.5   CONCLUSIONS

Foremost, the methods used for collecting and processing macroinvertebrate samples should provide data that adequately support the intended purpose of the biomonitoring program, although this can be extremely complicated to determine in environmental assessment programs (see Downes *et al.*, 2002). Consequently, when programs differ in the fundamental questions being addressed, differences in the methods used for collecting and processing samples of macroinvertebrates are expected. Furthermore, even when programs have similar goals, differences can develop because of individual preferences, perceived effects of specific steps and changes in personnel. Additionally, the methods used for collecting and processing benthic invertebrate samples for biomonitoring are composed of many steps and, at each step, differences among programs can develop.

Fundamental differences in program design currently exist between Canadian programs that are influenced by the EEM point-source studies and the broader scale, non-point-source studies that follow RBPs in the US. However, in general, when biomonitoring is undertaken to identify larger-scale, non-point-source situations in Canada, there are greater similarities than differences in the methods used between the two countries.

Over the last two to three decades, the methods used for evaluating the health of aquatic systems have advanced substantially (Karr, 1981; Wright *et al.*, 1984; Plafkin *et al.*, 1989; Barbour *et al.*, 1999) and numerous countries, states and provinces have well-designed monitoring programs. Over this same period, a better understanding of some of the inherent limitations in assessing biota for water-quality monitoring has also developed (Barbour *et al.*, 1996; Courtemanch, 1996; Vinson and Hawkins, 1996; Cao *et al.*, 1998, 2002). How do the biomonitoring methods used in Canadian and US programs that we have described in this chapter differ from those currently being developed and tested by the European Union (EU)? We compare North America (NA) methods to those biomonitoring programs that are being

developed by the EU for the Water Framework Directive (WFD) (http://europa.eu.int/
comm/environment/water/water-framework/index_en.html).

One of the first large projects to address the development of techniques for
assessing European streams and rivers using benthic macroinvertebrates was the
AQEM (The Development and Testing of an Integrated Assessment System for
the Ecological Quality of Streams and Rivers throughout Europe using Benthic
Macroinvertebrates) (Hering *et al.*, 2004). Currently, the STAR (Standardisation
of River Classification) effort is addressing calibration and standardization among
methods for determining ecological quality classifications. The field and laboratory
methods to which NA methods are compared are those protocols with the greatest
amount of methodological information listed on the STAR web site (http://www.eu-
star.at/frameset.htm) under the 'Protocol' link.

Just as there are many different methods used in NA, there also are different
methods used throughout the EU. However, there are numerous similarities in the
field methods employed on both continents. For example, the device most frequently
used for benthic sampling in both the US and Europe is a 30 cm wide kicknet fitted
with a mesh of approximately 500 μm. In general, the device is used to collect
composited samples. At least three different methods to standardize effort are used:
collecting from a known area, collecting for a fixed amount of time and collecting by
using a specific number of 'jabs' of the sampler. Most programs collect from a predetermined area, often between 1 to 2 m^2. Other programs, e.g. RIVPACS, (**R**iver
Invertebrate **P**rediction and **C**lassification **S**ystem) and some US state programs
limit the amount of time allotted for sample collection (e.g. 3 min). Some methods
(e.g. AQEM) distribute a fixed number of jabs (e.g. 20) of the net, trying to sample
from 0.0625 m^2 per jab to form a collection that represents a total area of 1.25 m^2.
Once the composited collection is formed, most programs clean the collection of
both extraneous inorganic and organic debris and then preserve it in the field with
5–10 % formalin or 70–80 % alcohol.

However, differences in defining the sampling reach and habitat to be sampled
vary greatly among programs in NA and Europe. In NA, some programs limit their
collection to a single riffle, while others collect from multiple habitats over a length
of up to 40 channel widths. The length over which the collection is made is often
determined by the geomorphology of the channel. Similarly, the reach length over
which collecting occurs in the AQEM protocol is also geomorphologically defined
but is a function of broadly categorized basin area. Unlike most methods, RIVPACS
protocols state that, in general, collecting should occur over a fixed length of stream,
and unlike other protocols, the length of stream over which collecting occurs tends
to decrease as stream width increases.

Most programs in NA and the EU process samples in the laboratory. However,
whether samples are sorted by eye or at higher magnification significantly influences
other aspects of sample processing. In NA, about 40 % of US programs and over
80 % of Canadian programs sort using approximately 10× or higher magnification.
Only about 20 % of US programs sort macroinvertebrates by eye. Many Canadian
programs collect using much finer mesh sizes than most other countries, and often

sort at higher magnification. Both the AQEM and RIVPACS protocols require sorting by eye – similar to many US programs (Carter and Resh, 2001); however, sorting at higher magnification has been introduced in the AQEM protocol as an option and is 'allowed if necessary'.

All programs subsample large collections when it would be prohibitively expensive to process them in their entirety. The AQEM protocol and most US programs subsample by randomly sorting a fixed count of organisms. In US and Canadian programs, subsampling is extremely important because of the high numbers of very small immature organisms found in most collections when they are sorted using magnifications of $10\times$ or greater, and in the case of some Canadian methods when samples are collected with very fine mesh nets. Conversely, the RIVPACS protocol subsamples only when necessary. Regardless as to whether subsampling is a formalized aspect of sample processing, most programs control sorting effort by limiting the total number of organisms sorted. The targeted number for the EU protocols is substantially higher ($\sim$ 700 individuals) than most US and Canadian protocols, which tend to have fixed count targets of 100 and 300 organisms, respectively.

Whether the sorting procedure includes a separate step for removing 'large–rare' organisms is a direct function of whether the sample is subsampled and the magnification used for sorting. Many Canadian programs size-fractionate samples in the laboratory using 1000–2000 μm sieves, which functionally sort all organisms that would be classified as large–rare organisms. RIVPACS protocol states that, in general, the entire sample is sorted by eye, which also would lead to the removal of all (or at least a representative number of) large–rare individuals. In the US, Carter and Resh (2001) found that about one-half of the programs included a special step in their protocol for sorting large–rare organisms. In contrast to other programs, the STAR version of the AQEM protocol states that large–rare individuals should not be removed.

There are many similarities among macroinvertebrate biomonitoring programs in Canada, the US and the EU. Assessing the significance of these differences will lead to a better understanding of anthropogenic effects on lotic ecosystems. Standardization of field and laboratory methods among programs may never occur and there are several good reasons why complete standardization may not be desirable (e.g. variation in purpose). However, an understanding of the significance of the differences may lead to a better understanding of the differences in interpretation about anthropogenic effects on lotic ecosystems.

ACKNOWLEDGEMENTS

We would like to thank all of the Canadian and US biologists and managers for completing the questionnaire on which this chapter is based. We also would like to thank L. Grapentine, F. Quinn, B. Ross, I. Sharpe, S. Sylvestre, H. Vaughn and D. Williamson for their suggestions and input into the legal basis for biomonitoring in Canada.

REFERENCES

Bailey, R. C., Norris, R. H. and Reynoldson, T. B., 2001. 'Taxonomic resolution of benthic macroin-vertebrate communities in bioassessments'. *Journal of the North American Benthological Society*, **20**, 280–286.

Barbour, M. T. and Gerritsen, J., 1996. 'Subsampling of benthic samples: a defense of the fixed-count method'. *Journal of the North American Benthological Society*, **15**, 386–391.

Barbour, M. T. and Gerritsen, J., 2006. 'Key features of bioassessment development in the United States of America'. In: *Biological Monitoring of Rivers: Applications and Perspectives*, Ziglio, G., Siligardi, M. and Flaim, G. (Eds). Wiley: Chichester, UK pp 351–368.

Barbour, M. T., Gerritsen, J., Snyder, B. D. and Stribling, J. B., 1999. 'Rapid bioassessment protocols for use in streams and wadable rivers: periphyton, benthic macroinvertebrates and fish', 2nd Edition, EPA 841-B-99002. Office of Water, US Environmental Protection Agency, Washington, DC, USA.

Cairns, J., Jr, 1979. 'Biological monitoring – concept and scope'. In: *Environmental Biomonitoring, Assessment, Prediction, and Management – Certain Case Histories and Related Quantitative Issues*, Cairns, J., Jr, Patil, G. P. and Waters, W.E. (Eds). International Co-operative Publishing House, Fairland, MD, USA, pp. 3–20.

Cao, Y., Larsen, D. P., Hughes, R. M., Angermeier, P. L. and Patton, T. M., 2002. 'Sampling effort affects multivariate comparisons of stream assemblages'. *Journal of the North American Benthological Society*, **21**, 701–714.

Cao, Y., Williams, D. D. and Williams, N. E., 1998. 'How important are rare species in aquatic community ecology and bioassessment?'. Limnology and Oceanography, **43**, 1403–1409.

Carter, J. L. and Resh, V. H., 2001. 'After site selection and before data analysis: sampling, sorting and laboratory procedures used in stream benthic macroinvertebrate monitoring programs by USA state agencies'. *Journal of the North American Benthological Society*, **20**, 658–682.

Caton, L. W., 1991. 'Improving subsampling methods for the EPA "Rapid Bioassessment" benthic protocols'. *Bulletin of the North American Benthological Society* **8**, 317–319.

Courtemanch, D. L., 1996. 'Commentary on the subsampling procedures used for rapid bioassess-ments'. *Journal of the North American Benthological Society*, **15**, 381–385.

Cuffney, T. F., Gurtz, M. E. and Meador, M. R., 1993. 'Methods for collecting benthic invertebrate samples as part of the National Water Quality Assessment Program', US Geological Survey Open-File Report 93-406. US Geological Survey: Raleigh, NC, USA.

Davies, P. E., 2000. 'Development of a national river bioassessment system (AUSRIVAS) in Australia'. In: *Assessing the Biological Quality of Fresh Waters. RIVPACS and Other Tech-niques*, Wright, J. F., Sutcliffe, D. W. and Furse, M. T. (Eds). Freshwater Biological Association: Ambleside, UK, pp. 113–124.

Davis, W. S., Snyder, B. D., Stribling, J. B. and Stoughton, C., 1996. 'Summary of state biological assessment programs for streams and wadeable rivers, EPA 230-R-96-007. Office of Policy, Planning, and Evaluation, US Environmental Protection Agency: Washington, DC, USA.

Downes, B. J., Barmuta, L. A., Fairweather, P. G., Faith, D. P., Keough, M. J., Lake, P. S., Mapstone, B. D. and Quinn, G. P., 2002. *Monitoring Ecological Impacts: Concepts and Practice in Flowing Waters*. Cambridge University Press: Cambridge, U.K.

Dumaresq, C., Hedley, K. and Michelutti, R., 2002. 'Overview of metal mining Environmental Effects Monitoring program'. *Water Quality Research Journal of Canada*, **37**, 213–218.

Environment Canada, 2003. 'National assessment of pulp and paper environmental effects mon-itoring data: a report synopsis', NWRI Scientific Assessment Report Series No. 2. National Water Institute: Burlington, ON, Canada.

Green, R. H., 1979. *Sampling Design and Statistical Methods for Environmental Biologists*. Wiley: New York, NY, USA.

Gurtz, M. E. and Muir, T. A. (Eds), 1994. 'Report of the interagency biological methods workshop', US Geological Survey Open-File Report 94-490. US Geological Survey: Raleigh, NC, USA.

Hawkins, C. P., Norris, R. H., Hogue, J. N. and Feminella, J. W., 2000. 'Development and evaluation of predictive models for measuring the biological integrity of streams'. *Ecological Applications*, **10**, 1456–1477.

Hering, D., Moog, O., Sandin, L. and Verdonshot, P. F. M., 2004. Overview and application of the AQEM assessment system'. *Hydrobiologia*, **516**, 1–20.

Hornig, C. E. and Pollard, J. E., 1978. 'Macroinvertebrate sampling techniques for streams in semi-arid regions', EPA-600/4-78-040. Office of Research and Development, Environmental Monitoring and Support Laboratory, US Environmental Protection Agency: Las Vegas, NV, USA.

Hurlbert, S. H., 1984. 'Pseudoreplication and the design of ecological field experiments'. *Ecological Monographs*, **54**, 187–211.

Karr, J. R., 1981. 'Assessment of biotic integrity of fish communities'. *Fisheries*, **6**, 21–27.

Karr, J. R., 1991. 'Biological integrity: a long-neglected aspect of water-resource management'. *Ecological Applications*, **1**, 66–84.

Lenat, D. R. and Resh, V. H., 2001. 'Taxonomy and stream ecology – The benefits of genus- and species-level identifications'. *Journal of the North American Benthological Society*, **20**, 287–298.

Marchant, R., 1989. 'A subsampler for samples of benthic invertebrates'. *Bulletin of the Australian Society of Limnology*, **12**, 49–52.

Merritt, R. W. and Cummins, K. W. (Eds), 1996. *An Introduction to the Aquatic Insects of North America*, 3rd Edition. Kendall/Hunt: Dubuque: IA, USA.

Moulton, S. R., Carter, J. L., Grotheer, S. A., Cuffney, T. F. and Short, T. M., 2000. 'Methods of analysis by the US Geological Survey National Water Quality Laboratory – processing, taxonomy, and quality control of benthic macroinvertebrate samples', US Geological Survey Open-File Report 00-212. US Geological Survey: Denver, CO, USA.

Mumme, S. P. and Pineda, N., 2002. 'Water Management on the US–Mexico Border: Mandate Challenges for Binational Institution', In: *The Future of the US–Mexican Border*, Environmental Change and Security Project. Woodrow Wilson International Center for Scholars: Washington, DC, USA (available in electronic format at http://ecsp.si.edu/tijuana-mp.htm).

Munkittrick, K. R., McMaster, M. E. and Courtenay, S. C., 2002. 'Introductory remarks'. *Water Quality Research Journal of Canada*, **37**, 3–6.

Palmer, M., Bernhardt, E., Chornesky, E., Collins, S., Dobson, A., Duke, C., Gold, B., Jacobson, R., Kingsland, S., Kranz, R., Mappin, M., Martinez, M., Micheli, F., Morse, J., Pace, M., Pascual, M., Palumbi, S., Reichman, O. J., Townsend, A. and Turner, M., 2004. 'Ecology for a Crowded Planet. Ecological Science and Sustainability for a Crowded Planet: 21st Century Vision and Action Plan for the Ecological Society of America' (available in electronic format at www.esa.org/ecovisions).

Plafkin, J. L., Barbour, M. T., Porter, K. D., Gross, S. K. and Hughes, R. M., 1989. 'Rapid bioassessment protocols for use in streams and rivers: benthic macroinvertebrates and fish', EPA 444/4-89-001. Office of Water, US Environmental Protection Agency: Washington, DC, USA.

Resh, V. H., 1979. 'Sampling variability and life history features: basic consideration in the design of aquatic insect studies'. *Journal of the Fisheries Research Board of Canada*, **36**, 290–311.

Resh, V. H., Rosenberg, D. M. and Feminella, J. W., 1985. 'The processing of benthic samples: responses to the 1983 NABS questionnaire'. *Bulletin of the North American Benthological Society*, **2**, 5–11.

Reynoldson, T. B., Bombardier, M., Donald, D. B., O'Neill, H., Rosenberg, D. M., Shear, H., Tuominen, T. M., and Vaughn, H. H., 1999. 'Strategy for a Canadian aquatic biomonitoring network', NWRI Contribution No. 99-249. National Water Research Institute, Canada Centre for Inland Waters, Environment Canada: Burlington, ON, Canada.

Reynoldson, T. B., Day, K. E. and Pascoe, T., 2000. 'The development of the BEAST: a predictive approach for assessing sediment quality in the North American Great Lakes'. In: *Assessing the Biological Quality of Fresh Waters. RIVPACS and Other Techniques*, Wright, J. F., Sutcliffe, D. W. and Furse. M. T. (Eds). Freshwater Biological Association: Ambleside, UK, pp. 165–180.

Rosenberg, D. M., Reynoldson, T. B. and Resh, V. H., 2000. 'Establishing reference conditions in the Fraser River catchment, British Columbia, Canada, using the BEAST (BEnthic Assessment of SedimenT) predictive model'. In: Assessing the Biological Quality of Fresh Waters. RIVPACS and Other Techniques, Wright, J. F., Sutcliffe, D. W. and Furse, M. T. (Eds). Freshwater Biological Association: Ambleside, UK, pp. 181–194.

Slack, K. V., Tilley, L. J. and Kennelly, S. S., 1991. 'Mesh-size effects on drift sample composition as determined with a triple net sampler'. *Hydrobiologia*, **209**, 215–226.

USEPA, 1993. SAB Report: evaluation of draft technical guidance on biological criteria for streams and small rivers', Biological Criteria Subcommittee of the Ecological Processes and Effects Committee, EPA-SAB-EPEC-94-003. Science Advisory Board: Washington, DC, USA.

USEPA, 2002. 'Summary of biological assessment programs and biocriteria development for states, tribes, territories, and interstate commissions: streams and wadeable rivers', EPA-822-R-02-048. US Environmental Protection Agency: Washington, DC, USA.

Underwood, A. J., 1997. *Experiments in Ecology: their Logical Design and Interpretation using Analysis of Variance.* Cambridge University Press, Cambridge, UK.

Vinson, M. R. and Hawkins, C. P., 1996. 'Effects of sampling area and subsampling procedure on comparisons of taxa richness among streams'. *Journal of the North American Benthological Society*, **15**, 392–399.

Walker, S. L., Hedley, K. and Porter, E., 2002. 'Pulp and paper environmental effects monitoring in Canada: an overview'. *Water Quality Research Journal of Canada*, **37**, 7–19.

Weigel, B. M., Henne, L. J. and Martinez-Rivera, L. M., 2002. 'Macroinvertebrate-based index of biotic integrity for protection of streams in west-central Mexico'. *Journal of the North America Benthological Society*, **21**, 686–700.

Williams, D. D., and Feltmate, B. W., 1992. *Aquatic Insects.* CAB International: Wallingford, Oxford, UK.

Wright, J. F., 2000. 'An introduction to RIVPACS'. In: *Assessing the Biological Quality of Fresh Waters. RIVPACS and Other Techniques*, Wright, J. F., Sutcliffe, D. W. and Furse, M. T. (Eds). Freshwater Biological Association: Ambleside, UK, pp. 1–24.

Wright, J. F., Moss, D., Armitage, P. D. and Furse, M. T., 1984. 'A preliminary classification of running water sites in Great Britain based on macroinvertebrate species and prediction of community type using environmental data'. *Freshwater Biology*, **14**, 221–256.

Wrona, F. J., Culp, J. M. and Davies, R.W., 1982. 'Macroinvertebrate subsampling: a simplified apparatus and approach'. *Canadian Journal of Fisheries and Aqualic Science*, **39**, 1051–1054.

2.6

Biological Monitoring of Rivers and European Water Legislation

Ana Cristina Cardoso, Angelo Giuseppe Solimini
and **Guido Premazzi**

2.6.1   INTRODUCTION

The determination of water quality is more often done through the assessment of physico-chemical and biological elements. Biological assessment relies on the fact that pollution of a water body will cause changes in the physical and chemical environment of that water and that changes will affect the ecosystem composition and functioning, thus causing changes that are measurable through the biological elements. By measuring the extent of the biological upset, the severity of the pollution can be evaluated.

This chapter provides an overview of the European water legislation relevant to river quality, with emphasis on the new Water Framework Directive (WFD). This

Biological Monitoring of Rivers   Edited by G. Ziglio, M. Siligardi and G. Flaim
© 2006 John Wiley & Sons, Ltd.

Directive is unique among the water legislation because it includes for the first time the nature of the information required to determine the ecological functioning (status) of an aquatic system.

2.6.2   THE EVOLUTION OF THE EUROPEAN WATER LEGISLATION

Water is one of the most comprehensively regulated areas of the European Union (EU) environmental legislation. In the last twenty years, the approaches to water protection have evolved in significant ways. Early European water policy began in the 1970s, with the adoption of political programmes, as well as legally binding legislation. Parallel to the political programmes, a first wave of legislation was adopted, starting with the 1975 'Surface Water Directive' and culminating in the 1980 'Drinking Water Directive'. This first wave of water legislation included water quality legislation on fish waters (1978), shellfish waters (1979), bathing waters (1976) and ground waters (1980). In the field of emission limit value legislation, the 'Dangerous Substances Directive' (1976) and its 'Daughter directives' (1982–1986) on various individual substances were adopted.

A second wave of water legislation followed a review of existing legislation and an identification of necessary improvements and gaps to be filled. This phase of water legislation included the 'Urban Waste Water Treatment Directive' (1991) and the 'Nitrates Directive' (1991). Other elements identified were revisions of the 'Drinking Water' and 'Bathing Water' Directives to bring them up to date (proposals for revisions being adopted in 1994 and 1995, respectively), the development of a 'Groundwater Action' Programme and a 1994 proposal for an 'Ecological Quality of Water' Directive. In addition, for large industrial installations, the 'IPPC (Integrated Pollution Prevention and Control) Directive' (adopted in 1996) covered water pollution as well.

There have been, at Member State, as well as at European level, basically two different approaches to tackle water pollution: the Water Quality Objective (WQO) approach and the Emission Limit Value (ELV) approach. The WQO defines the minimum quality requirements of water to limit the cumulative impact of emissions, both from point sources and diffuse sources. This approach, therefore, focuses on a certain quality level of water in which condition and use is not harmful for the environment and human health. This approach was mainly used in the first wave of water directives (1975 to 1980), such as the 'Surface Water Directive' (1975) or the 'Bathing Water Quality Directive' (1976). On the other hand, the ELV approach focuses on the maximum allowed quantities of pollutants that may be discharged from a particular source into the aquatic environment. This approach, in fact, looks at the end-product of a process (waste water treatment, discharges from industry, effect of agriculture on water quality, etc.) or what quantities of pollutants may go into the water. This was mainly used in the second wave of water legislation during the 1990s: the 'Urban Waste Water Treatment Directive' (1991), the 'Nitrates Directive' (1991) and the 'IPPC Directive' (1996).

Since then, the question of which approach is the most appropriate one has been the subject of long scientific and political debate. The issue fuelling the debate has always been on the dilemma between two possible choices: establishing quality standards for water or imposing emission standards for water contaminants. The quality standards define the desired quality of the environment whereby the introduction of certain contaminants is tolerated, while the emission standards (limit values) set maximum quantities for pollutants, which can be emitted, and therefore, may be stricter than the former. In the end, the result is an environmental policy – the WFD – which is based on a 'combined approach', requiring both the establishment of emission standards and the implementation of water quality objectives.

2.6.3 THE MOST RECENT INITIATIVE OF THE EUROPEAN COMMISSION: THE WATER FRAMEWORK DIRECTIVE

December 22, 2000 will remain a milestone in the history of water policies in Europe: on that date, the Water Framework Directive (WFD), or the Directive 2000/60/EC of the European Parliament and of the Council of 23 October 2000 establishing a framework for Community action in the field of water policy was published in the Official Journal of the European Communities and thereby entered into force.

This Directive is the result of a process of more than ten years of discussions and negotiations between a wide range of experts, stakeholders and policy makers. This process has stressed the widespread agreement on key principles of modern water management that form today the foundation of the WFD. The word 'framework' means that it sets out common principles and provides an overall framework for protection and action, rationalizing existing legislation to produce an integrated approach to water resources management in the Union.

The new European Water Policy, and its operative tool, the Water Framework Directive, is based on the 'combined approach' where ELVs and WQOs are used to mutually reinforce each other. This combined approach is in accordance with principles established in the Treaty as the precautionary principle, and the principle that environmental damage should be rectified at the source, as well as the principle that environmental conditions in the various regions shall be taken into consideration.

The major purpose of the WFD is to establish a framework for the protection of all waters (including inland surface waters, transitional waters, coastal waters and groundwater). This directive aims to (1) protect and improve the quality of aquatic ecosystems, (2) promote sustainable water use based on water management for the long term, and (3) ensure that the right amount of water is available, where and when it is needed. There are clear deadlines for each of the requirements, which add up to an ambitious overall timetable. The overall deadline is the objective of achieving good water status for all waters in the Union by 2015.

The WFD is based on a 'framework philosophy', in line with the principle of subsidiarity. It comes to set only the objectives to be fulfilled by Member States (i.e.

good quality of all waters), defining the organizational structure (river basin authorities) and mechanisms (existing legislation and further measures) to achieve them. As such, the WFD best exemplifies the new approach in the EU environmental policy, where environmental protection is married with subsidiarity through the division of objectives at a European level and measures at a national level. This Directive will replace many of the 'first-wave' legislation, such as the fish and shellfish water directives. This satisfies the call for deregulation (simplification) of the existing legislative frame. On the other hand, the European Commission (EC) considered that the public-health-related standards (such as those of the drinking and bathing directives) should not be affected. In the WFD, one can see elements from all different forces that have guided the reform of EU water policy, i.e. environmental protection, deregulation and subsidiarity. Moreover, elements of the economic instruments approach (i.e. introducing the cost recovery principle), quantitative concerns (i.e. in setting minimum flow objectives for rivers and abstraction limits for ground waters) and the quest for integration (i.e. river basin management with representation of all stakeholders) are all reflected in the WFD.

2.6.4   A NEW STRATEGY FOR WATER PROTECTION

A number of new strategies will result from the implementation of the WFD. These include the river basin management, the programmes of measures and monitoring, cost recovery and the public consultation.

The new approach to water management requires water to be managed on the basis of river basins, rather than according to geographical or political boundaries. This enables assessment of all activities, which may affect the watercourse, and their eventual control by measures, which may be specific to the conditions of the river basin. The WFD requires river basin management plans to be drawn up on a river-basin basis. It may be necessary to sub-divide a large river basin into smaller units, and sometimes a particular water type may justify its own plan.

Central to each river basin management plan will be a programme of measures to ensure that all waters in the river basin achieve good water status. The starting point for this programme is the full implementation of any relevant national or local legislation, as well as a range of Community legislation on water and related issues. If this basic set of measures is not enough to ensure that the goal of good water status is reached, the programme must be supplemented with whatever further measures are necessary. These might include stricter controls on polluting emissions from industry or agriculture, as well as from urban wastewater sources; land use planning might, in this context, be a key issue to be taken into account. In any particular situation, the more rigorous approach will apply.

The application of economic instruments, such as charges for use of water as a resource or for the discharge of effluents into watercourses, is a policy explicitly endorsed in the new directive. The 'polluter pays' principle must be applied, and economic assessment becomes an essential part of water-management planning. The principle of charges for water reflecting the true costs is a radical innovation at

European level. There is a danger in this proposal that water may become a too expensive commodity for many, and a general reduction in its beneficial use may result.

The WFD requires consultation to take place and the competent authorities should
arrange consultation mechanisms with the interested parties, such as the general
public, NGOs, farmers, water companies, etc. In this context, consultation with
all relevant parties might achieve a best cost-effectiveness and identify the best
combination of measures on a proportionate level. Therefore, preparation of those
parts of the WFD addressing information and consultation of all stakeholder groups
and the public should be subject to serious efforts. Integration of stakeholders and
the civil society in decision-making, by promoting transparency and information to
the public and by offering a unique opportunity for involving stakeholders in the
development of river basin management plans, is a central concept of the WFD.

2.6.5 ROLE OF MONITORING

Monitoring will be an essential part in the implementation of the WFD. Requirements for monitoring of surface water are established in Article 8 of the Directive
and described in Annex V. Monitoring programmes are required to establish a coherent and comprehensive overview of water status within each river basin district
characterized as in Article 5 of the Directive. The results of monitoring should permit
the classification of surface waters into five quality classes. The Directive gives a
general definition of ecological quality (Table 2.6.1) and specifies quality elements
(QEs) for the classification of ecological status including biological (Table 2.6.2),

Table 2.6.1 General definition of ecological quality for rivers, lakes, transitional and coastal
waters (from the WFD, Annex V, Table 1.2). Waters achieving a status below moderate shall be
classified as poor or bad. Waters showing evidence of major alterations to the values of the
biological quality elements for the surface water body type and in which the relevant biological
communities deviate substantially from those normally associated with the surface-water-body
type under undisturbed conditions, should be classified as poor

High status	Good status	Moderate status
There are no, or only very minor, anthropogenic alterations to the values of the physico-chemical and hydromorphological quality elements for the surface-water-body type from those normally associated with that type under undisturbed conditions. The values of the biological quality elements for the surface water body reflect those normally associated with that type under undisturbed conditions, and show no, or only very minor, evidence of distortion. These are the type-specific conditions and communities	The values of the biological quality elements for the surface-water-body type show low levels of distortion resulting from human activity, but deviate only slightly from those normally associated with the surface-water-body type under undisturbed conditions	The values of the biological quality elements for the surface-water-body type deviate moderately from those normally associated with the surface-water-body type under undisturbed conditions. The values show moderate signs of distortion resulting from human activity and are significantly more disturbed than under conditions of good status

Table 2.6.2 Definitions for high, good and moderate ecological status in rivers – biological quality elements (from the WFD, Annex V, Table 1.2.1)

Element	High status	Good status	Moderate status
Phytoplankton	The taxonomic composition of phytoplankton corresponds totally, or nearly totally, to undisturbed conditions. The average phytoplankton abundance is wholly consistent with the type-specific physico-chemical conditions and is not such as to significantly alter the type-specific transparency conditions. Planktonic blooms occur at a frequency and intensity which is consistent with the type-specific physico-chemical conditions	There are slight changes in the composition and abundance of planktonic taxa compared to the type-specific communities. Such changes do not indicate any accelerated growth of algae resulting in undesirable disturbances to the balance of organisms present in the water body or to the physico-chemical quality of the water or sediment. A slight increase in the frequency and intensity of the type-specific planktonic blooms may occur	The composition of planktonic taxa differs moderately from the type-specific communities. Abundance is moderately disturbed and may be such as to produce a significant undesirable disturbance in the values of other biological and physico-chemical quality elements. A moderate increase in the frequency and intensity of planktonic blooms may occur. Persistent blooms may occur during summer months
Macrophytes and phytobenthos	The taxonomic composition corresponds totally, or nearly totally, to undisturbed conditions. There are no detectable changes in the average macrophytic and the average phytobenthic abundances	There are slight changes in the composition and abundance of macrophytic and phytobenthic taxa compared to the type-specific communities. Such changes do not indicate any accelerated growth of phytobenthos or higher forms of plant life resulting in undesirable disturbances to the balance of organisms present in the water body or to the physico-chemical quality of the water or sediment. The phytobenthic community is not adversely affected by bacterial tufts and coats present due to anthropogenic activity	The composition of macrophytic and phytobenthic taxa differs moderately from the type-specific communities and is significantly more distorted than at good status. Moderate changes in the average macrophytic and the average phytobenthic abundance are evident. The phytobenthic community may be interfered with and, in some areas, displaced by bacterial tufts and coats present as a result of anthropogenic activities

	High status	Good status	Moderate status
Benthic invertebrate fauna	The taxonomic composition and abundance correspond totally, or nearly totally, to undisturbed conditions. The ratio of disturbance-sensitive taxa to insensitive taxa shows no signs of alteration from undisturbed levels. The level of diversity of invertebrate taxa shows no sign of alteration from undisturbed levels	There are slight changes in the composition and abundance of invertebrate taxa from the type-specific communities. The ratio of disturbance-sensitive taxa to insensitive taxa shows slight alteration from type-specific levels. The level of diversity of invertebrate taxa shows slight signs of alteration from type-specific levels	The composition and abundance of invertebrate taxa differ moderately from the type-specific communities. Major taxonomic groups of the type-specific communities are absent. The ratio of disturbance-sensitive taxa to insensitive taxa, and the level of diversity, are substantially lower than the type-specific level and are significantly lower than for good status
Fish fauna	Species composition and abundance correspond totally, or nearly totally, to undisturbed conditions. All the type-specific disturbance-sensitive species are present. The age structures of the fish communities show little sign of anthropogenic disturbance and are not indicative of a failure in the reproduction or development of any particular species	There are slight changes in species composition and abundance from the type-specific communities attributable to anthropogenic impacts on physico-chemical and hydromorphological quality elements. The age structures of the fish communities show signs of disturbance attributable to anthropogenic impacts on physico-chemical or hydromorphological quality elements, and, in a few instances, are indicative of a failure in the reproduction or development of a particular species, to the extent that some age classes may be missing	The composition and abundance of fish species differ moderately from the type-specific communities attributable to anthropogenic impacts on physico-chemical or hydromorphological quality elements. The age structure of the fish communities shows major signs of anthropogenic disturbance, to the extent that a moderate proportion of the type-specific species are absent or of very low abundance

hydromorphological (Table 2.6.3) and physico-chemical (Table 2.6.4) elements. It is important to note that the use of hydromorphological and physico-chemical indicators is in support to the use of biological indicators but cannot replace it. Therefore, a direct measure of biological quality elements is always necessary to validate any potential impacts suggested by non-biological indicators. The role of the different elements (hydromorphological, physico-chemical and biological) in the classification of the ecological status has been agreed in a guidance developed within the WFD Common Implementation Strategy ECOSTAT WG2A (Anon, 2004) (see below).

The results of monitoring should be expressed as ecological quality ratios (EQRs) for the purposes of classification of ecological status and based on the principle of 'one-out, all-out'. These ratios represent the relationship between the values of the biological parameters observed for a given body of surface water and the values for these parameters in the reference conditions applicable to that type of water body. The ratio is expressed as a numerical value between zero and one, with high ecological status represented by values close to one and bad ecological status by values close to zero. Frequency of monitoring depends on a specific determinant and quality element so that financial resources can be allocated in a cost-effective way. This should prevent a situation where countries have to monitor for chemical substances, even though they are known not to be present in the catchment, except where validation of the risk assessments is required. Therefore, a crucial aspect in the design of monitoring programmes is quantifying the temporal and spatial variability of quality elements. Parameters that are very variable in time and/or space may require more intensive sampling (and hence cost) than those that are less variable.

Three types of monitoring for surface waters are described in Annex V, namely, (1) surveillance, (2) operational, and (3) investigative monitoring.

Surveillance monitoring activities should provide information for the classification of surface water, for supplementing and validating the risk-assessment procedure (Annex II of the Directive), for an efficient and effective design of future monitoring programmes, and for the assessment of long-term changes in natural conditions, including those changes resulting from anthropogenic activity. Parameters indicative of all of the biological, hydromorphological and all general and specific physico-chemical quality elements are required to be monitored. The results of a surveillance monitoring programmes should be reviewed and used, in combination with the impact assessment procedure described in Annex II, to determine the requirements for monitoring programmes in river-basin-management plans.

Operational monitoring should be undertaken to establish the status of those water bodies identified by surveillance monitoring or impact assessment as being at risk of failing to meet their environmental objectives, and to assess any changes in the status of such bodies. The parameters monitored should be those indicative of the biological and hydromorphological quality elements most sensitive to the pressures to which the body is subject, or to all priority substances discharged and other substances discharged in significant quantities. The operational monitoring

Table 2.6.3 Definitions for high, good and moderate ecological status in rivers – hydromorphological quality elements (from the WFD, Annex V, Table 1.2.1)

Element	High status	Good status	Moderate status
Hydrological regime	The quantity and dynamics of flow, and the resultant connection to groundwaters, reflect totally, or nearly totally, to undisturbed conditions	Conditions consistent with the achievement of the values specified in Table 2.6.2 for the biological quality elements	Conditions consistent with the achievement of the values specified in Table 2.6.2 for the biological quality elements
River continuity	The continuity of the river is not disturbed by anthropogenic activities and allows undisturbed migration of aquatic organisms and sediment transport	Conditions consistent with the achievement of the values specified in Table 2.6.2 for the biological quality elements	Conditions consistent with the achievement of the values specified in Table 2.6.2 for the biological quality elements
Morphological conditions	Channel patterns, width and depth variations, flow velocities, substrate conditions, and both the structure and condition of the riparian zones, correspond totally, or nearly totally, to undisturbed conditions	Conditions consistent with the achievement of the values specified in Table 2.6.2 for the biological quality elements	Conditions consistent with the achievement of the values specified in Table 2.6.2 for the biological quality elements

Table 2.6.4 Definitions for high, good and moderate ecological status in rivers – physico-chemical quality elements (Environmental Quality Standards (EQSs)) (from the WFD, Annex V, Table 1.2.1)

Element	High status	Good status	Moderate status
General conditions	The values of the physico-chemical elements correspond totally, or nearly totally, to undisturbed conditions. Nutrient concentrations remain within the range normally associated with undisturbed conditions. The levels of salinity, pH, oxygen balance, acid neutralizing capacity and temperature do not show signs of anthropogenic disturbance and remain within the range normally associated with undisturbed conditions	Temperature, oxygen balance, pH, acid neutralizing capacity and salinity do not reach levels outside the range established so as to ensure the functioning of the type-specific ecosystem and the achievement of the values specified in Table 2.6.2 for the biological quality elements. Nutrient concentrations do not exceed the levels established so as to ensure the functioning of the ecosystem and the achievement of the values specified in Table 2.6.2 for the biological quality elements	Conditions consistent with the achievement of the values specified in Table 2.6.2 for the biological quality elements
Specific synthetic pollutants	Concentrations close to zero and at least below the limits of detection of the most advanced analytical techniques in general use	Concentrations not in excess of the standards set in accordance with the procedure detailed in Section 1.2.6 of the Directive without prejudice to Directive 91/414/EC and Directive 98/8/EC (EQSs)	Conditions consistent with the achievement of the values specified in Table 2.6.2 for the biological quality elements
Specific non-synthetic pollutants	Concentrations remain within the range normally associated with undisturbed conditions (background levels)	Concentrations not in excess of the standards set in accordance with the procedure detailed in Section 1.2.6 20 without prejudice to Directive 91/414/EC and Directive 98/8/EC (EQSs)	Conditions consistent with the achievement of the values specified in Table 2.6.2 for the biological quality elements

programme may be amended during the period of the river-basin-management plan to allow a reduction in frequency where an impact is found not to be significant or the relevant pressure is removed. Sampling frequency for surveillance and operational monitoring programmes should be critically assessed, especially considering the precision and accuracy of the estimates that they will provide.

Investigative monitoring should be carried out where the reason for an exceedance of environmental objectives is unknown, where surveillance monitoring indicates that the quality objectives for a water body are not likely to be achieved (and operational monitoring has not already been established), in order to assess the causes of a water body failing to achieve the environmental objectives, or to ascertain the magnitude and impacts of accidental pollution events.

2.6.6 COMMON IMPLEMENTATION STRATEGY

The implementation of the WFD raises challenges, which are widely shared by Member States and the European Commission. These include: an extremely demanding timetable, in particular in the nine preparatory years; the complexity of the text and the diversity of possible solutions to scientific, technical and practical questions; the problem of capacity building and an incomplete technical and scientific basis with a large number of fundamental issues in Annexes II and V, which need further elaboration and substantiation to make the transition from principles and general definitions to a successful practical implementation.

In order to respond to these challenges, the EC, together with Member States, jointly developed an informal programme of co-operation, known as the Common Implementation Strategy (CIS) (Anon, 2001). The EC, the Member States, the Candidate Countries and all relevant stakeholders take part in the CIS. The aim of this strategy is to allow, as far as possible, a coherent and harmonious implementation of the WFD in the EU. The focus is on methodological questions related to a common understanding of the technical and scientific implications of the WFD by developing guidance documents (e.g. 'Guidance on Monitoring' (Anon, 2003a); 'Guidance on Reference Conditions' (Anon, 2003b); 'Guidance on Intercalibration' (Anon, 2003c) and advice on operational methods.

A testing of the guidance documents developed during the first year of the WFD-CIS is taking place in the so-called 'pilot river basins'. The objectives are to gain practical experiences and to test for consistency and coherency of the guidances. Because of the diversity of circumstances within the EU, Member States may apply this guidance in a flexible way in answering to problems that will vary from one river basin to the next. This proposed guidance will, therefore, need to be tailored to specific circumstances. Following the testing phase, the final aim of this joint strategy is to produce 'The Manual for Integrated River Basin Management'. This manual would be used to prepare river basin management plans under the WFD.

2.6.7   FINAL CONSIDERATIONS

River quality in Europe has traditionally been determined using physico-chemical elements and benthic invertebrate fauna (FBA, 2000), often with classification computed by not considering background values for those elements. The classification requirements introduced by the WFD involve a considerable burden to the river-management authorities that are forced to develop monitoring programs, including all biological elements/parameters and to establish the appropriate protection/restoration measures. However, in practice, the data as required by the WFD is currently scarce and the application of the new concept of 'ecological status' is a largely unknown issue.

The WFD provides the basis for an overall management framework for surface waters and ground waters. Its implementation entails not simply the adoption of new standards, but requires the introduction of a modern regime of water management, based on river basin units. The WFD–CIS guidance documents will definitively give an important contribution to a harmonious and coherent implementation of the Directive. However, the lack of knowledge on how biological elements correlate with certain pressures, on their spatial and temporal variability, and the limitation of financial resources, may limit the effective implementation of the WFD, at least in the next few years.

REFERENCES

Anon, 2001. 'Strategic Document – Common Strategy on the Implementation of the Water Framework Directive', May 2, 2001 (available at: http://forum.europa.eu.int/Public/irc/env/wfd/library).

Anon, 2003a. 'Common Implementation Strategy for the Water Framework Directive (2000/60/EC). Guidance on Monitoring for the Water Framework Directive' (available at: http://forum.europa.eu.int/Public/irc/env/wfd/library).

Anon, 2003b. 'Common Implementation Strategy for the Water Framework Directive (2000/60/EC). Guidance on Establishing Reference Conditions and Ecological Status Class Boundaries for Inland Surface Waters' (available at: http://forum.europa.eu.int/Public/irc/env/wfd/library).

Anon, 2003c. 'Common Implementation Strategy for the Water Framework Directive (2000/60/EC). Towards a Guidance on Establishment of the Intercalibration Network and on the Process of the Intercalibration Exercise' (available at: http://forum.europa.eu.int/Public/irc/env/wfd/library).

Anon, 2004. 'Common Implementation Strategy for the Water Framework Directive (2000/60/EC). Overall Approach to the Classification of Ecological Status and Ecological Potential' (available at: http://forum.europa.eu.int/Public/irc/env/wfd/library).

FBA, 2000. *Assessing the Biological Quality of Freshwaters: RIVPACS and Other Techniques*, Wright, J. F., Sutcliffe, D. W. and Furse, M. T. (Eds), Proceedings of an International Workshop held in Oxford, UK, on 16–18 September, 1997. Freshwater Biological Association: Ambleside, UK.

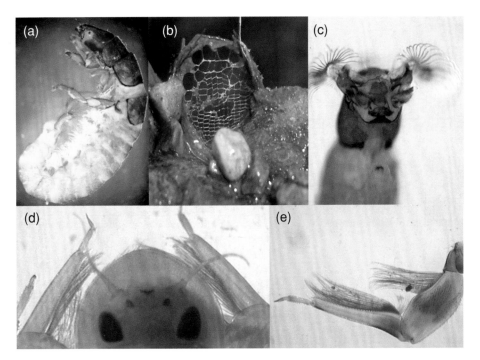

Plate 1 (Figure 1.5.16) Examples of different mechanisms for detritivore-filters (collectors) by using silk nets (a, b) and filter setae (c–e): Trichoptera *Hydropsyche* sp. lava (a) and net (b); head of a larva of Simuliidae, showing the filtering mandibular folding fans (c); Ephemeroptera of the family Oligoneuriidae showing the long filter bristles of the internal sides of the forelegs (d, e). See page 85

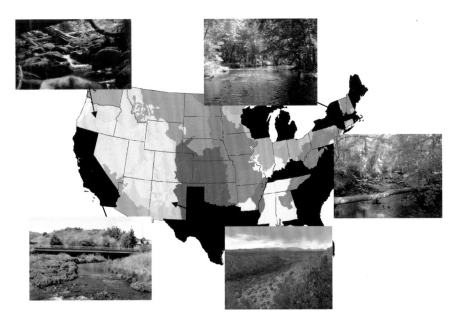

Plate 2 (Figure 3.5.1) Stream-type variation in the USA: State boundaries are delineated on the map by dark lines, while colours denote different ecoregions. See page 352

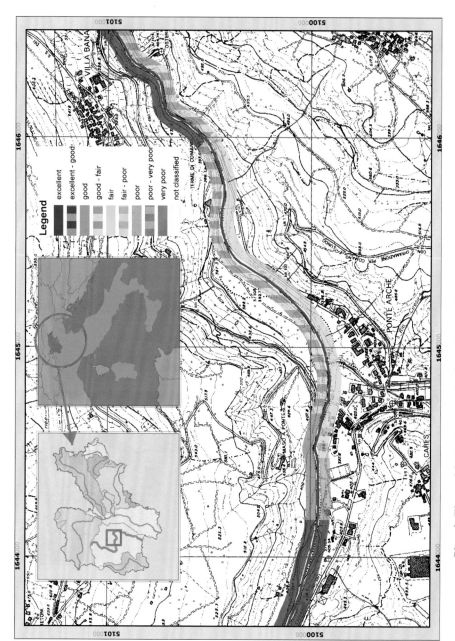

Plate 3 (Figure 4.3.1)   An example of an FFI map (River Noce, Trentino, Italy). See page 412

2.7

Ecotoxicological Experiments in River Pollution Assessment

Daren M. Carlisle and **William H. Clements**

2.7.1 INTRODUCTION

Presentations in this volume demonstrate the variety of methods used for biomonitoring rivers. Although approaches vary in the ecological attributes measured (e.g. fish vs. macroinvertebrate communities) and subsequent analysis, they share a commonality of being mostly descriptive. Descriptive studies can be used to identify variability in ecological attributes, which can then be correlated to potential anthropogenic disturbance. However, there are several questions that cannot be addressed solely with descriptive, biomonitoring data, and these are often paramount to the overall objectives of environmental monitoring and assessment. Addressing these questions requires some level of experimentation. We believe that experimental data are a necessary part of any environmental assessment and subsequent remediation program. The purpose of this chapter is to highlight ecotoxicological experiments that can be integrated with monitoring data in river pollution assessments. After reviewing the basic tenets of ecotoxicology and ecological experimental design, we describe an integrative approach that we have found effective for addressing these basic questions.

2.7.2 WHAT BIOLOGICAL MONITORING CANNOT REVEAL

Analysis and interpretation of biomonitoring data often generate additional questions. These questions may be crucial to understanding the causes (and potential sources) of impairment, the extent of ecological degradation and the most effective remedial options.

Question 1: What are the stressors responsible for the observed degradation in biological communities? Biomonitoring data are often used to determine the status of stream ecosystems. Identification of the causes of ecological degradation is a prerequisite to implementing the necessary remedial management actions. Because most biomonitoring is descriptive, establishing causal relationships between specific anthropogenic stressors and biological responses is often not possible, especially when multiple natural and anthropogenic stressors simultaneously exist. The best biomonitoring studies often can only muster support for the hypothesis that observed responses are a result of a presumed stressor (Suter *et al.*, 2002; Clements, 2004).

Question 2: What are the ecosystem consequences of anthropogenic stressors? Biological monitoring and assessments may reveal alterations in community structure or composition, but often lack information about important ecological processes that have high societal relevance. For example, indices of biotic integrity are popular tools in biological monitoring and assessment (Rosenberg and Resh, 1993) because they have been useful for detecting changes in ecological conditions. The underlying assumption when simplifying community data with additive metrics is that ecological information (and hence relevancy) is preserved in the final index. However, individual metrics are often selected based on their responsiveness to known stressors rather than on the ecological relevancy they impart to the final index. As a consequence, we may have useful tools for detecting change, but lack the ability to infer their ecological relevance. Additional observational and experimental work in the field and laboratory are often required to describe the full breadth of the ecological consequences of anthropogenic stressors.

Question 3: From a management standpoint, what level of cleanup is required to support and maintain target levels of ecological attributes (e.g. species diversity and ecosystem functioning)? The socio-economic costs of remediation require an informed estimate of specific contaminant levels necessary to restore ecological integrity. Taxon-specific chronic criteria exist for many chemicals, but are based on laboratory bioassays using surrogate species often not found in the ecosystem under study. The inability of single-species bioassays to predict ecological responses is well documented (Cairns, 1983, 1986).

2.7.3   THE DISCIPLINE OF ECOTOXICOLOGY

Ecotoxicology is the application of the principles of toxicology to ecological systems. Broadly defined, toxicology is the study of the effects of chemicals on organisms. Most toxicological principles are based on measured responses of cells, tissues, organs or individuals experimentally exposed to known toxicant levels. The dose–response relationship is a fundamental concept in toxicology because it summarizes the characteristics of exposure and the wide spectrum of possible responses. Indeed, Eaton and Klaassen (1996) assert that understanding the dose–response relationship is 'essential for the study of toxic materials' (p.18). Toxicologists generally model dose–response relationships for individuals and populations, but the goal of ecotoxicology is to identify dose (exposure)–responses for community and ecosystem endpoints (Figure 2.7.1).

2.7.4   EXPERIMENTAL APPROACHES IN ECOTOXICOLOGICAL RESEARCH

2.7.4.1   Background

Certain types of experimental approaches are more useful than others for investigating ecological responses to perturbations. Diamond (1986) distinguishes three types of experiments in ecological research: laboratory experiments, field experiments and natural experiments (Table 2.7.1). These approaches can be contrasted in terms of control over independent variables, site matching (e.g. pre-treatment similarity among experimental units), ability to follow a trajectory, spatiotemporal scale, scope, ecological realism and generality. Laboratory experiments rank high for control of independent variables and site matching, but are undesirable because of their limited scope, spatiotemporal scale, ecological realism and generality. Field experiments are conducted outdoors and often involve manipulation of natural communities. Relative to laboratory experiments, field experiments are more realistic and offer more possible manipulations. However, field experiments have less control

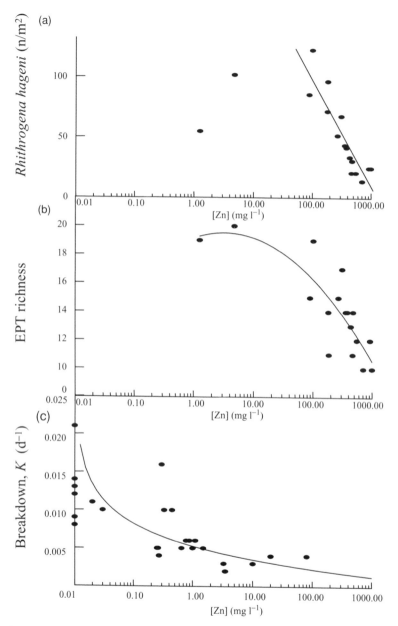

Figure 2.7.1 Concentration–response models across multiple levels of biological organization typically studied in ecotoxicology (a) population density of the mayfly (Ephemeroptera:Heptageniidae) *Rhithrogena hageni* in stream microcosms exposed to Zn; (b) taxa richness of mayflies, caddisflies and stoneflies in stream microcosms exposed to Zn; (c) leaf litter breakdown rate in streams along a gradient of Zn contamination (adapted from Niyogi *et al.*, 2001)

Table 2.7.1 Relative strengths and weaknesses of different types of experiments used in ecology. Table 1.1, p.4 from 'Overview: Laboratory experiments, field experiments and natural experiments' by Jared M. Diamond in COMMUNITY ECOLOGY edited by Jared M. Diamond and Ted. J. Case. Copyright © 1986 by Harper and Row, Publishers, Inc. Reprinted by permission of Pearson Education, Inc.

	Type of experiment			
	Laboratory	Field	Natural trajectory	Natural snapshot
Regulation of independent variables	Highest	Medium to low	None	None
Site matching	Highest	Medium to low	Medium	Lowest
Ability to follow trajectory	Yes	Yes	Yes	No
Maximum temporal scale	Lowest	Lowest	Highest	Highest
Maximum spatial scale	Lowest	Low	Highest	Highest
Scope (range of manipulations)	Lowest to low	Medium to high	Medium	Highest
Realism	Low/none	High	Highest	Highest
Generality	None	Low	High	High

than laboratory studies and may be confounded by pre-treatment differences among experimental units. Field experiments are usually conducted at a small spatiotemporal scale and lack generality (Diamond, 1986). Natural experiments differ from field experiments in that the variables of interest are not directly manipulated, but are instead used to select sites along gradients of the suspected variables. Natural experiments can be further distinguished as natural snapshot experiments, where a researcher compares sites that differ in a particular characteristic (e.g. presence or absence of a predator) and natural trajectory experiments, where a researcher makes comparisons before and after a perturbation. Ecotoxicologists typically employ the same basic experimental approaches as ecologists to investigate the effects of contaminants on ecosystems (Clements and Newman, 2002).

Standardized laboratory experiments, such as 96 h toxicity tests, have been the workhorse of aquatic toxicologists for many years (Cairns, 1983), but the regulatory and cultural inertia associated with single-species laboratory experiments have partially impeded implementation of more ecologically relevant approaches (Clements and Newman, 2002). Recent studies show that single species tests may not predict community-level responses to contaminants because of indirect effects and higher-order interactions (Schindler, 1987; Clements *et al.*, 1989; Gonzalez and Frost, 1994). Currently, there are no established protocols for community- and ecosystem-level experimentation in ecotoxicology. Recent reviews reveal an astonishing diversity of experimental conditions, communities, duration, spatiotemporal scale, experimental designs and test endpoints (Gillett, 1989; Gearing, 1989; Kennedy *et al.*, 1995; Pontasch, 1995; Shaw and Kennedy, 1996; Clements, 1997).

2.7.4.2 Microcosm and mesocosm experiments

Model systems are not typically employed in ecological risk assessment or used for establishing chemical criteria, but the value of microcosms and mesocosms to assess effects of contaminants on communities is widely recognized (see reviews by Gillett, 1989; Gearing, 1989; Graney *et al.*, 1989). Although the distinction between microcosms and mesocosms is not always obvious, microcosms are generally located indoors and are smaller in size than mesocosms. Microcosms are defined as controlled, laboratory systems that attempt to simulate a portion of nature. Odum (1984) defined mesocosms as 'bounded and partially enclosed outdoor experimental setups'. Because they are only partially enclosed, mesocosms generally have greater exchange with the natural environment. Despite these differences, a common feature of both microcosm and mesocosm experiments is that they can investigate the responses of numerous species simultaneously. Endpoints examined in microcosm and mesocosm experiments are therefore not restricted to simple estimates of mortality or growth, but generally include an array of structural and functional measures (e.g. community composition, species richness, primary productivity). Although microcosms and mesocosms have been used to assess impacts of contaminants on populations and communities, they have not played a major role in ecotoxicological research. Reviews of the major journals in aquatic and terrestrial toxicology reveal a surprisingly infrequent application of these tools. Notable exceptions include a few published symposia and special features that have focused on microcosm and mesocosm experiments (*Environmental Toxicology and Chemistry*, 1992, 11(1); *Ecological Applications*, 1997, 7(4)).

Unlike descriptive studies, microcosm and mesocosm experiments can be used to unambiguously test specific hypotheses (Daehler and Strong, 1996). Unfortunately, the degree of simplification required to obtain precise control often compromises the study's ecological realism. The seriousness of this issue depends on the specific study goals. If the primary objective is to obtain a mechanistic understanding of underlying processes, then realism may not be a significant issue (Peckarsky, 1998). We agree with Lawton (1996) that the best way to understand the operations of a complex ecological system is to construct a model and determine whether it functions as predicted. In summary, we feel that perturbations of model systems provide a powerful way to test hypotheses about the effects of anthropogenic perturbations on ecosystems.

The early focus of microcosm and mesocosm research was to predict the transport and fate of contaminants under controlled conditions. Various processes involved in contaminant transport through biotic and abiotic compartments, including volatilization, microbial degradation, biotransformation and food chain transfer, are readily quantified using microcosms and mesocosms. More recent applications of microcosms and mesocosms aim to evaluate ecological effects. Model systems have been used to support regulatory decisions regarding safe concentrations of pesticides (Giddings *et al.*, 1996), establish concentration–response relationships in community level experiments (Kiffney and Clements, 1996), validate single-species toxicity tests (Pontasch *et al.*, 1989), compare sensitivity of different endpoints (Carlisle and Clements, 1999), validate field responses (Niederlehner *et al.*, 1990) and evaluate interactions among multiple stressors (Barreiro and Pratt, 1994; Pratt and Barreiro, 1998). Mesocosms also provide an efficient way to compare effects among different communities (e.g. from different locations).

2.7.4.3   Field experiments

Although microcosm and mesocosm studies have contributed to our understanding of ecological responses to anthropogenic perturbations, spatial and temporal scaling issues may detract from the ecological realism of these approaches (Carpenter, 1996). Container effects in small model systems may change environmental conditions, alter exposure regimes and limit the duration of microcosm and mesocosm experiments. At least two types of field experiments, whole-system manipulations and manipulative experiments, overcome many of these limitations.

One solution to the limited spatiotemporal scale and lack of ecological realism of model ecosystems is the direct application of contaminants in the field. One of the first manipulations conducted in a riparian ecosystem examined the effects of clear cutting and herbicide applications on nutrient budgets in the Hubbard Brook Experimental Forest, New Hampshire (Likens *et al.*, 1970). Wallace and colleagues at Coweeta Hydrologic Laboratory exposed a small stream to the pesticide methoxychlor (Wallace *et al.*, 1982; Cuffney *et al.*, 1984). There are several limitations to ecosystem manipulations (Perry and Troelstrup, 1988), including the difficulty in replicating treatments, high costs and limited types of contaminants that may be investigated. The lack of replication appears to be a major shortcoming of whole ecosystem experiments. Control, randomization and replication are generally considered the major components of a legitimate experiment. Carpenter (1989) estimated that approximately ten replicate lakes were required to detect the effects of contaminants on primary production because of high natural variability. The luxury of this level of replication seems out of reach of most research programs. Even in situations such as the Experimental Lakes Area (Canada), where a large number of lakes are available for manipulation, it is difficult to locate true replicates (Schindler, 1998). As a compromise, sustained, long-term manipulations using unreplicated paired ecosystems may be a reasonable approach for assessing ecosystem responses (Carpenter, 1989; Schindler, 1998). The high cost of ecosystem manipulations has also limited their widespread use in ecotoxicology, but may be justified in some instances because well-designed experiments generate extensive data on responses at different levels of biological organization. Ecosystem experiments often involve multiple investigators and promote cost-effective, interdisciplinary research (Perry and Troelstrup, 1988).

There is a limit to the types of experiments that can be conducted in whole ecosystems. For example, experimental introduction of highly persistent compounds, such as PCBs and dioxins, would not (and should not) be allowed in most natural systems. Integration of smaller-scale studies (microcosms) with ecosystem experiments and taking advantage of unexpected environmental perturbations (Wiens and Parker, 1995) will be essential to understand effects of these persistent, highly toxic compounds.

Experiments that manipulate portions of communities can also overcome the limitations of laboratory methods. Several of the most significant contributions to the discipline of community ecology were derived from manipulative experiments conducted in rocky intertidal habitats (Connell; 1961; Paine, 1966). Rocky intertidal habitats are less complex than many ecosystems and lend themselves to simple experimental manipulation. Removing competitors or excluding predators is relatively simple in these essentially two-dimensional systems, where most of the organisms are either sessile or very slow moving. Transplanting major portions of communities between polluted and control sites can be a powerful manipulative approach. Courtney and Clements (2002) transferred benthic communities from an unpolluted tributary to a metal-contaminated stream, but to our knowledge there have been few similar attempts by ecotoxicologists.

2.7.4.4   Natural experiments

Natural experiments differ from other experimental methods in that the researcher does not directly control variables of interest. Large-scale monitoring studies that compare communities under varying influences of perturbation are analogous to Diamond's (1986) definition of natural experiments. Because treatments are not assigned randomly, however, these designs also suffer from some of the same limitations as field experiments. Natural trajectory experiments have been used when researchers took advantage of planned perturbations. If data are collected before a particular stress occurs, the before-after control-impact (BACI) design (Stewart-Oaten *et al.*, 1986) is a powerful 'quasi-experimental' approach. Based on experiences from the *Exxon Valdez* oil spill, Wiens and Parker (1995) reviewed quasi-experimental approaches for assessing the impacts of unplanned perturbations. They stress that experimental designs that treat the level of contamination as a continuous variable are generally more precise and offer the greatest opportunity to detect nonlinear responses.

2.7.5   AN INTEGRATED APPROACH TO DEMONSTRATE CAUSATION

Biological assessments and monitoring are frequently used to determine the status of stream ecosystems, but rarely indicate clear causes when impairment is revealed (Suter *et al.*, 2002). The need for attributing ecological impairment to specific natural and anthropogenic causes is obviously prerequisite to remediation. However, cause–effect relationships are notoriously elusive because potential causative factors are not

Table 2.7.2 Assembly rules for construction of an argument supporting causal inference (from Beyers, 1998; integration of Hill, 1965 and Suter, 1993). Reproduced by permission of the North American Benthological Society from Beyers, D. W., 1998, 'Causal inference in environmental impact studies', *Journal of the North American Benthological Society*, **17**, 367–373

Rule	Description
Strength	A large proportion of individuals are affected in the exposed areas relative to unexposed areas
Consistency	The association has been observed by other investigators at other times and places
Specificity	The effect is diagnostic of exposure
Temporality	Exposure must precede the effect in time
Biological gradient	The risk of effect is a function of magnitude of exposure (e.g. dose–response relationship)
Biological plausibility	A plausible mechanism of action links cause and effect
Experimental evidence	A valid experiment provides strong evidence of causation
Analogy	Similar stressor cause similar effects
Coherence	The causal hypothesis does not conflict with existing knowledge of natural history and biology
Biological indicator	Indicators of exposure to the stressor must be found in the organisms

controlled in most monitoring and assessment designs. Hill (1965) proposed several criteria for strengthening causal relationships in epidemiological studies, which have been repeated and modified by others (Suter, 1993; Beyers, 1998; Suter *et al.*, 2002). We repeat these rules as modified by Beyers (1998) in Table 2.7.2, recognizing that they may not be universally applicable to all types of anthropogenic stressors.

Suter *et al.* (2002) describe a system for causal inference that emphasizes the integration and weighing of case-specific information with evidence from other sources. Temporal or spatial associations of cause and effect, relevant field or laboratory experiments and diagnostic evidence (e.g. biomarkers) are examined to construct conceptual models of alternative causal agents. Exclusion of least likely alternatives, diagnostic protocols and strength of evidence are then used to identify the most probable causes of observed impairment. Norton *et al.* (2002) and Corimer *et al.* (2002) apply this system to diagnose the causes of impairment in an Ohio river.

Clements *et al.* (2002) detail an integration of monitoring and experimental methods to infer the effects of metal pollution on stream invertebrate communities in a Rocky Mountain stream. Monitoring data revealed that mayfly (Ephemeroptera) taxa richness and the abundance of a mayfly family (Heptageniidae) increased after remediation significantly reduced metal inputs to the Arkansas River, Colorado. Microcosm experiments were used to develop exposure–response relationships for metal mixtures and population- and community-level measures. Field experiments demonstrated that filter-feeding caddisflies accumulated metals. Field experiments also showed that abundance of mayflies, and heptageniids in particular, were disproportionately reduced after colonized substrates from an unpolluted site were transferred to a polluted site. Natural experiments showed that species richness and abundance of heptageniids were significantly lower in stream segments downstream of metal inputs than unpolluted upstream segments. A large-scale spatial gradient of dissolved metals was also used in a natural experiment to show that the above patterns observed in the Arkansas River could be generalized throughout the southern Rocky Mountains. This integration of monitoring and ecotoxicological experimentation resulted in the development of robust indicators of metal contamination for a large geographic region. Similar applications are needed for other stressors and geographic areas.

2.7.6 MONITORING QUESTIONS REVISITED

We have explored the tools available for ecotoxicological investigations in river pollution assessment. We assert that integration of observational and experimental methods, across scales of time, space, and biological organization strengthens weight-of-evidence approaches. In the following sections, we briefly review several studies that have embraced these principles and, as a result, proved insightful to scientists and managers. We show how the basic three questions from monitoring studies can be convincingly addressed using integrated methods.

Question 1: Stressor identification?   Distinguishing natural from anthropogenic variation in ecosystems is one of the most challenging aspects of interpreting results of biomonitoring studies (Clements and Kiffney, 1995; Growns *et al.*, 1995). Estimating natural background variation is especially important where anthropogenic disturbance is subtle (Clements and Kiffney, 1995). Accounting for natural variation often requires extensive background data, such as results obtained from long-term biomonitoring projects or large-scale studies involving several watersheds. In Rocky Mountain streams, we found that natural factors such as seasonality, stream size and alkalinity, are often confounded with heavy-metal concentrations (Clements and Kiffney, 1995). Despite the daunting complexity of natural and anthropogenic influences on communities, some experiments have successfully identified the relative importance of multiple stressors and their possible interactions.

Ecotoxicological experiments can be designed to quantify the influence of natural environmental conditions on biological responses to anthropogenic stressors. Factorial designs using microcosms or mesocosms can be used to compare the impacts of a stressor on communities collected from different seasons or obtained from different locations. Barreiro and Pratt (1994) demonstrated that communities established under low nutrient conditions were more susceptible to chemical stress and required longer time to recover than communities from enriched environments. Kiffney and Clements (1996) used microcosms to compare the responses of benthic macroinvertebrate communities collected from two stream sizes to heavy metals. Because communities from both stream sizes were exposed to the same metal concentrations, the experiment provided an opportunity to estimate differences in sensitivity between locations. Their results showed that headwater communities were more sensitive to heavy metals than communities from a lower-elevation stream, which suggests that criterion values protective of low-elevation communities may not be protective of those from high elevations (Kiffney and Clements 1996). This pattern was reversed for diatom assemblages. Because headwater streams were naturally dominated by early successional species (*Achnanthes minutissima*), which are also tolerant of metals, these communities showed little response to metals in experimental streams relative to communities from larger streams (Medley and Clements, 1998).

Perhaps the most important contribution of microcosm and mesocosm experiments, which cannot be easily investigated in ecosystem manipulations or natural experiments, is the opportunity to measure stressor interactions. Using simple factorial designs, researchers can investigate effects of multiple stressors simultaneously and estimate the potential interaction among stressors (Genter *et al.*, 1988; Genter, 1995; Courtney and Clements, 2000). Genter (1995) used stream microcosms to quantify the interactive effects of acidification and aluminum on periphyton communities, and to measure the indirect effects of heavy metals through snail grazing (Genter *et al.*, 1988). Studies where direct and interactive effects of multiple stressors are investigated simultaneously require a degree of control that is not possible in field studies. The opportunity to examine interactions among multiple stressors in microcosm experiments and to develop mechanistic explanations for these interactions will greatly improve our ability to predict responses in natural systems, and identify the relative importance of stressors for management purposes.

Habitat degradation is widely cited as a cause for stream impairment (USEPA, 2002), but the relative importance of habitat and chemical stressors to stream impairment is rarely quantified. Courtney and Clements (2002) allowed substrate-filled trays to colonize in an unpolluted tributary, and then placed these trays in the Animas River (Colorado) to measure the direct effects of aqueous contamination. A substrate with a contaminated biofilm from the Animas River was placed in the unpolluted tributary, and colonization was compared with a clean substrate placed adjacent to the contaminated substrate. The effects of chronic exposure to the contaminated substrate were studied with a mayfly-growth experiment in the laboratory. Their results showed that although several taxa were sensitive of aqueous metal concentrations, some were sensitive to both aqueous exposure and degraded substrate. Chronic exposure to the contaminated substrate reduced the growth rates of grazing mayflies. This approach provides a useful example of how to experimentally isolate potential causes of stream degradation. There are, however, surprisingly few studies that have attempted to decouple the potential interacting effects of habitat degradation and chemical contamination in streams.

Question 2: Integration of endpoints across biological organization levels? Because the ultimate goal of environmental regulation is protection of ecological integrity, ecological relevance is an essential consideration in endpoint selection. Contaminant effects are manifested at several levels of biological organization, and responses at each level provide different types of information concerning overall ecological effect. Responses at lower levels of organization are more closely linked to exposure, whereas those at higher levels are better indicators of ecological significance (Clements, 1997).

A significant contribution of Wallace's experiments (Wallace *et al.*, 1982) was the establishment of a relationship between structural and functional characteristics of headwater streams under anthropogenic stress. Wallace and colleagues found that functional measures were relatively sensitive to chemical stress. Application of methoxychlor resulted in significant alterations in detritus processing in the treated stream. These results suggest that toxic effects on sensitive populations can have cascading impacts on ecosystem dynamics, such as organic matter processing and export of particulate material. The inhibition of ecosystem processes was an indirect effect of toxic exposure, and led to far greater ecological degradation than expected based on direct toxicity alone (Wallace *et al.*, 1989).

In another integrative study, Fairchild *et al.* (1992) measured the effects of pyrethroid insecticides on population- (bluegill survival, growth, reproduction), community- (zooplankton and macroinvertebrate community structure) and ecosystem-processes (production and respiration). Although direct mortality to fish and invertebrates was predictable from laboratory tests, an understanding of factors that regulate ecosystem structure and function was necessary to predict indirect effects such as competition and increased grazing pressure.

These present authors (Carlisle and Clements, 2003) integrated experimental and observational approaches to link metal effects across scales of ecological organization in Rocky Mountain streams. Populations of species within the mayfly family Heptageniidae are predictably reduced in the presence of metal mixtures. Although this information alone is ecologically meaningful, the ramifications of reduced heptageniid populations for the overall community and ecosystem are unknown. Carlisle and Clements (2003) used natural and microcosm experiments to measure functional attributes of heptageniid populations along a metal-pollution gradient. Secondary production and seasonal food consumption were estimated for three heptageniid species. Secondary production is an integrative measure of the ultimate reproductive output and energetic flows through a population (Benke, 1993). When population production is coupled with detailed dietary information, consumption can be estimated using bioenergetic principles (Benke *et al.*, 1988). This approach has been used to quantify the trophic roles of a species in food webs (Benke and Wallace, 1997). We found that heptageniids are responsible for a disproportionate amount of total energy flow through invertebrate populations in unpolluted Colorado streams. The individual growth rates of some species were also reduced in metal-polluted streams. By simultaneously measuring individual and population measures of structure and function, the following patterns emerged. At Zn concentrations that exceeded the chronic criterion concentration (defined as the concentration protective of aquatic organisms), population abundances of all heptageniids are reduced, presumably by direct toxicity to individuals. Those remaining individuals are forced to divert energy from growth to detoxify metals, and consequently grow slower. Slower growth

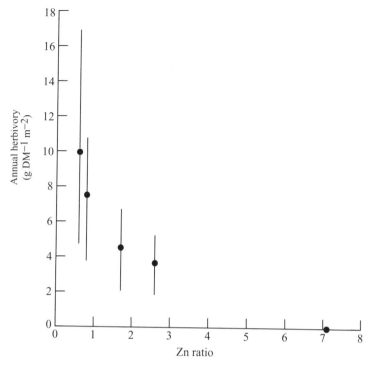

Figure 2.7.2 Total annual algal consumption by grazing mayflies (Ephemeroptera) in five Rocky Mountain streams, central Colorado, 1998. The study design was a natural experiment where similar streams were compared along a gradient of metal contamination: DM, dry mass; the Zn ratio is the annual mean ratio of observed dissolved Zn concentration to the hardness-adjusted criterion value

rates are also likely caused by reduced food quality and quantity. As a result of lower population abundance and individual growth, total resource consumption by grazing mayflies is reduced by as much as 50 % (Figure 2.7.2). In moderately contaminated streams, metal-tolerant species replace sensitive heptageniids, but they do not replace their role in food-web energetics. Consequently, stream ecosystems were depleted of energy as the heptageniid assemblage diminished. By integrating these functional attributes with known structural responses to metals, the relevance of heptageniid losses to the ecosystem are more clear.

We suggest that without establishing causal linkages between levels of biological organization, molecular and biochemical responses will remain indicators of exposure with little or no ecological relevance. Similarly, without understanding the mechanisms responsible for observed ecological changes in communities and ecosystems, responses at higher levels of organization often cannot be linked to specific contaminants. Despite the importance in doing so, few studies have simultaneously investigated the effects of contaminants at different levels of organization.

Question 3: Ecologically relevant concentration–response relationships?　One important use of concentration–response relationships is to identify contaminant levels protective of specific endpoints. Since communities and ecosystems are the targets of our protective strategies, concentration–response relationships at these levels would allow the establishment of ecologically relevant remediation goals. Concentration–response relationships for communities and ecosystems are virtually nonexistent, and 'no-observed effect levels' (NOELs) established for single species in laboratory bioassays cannot be applied to natural systems. Instead, the direct effects of chemicals on community responses could be quantified in mesocosm experiments.

An important application of microcosm and mesocosm research is to establish concentration–response relationships between contaminants and community level endpoints. If treatments are selected to represent a range of potential responses, we can predict effects at various chemical concentrations (e.g. the concentration that results in 20 % reduction in species richness). Clements (2004) described a series of experiments in which indigenous macroinvertebrate communities were exposed to metal mixtures representative of contamination throughout Colorado. After a 10-d exposure, the EC_{10} (concentration at which endpoint changed 10 % from the control) values for a variety of populations, community richness and community respiration were estimated. Expressed as hardness-adjusted chronic criterion units, the lowest EC_{10} values were 4.9 for *Epeorus longimanus* and 5.0 for community respiration. Clements therefore recommended that a chronic criterion unit of 5.0 was probably safe for benthic communities. Managers charged with metal remediation in Colorado streams now have a clean-up level that is ecologically relevant.

Because most microcosm and mesocosm experiments involve exposure of numerous species simultaneously, regression approaches can be used to estimate species-specific sensitivity to particular contaminants. The slopes of concentration–response relationships for individual taxa provide an objective estimate of tolerance and can be used to develop biotic indices. These population and community responses observed in mesocosms could then be verified by using routine field biomonitoring.

2.7.7　SYNTHESIS

The most obvious way to establish causality is by experimentation, but as discussed in this chapter there are significant limitations to all experimental approaches employed in ecotoxicology (Diamond, 1986). Indirect effects, stressor interactions and potential artifacts introduced by the experimental system complicate interpretations of experimental studies. Even complex factorial designs which investigate stressor interactions are limited in the number of variables that can be manipulated. Because the goal of many experiments is to demonstrate the importance of a single factor (e.g. the effects of a specific chemical), the connection between the experiment and the natural system is often lost. Furthermore, while a well-designed experiment with sufficient power can demonstrate the statistical significance of a single factor, the

importance of this factor relative to other unmanipulated variables remains unknown
without supporting comparative data. Consequently, a variety of experimental and
observational approaches at different temporal and spatial scales and at different
levels of biological organization must be employed (Figure 2.7.3). We assert that
thoughtful integration of multiple approaches produces the most credible evidence
for causation. Indeed, there is much to be learned by comparing patterns observed in
nature to those predicted from theory, and a successful research program should combine theory, observations, and experiments. Werner (1998) describes the advantages

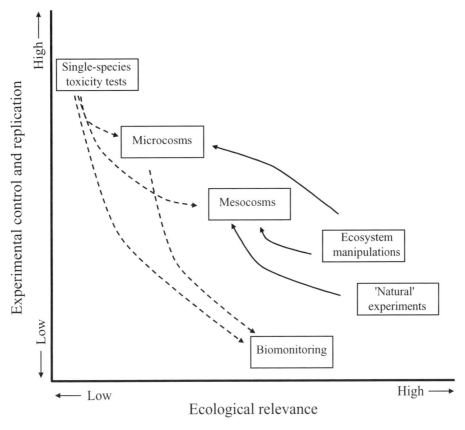

Figure 2.7.3   The relationship between ecological relevance, and experimental control and replication, in ecotoxicological assessments presented as two continua. Small-scale laboratory and microcosm experiments lack ecological realism but are easily replicated and controlled. Ecosystem
manipulations and natural experiments have greater ecological relevance but lack rigorous control
and reliable replicability. Environmental assessments that integrate experimental approaches at
different spatial and temporal scales are optimal for determining causation. For example, underlying causative mechanisms responsible for changes observed in unreplicated observational studies
can be rigorously tested in small-scale studies (represented by continuous lines). Similarly, the
ecological relevance of laboratory-based conclusions can be determined by conducting studies at
larger spatial and temporal scales (represented by dashed lines)

of a research program that integrates experimental techniques with theoretical and comparative approaches for understanding basic ecological patterns. The same argument applies to ecotoxicology.

REFERENCES

Barreiro, R. and Pratt, J. R., 1994.' Interaction of toxicants and communities: the role of nutrients'. *Environmental Toxicology and Chemistry*, **13**, 361–368.

Benke, A. C. 1993. 'Concepts and patterns of invertebrate production in running waters', Edgardo Baldi Memorial Lecture. *Verhandlungen der Internationalen Vereinigung für Theoretische und Angewandte Limnologie*, **25**, 15–38.

Benke, A. C. and Wallace, J. B., 1997. 'Trophic basis of production among riverine caddisflies: implications for food web analysis'. *Ecology*, **78**, 1132–1145.

Benke, A. C., Hall, C. A. S., Hawkins, C. P., Lowe-McConnell, R. H., Stanford, J. A., Suberkropp, K. and Ward, J. V., 1988. 'Bioenergetic considerations in the analysis of stream ecosystems'. *Journal of the North American Benthological Society*, **7**, 480–502.

Beyers, D. W., 1998. 'Causal inference in environmental impact studies'. *Journal of the North American Benthological Society*, **17**, 367–373.

Cairns, J., Jr, 1983. 'Are single species toxicity tests alone adequate for estimating environmental hazard?'. *Hydrobiologia*, **100**, 47–57.

Cairns, J., Jr, 1986. 'The myth of the most sensitive species'. *BioScience*, **36**, 670–672.

Carlisle, D. M. and Clements, W. H., 1999. 'Sensitivity and variability of metrics used in biological assessments of running waters'. *Environmental Toxicology and Chemistry*, **18**, 285–291.

Carlisle, D. M. and Clements, W. H., 2003. 'Growth and secondary production of aquatic insects along a gradient of Zn contamination in Rocky Mountain streams'. *Journal of the North American Benthological Society*, **22**, 582–597.

Carpenter, S. R., 1989. 'Replication and treatment strength in whole-lake experiments'. *Ecology*, **70**, 453–463.

Carpenter, S. R., 1996. 'Microcosm experiments have limited relevance for community and ecosystem ecology'. *Ecology*, **77**, 677–680.

Clements, W. H., 1997. 'Effects of contaminants at higher levels of biological organization in aquatic ecosystems'. *Reviews in Toxicology*, **1**, 107–146.

Clements, W. H., 2004. 'Small-scale experiments support causal relationships between metal contamination and macroinvertebrate community responses'. *Ecological Applications*, **14**, 954–967.

Clements, W. H. and Kiffney, P. M., 1995. 'The influence of elevation on benthic community responses to heavy metals in Rocky Mountain streams'. *Canadian Journal of Fisheries and Aquatic Sciences*, **52**, 1966–1977.

Clements, W. H. and Newman, M. C., 2002. *Community Ecotoxicology*. Wiley: New York, NY, USA.

Clements, W. H., Cherry, D. S. and Cairns, J., Jr, 1989. 'The influence of copper exposure on predator–prey interactions in aquatic insect communities'. *Freshwater Biology*, **21**, 483–488.

Clements, W. H., Carlisle, D. M., Courtney, L. A. and Harrahy, E. A., 2002. 'Integrating observational and experimental approaches to demonstrate causation in stream biomonitoring studies'. *Environmental Toxicology and Chemistry*, **21**, 1138–1146.

Connell, J. H., 1961. 'The influence of interspecific competition and other factors on the distribution of the barnacle *Chthamalus stellatus*'. *Ecology*, **42**, 710–723.

Corimer, S. M., Norton, S. M., Suter II, G. W., Altfater, D. and Counts, B., 2002. 'Determining the causes of impairments in the Little Scioto River, Ohio, USA: Part 2. Characterization of causes'. *Environmental Toxicology and Chemistry*, **21**, 1125–1137.

Courtney, L. A. and Clements, W. H., 2000. 'Sensitivity to acidic pH in benthic invertebrate assemblages with different histories of exposure to metals'. *Journal of the North American Benthological Society*, **19**, 112–127.

Courtney, L. A. and Clements, W. H., 2002. 'Assessing the influence of water quality and substratum quality on benthic macroinvertebrate communities in a metal-polluted stream: an experimental approach'. *Freshwater Biology*, **47**, 1766–1778.

Cuffney, T. F., Wallace, J. B. and Webster, J. R., 1984. 'Pesticide manipulation of a headwater stream: invertebrate responses and their significance for ecosystem processes'. *Freshwater Invertebrate Biology*, **3**, 153–171.

Daehler, C. C. and Strong, D. R., 1996. 'Can you bottle nature? The roles of microcosms in ecological research'. *Ecology*, **77**, 663–664.

Diamond, J. M., 1986. 'Overview: laboratory experiments, field experiments, and natural experiments'. In: *Community Ecology*, Diamond, J. M. and Chase, T. J. (Eds). Harper and Row: New York, NY, USA, pp. 3–22.

Eaton, D. L. and Klaassen, C. D., 1996. 'Principles of toxicology'. In: *Casarett and Doull's Toxicology: The Basic Science of Poisons*, Klaassen, C. D., Amdur, M. O. and Doull, J. (Eds). McGraw-Hill: New York, NY, USA, pp. 13–33.

Fairchild, J. F., La Point, T. W., Zajicek, J. L., Nelson, J. K., Dwyer, F. J. and Lovely, P. A., 1992. 'Population-, community- and ecosystem-level responses of aquatic mesocosms to pulsed doses of a pyrethroid insecticide'. *Environmental Toxicology and Chemistry*, **11**, 115–129.

Gearing, J. N., 1989. 'The role of aquatic microcosms in ecotoxicologic research as illustrated by large marine systems'. In: *Ecotoxicology: Problems and Approaches*, Levin, S. A., Harwell, M. A., Kelly, J. R. and Kimball, K. D. (Eds). Springer-Verlag: New York, NY, USA, pp. 409–470.

Genter, R. B., 1995. 'Benthic algal populations respond to aluminum, acid, and aluminum-acid mixtures in artificial streams'. *Hydrobiologia*, **306**, 7–19.

Genter, R. B., Colwell, F. S., Pratt, J. R., Cherry, D. S. and Cairns, J., Jr, 1988. ' Changes in epilithic communities due to individual and combined treatments of zinc and snail grazing in stream mesocosms'. *Toxicology and Industrial Health*, **4**, 185–201.

Giddings, J. M., Biever, R. C., Annunziato, M. F. and Hosmer, A. J., 1996. 'Effects of diazinon on large outdoor pond microcosms'. *Environmental Toxicology and Chemistry*, **15**, 618–629.

Gillett, J. W., 1989. 'The role of terrestrial microcosms and mesocosms in ecotoxicologic research'. In: *Ecotoxicology: Problems and Approaches*, Levin, S. A., Harwell, M. A., Kelly, J. R. and Kimball, K. D. (Eds). Springer-Verlag: New York, NY, pp. 1–67.

Gonzalez, M. J. and Frost, T. M., 1994. 'Comparisons of laboratory bioassays and a whole-lake experiment: rotifer responses to experimental acidification'. *Ecological Applications*, **4**, 69–80.

Graney, R. L., Giesy, Jr, J. P. and DiToro, D., 1989. 'Mesocosm experimental design strategies: advantages and disadvantages in ecological risk assessment'. In: *Using Mesocosms to Assess the Aquatic Ecological Risk of Pesticides: Theory and Practice*, Voshell, Jr, J. R. (Eds). Entomological Society of America: Lanahm, MD, USA, pp. 74–88.

Growns, J. E., Chessman, B. C., Mcevoy, P. K. and Wright, I. A., 1995. 'Rapid assessment of rivers using macroinvertebrates: Case studies in Nepean River and Blue Mountains, NSW'. *Australian Journal of Ecology*, **20**, 130–141.

Hill, A. B., 1965. 'The environment and disease: association or causation'. *Proceedings of the Royal Society of Medicine*, **58**, 295–300.

Kennedy, J. H., Johnson, Z. B., Wise, P. D. and Johnson, P. C., 1995. 'Model aquatic ecosystems in ecotoxicology research: considerations of design, implementation, and analysis'. In: *Handbook *of Ecotoxicology*, Hoffman, D. J., Rattner, B. A., Burton, Jr, G. A. and Cairns, Jr, J. (Eds). CRC Press: Boca Raton, FL, USA, pp. 117–162.

Kiffney, P. M. and Clements, W. H., 1996. 'Effects of metals on stream macroinvertebrate assemblages from different altitudes'. *Ecological Applications*, **6**, 472–481.

Lawton, J. H., 1996. 'The Ecotron facility at Silwood Park: The value of "big bottle" experiments'. *Ecology*, **77**, 665–669.

Likens, G. E., Bormann, F. H., Johnson, N. M., Fisher, D. W. and Pierce, R. S., 1970. 'Effects of forest cutting and herbicide treatment on nutrient budgets in the Hubbard Brook Watershed-ecosystem'. *Ecological Monographs*, **40**, 23–47.

Medley, C. N. and Clements, W. H., 1998. 'Responses of diatom communities to heavy metals in streams: The influence of longitudinal variation'. *Ecological Applications*, **8**, 631–644.

Niederlehner, B. R., Pontasch, K. W., Pratt, J. R. and Cairns, Jr, J., 1990. 'Field evaluation of predictions of environmental effects from a multispecies-microcosm toxicity test'. *Archives of Environmental Contamination and Toxicology*, **19**, 62–71.

Niyogi, D. K., Lewis, Jr, W. M. and McKnight, D. M., 2001. 'Litter breakdown in mountain streams affected by mine drainage: biotic mediation of abiotic controls'. *Ecological Applications*, **11**, 506–516.

Norton, S. B., Corimer, S. M., Suter II, G. W., Subramanian, B., Lin, E., Altfater, D. and Counts, B., 2002. 'Determining the causes of ecological impairment in the Little Scioto River, Ohio, USA: Part 1. Listing candidate causes and analyzing evidence'. *Environmental Toxicology and Chemistry*, **21**, 1112–1124.

Odum, E. P., 1984. 'The mesocosm'. *BioScience*, **34**, 558–562.

Paine, R. T., 1966. 'Food web complexity and species diversity'. *American Naturalist*, **100**, 65–75.

Peckarsky, B. L., 1998. 'The dual role of experiments in complex and dynamic natural systems'. In: *Experimental Ecology: Issues and Perspectives*, Resetarits, Jr, W. J. and Bernardo, J. (Eds). Oxford University Press: New York, NY, USA, pp. 311–324.

Perry, J. A. and Troelstrup, Jr, N. H., 1988. 'Whole ecosystem manipulation: a productive avenue for test system research?'. *Environmental Toxicology and Chemistry*, **7**, 941–951.

Pontasch, K. W., 1995. 'The use of stream microcosms in multispecies testing'. In: *Ecological Toxicity Testing: Scale, Complexity, and Relevance*, Cairns, Jr, J. and Niederlehner, B. R. (Eds). CRC Press: Boca Raton, FL, USA, pp. 169–191.

Pontasch, K. W., Niederlehner, B. R. and Cairns, Jr, J., 1989. 'Comparisons of single-species, microcosm and field responses to a complex effluent'. *Environmental Toxicology and Chemistry*, **8**, 521–532.

Pratt, J. R. and Barreiro, R., 1998. 'Influence of trophic status on the toxic effects of a herbicide: a microcosm study'. *Archives of Environmental Contamination and Toxicology*, **35**, 404–411.

Rosenberg, D. M. and Resh, V. H., 1993. *Freshwater Biomonitoring and Benthic Macroinvertebrates*. Chapman & Hall, New York, NY, USA.

Schindler, D. W., 1987. 'Detecting ecosystem responses to anthropogenic stress'. *Canadian Journal of Fisheries and Aquatic Sciences*, **44**, (Supplement 1), 6–25.

Schindler, D. W., 1998. 'Replication versus realism: The need for ecosystem-scale experiments'. *Ecosystems*, **1**, 323–334.

Shaw, J. L. and Kennedy, J. H., 1996. 'The use of aquatic field mesocosm studies in risk assessment'. *Environmental Toxicology and Chemistry*, **15**, 605–607.

Stewart-Oaten, A., Murdoch, W. W. and Parker, K. R., 1986. 'Environmental impact assessment: "pseudoreplication" in time?'. *Ecology*, **67**, 929–940.

Suter, G. W., II, 1993. *Ecological Risk Assessment*. Lewis publishers: Boca Raton, FL, USA.

Suter, G. W., II, Norton, S. B. and Corimer, S. M., 2002. 'A methodology for inferring the causes of observed impairments in aquatic communities'. *Environmental Toxicology and Chemistry*, **21**, 1101–1111.

USEPA, 2002. National Water Quality Inventory: 2000 Report, EPA 841-R-02-001. Office of Water, United States Environmental Protection Agency: Washington, DC, USA.

Wallace, J. B., Webster, J. R. and Cuffney, T.F., 1982. 'Stream detritus dynamics: regulation by invertebrate consumers'. *Oecologia*, **53**, 197–200.

Wallace, J. B., Lugthart, G. F., Cuffney, T. F. and Schurr, G. A., 1989. 'The impact of repeated insecticidal treatments on drift and benthos of a headwater stream'. *Hydrobiologia*, **179**, 135–147.

Werner, E. E., 1998. 'Ecological experiments and a research program in community ecology'. In: *Experimental Ecology: Issues and Perspectives*, Resetarits, Jr, W. J. and Bernardo, J. (Eds). Oxford University Press: New York, NY, USA, pp. 3–26.

Wiens, J. A. and Parker, K. R., 1995. 'Analyzing the effects of accidental environmental impacts: approaches and assumptions'. *Ecological Applications*, **5**, 1069–1083.

Section 3

Various Experiences in Biological Monitoring

3.1
Monitoring of Alpine Rivers: The Italian Experience

Francesca Ciutti and **Giovanna Flaim**

3.1.1   INTRODUCTION

Because of its geography, Italy's hydrological network has few large rivers (Po, Adige, Tevere). Most water courses are of the 'alpine typology' – fairly short streams, with torrential flow, originating in the Alps or Appenines. Anthropogenic impacts on stream integrity are those experienced in all industrialized nations: civil, agricultural and industrial pollution. In addition, streams in mountain habitats are subject to a series of unique impacts, mostly tied to alterations of their physical and morphological components.

Biological Monitoring of Rivers   Edited by G. Ziglio, M. Siligardi and G. Flaim
© 2006 John Wiley & Sons, Ltd.

3.1.1.1   Special impacts

Mountain streams, with their combination of flow and altitude, have a great potential for hydroelectric power production, explaining why almost 20 % of energy
production in Italy originates from hydroelectric power plants (APAT, 2003) and
nearly all of it is produced in the Alpine regions. The presence of hydroelectric
reservoirs influences downstream ecosystems in several ways by causing physical
alterations due to the discharge regime and changes in water quality during storage
before release downstream (Crisp, 1995; Petts, 1984). In particular, the presence of
peaking hydroelectric power plants causes artificial flow fluctuations and highly unnatural discharge phenomena in terms of flow magnitude, duration, sequence and frequency. This prevents the establishment of a stable and diversified biotic community
(Cushman, 1985; Moog, 1993; Gore, 1994).

In alpine regions, flood control measures are in most cases indispensable, especially in urban areas. Near human settlements, the main strategy for control is
twofold: increase resistance and control of catchment erosion processes, coupled
with channel works to check the high kinetic energy of fast-moving-streams. In particular, channel works consist of two classes of structure: cross-sectional structures,
such as check dams, sills and groins, which serve to check the high kinetic energy of
torrents, and longitudinal, structures such as embankments and revetments, which
serve to increase the scouring resistance of the channel (Kostadinov, 1995). The
physical simplification of stream morphology causes a consequent reduction in self-
purification processes in streams and interruption in water flow, both transversally
and longitudinally.

Most alpine streams are also affected by sewage pollution. Tourism is a major
source of revenue in mountain areas and during peak summer and winter tourist
seasons with their greatly increased nutrients loads, raw effluents frequently enter
waterways directly, bypassing undersized treatment plants.

In addition, in typical trout streams, high fishing pressure has resulted in using a
heavy hand in management and at least until very recently massive restocking has
altered equilibrium, especially in 'hard-to-find' natural conditions. During the last
few decades, deliberate or fortuitous introduction of allochtonous species has often
wrecked havoc in fish communities.

The status of watercourses in Italy's alpine environment more or less mirrors other
alpine realities with anthropogenic impacts such as land use and alterations in river
bed morphology and hydrology making pristine or nearly pristine ecosystems rare.
Martinet and Dubost (1992) observe that in general only 10 % of alpine water courses
can be considered 'near-natural'. In Austria, a recent study has evidenced that only
6 % of water ways maintain a natural/'nature-like' character (79 % are moderately
to heavily impacted) (Muhar *et al.*, 2000).

Another aspect that must be mentioned is that at least for the Italian situation, national monitoring programmes do not normally include mid-to-high altitude stations,
notwithstanding that these sites are subject to important impacts (water harness,
tourism pressure, climate change). Even though headwaters show high sensitivity

and vulnerability to environmental changes, on both the local and global scale they are studied only within specific and sporadic research projects (Maiolini and Lencioni, 2001; Brittain *et al.*, 1998). In any case, the monitoring methods normally used have to be calibrated against these specific environments because often there is an altitudinal limit for method application.

3.1.2 MONITORING EXPERIENCE IN ITALY

3.1.2.1 Water protection legislation

Up to the early 1970s, water protection laws in Italy considered running waters only as receptacles for waste (Law 10, May 1976, no. 319 – Merli) and defined maximum concentration limits in effluents without considering the flow volume of the receiving watercourse. As river monitoring evolved towards an 'ecosystem' approach, development of methodologies for water quality classification based on the aquatic community was promoted in Italy as in other European and 'extra-European' countries. Biological river monitoring enhanced the use of biotic indices and their territorial application. Simultaneously, there was also an evolution in environmental legislation which only slowly adapted to the new view of running waters, in part because the biological approach to monitoring was only slowly accepted by technical personnel responsible for environmental surveillance.

With respect to the 'Merli Law', a radical change was bought about in 1999 with Decree 152/1999 (Decreto Legislativo, 1999) and subsequent amendments that adapted an integrated approach for the protection of river ecosystems by considering biological, chemical, physical and morphological characteristics. In particular, this Decree 152/99 introduced an innovative concept in water surveillance by defining the 'ecological state of surface waters' (SECA) as 'expressions of the complexity of aquatic ecosystems, and the physical and chemical nature of water and sediments, of the characteristics of water flow and the morphology of the water body, while considering as priority the biological components of the system'. With the passage of Decree 152/99 and subsequent amendments, the 'Indice Biotic Esteso' (IBE) (Extended Biotic Index) method becomes an official method for the determination of water quality in running waters, together with the more traditional chemical, physical and bacteriological parameters. Some of these parameters (ammonium N, nitrate N, dissolved oxygen, BOD_5, total phosphorus and *Escherichia coli*) form the LIM (macro-descriptor pollution level), as described in Table 3.1.1. The definition of the ecological state of running waters (SECA) is determined by an integration of information derived from the LIM and IBE with the lower value determining the quality class for a specific water course (Table 3.1.2).

An overview of water quality in Italy according to the principles of Decree 152 (APAT, 2003) was taken using the 2002 monitoring programme based on available streams and sampling sites for LIM (578 sites), IBE (551 sites) and SECA (513 sites). The results show that the majority of Italian waterways are in II and III quality

Table 3.1.1 Pollution level expressed as macrodescriptors, from Table 7, Annex, 1 of Italian Water Protection Decree 152/99

Parameter	Level 1	Level 2	Level 3	Level 4	Level 5
100-DO (% saturated)[a]	$\leq 10^b$	≤ 20	≤ 30	≤ 50	> 50
BOD_5 (O_2 mg/l^{-1})	< 2.5	≤ 4	≤ 8	≤ 15	> 15
COD (O_2 mg/l^{-1})	< 5	≤ 10	≤ 15	≤ 25	> 25
NH_4 (N mg/l^{-1})	< 0.03	≤ 0.10	≤ 0.5	≤ 1.50	> 1.50
NO_3 (N mg/l^{-1})	< 0.3	≤ 1.5	≤ 5.0	≤ 10.0	> 10.0
Total phosphorous (P mg/l^{-1})	< 0.07	≤ 0.15	≤ 0.30	≤ 0.60	> 0.60
Escherichia coli (UFC (100/mL)$^{-1}$	< 100	≤ 1000	≤ 5000	$\leq 20\,000$	$> 20\,000$
Score given for every measured parameter (75° percentile during monitoring period)	80	40	20	10	5
Pollution level according to macrodescriptors	480–560	240–475	120–235	60–115	< 60

[a] Measurement must be made in absence of vortices, the datum of deficit or surplus has to be considered an absolute value.
[b] In the absence of eutrophication.

class (86 % for LIM and 70 % for IBE), while only 3 % of the sites are in class I for chemical quality but 13 % for biological quality. For overall ecological quality (SECA), the vast majority of sites (79 %) fall between class II and III (Table 3.1.3). Looking only at the SECA data and assuming the absence of micro-pollutants, we see that 20 % of the monitored sites are below the environmental objectives foreseen for 2008 (sufficient ecological state), while a good 61 % are below the environmental objectives foreseen for 2016 (good ecological state).

Decree 152/99 therefore anticipates the principles outlined in the Water Framework Directive (WFD) 2000/60/CE (European Parliament, 2000) for the definition of quality objectives, that also re-enforce the importance of the biological component: this document represents the reference text for Member States in the European Union for the next decades and underlines the importance of evaluating environmental status both through the biological components and other non-biological factors. In particular, the WFD introduces innovative principles, referable to a reference condition which should be defined for each water type in each eco-region. This reference community is needed to evaluate the distance of the observed state from the reference condition, thus defining and programming quality objectives.

Table 3.1.2 Ecological state of water courses (SECA), from Table 8 of Italian Water Protection Decree 152/99; the lower value between IBE and macrodescriptors is considered

Class	1	2	3	4	5
IBE	≥ 10	8–9	6–7	4–5	1, 2, 3
Pollution level based on macrodescriptors	480–560	240–475	120–235	60–115	< 60

Table 3.1.3 Number of samples sites and their distribution among quality classes according to the LIM, IBE and SECA

Class value	LIM		IBE		SECA	
	Sampling stations	%	Sampling stations	%	Sampling stations	%
I	18	3.1	74	13.4	9	1.8
II	304	52.6	195	35.4	189	36.8
III	191	33.0	190	34.5	211	41.1
IV	49	8.5	74	13.4	88	17.2
V	16	2.8	18	3.3	16	3.1

3.1.2.2 Biological monitoring

Different impacts incumbent upon waterways require adequate instruments for their evaluation. A knowledge of the structure and composition of the biological community and of the interacting factors involved in river functionality are therefore needed for a definition of ecological integrity.

Stream macroinvertebrates have the longest history as biological monitoring tools for running waters in Italy. Ghetti (1986) proposed the first version of the (IBE – Extended Biotic Index) method, essentially an adaptation of the Trent Biotic Index (Woodiwiss, 1964), for the Italian situation and provided a first sampling protocol. Since then, the method has been constantly refined on the basis of feedback from workers operating in the field (Sittoni and Flaim, 1985; Ghetti, 1995, 1997; APAT-CNR, 2004). This has gone hand in hand with an evolution and amelioration of the taxonomic guides available for the determination of the Italian macroinvertebrate fauna – very varied because of latitude and altitude ranges that characterize the peninsula. Worthy of mention are Ruffo's series (1977–1985), Sansoni's photographic atlas (1988) and the Campaioli *et al.* series (1994–1999) which have given operators valid instruments for taxa identification and consequently for the diffusion of the IBE method.

Until the beginning of the 1990s, the IBE method was applied sporadically at the national level (Manzini and Spaggiari, 1989) with some provinces and fishery management plans including biologial water quality data based on biotic indexes (Vittori, 1982). The first widespread and rational application on the national scale started in 1992 with Law 130, the implementation of the Council Directive 78/659/EEC on the quality of fresh waters needing protection or improvement to support fish life. This Law encouraged a more in-depth analysis of water quality and invited the regions to promote the realization of monitoring campaigns of biological monitoring of water courses with the IBE method (Spaggiari, 1999).

Fish as quality indicators have only been used in fisheries management plans with investigation of community composition, population structure and analysis of the main growth parameters, and therefore limits the application essentially for fisheries management plans (stocking plans, fishing limits, protection of endangered species).

A first quality index based on fish has been attempted by Zerunian (2004, 2005) and Forneris *et al.* (2005), while Scardi *et al.* (2004) have experimented with neural network as predictive models.

Presently, methods that utilize other biological components considered in the WFD are being developed; some examples are components of aquatic vegetation and in particular, diatoms (Dell'Uomo, 2004) and macrophytes (Minciardi *et al.*, 2003). The development and application of biological methods for environmental quality determination has shown that algal indicators such as diatoms are capable of underlining the different trophic levels in the aquatic community. In particular, epilithic diatoms are known to be excellent quality indicators for running waters and represent a reference group for the study of the aquatic plant community. In many European countries, methods have already been developed using diatoms for quality assessment (Whitton *et al.*, 1991; Whitton and Rott, 1996; Prygiel *et al.*, 1999). In Italy, the Eutrophication-Pollution Index with Diatoms (EPI-D) method (Dell'Uomo, 2004) was originally calibrated for streams in the Central Apennines. Besides organizing workshops on the use of this method, the Province of Trento is also testing the applicability of the index in other settings. The goal is to provide guidelines for the standardization of the method on a national scale, to update the floristic list of Italian diatoms present in Italian running waters, to calibrate the point scale assigned to each species and to test its application in high-elevation environments (Cappelletti *et al.*, 2003; Ciutti *et al.*, 2003; Ciutti, 2005).

Conscious that river ecosystems have to be considered holistically, a further step forward has been the development of the Index of Fluvial Functionality (IFF) that integrates traditional biological methods of investigation with an approach that goes beyond the analysis of the aquatic community and considers the river as a continuum with the riparian ecotone and the surrounding territory (Siligardi *et al.*, 2000). The IFF method has been applied in the Province of Trento for territorial planning and represents a valid environmental instrument that gives a complete picture of restoration phases (Siligardi and Cappelletti, 2006; Chapter 4.3 in this text).

3.1.3   THE MONITORING SITUATION IN AN ALPINE ENVIRONMENT

The Autonomous Province of Trento (Northern Italy – Central Alps), with its 6338 km^2 and its 5419 km long hydrographic network ('river density' of about 0.9 km per km^2), has a long history of biological monitoring and has always favoured the development and application of analytical methods capable of evaluating different types of impacts on stream quality. The use of methods that consider not only biological components (macroinvertebrates, diatoms, fish) but also those factors that interact to determine river functionality for a more holistic evaluation of ecological integrity have always had priority. Without a doubt, stream monitoring with macroinvertebrates in Trento is a consolidated experience; most water courses have a continuous series of macroinvertebrate data from 1985. Punctual and continuous sampling, together with chemical and bacteriological indicators, permit establishing

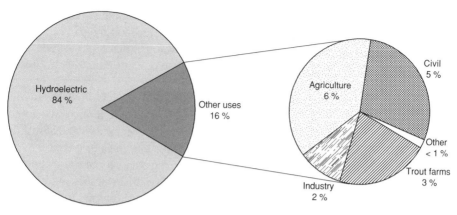

Figure 3.1.1 Water use permits released according to use for the Province of Trento Italy, 2002. Data expressed as percentage of total use (667 m^3 s^{-1}) and refers to first use only; cascading uses are not considered. Data from Ufficio Pianificazioni e Rivelazioni Idriche – PAT. Reproduced by permission of Provincia Autonoma di Trento: Trento, Italy

trends, evidence degraded situations, test law fulfilment and verify the efficiency of restoration plans.

The application of Law 130 confirmed the presence of some high-altitude areas, such as the Adamello Brenta Natural Park, that are still in a pristine state (Corradini *et al.*, 2005). Unfortunately however, many high-altitude sites show critical situations: many sites are subject to regimented water flow combined with heavy tourist pressure. The problem is accentuated in winter when high tourist presence coincides with low water flow.

Even though in the past decade biological quality has improved in lowland stretches, confirming the efforts of water quality improvement plans and the validity of sewage effluent treatment, improvement has not been as much as hoped because of the chronic shortage of hydraulic loading which depresses the possibility of self-purification processes (Provincia Autonoma di Trento, 2000, 2001). As clearly shown in Figure 3.1.1, hydroelectric exploitation of water is very strong and there is a serious need of rigorous policies based on minimum-flow criteria to protect river ecosystems.

In oligotrophic environments, even the presence of trout farms have an impact on receiving waters. In the Province of Trento, there are about 50 trout farms and one of the few surveys carried out has shown that water quality downstream from the farms is degraded both biologically and chemically (Pontalti and Baruchelli, 1988).

3.1.4 MONITORING IN AN ALPINE ENVIRONMENT – SPECIAL CONSIDERATIONS

As already underlined, sites that are found at high altitudes are not normally part of monitoring programmes for Decree 152/99, even though these sites require special attention. Even if some areas have a more dense monitoring network and include minor watercourses, sampling sites are almost always placed at the end of a catchment or downriver from populated areas. We feel, however, that high-altitude sites, for their intrinsically peculiar and fragile nature, should be monitored with greater attention.

At the same time, the environmental conditions typical of high-altitude environments (low trophism, instable morphology) are such that some biotic indices (i.e. EBI) are not applicable above the tree line and, for example, near the source of springs (Ghetti, 1997). It is known that macroinvertebrate taxa and density increase along a gradient with increasing distance from the source in glacial streams as a function of various factors, including water temperature and river bed stability (Milner and Petts, 1994). Presently, there are few studies regarding the altitudinal limits for applying the IBE index in alpine streams of different origin – glacial melt waters and snow pack melt waters and underground springs. Generally speaking, there seems to be no neat altitudinal limit for IBE but its validity depends on stream typology and on the orographic conditions of the watershed; in some cases, it was efficient up to relatively high altitudes, while in others extreme fluvial conditions limiting its use were evident below the tree line (Lencioni *et al.*, 2001, 2002). These aspects need to be better defined.

The same WFD in its system A for the classification of water courses indicates that high-altitude environments are those more than 800 m above sea level. Actually, a first analysis of the high-elevation diatom community (> 800 m above sea level) shows that it would be useful to further divide elevation as > 800 and > 1800 m above sea level and consider distance from the source as a determining factor in defining community structure in algae (Rimet *et al.*, 2004).

Qualitative or semi-quantitative evaluation methods (IBE, EPI-D) are frequently not adequate to describe physical and morphological types of impacts that often have a greater affect on the number of organisms than the relative community composition. This means that the influence of hydraulic loading variation and hydroelectric harnessing can be underestimated if quantitative methods are not used. As an example, the evaluation of 'drop-down' of hydroelectric reservoirs often has an underestimated impact on alpine water courses. The progressive deposition of sediment in basins causes the blockage of outlets, directly influencing the safety of reservoirs; the least expensive and often used maintenance solution is flushing of the sediments through the bottom outlet. In a study regarding the impact of hydroelectric reservoir maintenance operations on the downstream macroinvertebrate community, quantitative sampling showed a reduction in the number of individuals up to 87 % in the five days following 'drop-down' with respect to the previous situation. This showed a strong impact on the downstream community which was not seen in the slight reduction in the number of taxa (Ciutti *el al.*, 2000). Therefore, tools which evaluate ecosystem functionality, such as the Index of Fluvial Functioning (IFF) (Siligardi *et al.*, 2000), are useful instruments for underlining critical points in the river ecosystem taken as a whole, going beyond the simple analysis of the aquatic community.

3.1.5 FUTURE PROSPECTS

3.1.5.1 The water framework directive

In Italy, as in other European countries, the adaptation of available methods and the development of new methodologies able to evaluate other biological components is given priority for the future application of the WFD. In particular, the 2000/60/CE Directive indicates 'high ecological status' as the quality that all water bodies should attain. It is therefore indispensable to define a reference condition for each stream type in each ecoregion in order to assess distance between the observed state of affairs and the corresponding ideal state.

Defining and finding reference sites is a problem already considered in several EU projects (REFCOND, 'Development of a protocol for identification of reference conditions'; AQEM, 'The development and testing of an integrated assessment system for the ecological quality of streams and rivers throughout Europe using benthic macroinvertebrates'; STAR, 'Standardization of River Classifications'; PAEQANN, 'Predicting Aquatic Ecosystems Quality using Artificial Neural Networks' (Lek and Guégan, 1999)) and only partially resolved because in many situations reference conditions can only be inferred through historical data analysis that is not always available (Rimet *et al.*, 2005). Finding reference conditions for certain water course typologies is a very serious problem: for alpine-type streams in particular, pristine conditions can be found in protected areas, such as natural parks and at high elevations, but in lowlands, land use and urban settling have so altered the territory that it is more or less impossible to find a reference condition (Füreder *et al.*, 2002).

3.1.5.2 Worker formation

A fundamental step in water quality monitoring is the dissemination of an ecological culture together with the technical know-how for the correct application of evaluation methods among technicians. In this sense, an important role in Italy has been played by the Autonomous Province of Trento and the Centro Italiano Studi di Biologia Ambientale (CISBA). Under the aegis of the national Environmental Protection and Technical Services Agency (APAT), biologists working for local provincial and regional agencies are trained in environmental monitoring techniques. Practical one-week-long training courses on the proper use of the IBE, held from 1980 to the present, but also other methods such as algal indicators (EPI-D) and IFF, have had a significant role in creating an ecological mind-set among workers and have also been important moments of networking.

The Environmental Protection Agency of Trento remains in this sense the national reference point in Italy for the training and development of biological monitoring methods for water quality assessment. In fulfilment of this role, the Province has recently started a Suitability Assessment Licensing Procedure (Processo di Accertamento dell'Idoneità) (PAI) to guarantee the correct application of biological methods.

For IBE certification, a technician must pass a written/field PAI–IBE test to qualify as a certified operator included in a national list managed by the APAT.

A workshop on the 'Use of Biotic Indexes to evaluate the quality of freshwater streams: a comparison among four different methods (IBE, BBI, BMWP, RIVPACS)' was organized by the University of Trento, the Istituto Agrario of San Michele, TECHWARE (TECHnology for WAter REsources – Brussels, Belgium) and CISBA (Centro Italiano Studi di Biologia Ambientale) and gave rise to this book. This workshop provided an interesting comparison of four widely used methods tested simultaneously in different streams differing in morphology and level of organic pollution (Flaim *et al.*, 2000). The workshop was set up according to the scheme used by the national training courses and participants were experienced workers from several European countries. The good instructor/student ratio and the possibility of applying different methods created a valuable exchange of experience and information among workers. Among the interesting observations that emerged from the workshop was how all four methods and groups found the same dominant organisms but the presence of the rarer macroinvertebrate taxa was not uniform, nor was the consideration given them among working groups and methods. In the final analysis, however, all four methods agreed on the quality evaluation when tested on different sites.

In this context, it does not seem necessary to limit or standardize the biological methods used, but instead spend more effort in homogenizing the end results, i.e. each method should have the same number of quality classes and clarify what exactly is meant by 'excellent', 'good', 'fair' or 'poor' water quality.

In addition, one should not forget that an important impulse in the application of biological monitoring methods derives from the availability of application manuals and taxonomic guides for emerging methods such as diatoms and macrophytes, besides the already proven macroinvertebrates.

REFERENCES

APAT, 2003. 'Annuario dei dati ambientali', SISTAN Sistema statistico nazionale. Agenzia per la protezione dell'ambiente e per i servizi tecnici: Rome, Italy.

APAT-CNR, 2004. Indice Biotico Esteso (IBE), APAT Rapporti, 29/2003. Agenzia per la protezione dell'ambiente e per i servizi tecnici and Istituto di Ricerca sulle Acque – Consiglio Nazionale delle Ricerche: Rome, Italy.

Brittain, J. E., Lencioni, V. and Maiolini, B., 1998. 'Arctic and alpine stream ecosystem research (AASER)'. In: *The Arctic and Global Change*, Casacchia, R., Koutsileos, H., Morbidoni, M., Petrelli, P. D., Pettersen, M. R., Salvatori, S., Sparapani, R. and Stolz Larsen, E. (Eds). Ny Alesund Science Managers Committee: Ravello, Italy, pp. 123–126.

Campaioli, S., Ghetti, P. F., Minelli, A. and Ruffo, S., 1994–1999. *Manuale per il Riconoscimento dei Macroinvertebrati delle Acque Dolci Italiane*, Vols I and II. Provincia Autonoma di Trento: Trento, Italy.

Cappelletti, C., Ciutti, F. and Torrisi, M., 2003. 'Diatomee epilitiche e qualità biologica del torrente Noce (Trentino)'. In: *Atti del Seminario di Studi "I Nuovi Orizzonti dell'Ecologia"*, Trento,

18–19 April, 2002, CISBA, Baldaccini, G. N. and Sansoni, G. (Eds). Provincia Autonoma di Trento: Trento, Italy, pp. 177–181.

Ciutti, F., 2005. 'Monitoraggio dei corsi d'acqua con indicatori algali (Diatomee)'. *Annali Istituto Superiore Sanità*, in press.

Ciutti, F., Cappelletti, C., Monauni, C. and Pozzi, S., 2000. 'Effetti dello svaso di un bacino idroelettrico sulla comunità dei macroinvertebrati'. *Rivista Idrobiologia*, **39**, 165–184.

Ciutti, F., Cappelletti, C. and Corradini, F., 2003. 'Applicazione dell'indice EPI-D a un corso d'acqua delle Alpi (Torrente Fersina): osservazioni sulla metodica di determinazione delle abbondanze relative'. *Studi Trentini Scienze Naturali Acta Biologica*, **80**, 95–100.

Corradini, F., Cappelletti, C. and Flaim, G., 2005. 'Qualità delle acque idonee alla vita dei pesci: otto anni di applicazione del D. Lgs. 130/92 in Trentino'. *Biologia Ambientale*, **19**, 197–200.

Crisp, D. T., 1995. 'The ecological basis for the management of flows regulated by reservoirs in the United Kingdom', In: *The Ecological Basis for River Management*, John Wiley & Sons Ltd. Chichester, England. Harper, D. M. and Ferguson, A. D. J. (Eds). Wiley: Chichester, UK, pp. 93–100.

Cushman, R. M., 1985. 'Review of ecological effects of rapidly varying flows downstream from hydroelectrical facilities'. *North American Journal of Fisheries Management*, **5**, 330–339.

Decreto Legislativo, 1999. 'Testo aggiornato del decreto legislativo, 11 May, 1999, no. 152, recante: "Disposizioni sulla tutela delle acque dall'inquinamento e recepimento della direttiva 91/271/CEE concernente il trattamento delle acque reflue urbane e della direttiva 91/676/CEE relativa alla protezione delle acque dall'inquinamento provocato dai nitrati provenienti da fonti agricole", a seguito delle disposizioni correttive ed integrative di cui al decreto legislativo 18 August, 2000, no. 258'. Gazzetta Ufficiale, no. 246, del 20/10/2000, Supplemento Ordinario no. 172.

Dell'Uomo, A., 2004. 'L'indice diatomico di eutrofizzazione/polluzione (EPI-D) nel monitoraggio delle acque correnti', Linee guida. Agenzia per la Protezione dell'Ambiente e per i servizi Tecnicin (APAT): Rome, Italy.

European Parliament, 2000. 'Directive 2000/60/EC of the European Parliament and of the Council establishing a framework for Community action in the field of water policy'. *Official Journal of the European Communities*, **L327**.

Flaim, G., Ziglio, G., Siligardi, M., Ciutti, F., Monauni, C. and Cappelletti, C., 2000. 'Report on the course: A comparison among four different European biotic indexes (IBE, BBI, BMWP, RIVPACS) for river quality evaluation', International Symposium of the Learning Society and the Water Environment'. UNESCO International Hydrological Programme, Paris, France, pp. 454–458.

Forneris, G., Merati, F., Pascale, M. and Perosino, G. C., 2005. 'Proposta di indice ittico (II) per il bacino occidentale del Po'. *Biologia Ambientale*, in press.

Füreder, L., Vacha, C., Amorosi, K., Bühler, S, Hansen, C. M. E. and Moritz, C., 2002. 'Reference conditions of alpine streams: physical habitat and ecology'. *Water, Air and Soil Pollution: Focus*, **2**, 275–294.

Ghetti, P. F., 1986. *I Macroinvertebrati nell'Analisi di Qualità dei Corsi d'Acqua. Manuale di Applicazione – Indice Biotico (IBE) Modificato*. Provincia Autonoma di Trento: Trento, Italy.

Ghetti, P. F., 1995. *Indice Biotico Esteso* (IBE). Notiziario dei Metodi Analitici IRSA–CNR: Rome, Italy.

Ghetti, P. F., 1997. *Manuale di Applicazione: Indice Biotico Esteso. I Macroinvertebrati nel Controllo della Qualità degli Ambienti di Acque Correnti*. Agenzia provinciale per la protezione dell'ambiente, Provincia Autonoma di Trento: Trento, Italy.

Gore, J. A., 1994. 'Hydrological change'. In: *The Rivers Handbook*, Vol. 2, Calow, P. and Petts, G. E. (Eds). Blackwell Science: Oxford, UK, pp. 35–54.

Kostadinov, S., 1995. 'The ecological basis for torrent control in mountainous landscapes'. In: The Ecological Basis for River Management, Harper, D. M. and Ferguson, A. J. D. (Eds). Wiley: Chichester, UK, pp. 51–57.

Law 10, May 1976, no. 319. 'Norme per la tutela delle acque dall'inquinamento', Gazzetta Ufficiale del 29 May, 1976, no. 141.

Lek, S. and Guégan, J F., 1999. 'Artificial neural networks as a tool in ecologae modelling: An introduction'. *Ecological Modelling*, **120**, 65–73.

Lencioni, V., Boscaini, A., Franceschini, A. and Maiolini, B., 2001. 'Distribuzione di macroin-vertebrati bentonici in torrenti d'alta quota sulle Alpi italiane: stato delle conoscenze e recenti risultati', Atti S.It.E 25, CD-ROM, no. 42.

Lencioni, V., Maiolini B. and Margoni, S., 2002. 'Il limite altitudinale di applicazione degli Indici IBE (Indice Biotico Esteso) e IFF (Indice di Funzionalità Fluviale) in due sistemi fluviali alpini (Amola e Cornisello, Trentino). Studi Trentini Scienze Naturali', *Acta Biologica*, **78**, 81–90.

Maiolini, B. and Lencioni, V., 2001. 'Longitudinal distribution of macroinvertebrate community assemblages in a glacially influenced system in the Italian Alps'. *Freshwater Biology*, **46**, 1625–1639.

Manzini, P. and Spaggiari, R., 1989. 'Le indagini sulla qualità biologica dei corsi d'acqua ital-iani'. In: *Atti del Convegno 'La Qualità delle Acque Superficiali. Criteri per una Metodologia Omogenea di Valutazione'*, Siligardi, M. (Ed.). Riva del Garda (TN), Provincia Autonoma di Trento: Trento, Italy, pp. 270–278.

Martinet, F. and Dubost, M., 1992. 'Die letzen naturnahen Alpenflüsse'. CIPRA, Kleine Schriften 11: pp. 10–58.

Milner, A. N. and Petts, G. E., 1994. 'Glacial rivers: physical habitat and ecology'. *Freshwater Biology*, **32**, 295–307.

Minciardi, M. R., Rossi, G. L., Azzollini, R. and Betta, G., 2003. 'Linee guida per il biomon-itoraggio di corsi d'acqua in ambiente alpino'. ENEA: Turin, Provincia di Torino, Torino, Italy.

Moog, O., 1993. 'Quantification of daily peak hydropower effects on aquatic fauna and management to minimize environmental impacts'. *Regulated Rivers: Research and Management*, **8**, 5–14.

Muhar, S., Schwarz, M., Schmutz, S. and Jungwirth, M., 2000. 'Identification of rivers with high and good quality: Methodological approach and applications in Austria'. *Hydrobiologia*, **422/423**, 343–358.

Petts, G. E., 1984. *Impounded Rivers: Perspectives for Ecological Management*. Wiley: Chichester, UK.

Pontalti, L. and Baruchelli, G., 1988. 'Impatto ambientale e qualità delle acque utilizzate dalle troti-colture trentine, Stazione Sperimentale Agraria Forestale di S. Michele all'Adige'. *Esperienze e Ricerche*, **17**, 159–186.

Provincia Autonoma di Trento, 2000. 'Qualità delle acque superficiali. Monitoraggio dei corsi d'acqua principali in provincia di Trento. Elaborazione complessiva per il decennio 1990–1999'. Agenzia Provinciale per la Protezione dell'Ambiente: Trento, Italy.

Provincia Autonoma di Trento, 2001. 'Valutazione della qualità ambientale dei principali corpi idrici trentini recettori di acque reflue depurate. Indagini 1991–2000'. Servizio Opere Igienico-Sanitarie: Trento, Italy.

Prygiel, J., Coste, M. and Bukowska, J., 1999. 'Review of the major diatom-based techniques for the quality assessment of rivers. State of the art in Europe'. In: *Use of Algae for Monitoring Rivers, IIII*, Prygiel, J., Whitton, B. B. and Bukowska, J. (Eds). Agence de l'Eau Artois-Picardie: Douai, France, pp. 224–238.

Rimet, F., Bertuzzi, E., Cantonati, M., Cappelletti, C., Ciutti, F., Cordonier, A., Coste, M., Gomà, J., Tison, J., Tudesque, L., Vidal, H., Huck, V. and Ector, L., 2004. 'Distribution of diatom assemblages in high altitude watercourses in Western Europe: implications for the System A of the Water Framework Directive', 18 Tagung Deutchsprachiger Diatomologen, Limnologische Station Iffeldorf, Germany, 26–28 March, Abstracts, p. 11.

Rimet, F., Ciutti F., Cappelletti C. and Ector, L., 2005. 'Ruolo delle Diatomee nell'applicazione della Direttiva Europea Quadro sulle acque'. *Biologia Ambientale*, **19**, 87–93.

Ruffo, S. (Ed.), 1977–1985. *Guide per il Riconoscimento delle Specie Animali delle Acque Interne Italiane*, Collana del Progetto Finalizzato 'Promozione della Qualità dell'Ambiente'. CNR: Rome, Italy.

Sansoni, G., 1988. *Atlante per il Riconoscimento dei Macroinvertebrati dei Corsi d'Acqua Italiani*. Centro Italiano Studi di Biologia Ambientale, Provincia Autonoma di Trento: Trento, Italy.

Scardi, M., Cataudella, S., Ciccotti, E., Di Dato, P., Maio, G., Marconato, E., Salviati, S., Tancioni, L., Turin, P. and Zanetti, M., 2004. 'Previsione della composizione della fauna ittica mediante reti neurali artificiali'. *Biologia Ambientale*, **18**, 25–31.

Siligardi, M. and Cappelletti, C., 2006. 'A new approach to evaluating fluvial functioning (FFI): towards a landscape ecology'. In: *Biological Monitoring of River: Applications and Perspectives*, Ziglio, G., Siligardi, M. and Flaim, G. (Eds). Wiley: Chichester, UK, pp. 401–418.

Siligardi, M., Bernabei, S., Cappelletti, C., Chierici, E., Ciutti, F., Egaddi, F., Franceschini, A., Maiolini, B., Mancini, L., Minciardi, M. R., Monauni, C., Rossi, G., Sansoni, G., Spaggiari, R. and Zanetti, M., 2000. IFF Indice di Funzionalità Fluviale, Manuale ANPA, November, 2000. Agenzia Nazionale per la Protezione dell'Ambiente: Rome, Italy.

Sittoni, L. and Flaim, G., 1985. 'Indagine biologica preliminare delle acque correnti nel Trentino: problemi nell'applicazione dell'indice di qualità', Atti del convegno 'Esperienze e confronti nell'applicazione degli indicatori biologici in corsi d'acqua Italiani', 6–7 September, San Michele all'Adige, Trento, Italy, pp. 171–196.

Spaggiari, R., 1999. 'La qualità ecologico-biologica dei fiumi alle soglie del duemila'. In: *Atti Seminario di Studi 'I Biologi e l'ambiente . . . . Oltre il duemila'*, Baldaccini, G. N. and Sansoni, G. (Eds), 22–23 November, 1996, Venice, Italy. CISBA: Raggio Emilia, Italy, pp. 55–63.

Vittori, A., 1982. *Carta Ittica della Provincia di Trento: Documento Fondamentale della Provincia Automona di Trento per una Razionale Politica di Gestione delle Acque*. Stazione Sperimentale di San Michele all'Adige: Trento, Italy.

Whitton, B A. and Rott, E. (Eds), 1996. *Use of Algae for Monitoring Rivers, II*, Proceedings of International Symposium, Innsbruck, Austria, 17–19 September, 1995. Institut für Botanik, University of Innsbruck: Innsbruck, Austria.

Whitton, B. A., Rott, E. and Friederich, G., (Eds), 1991. *Use of Algae for Monitoring Rivers*, Proceedings of International Symposium, Düsseldorf, Germany, 26–28 May, 1991. Institut für Botanik. University of Innsbruck: Innsbruck, Austria.

Woodiwiss, F. S., 1964. 'The biological system of stream classification used by the Trent River Board'. *Chemistry and Industry*, **14**, 443–447.

Zerunian, S., 2004. 'Proposta di un Indice dello Stato Ecologico delle Comunità Ittiche viventi nelle acque interne italiane'. *Biologia Ambientale*, **18**, 25–30.

Zerunian, S., 2005. 'Ruolo della fauna nell'applicazione della Direttiva Quadro'. *Biologia Ambientale*, **19**, 61–69.

3.2

Biological Monitoring of North European Rivers

Leonard Sandin and **Nikolai Friberg**

3.2.1   INTRODUCTION

The historical as well as present day monitoring and bioassessment of running waters within the Nordic countries (Denmark, Finland, Iceland, Norway and Sweden) is varied, but with many similarities. Friberg and Johnson (1995) summarized running water monitoring using benthic macroinvertebrates within the Nordic countries and the national sampling and assessment methods for fish, macrophytes and macroinvertebrates within the five countries were collected by Skriver (2001). A nationwide monitoring programme for running waters was set up in Denmark in 1987. The programme initially contained some 200 streams, sampled for benthic macroinvertebrates twice a year. This programme was later (1998) expanded and presently includes 1053 stations and 80 intensively studied sites where macroinvertebrates, macrophytes and fish are sampled annually. In Finland, an investigation was carried out on the suitability of benthic macroinvertebrates for the bioassessment of rivers (Nyman *et al.*, 1986). Since 1999, 16 larger rivers are sampled annually for benthic macroinvertebrates and 20 rivers are sampled for fish. Norway has had monitoring of certain rivers since the beginning of the 1980s. These programmes have focused on specific stressors such as acidification, heavy metals or organic pollution. The national monitoring programme of rivers includes biology (fish, benthic macroinvertebrates and phytobenthos) in the River Atna and River Vikedal. The national programme also includes twenty one rivers where the effects of liming are investigated (at both limed and non-limed sections). These rivers are sampled for benthic macroinvertebrates, but also fish and phytobenthos are monitored in some of them. There is also a biological monitoring of acidification problems, where five rivers are monitored using either benthic macroinveretbrates or fish (in one case, both). National biological monitoring of running waters in Sweden started in the mid 1980s with five streams and in 1995 the first national survey of running waters including benthic macroinvertebrates was carried out (Wilander *et al.*, 1997; Sandin, 2003). Biological quality elements (fish and benthic macroinvertebrates) are also included in the national monitoring of temporal reference streams and rivers in Sweden (15 streams) since 2000. There is currently no national biological monitoring of running waters in Iceland due to the minor anthropogenic influence in the country, but there is large ongoing research on the Myvatn-Laxá river ecosystems, studied since 1977 (e.g. Gíslason *et al.*, 1998, 2000, 2001).

Three of the Nordic countries have published national assessment systems for running waters, i.e. the Danish Stream Fauna Index (DSF1) (Skriver *et al.*, 2000) in Denmark, the Classification of Environmental Quality of Freshwaters (SFT, 1997) in Norway and the Environmental Quality Criteria (Swedish Environmental Protection Agency, 2000) in Sweden. Macroinvertebrates have been the most commonly used organism group for assessment of running waters in all five countries and the sampling techniques is similar (kick-sampling using a 0.5 mm mesh sized net) in most cases (Johnson *et al.*, 2001). This similarity is in part due to the international (ISO) and Nordic standardization work that took place in the late 1980s. The sampling methods for macroinvertebrates used in the Nordic countries were intercalibrated at two streams in south-central Sweden in the autumn of 1998 (Johnson *et al.*, 2001). Four countries used some sort of standardized kick-sampling, whereas Iceland took their samples by collecting individual stones as discrete sampling units. No significant differences were found among the countries using kick-sampling at any of the streams (one reference and one impacted) regarding number of taxa, number of EPT (Ephemeroptera, Plecoptera and Trichoptera) taxa, ASPT (Armitage *et al.*, 1983), DSFI (Skriver *et al.*, 2000), or for any of the Norwegian acid indices (Raddum et al., 1988; Bækken and Aanes, 1990), or Shannon's diversity index, despite the fact that the sampling methods differed to some degree (habitats included and time for taking each sample).

Macrophyte monitoring is currently only taking place in Denmark (80 streams) and Norway (6 rivers) and none of the Nordic countries have so far developed an assessment system for macrophytes in running waters. Focus has instead been on evaluating the community composition of plants in the stream or river. For fish monitoring in rivers and streams, the main focus has been on salmonids and other economically important species, where in the year 2000, ca. 100 streams across the Nordic countries were intensively monitored in a standardized way (Malmquist *et al.*, 2001). The sampling method is similar among the countries and includes pulsating direct current equipment and successive removal of fish for density estimates. Malmquist *et al.* (2001) suggested eleven fish species which could serve as indicators of human disturbance to running waters in the Nordic countries. National long-term monitoring using phytobenthos has only taken place in Norway and only Sweden has a national assessment system of running waters that includes phytobenthos.

3.2.2 STRESSORS/IMPACTS TO RUNNING WATERS IN THE NORDIC COUNTRIES

All of the Nordic countries are characterized by a vast freshwater resource. There is a gradient in environmental impacts from Denmark in the south to the northern parts of Norway and Finland, largely reflecting population density (Table 3.2.1). Denmark and southern Sweden are intensively farmed and populated and as a result heavily impacted locally by human activities. The main threats to freshwater biological communities in this part of the Nordic countries are organic pollution, excessive nutrient loading, toxic compounds and effects of drainage on the physical features, especially of streams and wetlands. A survey from 1996 among the regional water authorities

Table 3.2.1 Major potential threats to the macroinvertebrate communities in Nordic streams and rivers (excluding Iceland): DK, Denmark; N, Norway; S, Sweden; SF, Finland

Threat	Local	Trans-boundary	Area of concern
Organic pollution	×	—	DK (others)
Habitat (physical) modifications	×	—	All
Toxic compounds	×	(×)	All
Catchment scale changes	×	—	All
Acidification	(×)	×	N, S, SF
Liming	×	—	S
Introduction of alien species	—	×	All

in Denmark revealed that ca. 25 % of the streams not fulfilling their quality objectives could be attributed to sewage input from scattered dwellings (Windolf, 1997). The same survey also revealed that habitat degradation (channelization, dredging and weed cutting), could explain another 25 % of the non-compliance with the quality objectives. Comparing the ecological quality criteria in Denmark with Sweden reveals large differences, where the main parts of sites sampled in the Swedish national stream survey had a DSFI value of 6 or 7 (indicating a high ecological quality), whereas the main part of the Danish sites sampled in the NOVA programme had DSFI index values of 4 or 5 (indicating a good or moderate quality) (Figure 3.2.1). In other parts of the Nordic countries these threats are mostly restricted to urban areas and areas with concentrated human activities such as agriculture, mining and industry, one exception being hydroelectric power plants which occur throughout the Nordic countries on almost all major river systems. In contrast to the more locally restricted pollutants, the major threat to freshwater invertebrates in Norway, Sweden and Finland has been long-range transboundary pollution with acidifying substances (Table 3.2.1). To counteract the effects of acidification, extensive liming programmes have been initiated, especially in Sweden.

Denmark and Sweden has the most developed national monitoring systems among the Nordic countries and a more detailed description of these programmes is described below.

3.2.3   DENMARK

3.2.3.1   Background history of national monitoring

Routine biological assessment of running waters has been performed in Denmark since the beginning of the 1970s. The reason for introducing monitoring activities was that most Danish streams became severely degraded, especially during the first part of the 20th Century. River valleys were in increasing numbers used as arable land and the majority of streams became physically modified, i.e. straightened and channelized. At the same time, Danish streams became increasingly polluted by easy degradable organic matter from untreated sewage and various agricultural point sources. Sewage treatment plants were established in the 1950s and 1960s and the fight against water pollution was intensified with the introduction of the Environmental Protection Act in 1974. This Act stipulates a planning system requiring specific quality objectives to be set for individual streams. Fulfilment of quality objectives is assessed using only macroinvertebrates (see below). As part of the Danish Action Plan on the Aquatic Environment, a nationwide monitoring programme was established in 1988. Three major programme revisions have taken place since 1988. The first revision took place in 1992 for the period 1993–1997, the second in 1997 for the period 1998–2003 (named NOVA 2003) and the third in 2003 for the period 2004–2009 (named NOVANA).

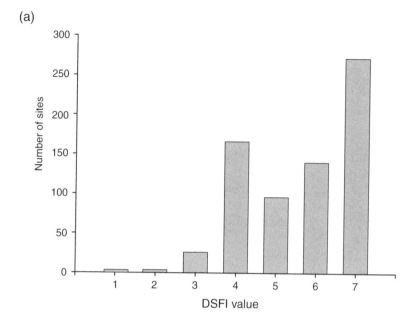

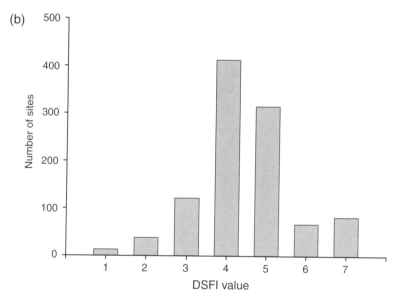

Figure 3.2.1 Ecological quality classification according to the Danish Stream Fauna Index (DSFI) (see text), where a DSFI value of 1 indicates very low ecological quality and a value of 7 a very high quality for (a) the 707 sites sampled in the 2000 national survey of lakes and streams in Sweden, and (b) the 1051 sites sampled for macroinvertebrates in the Danish 2002 NOVA monitoring of running waters

3.2.4 THE NOVANA PROGRAMME

3.2.4.1 Background

The design of the programme and selection of monitoring sites were influenced by several demands and constraints. The programme should be in compliance with EU directives relating to streams and riparian areas. The new monitoring programme must also cover monitoring of river freshwater habitats and species in relation to the European Habitat Directive (Council of the European Communities, 1992). In addition to the international demands, several national considerations were included when developing the NOVANA monitoring strategy. Nature elements should be prioritized in the new programme and as a consequence more components of the biota (macrophytes, riparian vegetation and fish) should be monitored at all sites in addition to macroinvertebrates. The latter should, in addition, be identified to species level to give a better understanding of diversity patterns in Danish streams. Another more site-specific demand to the new programme was that the river Skjern should be monitored. The lower part of the river Skjern was re-meandered in 2000–2001. Pre-restoration monitoring and post-restoration monitoring two years after the restoration were carried out as a 'stand-alone' project (Svendsen and Hansen, 2000) whereas the long-term monitoring activities should be covered by NOVANA. An important consideration of the design of the new monitoring network was the re-use of old stations. The first two monitoring programmes have yielded a substantial number of time series of macroinvertebrates, water chemistry and discharge, dating back to at least 1989. The aim when selecting the new monitoring stations was consequently to maximize the overlap with previous programmes.

3.2.4.2 Strategy of the NOVANA programme

The most important criteria when selecting the monitoring stations was that they should be geographically representative. Despite the small area of Denmark (43 000 km^2) there are distinctive geological and geographical differences between the different parts of the country. These differences affect in-stream hydromorphological features as well as the hydrological regime. In addition to the natural differences, human impact on the different geographical regions is also highly variable. The eastern part of the country, especially the greater Copenhagen area, is more densely populated, whereas farming, and the number of livestock, is more intense in the western part. In order to ensure that all major impacts are covered in the NOVANA programme, seven monitoring station types were predefined and denoted agricultural stations (AGI–III), point source stations (PSI–II), habitat stations (HABs) and reference stations (REFs). Three types are directly related to agricultural activities (AGI–IIIs) and two indirectly (PSI and HABs). This focus on agriculture reflects the fact that arable farmland covers ca. 63 % of Denmark's total area. AGI to AGIII are defined so that they cover a gradient with respect to possible impact by agricultural activities with AGI stations having the highest risk of being impacted. In agricultural and rural catchments, scattered dwellings with no or limited sewage treatment systems dominate PSI. HABs are in the agricultural landscape streams that are physically degraded but otherwise unpolluted. Despite comprehensive efforts to reduce point-source discharges into streams, there are still locally situated important point sources and PSII stations are in these streams. Reference stations (REFs) were included in order to have an absolute reference point and to comply with the Water Framework Directive (WFD). An important part of the monitoring strategy was to divide the programme into an intensive and extensive programme. The aim of the intensive network stations is to provide in-depth knowledge of elements less thoroughly investigated in the extensive network stations, including methodological issues, year-to-year variability, etc. The extensive network aims at providing knowledge on the overall status of Danish streams and rivers.

3.2.4.3 Selection of monitoring network

In total, 800 extensive (1 per 53 km^2) and 50 intensive stations were selected as the total number was within the economical scope of the programme. The counties submitted information on 1210 potential monitoring stations that fulfilled the selection criteria to the National Environmental Research Institute (NERI). In order to fulfil the constraints imposed on the selection of stations (e.g. the river Skjern), a total of 80 stations were pre-selected. The remaining 1130 stations were subjected to a random selection procedure constrained by stream size, impact type and regional distribution. The stations in the intensive network were selected along the continuum in twelve large river systems distributed across Denmark. All of the intensive stations were selected as agricultural type III with little impact on the riparian areas or reference sites. This selection enabled monitoring of the interaction between the stream and the riparian areas.

3.2.4.4 Elements in the new monitoring programme

Elements in the new monitoring programme include hydromorphology, water chemistry and the biological indicators (macrophytes, macroinvertebrates and fish). Small impacted and reference streams (636 in total) will be sampled for all elements once during the six-year monitoring period (2004–2009) and larger streams twice (164 in total). Intensive stations are monitored each year in addition to 250 stations in which only macroinvertebrates are sampled. The latter stations are a sub-set of 1051 NOVA 2003 stations in order to provide time-series data on the macroinvertebrate community, the only new feature being that both hydromorphology and physico-chemistry are now measured.

Catchment characteristics, such as land use, geology and geomorphology, are collected in a standardized way for each station using the site protocol. The quality of the physical conditions in the streams is monitored using the Danish Habitat Quality Index (Pedersen and Baattrup-Pedersen, 2003). On intensive stations, information on water level fluctuations and discharge regime is monitored by continuous water level registration and the discharge is measured eight times per year. Basic physico-chemical parameters, such as temperature and pH, are monitored along with important macro-ions and nutrients. The loading of organic matter from scattered households and sewage treatment plants requires monitoring of the biochemical oxygen demand (BOD) concentrations.

Macroinvertebrates are still a key parameter in Danish stream monitoring. In the new monitoring programme, identification to the species level is required for most taxonomic groups. Samples are multi-habitat kick-samples (12 kicks in 3 transects) covering the riffle, glide/run, pool and edge habitats in the streams using the Danish Stream Fauna Index sampling procedure (Skriver *et al.*, 2000). Macrophytes are surveyed in 200 plots laid out in transects across the stream on both extensive and intensive stations. Plant species are scored in the plots depending on coverage and depth and dominant substratum are also registered for each plot. Along with vegetation in the stream bank, vegetation (up to 2 m from the stream) is also surveyed quantitatively in a number of transects at each station. On intensive stations, the spatial distribution of macrophyte species and substrata is surveyed on a 20-m sub-reach using a grid net strategy (Wright *et al.*, 1981). Fish densities are monitored on a 100 m reach using multiple-run 'electro-fishing'. Densities of all species are estimated. However, trout (*Salmo trutta*) is the only useful indicator in small Danish streams and is therefore the primary target species. In species monitoring such as the river Skjern population, estimates of salmon and houting are obtained by the capture/recapture method (Ricker, 1975; Schneider, 2000). On intensive stations, the fish monitoring is carried out for the entire river system (Pedersen and Baattrup-Pedersen, 2003).

Two different strategies are used for monitoring of the vegetation on the riparian areas. On both extensive and intensive stations, vegetation is categorized into a number of vegetation types. This is done by determining the seven most dominant species in 30 plots of 100 m^2 on each side of the stream. By registering the presence of certain wetland species, each plot can be assigned to a vegetation type (Nygård *et al.*, 1999). In addition, permanent transects are established perpendicularly to the stream at the intensive stations. Plant communities are then monitored quantitatively in randomly selected plots each year at a distance of 2–10 m from the stream. Important explanatory variables such as soil types and groundwater level are also monitored.

3.2.4.5 Assessment system

The first biomonitoring method in Denmark used benthic macroinvertebrates collected by hand net with a 1 mm mesh size and identified in the field to genus, family or order (Ministry of Agriculture, 1970). The outcome of the assessment was therefore very dependent on the taxonomic skills and experience of the person undertaking the field sampling. How a given macroinvertebrate species composition was to be interpreted in terms of pollution degree was, furthermore, only very roughly described in the guidelines. Consequently, regional water authorities developed a range of modifications that made inter-regional comparisons impossible. In addition, quality assessment varied over time, as surveyors became increasingly critical as water quality improved during the late 1970s and early 1980s which made temporal comparisons very difficult (Windolf *et al.*, 1997). To overcome these shortcomings a biotic index (the Viborg Index) (Andersen *et al.*, 1984) was developed, based on the principles used in the Trent Biotic Index (Woodiwiss, 1964). The Viborg Index has never been a national standard in Denmark, but was used by some Danish counties up until 1992. A working group appointed by the Danish Environmental Protection Agency used the Viborg Index as a template for a new official Danish biotic index. Testing the index showed that it could be used after minor modifications and the revised index was named the Danish Fauna Index (DFI). The DFI has never been used officially as the standard method, but was used during the period 1993–1997 for biological assessment of the ca. 220 sites included in the Nationwide Monitoring Programme under the Action Plan on the Aquatic Environment (Kirkegaard *et al.*, 1992). In 1996, a new working group was established, after the county biologists had argued that the DFI had more disadvantages than benefits. After thorough analysis of various modifications, the working group recommended a very minor change of DFI as well as a revision of the scale and the use of Arabic numerals 1 (heavily polluted) to 7 (unpolluted) instead of Roman numerals which had been used in the previous indices (Friberg *et al.*, 1996). The new index was named the Danish Stream Fauna Index (DSFI) and became the national standard in 1998 (Danish Environmental Protection Agency, 1998).

The Danish Stream Fauna Index (DSFI) manual includes both the sampling procedure and the index itself. The sampling procedure is standardized, and includes in principle all habitats at the site. Sampling is undertaken by using a standard handnet with a 25 × 25 cm opening and a tapering netbag with a mesh size of 0.5 mm. The index calculation and sampling method is thoroughly described in Skriver *et al.* (2000). Species sensitive or tolerant towards low oxygen concentrations are primarily included in the index. However, other impacts such as habitat degradation have become more evident in recent years and it has, as a consequence, been tested if the DSFI was a sensitive indicator towards stressors other than organic pollution. In a comparison of physically homogenous and heterogeneous reaches in six streams, Olsen and Friberg (1999) found that the DSFI values were higher in the heterogeneous reaches. This result indicated that the DSFI is at least partly sensitive towards habitat degradation. In another study, Skriver (1999) found a strong correlation between the DSFI value and the number of 'red-listed' macroinvertebrate species, hence suggesting that the DSFI could be used as an overall indicator for in-stream biodiversity.

No specific assessment system has been developed for macrophytes. Up until now, macrophyte data collected as part of the NOVA 2003 monitoring programme has been analysed using various types of multivariate statistics or with simple metrics

such as taxon richness. The Mean Trophic Rank (MTR) (Holmes *et al.*, 1999) has
been applied in some Danish streams with limited success (Baattrup-Pedersen *et al.*,
2004). The MTR differentiates poorly among sites varying in ecological quality,
as the range of nutrient concentrations in Danish streams is very narrow. As for
macrophytes, no specific assessment system is used for fish and they have not yet
been used in an ecological context. The focus has until now been primarily on
assessing fish stocks.

3.2.5   SWEDEN

3.2.5.1   Background history of national monitoring

Regional monitoring of Swedish lakes and watercourses had already started in some
areas in the early 1950s. These programmes were, however, of varying quality and
the Swedish Environmental Protection Agency therefore issued general guidelines
on local monitoring of freshwater systems in 1986. The national monitoring of freshwaters was initiated in the 1960s using a combination of both chemical and biological measures. The national biological monitoring of streams and rivers, however,
started in Sweden with the programme 'Integrated monitoring of 18 small catchment
areas' in the mid-1980s. Here, five streams were monitored temporally (Wiederholm,
1992), together with seven limed streams where biological monitoring started in 1989
(Bergquist, 2000). The national monitoring programme for lakes and watercourses
was later redesigned following the suggestions from Wiederholm (1992).

   This included a nested sampling design with three tiers or levels as described
by Wiederholm and Johnson (1997) with (i) national lake and watercourse surveys
taking place every fifth year in order to determine spatial patterns, using physico-
chemical and biological indicators, (ii) temporal references including annual monitoring of reference lakes and watercourses to determine 'among-year' variability and
trends of indicator metrics (including macroinvertebrate and fish), and (iii) integrated
temporal reference sites with annual monitoring of physico-chemical and biological indicators of lakes and watercourses (including macroinvertebrates, periphyton,
fish and macrophytes). However, not all the suggestions included in this sampling
scheme have been implemented into the national Swedish monitoring programme.

   The new programme started with the national survey of lakes and streams in 1995
(where some 700 streams were sampled for benthic macroinvertebrates; see Sandin,
2003). The national survey was repeated in 2000 including the same 700 streams,
once again sampled for benthic macroinvertebrates (Wilander *et al.*, 2003). The
temporal reference programme for running waters was started in the year 2000 and
includes 50 medium-sized streams, sampled for water chemistry. Fifteen of these
are intensively studied and includes annual sampling of benthic macroinvertebrates,
as well as fish. The national programme also includes 12 limed streams, sampled
annually for benthic macroinvertebrates since 1989 in most cases and fish, where
sampling started in the early or mid-1990s.

The national freshwater monitoring programme is currently being revised. The main aims of the revision is to better follow up all of the 15 Swedish environmental quality objectives (such as 'Natural Acidification Only' and 'Flourishing Lakes and Streams' (Swedish Government Bill 2000/01:130), and measure up to the demands of the EC Water Framework Directive (WFD) (European Commission, 2000). The prioritized questions for the freshwater monitoring programme are biological diversity and organic pollutants. This new programme will be finalized in 2005 and in operation in 2006.

3.2.5.2   Selection of monitoring sites

The random stratification of sites sampled in the national survey of streams and watercourses in 1995 and 2000 is described in Sandin (2003). Streams for sampling of benthic macroinvertebrates were randomly selected from the Swedish Hydrological and Meteorological Institute's watercourse and catchment register. This register contained 3198 streams with a catchment area between 15 and 250 km^2 (which were chosen because they will most probably be permanent and suitable for sampling through wading). Out of the 3198, 600 with a catchment area between 15 and 50 km^2 and 600 with a catchment area between 50 and 250 km^2 were randomly chosen. The streams were then evaluated for accessibility (with a road within 600 meters) and then stratified according to size; 350 within 15–50 km^2 catchments and 350 within 50–250 km^2 catchments were finally chosen for sampling. The randomly chosen sampling coordinates for each stream constitute the point where the stream to be sampled flowed into a larger stream or lake. This point was *a priori* deemed to be not suitable for sampling and site selection for sampling was constrained to within 100–600 m upstream of the sampling coordinates. Accordingly, a 50-m reach (sampling site) of homogenous substratum (preferably hard-bottom), flow and riparian vegetation, was chosen in each stream and sampled for benthic macroinvertebrates at the same time that water chemistry samples were taken and the physical habitats were evaluated.

The criteria for including sites into the temporal reference programme takes into account a stratification among the six main ecoregions of Sweden (Gustafsson and Ahlén, 1996), and finding streams or rivers not affected by anthropogenic factors. The streams or rivers monitored should also be representative for the region where they are situated. Some of the sites included in the programme have been part of an earlier monitoring programme, initiated in the early 1960s, whereas others have been chosen more specifically for monitoring of the biological quality elements (fish and benthic macroinvertebrates; eleven streams were also sampled for phytobenthos in 2001). All biologically sampled water courses are relatively small, with only two out of the fifteen having a catchment area larger than 100 km^2.

The sampling of biological quality elements for the national monitoring of freshwaters is done according to the Handbook for Environmental Monitoring ('Handbok för Miljöövervakning' (http://www.naturvardsverket.se)). The handbook currently contains seven biological investigation types for water courses (i.e. two methods for sampling benthic macroinvertebrates, as well as methods for monitoring fish, macrophytes, phytobenthos, crayfish and large mussel species).

The method used in the national monitoring programme for sampling of benthic macroinvertebrates is done in accordance with the European standard SS-EN 27 828 (Anon, 1994). Five kick-samples are taken in an area of 10 m at a homogeneous stretch, preferably in a riffle section using a hand-net (mesh size, 500 µm). Each sample comprises a fixed distance of 1 m (25 cm width) and a defined sampling time of 1 min. The sampled material is conserved in ethanol and brought back to the laboratory for sorting and identification. Sampling of phytobenthos is done by sampling stones (diameter preferably above 10 cm). Periphyton is scrubbed off (using a toothbrush or similar device) from at least five large stones. The material is placed in 1 l of water and two sub-samples are taken out, one for live analyses (no fixation, stored dark and cool) and one fixed with formalin for diatom analysis. 'Electro-fishing' by wading is the standardized sampling method and fishing is carried out by using pulsating direct current equipment, utilizing a fishing rod with a ring diameter of 20–30 cm and a net with a mesh size of 4 mm. Sampling is carried out by successive removal of fish (at least three runs), where the total area sampled should be 200–300 m^2. To obtain a robust estimate of the ecological quality of a stream or river, a number of sampling sites within each watercourse must be sampled (at least three, but preferably many more in large rivers).

The macrophyte survey methods in Sweden are in the stage of development. A sampling method was described in 2003, but has so far only been tested in some research projects. The method is based on the use of transects, where 6–10 transects are placed between stream banks at each locality, i.e. a stream reach of 100 m. Along each transect, 0.25 × 0.25 m quadrates are placed side-by-side transverse to the current, from one bank to the other of the water course. In each quadrate, the presence of bryophytes and vascular plant species is recorded. On each locality, the relative frequencies of species in the quadrates are calculated (only quadrates with plants are considered). If the spatial extension of macrophytes is very sparse on a site, more quadrates should be obtained to ensure accurate calculations of the frequencies.

3.2.5.3   Assessment system

Sweden has developed Ecological Quality Criteria for lakes and watercourses (Swedish Environmental Protection Agency, 2000) to assess the ecological quality of, e.g. watercourses in the country. The criteria for biological quality elements in running waters include phytobenthos, benthic macroinvertebrates and fish. The ecological quality criteria are divided into two parts: (i) assessment of current conditions, and (ii) assessment of deviation from a reference value (indicates environmental impact of human activities). The reference condition approach ideally assumes that the references are unaffected by any human activity. In practice, however, reference values are usually based on observations made in areas with a slight human impact. The extent of human impact is calculated by measuring the deviation from the reference value, given as a ratio of the measured value to the reference value and divided into a five-point scale.

The periphyton or diatom ecological quality criteria include two French indices for assessing running waters, namely the ISP (Indice de polluo-sensibilité) (CEMAGREF, 1982) and the IDG (Indice diatomique génerique) (Rumeau and Coste, 1988). With phytobenthos, it is only possible to assess the current conditions, since not enough background data exist to determine reference values for diatoms in Sweden. The benthic macroinvertebrate assessment system includes four indices to assess ecological quality (both current conditions and deviation from reference). These are Shannon's diversity index (Shannon, 1948), Average Score per Taxon (Armitage *et al.*, 1983), the Danish Stream Fauna Index (Skriver *et al.*, 2000), and a Swedish index to assess acid conditions (Henriksson and Medin, 1986). The Swedish fish assessment system (FIX) is an Index of Biotic Integrity (IBI) (Karr, 1981) type of multimetric index. The reference values represent 'typical' values in relation to the Swedish national 'electro-fishing' database, rather than values for the 'pristine state', since it is argued that it is not possible to distinguish such a state regarding the effects of, e.g. land use, introduction of non-native species, etc. The method could also be used to assess the current status as described above. The assessment of running waters is based on seven parameters, which are weighted together to give an overall index: (i) number of native fish species, (ii) biomass of native fish species, (iii) number of individuals of native fish species, (iv) proportions of salmonids based on numbers, (v) reproduction of native salmonids, (vi) presence of species and stages sensitive to acidification, and (vii) proportion of alien species based on number.

3.2.6 CONCLUSIONS

The monitoring and assessment of running waters in the Nordic countries have both similarities and differences: similar since, e.g. monitoring so far is mainly based on macroinvertebrates, and different, e.g. in terms of the extent of the national monitoring programmes. However, the Nordic countries have been working together towards a common goal of comparability in assessment and monitoring, e.g. (i) through the intercalibration of macroinvertebrate sampling (Johnson *et al.*, 2001), (ii) building the first steps towards developing a common Nordic monitoring strategy (Skriver, 2001), and (iii) developing a common predictive assessment system (NORDPACS) (Johnson *et al.*, 2001), similar to the RIVPACS system (Moss *et al.*, 1987) in the UK. With the Water Framework Directive, the monitoring programmes in the Nordic countries will become more comprehensive (e.g. to a greater extent include more biological monitoring elements), and in some cases also partly change its aim (e.g. in Sweden focus more on sites affected by human impact). Nordic rivers and streams make up a substantial proportion of running waters within Europe, where the Nordic countries (including Iceland) can be divided into eight main ecoregions (Nordic Council of Ministers, 1984), with a largely east–west distribution. These regions have similar threats to its fauna and flora, irrespective of country. The monitoring and assessment of running waters in the Nordic countries can thus benefit from work carried out in the same ecoregion in other countries (e.g. as regards defining reference conditions, relationships among organism groups, and the suitability of sampling methods for different monitoring purposes). It is, therefore, important that common methods and strategies for monitoring and assessment of running waters in the Nordic countries are developed. Some recommendations regarding future co-operation among the Nordic countries with respect to fish, macrophyte and macroinvertebarte biological assessment of running waters is given by Skriver (2001). Future work should also ideally include development of assessment systems for habitat/riparian vegetation (e.g. similar to the Danish Habitat Quality Index (Pedersen and Baattrup-Pedersen, 2003)) and integrating the Baltic states (Estonia, Latvia and Lithuania) into the Nordic freshwater monitoring and assessment work, since these countries are similar to the Nordic countries, both in terms of water resources and tradition in running water research.

ACKNOWLEDGEMENTS

We thank Ann Kristin Schartau, Norwegian Institute for Nature Research, and Gunnar Raddum, University of Bergen, for supplying information on the national monitoring of freshwaters in Norway. This chapter was written while Leonard Sandin was working at the National Environmental Research institute (NERI) with a mobility scholarship from the Nordic Academy for Advanced Study (NorFA).

REFERENCES

Andersen, M. M., Riget, F. F. and Sparholt, H.,1984. 'A modification of the Trent Index for use in Denmark'. *Water Research*, **18**, 145–151.

Anon, 1994. 'Water quality – methods for biological sampling – guidance on handnet sampling of aquatic benthic macroinvertebrates, EN 27 828. European Committee for Standardization, Brussels, Belgium.

Armitage, P. D., Moss, D., Wright, J. F. and Furse, M. T., 1983. 'The performance of a new biological water quality score system based on macroinvertebrates over a wide range of unpolluted running-water sites'. *Water Research*, **17**, 333–347.

Baattrup-Pedersen, A., Friberg, N., Pedersen, M. L., Skriver, J., Kronvang, B. and Larsen, S. E., 2004. 'Implementation of the Water Framework Directive in Danish streams', NERI Technical Report No. 499. National Environmental Research Institute: Silkeborg, Denmark.

Bækken, T. and Aanes, K. J., 1990. 'Use of macroinvertebrates to classify water quality', Report No. 2A, Acidification. The Norwegian Institute for Water Research (NIVA): Olso, Norway.

Bergquist, B. (Ed.), 2000. 'Kalkade vattendrag – miljökvalitet och biologisk mångfald; utvärdering av IKEU-programmets sex första år'. Swedish Environmental Protection Agency Report No. 5076. Swedish Environmental Protection Agency: Stockholm, Sweden (in Swedish).

CEMAGREF, 1982. 'Etude des méthodes biologiques quantitatives d' appréciation de la qualité des eaux', Rapport Division Qualité des Eaux Lyon. Agence Financière de Bassin Rhône–Méditerranée: Lyon, France.

Council of the European Communities, 1992. 'Council Directive 92/43/EEC of 21 May 1992 on the conservation of natural habitats and of wild fauna and flora'. Council of European Communities, Brussels, Belgium (online: http://www.ecnc.nl/doc/europe/legislat/habidire. html).

Danish Environmental Protection Agency, 1998. 'Biological assessment of watercourse quality', Guidelines, No. 5. Ministry of Environment and Energy, Danish Environmental Protection Agency: Copenhagen, Denmark.

European Commission, 2000. 'Establishing a framework for Community action in the field of water policy'. Directive 2000/60/EC, October 2000. European Commission: Brussels, Belgium.

Friberg, N. and Johnson, R. K. (Eds), 1995. 'Biological monitoring of streams'. Nordic Council of Ministers Report, TemaNord 1995:640. Nordic Council of Ministers: Oslo, Norway.

Friberg, N., Larsen, S. E., Christensen, F., Rasmussen, J. V. and Skriver, J., 1996. *Dansk Fauna Indeks: Test og Modifikationer*, NERI Technical Report, No. 181. National Environmental Research Institute: Silkeborg, Denmark (in Danish).

Gislason, G. M., Ólafsson, J. S. and Adalsteinsson, H., 1998. 'Animal communities in Icelandic rivers in relation to catchment characteristics and water chemistry'. *Nordic Hydrology*, **29**, 129–148.

Gislason, G. M., Ólafsson, J. S. and Adalsteinsson, H., 2000. 'Life in Glacial and Alpine Rivers in Central Iceland in Relation to Physical and Chemical Parameters'. *Nordic Hydrology*, **31**, 411–422.

Gislason, G. M. Adalsteinsson, H., Hansen, I., Ólafsson, J. S. and Svavarsdóttir, K., 2001. 'Longitudinal changes in macroinvertebrate assemblages along a glacial river system in central Iceland'. *Freshwater Biology*, **46**, 1737–1751.

Gustafsson, L. and Ahlén, I. (Eds), 1996. *Geography of Plants and Animals*, National Atlas of Sweden. Almqvist and Wiksell International: Stockholm, Sweden.

Henriksson, L. and Medin, M., 1986. 'Biologisk bedömning av försurningspåverkan på Lelångens tillflöden och grundområden 1986', Aquaekologerna, Report to the County board of Älvsborgs län Aquaekologerna: Hyssna, Sweden (in Swedish).

Holmes, N. H. T., Newman, J. R., Dawson, F. H., Chadd, S., Rouen, K. J. and Sharp, L., 1999. 'Mean Trophic Rank: A User's Manual', Research and Development Technical Report. Environment Agency: Bristol, UK.

Johnson, R. K., Aagaard, K., Aanes, K. J., Friberg, N., Gislason, G. M., Lax, H. and Sandin, L., 2001. 'Macroinvertebrates'. In: *Biological Monitoring in Nordic Rivers and Lakes*, Skriver, J. (Ed.), Nordic Council of Ministers Report, TemaNord 2001:513. Nordic Council of Ministers: Oslo, Norway, pp. 43–51.

Karr, J. R., 1981. 'Assessment of biotic integrity using fish communities'. *Fisheries*, **6**, 21–27.

Kirkegaard, J., Wiberg-Larsen, P., Jensen, J., Iversen, T. M. and Mortensen, E., 1992. Biologisk bedømmelse af vandløbskvalitet – Metode til anvendelse på vandløbsstationer i Vandmiljøplanens Overvågningsprogram, Technical Guidelines No. 5. National Environmental Research Institute: Silkeborg, Denmark (in Danish).

Malmquist, H. J., Appelberg, M., Dieperink, C., Hesthagen, T. and Rask, M., 2001. 'Fish'. In: *Biological Monitoring in Nordic Rivers and Lakes*, Skriver, J. (Ed.), Nordic Council of Ministers Report, TemaNord 2001:513. Nordic Council of Ministers: Oslo, Norway, pp. 61–71.

Ministry of Agriculture, 1970. Vejledning om fremgangsmåden ved bedømmelse af recipienters forureningsgrad. Ministry of Agriculture: Copenhagen, Denmark (in Danish).

Ministry of Interior, 1987. Document No. 46 of 19 October 1987. Ministry of Interior: Copenhagen, Denmark (in Danish).

Moss, D., Furse, M. T., Wright, J. F. and Armitage, P. D., 1987. 'The prediction of the macro-invertebrate fauna of unpolluted running-water sites in Great Britain using environmental data'. *Freshwater Biology*, **17**, 41–52.

Nordic Council of Ministers, 1984. "Naturgeografisk regionindelning av Norden', Nordiska min-isterrådet 1984. Nordic Council of Ministers: Oslo, Norway (in Swedish, Danish, Norwegian).

Nygaard B., Mark, S., Baattrup-Pedersen, A., Dahl, K., Ejrnæs, R., Fredshavn, J., Hansen, J., Lawesson, J., Münier, B., Møller, P. F., Risager, M., Rune, F, Skriver, J. and Søndergaard, M., 1999. 'Nature quality – development of criteria and methods', Faglig Rapport fra DMU No. 285.

Nyman, C., Anttila, M. -E., Lax, H. -G. and Sarvala, J., 1986. 'The bottom fauna of rapids as a measure of the quality of running waters', NBWE, Publication Series 3. National Board of Waters and the Environment: Helsinki, Finland, pp. 1–76, (in Finnish, with an English summary).

Olsen, H. M. and Friberg, N., 1999. 'Biological stream assessment in Denmark: The importance of physical factors'. In: *Biodiversity in Benthic Ecology*, Friberg, N. and Carl, J. (Eds), Proceedings from Nordic Benthological Meeting in Silkeborg, Denmark, 13–14 November 1997, NERI Technical Report No. 266. National Environmental Research Institute: Silkeborg, Denmark, pp. 89–95.

Pedersen, M. L. and Baattrup-Pedersen A., (Eds) 2003. 'Ecological monitoring in streams and riparian areas in NOVANA 2004–2009', Technical Guidelines No. 21. National Environmental Research Institute (NERI): Silkeborg, Denmark (online: http://tekniske-anvisninger.dmu.dk) (in Danish).

Raddum, G., Fjellheim, A. and Hesthagen, T., 1988. 'Monitoring of acidity by the use of aquatic organisms'. *Verhandlungen der Internationale Vereinigung für Limnologie*, **23**, 2291–2297.

Ricker, W. E., 1975. 'Computation and interpretation of biological statistics of fish populations in fresh waters', Bulletin 191. Department of the Environment, Fisheries and Marine Services: Ottawa, Canada.

Rumeau, A. and Coste, M., 1988. 'Initiation à la systématique des diatomées d'eau douce'. *Bull. Fr. Pêche Pisci*, **309**, 1–69.

Sandin, L., 2003. 'Benthic macroinvertebrates in Swedish streams: community structure, taxon richness, and environmental relations'. *Ecography*, **26**, 269–282.

Schneider, J. C., 2000. 'Manual of Fisheries Survey Methods II: with periodic updates'. DNR Special Report No. 25. Michigan Department of Natural Resources (DNR): Ann Arbor, MI, USA.

Shannon, D. E., 1948. 'A mathematical theory of communication'. *Bell Systems Technological Journal*, **27**, 379–423.

Skriver, J., 1999. 'Danish Stream Fauna Index (DSFI) as an indicator of rare and threatened benthic macroinvertebrates'. In: *Biodiversity in Benthic Ecology*, Friberg, N. and Carl, J. (Eds), Proceedings from Nordic Benthological Meeting in Silkeborg, Denmark, 13–14 November 1997, NERI Technical Report No. 266. National Environmental Research Institute: Silkeborg, Denmark, pp. 97–105.

Skriver, J., (Ed.), 2001. *Biological Monitoring in Nordic Rivers and Lakes*, Nordic Council of Ministers Report, TemaNord 2001:513. Nordic Council of Ministers: Oslo, Norway.

Skriver, J., Riis, T., Carl, J., Baattrup-Pedersen, A., Friberg, N., Ernst, M. E., Frandsen, S. B., Sode, A. and Wiberg-Larsen, P., 1999. 'Biologisk vandløbskvalitet (DVFI)', Udvidet biologisk program, NOVA 2003. Afdeling for Vandløbsøkologi og Afdeling for Sø- og Fjordøkologi: Silkeborg, Denmark (in Danish).

Skriver, J., Friberg, N. and Kirkegaard, J., 2000. 'Biological assessment of water course quality in Denmark: Introduction of the Danish Stream Fauna Index (DSFI) as the official biomonitoring method'. *Verhandlungen der Internationale Vereinigung für Limnologie*, **27**, 1822–1830.

SFT, 1997. 'Klassifisering av miljøkvalitet i ferskvann', Veiledning 1997:04: Oslo, Norway.

Svendsen, L. M. and Hansen, H.O. (Eds), 2000. 'The river Skjern restoration project – the monitoring programme', Arbejdsrapport fra DMU No. 139. Department of Streams and Riparian areas, National Environmental Research Institute: Silkeborg, Denmark (in Danish).

Swedish Environmental Protection Agency, 2000. 'Ecological quality criteria: lakes and watercourses', Report No. 5050. Swedish Environmental Protection Agency: Stockholm, Sweden.

Wiederholm, T. (Ed.), 1992. 'Freshwater environmental monitoring in Sweden', Proposals from a Working Group, Report No. 4111. Swedish Environmental Protection Agency: Stockholm, Sweden.

Wiederholm, T. and Johnson, R. K., 1997. 'Monitoring and assessment of lakes and watercourses in Sweden'. In: *Monitoring Tailor Made II. Information Strategies in Water Management*, Ottens, J. J., Claessen F. A. M., Stoks, P. G., Timmerman, J. G. and Ward, R. C. (Eds), Proceedings from an International workshop, September 1996, Nunspeet, The Netherlands, pp. 317–329.

Windolf, J. (Ed.), 1997, 'Freshwater monitoring – streams and springs', NERI Technical Report No. 214. National Environmental Research Institute: Silkeborg, Denmark.

Wilander, A., Johnson, R. K., Goedkoop, W. and Lundin, L., 1998. 'Riksinventering 1995. En synoptisk studie av vattenkemi och bottenfauna i svenska sjöar och vattendrag', Report No. 4813. Statens Naturvårdsverk: Stockholm, Sweden (in Swedish).

Wilander, A., Johnson, R. K. and Goedkoop, W., 2003. 'Riksinventering 2000. En synpotisk studie av vattenkemi och bottenfauna i svenska sjöar och vattendrag', SLU Publication 2003:1. Department of Environmental Assessment: Stockholm, Sweden (in Swedish).

Woodiwiss, F. S., 1964. 'The biological system of stream classification used by the Trent River Board'. *Chemistry and Industry*, **14**, 443–447.

Wright, J. F., Hiley, P. D., Ham, S. F. and Berrie, A. D., 1981. 'Comparison of three mapping procedures developed for river macrophytes'. *Freshwater Biology*, **11**, 369–379.

3.3

Biological Monitoring of Mediterranean Rivers with Special Reference to Greece

Konstantinos C. Gritzalis

Biological Monitoring of Rivers   Edited by G. Ziglio, M. Siligardi and G. Flaim
© 2006 John Wiley & Sons, Ltd.

3.3.1   INTRODUCTION

The continuous and rapid development of human activities during the last two centuries has strongly and dramatically affected the different types of the ecosystems on the earth. The anthropogenic activities resulting in environmental pressures and changes, in many cases take place in the vicinity of populated areas but sometimes, policy makers and responsible authorities in order to satisfy numerous and different social needs interfere with ecosystems in remote locations (e.g. establish reservoirs). The concept of exploitation in size, natural rhythm and area leads to ecosystems, especially aquatic ones, towards degradation in terms of quality and quantity, to the detriment of the global ecosystem.

The significance of an excellent aquatic system for human health has been known since many centuries ago as it was described by the Greek physician Galenus (129–201 AD), for the Mediterranean rivers and coastal zones (Gritzalis, 1998). Since then, aquatic ecosystems have been stressed with different scale pressures. On a global scale, many countries remain undecided about environmental policy, but some others such as the Member States of the European Union (EU), have created appropriate administrative mechanisms based on Directives such as the Water Framework Directive (WFD) (EU, 2000) in order to protect all types of water bodies and to restore health in degraded ones within the next few years, through assessment tools and biomonitoring.

Beyond doubt, such situations give rise to interactions among science, politics, public administration and the general public within the framework of the assessment of biological condition as it pertains to ecological integrity. In such cases, the level of public concern for environmental protection and health plays a decisive role regarding policy formulation and implementation (Moog and Chovanec, 2000). A main target of the WFD is the use of biotic elements such as fishes, macroinvertebrates and aquatic flora in the assessment of freshwater quality in streams. Apart from the membership of the EU, some South European countries are very familiar with this philosophy of freshwater quality assessment and monitoring. Some others, however, have been involved in this practice only during recent years at the research level and biological monitoring has not yet been adopted and applied at a national or even at a regional level, while the monitoring regarding the water quality in streams and rivers is based only on physico-chemical parameters (e.g. Greece, Cyprus, etc.).

The aim of this contribution is to provide some basic structural elements of the biological methods for such purposes, focusing mainly on recent approaches. Finally, through EU-funded programmes which fulfil the demands for the implementation of the WFD, some recent case studies in Greek streams and rivers are described.

3.3.2   ECOLOGICAL INTEGRITY AND ASSESSMENT APPROACHES

In order to detect and evaluate the effects of various types of pollutants emanating from anthropogenic activities on the quality of freshwaters, a number of impacts should be considered – either isolated or in combinations. A large array of factors such as the different types of pollution (see indicative list of the main pollutants in Annex VIII of the EU, 2000, WFD), its periodicity regarding duration and season-ality, its size, point source or diffused pollution, stream modifications, catchment area alterations (generally non-acceptable and sustainable uses) and finally the rela-tion with other natural environmental differences and scale characteristics (climatic conditions, water quantity, geomorphology, vegetation, etc.) constitute a complex issue.

Another issue, similarly complex, is that of ecological integrity. This later concept has been established in the Austrian Water Laws since 1985, reflecting the modern philosophy of water management of Austrian rivers (Moog and Chovanec, 2000). Disruptive facts of this integrity, such as river corridor alterations, modifications and straightening, can be seen in quantitative and qualitative changes in the assemblage of species (Karaouzas and Gritzalis, 2002) which can eventually lead to either the disappearance of naturally occurring species or the appearance of atypical ones (Moog, 1995). In Figure 3.3.1, a representative example is given by recent surveys

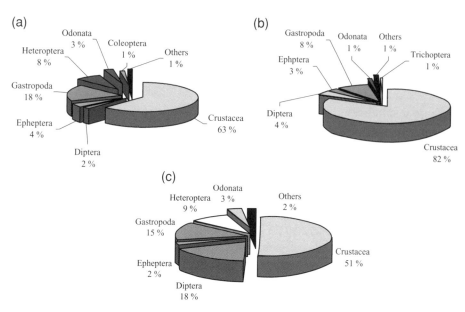

Figure 3.3.1 Distribution of macroinvertebrate major taxa groups in samples collected a modified stretch of the Pamisos River (close to the City of Messini, SW Greece), by the use of three different European sampling methods: (a) AQEM; (b) RIVPACS; (c) IBE. Data taken from the European Research Project 'Standardization of River Classifications. Framework for Calibrating Different Biological Survey Results against Ecological Quality Classifications to be Developed for the Water Framework Directive (STAR, www.eu-star.at), 2001–2005', and from I. Karaouzas, 'The Freshwater Quality Assessment of the Pamisos River (SW Greece) Using Different European Sampling Methods', MSc Thesis, Brunel University, Stanmore, UK. Reproduced by permission of I. Karaouzas

in Greece for the EU-funded[1] STAR[2] project purposes, illustrating the dominance of Crustacea in comparison with other major ones by the use of three European methods.

Additionally, the application of biotic indices and scores in such degradated sites may cause misidentification of the pollution level (Vourdoumpa and Gritzalis, 2000). Uses that are acceptable, i.e. not harmful and facilitating the sustainability of aquatic environments, are required as they keep in balance the capacity of the ecosystem and hence preserve the ecological integrity (Moog, 1995). The necessity of the ecological integrity assessment has been widely recognized and several scientists have been involved (Cairns, 1995; Harper *et al.*, 2000; Buffagni and Comin, 2000; Verdonschot, 2000; Moog and Chovanec, 2000).

Environmental degradation in aquatic systems and the unavoidable 'tipping' mechanism results in degradation of ecological integrity. The need to assess the effects of such degradation on freshwater quality led to the development of a system (saprobic) based on different taxonomic groups (presence of species in relation to organic pollution) in the first decade of the past century (Kolkwitz and Marsson, 1902, 1908, 1909). Since then, a large number of biological basis tools (more than 70) have been developed and the process is still evolving worldwide. Especially during the last few decades, the number of study methods for the quality of running waters based on the analysis of biocenosis of their natural ambient and utilizing taxonomic and pollution tolerance data has increased. Generally, these methods have been grouped into the following three principal categories (Metcalfe, 1989; Ghetti, 1997; Verdonschot, 2000):

(a) Saprobic Indices

(b) Biotic Indices and Scores

(c) Diversity Indices

Moreover, Ghetti (1997), refers to a fourth group constituted by comparative indices (similarity indices), which could have the same value applied alone or in combination with the same criteria described for the saprobic, biotic and diversity indices.

Finally, the relatively recent emergence of rapid assessment methods and multimetrics or multimeasures have been adopted by numerous scientific teams, especially in the USA, where a vast majority of the water resources agencies (90 %) use a multimetric approach (Barbour and Yoder, 2000). The same philosophy was

---

[1] European Commission under the 5th Framework Programme and contributing to the implementation of the Key Action 'Sustainable Management and Quality of Water' within the Energy, Environment and Sustainable Development Programme. Contract No: EVK1-CT 2001-0089.

[2] 'Standardization of River Classifications: Framework method for calibrating different biological survey results against ecological quality classifications to be developed for the Water Framework Directive', (www.eu-star.at).

the basis of the successful development of the EU-funded[3] AQEM[4] project (AQEM Consortium, 2002; Hering *et al.*, 2003, 2004). The aim of the AQEM project was the development of a framework for assessing the quality of streams by the use of benthic macroinvertebrates, thus contributing the fulfilling requirements of the E.U. WFD.

3.3.3 BIOLOGICAL QUALITY ELEMENTS

The assessment of running waters is based upon many taxonomic groups: macroinvertebrates, diatoms, plankton, fishes, macrophytes, periphyton and aquatic vegetation, generally known as Biological Quality Elements (BQEs). There are certain methods based on communities consisting solely of fishes or plankton or diatoms, etc., while a number of combinations of different taxonomic groups (e.g. macroinvertebrates and fishes, periphyton and diatoms, etc.) constitute other assessment methods (see relevant, inclusive table by Ghetti, 1997). In general, water quality assessment methods based partially or totally on the total of the BQE groups, measure actual effects on biota, whereas physical and chemical methods must eventually be interpreted on a biological basis. In order to assess the freshwater quality on the basis of ecosystem health, the response of the entire aquatic community to stress must be studied (Metcalfe, 1989). This approach, especially for biomonitoring aims, beyond the EU 2000/60 WFD requirements, could be inconvenient, cost ineffective and time-consuming and in many cases inapplicable due to the natural structure and conditions of the biota of the study area or region.

The natural environmental conditions and history of a study area play an important role as regards the selected approach towards the assessment of the water quality by means of all the BQEs. This has been derived from many cases in Greece, especially from the recent studies of the EU-supported research Project STAR (2001–2005). During the research surveys for the purposes of the STAR project in Greece, among the 35 sampling sites and in different sampling seasons, some permanent, in terms of flow regime, streams and characterized by excellent recent and past environmental conditions, have been proved to be 'fishless' (e.g. the upper part of Tsouraki stream in Peloponnesus, Tsivdoghianni and Aspropotamos streams in Samothraki and the Andros Aegean Islands, respectively). Additionally, in one site the collected fish fauna was comprised only by a hybrid (*Salmo sp.*) population (probably this has been due to the operation of a fish farm). Another difficulty also arises in cases where the fish fauna is represented almost exclusively by one endemic species such as the *Ladigesocypris ghigii*, found only in streams of Rhodes island, Greece, (Gianferrari, 1927; Stoumboudi *et al.*, 2002), where the proportion of ca. 100 % of the population

---

[3] 5th Framework Programme, Energy, Environment and Sustainable Development, Key Action Water, Contract No: EVK1-CT1999-00027.

[4] 'The Development and Testing of an Integrated Assessment System for the Ecological Quality of Streams and Rivers throughout Europe using Benthic Macroinvertebrates', (www.aqem.de).

and very rarely a few individuals of *Anguilla anguilla* could be collected during different surveys (M. Stoumboudi, personal communication).

Among the endemic species of the fish fauna found in Greece, another special case should be made out of the threatened and endemic fish *Pungitius hellenicus*, Stephanidis, 1971, distributed at the Spercheios river system and surrounding area in Central Greece (Keivany *et al.*, 1999). This rare and endemic species, contrary to the *Ladigesocypris ghigii* is a part of a fish population comprised of many other species and in different proportions (Ch. Daoulas, personal communication). A certain degree of distribution of endemic species is a phenomenon which gives rise to the following, complicated issue. 'In the case of endemic and rare species existence, a fish-fauna based assessment method of water quality should or should not change the database components, elaboration, evaluation and finally the total appraisal concept according to the study area, in order to assess the ecological status?'. There are also cases where the tolerance levels of endemic and rare species (especially in a small confined area) for different types of pollutants needs to be revised or are unknown. In addition, gaps regarding the ecological information and the stream or river type characteristics (e.g. river zonation preferences) will appear, while the distribution of these species is restricted in a very few stream types. In such cases, species belonging to the same subfamily or genus with similar life cycle strategy and habits could be used as an alternative for the endemic and rare species, but this entails an automatic rejection of the species level use in the quality assessment and monitoring procedures.

In terms of the conservation of the biodiversity, the endemic, native and the rare fishes are very important for the world natural heritage and should remain one of the main focus and aims of European and other environmental policies and strategies. Furthermore, such populations should remain a tool for quality assessment and they should serve as an argument for the elimination and avoidance of additional unsustainable and non-acceptable uses and activities within the river basin of their ranges. Moreover, they could also be used for conservation and preservation purposes. However, beyond these special cases, the major advantage of using fish fauna is that knowledge of autecology, particularly regarding needs in terms of the physical environment, is easy to obtain, at least in Central Europe (Moog, 1995). In Greece, however, such satisfying data are obtained (ca. 500 references were collected recently by the Institute of Inland Waters of the Hellenic Centre for Marine Research (HCMR), mostly in its western part and Peloponnesus. A major disadvantage, especially in the rhithral zones, is the paucity of species leading to a necessarily less sensitive response curve to environmental stress. Another problem is presented by the fact that the mobility of fish species confounds the spatial sensitivity of many assessment methods (Moog, 1995).

The periphyton, macrophytes and river phytoplankton constitute three very different groups of autotrophs occurring in stream and rivers (Allan, 1995) which, similarly to the fishes, lack some important attributes. This holds true, especially in cases where the lower reaches of streams and rivers are close to the river mouth and the floodplain is very wide, representative of a typical, wide, alluvial depositional stretch. This formation, in combination with the generally dry climatic conditions in the Mediterranean area, does not support well-developed community groups such as the vegetation elements described above. A similar problem often arises at the lowest reaches of streams and rivers, especially in Greece, in cases where the flow regime could be characterized as unpredictable. This is because the water flow quantity varies depending on the needs for energy, water supply and agricultural purposes, which are served by the establishment and operation of various reservoirs. Such situations affect the macroinvertebrate community close to the river mouth where freshwater benthic organisms (Chironomidae, Elmidae, Simuliidae, etc.) are occasionally re-placed by marine or brackish ones (e.g. Thyssanopods, Polychaeta, etc.) (Gritzalis *et al.*, 1993). The conditions for macrophytes and other types of vegetation com-munities are also affected, being constricted to a few species, generally dominated by Spermatophyta (*Scirpus sp., Carex sp., Juncus sp., Phragmites australis* (Cav.) Trin. Ex Steudel, *Salix sp.*), Pteridophyta (*Equisetum sp.*), Chlorophyta (*Cladophora agg.*), etc. In another case of a Mediterranean regulated river, Bertrand *et al.* (2001) have also pointed out the effects on algal community composition caused by flow and the heavy link of hydrological disturbance and freshwater needs.

Among all of the BQE groups, the macroinvertebrates are the basis of the majority of all methodologies developed for ecological assessment worldwide. Like any other tool, they have positive and negative attributes for such uses. Among their advan-tages are that they are ubiquitous and abound, easily collected and preserved. Their sedentary nature and relatively long life span provide data regarding environmental quality. Qualitative sampling and analysis are also well developed. The taxonomy of the vast majority of macroinvertebrate communities has reached very high levels and a satisfactory number of 'key books' are available. These organisms are also appro-priate for experimental studies of perturbation (Metcalfe, 1989; Resh *et al.*, 1996).

After taking into account the above advantages versus the disadvantages (some macroinvertebrates are not sensitive to some perturbations, seasonal variation could complicate interpretations or comparisons, etc.) and methods to overcome the latter (Resh *et al.*, 1996), the preference among scientists has been towards the use of such taxa as indicators for the ecological assessment and biomonitoring of running waters. Particularly in Mediterranean streams and rivers, their use is recommended more than other biological elements because notwithstanding the abnormalities of flow regime (alterations and modifications, e.g. abstraction for agricultural purposes or reservoir establishment) water courses are easily recolonizable by macroinvertebrates. Both in terms of taxonomic diversity and number of individuals, this community also has a great capacity to recover rapidly from severe droughts and floods (Pires *et al.*, 2000).

3.3.4 LEVEL OF IDENTIFICATION

The identification to different taxonomic levels and the enumeration of collected individuals in macroinvertebrate samples is essential for all biological assessment methods. As a consequence of that concept, and according to guidelines on stream biomonitoring that were published for cases in point by the Danish Ministry of Agriculture, collected benthic macroinvertebrates were identified in the field to the highest possible level, which in most cases was genus, family or order (Skriver *et al.*, 2000). In general, there are differences regarding the required taxonomical level among the developed assessment tools for freshwater quality worldwide. The most utilized indices in Europe for quality studies based on the analysis of the macroin- vertebrates community[5] do not require high taxonomic identification levels, e.g, the Belgian Biotic Index (BBI) (De Pauw and Vanhooren, 1983; NBN, 1984) requires identification to order, family and genus, while the Biological Monitoring Working Party (BMWP) and Average Score Per Taxon (ASPT) (Armitage *et al.*, 1983) scores (United Kingdom), as well as the Spanish modification of BMWP score (Alba- Tercedor and Sánchez-Ortega, 1988), require identification only to family level.

The selection of the appropriate taxonomic level of identification is made and tested by the application of various taxa communities identified to family, genus or species level for several indices as well as on aquatic organisms assemblages and other aims (Furse *et al.*, 1984; Graca *et al.*, 1995; Guerold, 2000; Reynoldson and Wright, 2000).

Guerold (2000) used three taxonomic levels (family, genus and species) of Ephemeroptera, Plecoptera and Trichoptera communities collected under different quality water conditions. He calculated richness, four diversity indices and two sim- ilarity indices. He reached the conclusion that certain underestimations or overesti- mations are possible and the use of family-level identification results in very unsafe interpretations of the metric values, particularly when it comes to the assessment of the water quality and of the changes in the macroinvertebrate community.

In general, some taxonomic groups appeared in huge numbers when the habitat and other conditions supported a vast population development. In Figure 3.3.2, an example is given where the macroinvertebrate taxa composition expressed at the family level is representative of such cases. This has been recorded during surveys for the AQEM and STAR programmes where very high densities of Asellidae (ca. 100 % of *Asellus aquaticus* and Physidae (*Physa spp.*) were collected in extremely polluted sites rich in vegetation.

In addition, in modified rivers, slightly or moderately polluted and choked with aquatic flora, many hundreds of Gammaridae, Ephemeroptera and Hydracarina in- dividuals have been recorded from STAR project surveys. It is widely accepted that the successful identification in the highest, where possible, levels (genus or species) of Chironomidae, Oligochaeta, Hydracarina and other 'difficult' groups is time- consuming, requires the appropriate 'key books' and in most cases depends on the assistance of taxonomic experts. The lack of appropriate identification books for ev- ery country or region (e.g. in Greece, the majority of the 'key books' used originates from Central and North Europe) for highest taxonomic levels identification, is a common hindrance to specialized studies. Nonetheless, species level identification,

---

[5] For an exhaustive review of these indices, see Ghetti, 1997.

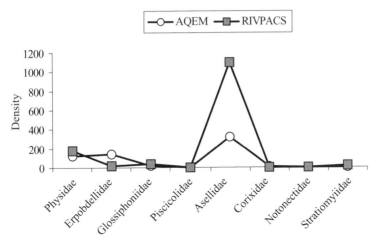

Figure 3.3.2 Density (expressed as individuals per sample) of macroinvertebrate taxa composition collected by the use of AQEM and RIVPACS methods on a polluted site (Lygkos River, Northern Greece). Data taken from the European Research Project 'Standardization of River Classifications. Framework for Calibrating Different Biological Survey Results against Ecological Quality Classifications to be Developed for the Water Framework Directive (STAR, www. eu-star.at), 2001–2005'.

especially for conservation studies under normal conditions, is required (Furse *et al.*, 1984). With regard to the required identification level for the reference sites, there is a long debate among scientists, well described by Reynoldson and Wright and the contributors of the working group (2000). The final conclusion was that the decision will largely depend on the objectives of the study, and the genus/species level identification is not always necessary. Finally, as regards the location of identification, this can be either the sampling site or the laboratory. The former is preferred when there is no need for highest-level identification, as it is time-saving. Irrespective of the location of identification, sorting in the field is sometimes recommended, e.g. for the fragile organisms.

3.3.5 HISTORICAL DATA

Historical data and generally the natural history of an area with the ecological causes of species abundance, richness and distribution play an important role in freshwater quality assessment because the population balance and distribution range changes, affected by different factors beyond human activities.

The existence of historical data is one of the most important and significant elements for developing an assessment tool for streams and rivers quality. This information is very helpful in describing and establishing reference conditions (Ehlert *et al.*, 2002; Nijboer *et al.*, 2004), while it is also suitable for restoration and conservation purposes.

In addition to the elaboration of macroinvertebrate historical data, some parameters and cases should be examined, because those data have been usually gathered
by different sampling methods. The past and recent degradation status, the cases
of migration, the case of an extreme climatic phenomenon (e.g. the appearance of
short or long dry periods, especially for the Mediterranean area), the case of gaps
regarding the sampling seasonality and every lack of foresight should all be taken
into account. This is necessary as a healthy, widely accepted ecological target is the
restoration of an ecosystem in scales and rhythms of a previous normal continuous
processing where possible. As a general rule, the criteria for an elaboration of the
historical data information should match those employed for the present use and aim.

3.3.6   REFERENCE CONDITIONS

The concept of reference conditions (species, river types, community, sample, etc.)
is indispensable to every approach of scientific research in different fields.

In cases such as ecological quality, the WFD (EU, 2000), requires the establishment
of type-specific reference conditions for surface water, body types. For running
waters, this is an area which has long been under discussion, especially during the
last few years with several thoughts, approaches, criteria (Hughes, 1995; Gibson
et al., 1996; Reynoldson and Wright, 2000; Ehlert *et al.*, 2002; Bonada *et al.*, 2002;
Nijboer *et al.*, 2004) and with an appropriate guidance-explanation (Wallin *et al.*,
2003) on the definition of the WFD.

Unimpaired stretches of stream generally have a defined length in relation to
the total length which could vary depending on the region. There is also no specific
relation between the reference site and the distance to sources or other morphological
characteristics (stream order, slope of the thalweg or valley floor, etc.). However, as
illustrated by the relevant surveys of the AQEM and STAR projects in Greece, the
stretch defined by the source and the reference site tends to be a multiple of the
headwaters. This is particularly the case for the mainland streams and less for those
of the Aegean Islands.

Regarding headwaters, a term frequently used in many cases of ecological studies,
these have been defined as the first 2.5 km of a watercourse from its furthest upstream
source (Furse, 2000). Undoubtedly, these stretches play an instrumental role in the
conservation of the macroinvertebrates species, thanks to their ability to support
many specialized taxa in whole river basins. Their size, however, is not a restrictive
measure for channel modification, agricultural pollution, acidification and drought
(Furse, 2000). All in all, the reference conditions fulfil the headwaters traits and they
are strongly linked with the conservation value.

In Greek streams and rivers, the conservation value is well supported not only by
the headwaters but also by a satisfying number of larger unimpaired stretches.

Concerning the establishment of type-specific reference conditions for running
waters, some consortia in Mediterranean countries (from Portugal, France, Italy and
Greece) involved in the EU-funded research projects AQEM and STAR and in Spain

for the purposes of the GUADALMED project (Bonada *et al.*, 2002) have worked towards this target based on several criteria. In Greece, with regard to the criteria used for reference site selection, the Institute of Inland Waters of the HCMR, which was involved with the AQEM and STAR projects, followed a strategy identical to the one adopted by the relevant consortia.

The basic principles of this approach are that the reference conditions must be politically reasonable and palatable, they must consider important aspects of 'natural' conditions and they must ensure that minimal anthropogenic disturbances have been taken into account. Additionally, the consortia of the AQEM and STAR projects have focused upon the main elements and parts affecting an ideal reference stream in a direct or indirect fashion. Among these criteria and their level of impairments where it was possibly acceptable (lowest values for some factors), the land-use practices in the catchment area, river channel and habitats, riparian vegetation and floodplain, hydrologic conditions and regulation, physical, chemical and biological conditions all came under scrutiny (AQEM Consortium, 2002; Hering *et al.*, 2003).

3.3.7 DEGRADATION FACTORS, SAMPLING NETWORK AND PERIODS

Even under the assumption of an ideal and undisturbed ecosystem where no degradation factors from anthropogenic activities exist, a number of natural ones affect the abiotic characteristics and the biota throughout long time intervals. Within this dynamic process of the earth landscapes, it is generally acceptable, that the water quality and quantity even under natural conditions varies within the same country or region, let alone among different countries and large regions such as those of Mediterranean Europe.

On the other hand, the types and the rhythm of development (industrial, agricultural, etc.) in relation to the adopted environmental policy, create unquestionable degradation scales regarding the water quality and quantity.

The results of such situations are due to some principal degradation factors and could be summarized as follows:

- Organic pollution

- Inorganic pollution

- Thermal pollution

- Sediment pollution

- Acidification

- River modification

- All possible combinations of the above factors (often referred to as general degradation).

In order to estimate the freshwater quality based upon the BQEs or upon the most suitable of those elements for this purpose, a strategy regarding the procedure is essential. As the construction of a sampling sites' network requires consideration of the river monitoring targets and aims, the most crucial questions which must be answered regarding the selection, the density of the network and the sampling period are the following: *what to assess* and *where and when to sample?*

The establishment of a representative and cost-effective sampling network regarding the sites and the seasons for freshwater quality assessment and biomonitoring would be an easy task if the degradation factor falls into one of the above mentioned categories. In cases where different combinations of pollutants within different affecting periods, originating from point or non-point sources with different residence times, create a mixture of degradation factors, the complexity of the issue increases, thus requiring special study in order to avoid potential misidentification of the quality level. Concerning the sampling period, as it has been observed during recent and past studies in Greek streams and rivers, the line differentiating between different season periods is not always clearly visible. Conditions in areas that are mountainous or away from coastal zones may be quite different from those in some other stream types located in lowlands, close to the sea or on islands.

3.3.8   HYDROMORPHOLOGY

The coexistence of anthropogenic activities and running waters entails an array of modifications on river corridors and habitats. In numerous cases, the lack of a sustainable and rational landscape use beyond the effect of pollutants has resulted in elimination of the natural running waters features, with further effects on the biota balance.

Following the dramatic effect upon the physical character and the habitat quality of rivers, a development of an evaluation method emerged in the United Kingdom during the last decade of the previous century as a result of collaboration among a wide spectrum of research teams, in order to facilitate the assessment of such situations. This derived method, 'River Habitat Survey' (RHS), is a valuable tool for river managers as it provides useful information, necessary in an attempt to sustain and enhance biodiversity. The collected data came from a huge number of sites and constitute a useful database. The RHS method employs some distinct components based on the physical structure of the river. First, the area under survey is established in a length of 500 m. Afterwards, a standard method for field survey and the map-based attributes acquisition is followed where the recorded data are derived not only from the 1 m width of ten spot checks established vertically every 50 m within the 500 m, but from a wider area and from the inter-spot sections. An amount of very important biotic and abiotic information is gathered, focusing on physical attributes and land uses of the river bed, bankface, banktop, river corridor, etc. (see Appendix 1 in Raven *et al.*, 1998). Finally, the collected results of the recent surveys are input into a computer database for the purposes of comparison with information obtained from other surveyed sites. The remaining two components deal with the Habitat

Quality Assessment (HQA) (quantification of the variety of natural features) and Habitat Modification Score (HMS) (quantification of the type and extend of artificial features) (Raven *et al.*, 1998). Generally, the concept and the history of the RHS system and other similar methods, such as the Index of Fluvial Functioning (IFF) (Siligardi *et al.*, 2000), in relation with the biotic scores and indices is very young and is not yet widely applied. According to Buffagni and Kemp (2002), only four European countries have relatively well-developed national programmes concerning the hydromorphological river assessment, suitable for application under the WFD (the French 'PSEQ', the Austrian nationwide method, the German 'Leitbild' and the RHS from the United Kingdom).

The biotic indices are suitable for the place where they have been developed, mainly because they are based on the regional biological reference conditions. The same applies to the hydromorphological features in the respective cases. The need for such modification when applying the RHS system in Southern European streams and rivers has been successfully described and presented by Buffagni and Kemp (2002) in their work. The latter especially focus upon and plan to investigate further in the future the important role of the secondary channels and other additional features regarding the substrate and flow types, in the application of the RHS to Southern European Rivers.

Generally, hydromorphology in the Mediterranean regions is a dynamic process and further studies are needed for a well-developed and representative system.

3.3.9   CASE STUDY

The following case study is divided in two sub-cases dealing with three different stream types, in accordance with system A of the WFD 2000/60 (EU, 2000). The basic distinction between the two cases is that, in the first one (Neda River) the study area is located on the mainland, while in the other (Fonias River and Tsivdoghianni Stream) it is located on a small Aegean island of Northern Greece.

3.3.9.1   Study areas

Neda River

The Neda River (Figure 3.3.3) is situated in the upper part of the Messinia Prefecture (Western Peloponnesus, Greece). The upper reaches are located on the mountain Lycaeon and finally the river meets the sea at the Kyparissia gulf in the Ionian Sea. Regarding the flow regime, this characterized as a perennial river, ca. 34.2 km long while its drainage basin encompasses ca. 289.6 km^2 (HMGS, 1978; NIGMR, 1975; IGME, 1982). The hydrological regime of its tributaries is a typical intermitted one, with very low flow periods during a few months (late November–early March) of the year.

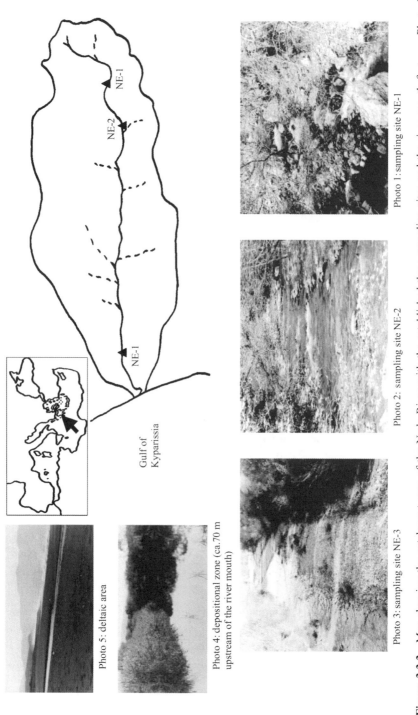

Figure 3.3.3 Map showing the catchment area of the Neda River with the established three sampling sites and the river mouth features: Photo 1, sampling site NE-1; Photo 2, sampling site NE-2; Photo 3, sampling site NE-3; Photo 4, depositional zone (ca. 70 m upstream of the river mouth); Photo 5, deltaic area

According to descriptors referred to the system A of the WFD 2000/60 (EU, 2000), this river belongs to the mid-altitude (200 to 800 m) type, while the size typology based on catchment area places this river in the medium ones (100 to 1000 km^2). From the geological point of view, the river basin of Neda is a typical calcareous one, dominated by carbonate rocks (ca. 70 %) and flysch and mollase (ca. 30 %) (NIGMAR, 1975; IGME, 1982). It belongs to ecoregion number 6 (Hellenic Western Balkans) (Illies, 1978; EU, 2000).

Fonias River and Tsivdoghianni Stream

Studies were conducted at the north-eastern part of the Samothraki Island drainage basin areas of the Fonias River and the Tsivdoghianni Stream (Figure 3.3.4). This ellipsoid-shaped island is located approximately 24 miles off the part of Alexandroupolis (NE Aegean Sea). The island surface covers an area of ca. 180 km^2, consisting of a massive mountain chain (Saos) where the highest top (Feggari) has an altitude of 1611 m (HAGS, 1969; IGSR, 1972). Regarding the hydrological regime of these two water sources, the Fonias River is a perennial river while at the Tsivdoghianni Stream the last part of ca. 2.4 km is intermitted during the summer season due to use of water for agricultural purposes. The whole island according to Illies (1978) and the EU (2000) belongs to the Eastern Balkans ecoregion (number 7). In contrast to the previous case study, the Fonias River and Tsivdoghianni Stream have a siliceous geological substrate. Based on the descriptors referred to in the system A of the WFD 2000/60 (EU, 2000), the Fonias River is characterized as small (10 to 100 km^2), and as a high-altitude (> 800 m) river. Analogously, the Tsivdoghianni

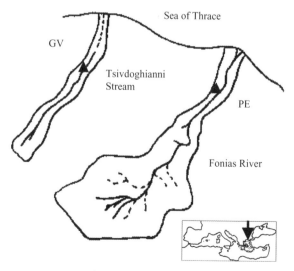

Figure 3.3.4 Map showing the catchment areas of the Tsivdoghianni Stream and the Fonias River, on Samothraki island, NE Aegean Sea, with the sampling sites GV and PE, respectively

Stream, is characterized as a small (10 to 100 km^2) – actually its drainage area is less than 10 km^2 – and mid-altitude (200 to 800 m) river. The Fonias River is 9.8 km long and the Tsivdoghianni Stream is 5.7 km long, while their catchment surface areas are 12.4 km^2 and 5.4 km^2, respectively. The two river basins are parallel to each other, while the distance between the river mouths is ca 5.5 km. In relation to the size of the river and the island, it is remarkable that at the last part of the Fonias River there is a mature, highly vegetated, delta formation. Within this triangular area (encompassing ca. 1.85 km^2), the dominant plants are *Platanus orientalis*, *Pteridium aquilinum* and *Juncus heldreichianus*.

3.3.9.2   Methods and materials

Protocols

Freshwater quality assessment requires a protocol that provides different kind of information, tailor-made to the needs of the particular research that it supports. For instance, in Spain, for the evaluation of the Ecological Status of Mediterranean Rivers a rapid protocol (RBP) within the GUADALMED project has been designed (PRECE – **P**rotocolo **R**ápido de **E**valuación de la **C**alidad **E**cológica). This protocol requires the determination of three indices: the IBMWP (former BMWP' for the assessment of biological quality of water), the index of riparian quality evaluation (QBR), and the physical habitat index (IHF). Methods for measuring the physico-chemical variables of water and river discharge are also included (Jáimez-Cuéllar *et al.*, 2002).

In the case of the AQEM project where many consortia were involved, the protocol (www. aqem.de) employs an extensive data set which provides both site- and sample-related information. The latter includes information on river and floodplain morphology, hydrology and vegetation, ensuring that the site can be precisely relocated in the field. Finally, documentation of the biological sampling process is also achieved. The AQEM site protocol contains 73 data fields which should be recorded (24 basic and 49 additional data), where the majority of these should be recorded in the field and the rest in the laboratory (AQEM Consortium, 2002). The steps described below are required by the AQEM and RHS protocols and have been applied for the case studies mentioned above.

Site selection and pre-classification

The establishment of the sampling sites for all studied cases was carried out under same criteria. The most important of these was that every investigated stretch should be a representative part of the river and should not be affected by certain factors such as bridges, weirs or other constructions, as the main target was the freshwater quality assessment from organic and inorganic pollutants. For this purpose, at the Neda River three sampling sites were established (NE-1, NE-2 and NE-3), starting from upstream to downstream (Figure 3.3.3). The distance from the source was for the NE-1 7.2 km, for the NE-2 10.8 km and for the NE-3 30.1 km. The last sampling site (NE-3, Photo 3 in Figure 3.3.3) was established in a stretch were no trace of salinity (> 0.06%) was found as this site was close to the river mouth. Remarkable is the fact that the dynamic effect of the sea waves (see Photo 5 in Figure 3.3.3) in relation to the relatively low yearly discharge of the river and with the river mouth dominant sediment, results into a restricted delta area (a very few km^2), where the direction of the last part of the narrow river corridor (70 to 120 m) undergoes occasional changes. Before the delta formation, there is a depositional zone ca. 60 m long due to anthropogenic activities (embankment) and natural processes (Photo 4 in Figure 3.3.3). Within this stretch, during the sampling site establishment, beyond the increased salinity levels caused by the sea-spray, a rough macroinvertebrate sampling revealed poor biodiversity (a few species of Atyidae, Chironomidae, Gammaridae, etc.).

For the Fonias River and Tsivdoghianni Stream, one sampling site (PE) and (GV) for each water source was established (Figure 3.3.4). Finally, based on the criteria for the definition of the reference conditions and the pressures on the catchment area, the sampling sites of the Neda River, NE-1, NE-2 and NE-3 were pre-classified as 'Good', 'Good' and 'Moderate', respectively, with the sites PE and GV of the Fonias River and Tsivdoghianni Stream as 'Reference' and 'Good', respectively.

Sampling equipment and techniques

Physical and chemical parameters   Dissolved oxygen, water temperature, pH and conductivity were measured by a portable water quality meter and current velocity was measured to assess water discharge ($1 \, s^{-1}$).

For every site, water samples (ca. 500 ml) were analysed (total hardness $[CaCO_3]$, magnesium $[Mg^{2+}]$, calcium $[Ca^{2+}]$, sodium $[Na^+]$, bicarbonate alkalinity $[HCO_3^-]$, potassium $[K^+]$, carbonate alkalinity $[CO_3^{2-}]$, sulfate $[SO_4^{2-}]$, chloride $[Cl^-]$, silicate $[SiO_2]$, and conventional pollutants such as nitrite $[NO_2^-]$, nitrate $[NO_3^-]$, ammonia $[NH_3^+]$, orthophosphate $[PO_4^{2-}]$ and total phosphorous $[TP]$).

Biological parameters – macroinvertebrate sampling devices and techniques   In the case studies of the Neda and Fonias Rivers and the Tsivdoghianni Stream, the macroinvertebrate sampling device used was a typical 500 µm mesh hand-net adopted for the AQEM project implementation. The sampling protocol is detailed in the AQEM Consortium (2002) and Hering *et al.* (2004). The collected material is preserved in 70 % ethanol and then transferred to the laboratory in plastic vials for sorting and identification to the highest where possible level by appropriate 'key-books'.

Hydromorpholigical and habitat features   The River Habitat Survey (RHS) system (Raven *et al.*, 1998) was applied at each site of the established sampling network of the investigated rivers and stream, in addition to the South Europe partly modified (RHS) protocol (Buffagni and Kemp, 2002). All surveys were applied during summer. The use of protocols in a 50 site network of all over Greece showed that late April to late September is the most appropriate period to assemble the full inventory of features described by the protocols.

AQEM software   The AQEM software (AQEM Consortium, 2002) performs all calculations necessary for applying the AQEM system, as follows:

(a) Calculation of the Ecological Quality Class of a sampling site, based upon a macroinvertebrate taxa list, by performing the stream-type specific calculations specified in Chapter 11 and Annex 7 of the appropriate Manual.[6]

(b) Calculation of a large number of additional metrics, which are helpful for further data interpretation. It should be noted, however, that the AQEM software is not designed for data storage.

3.3.9.3   Results and discussion

Very important entries in the AQEM protocol (AQEM Consortium, 2002) are the land-use practises in the catchment and sub-catchment area (area upstream of the sampling site).

In Table 3.3.1, the characteristics in 10 % steps of the Neda River catchment area and of the three sub-catchment ones are illustrated. With regard to the total catchment area, the dominant categories are macchie, crop land (mainly olive trees), deciduous native forest, and mixed native forest. Other categories such, as pasture, reeds, clear-cutting, olive oil presses, small villages' inhabitants, etc., are present in the whole river basin, especially in its final reach. The sub-catchment areas are dominated by native deciduous forest and macchie, with the scant presence of some other categories (Table 3.3.1). Generally, geomorphological formation and land use practices portray a typical mid-sized type, where low-level anthropogenic activities are compatible with slight environmental degradation.

Table 3.3.1 also shows the characteristics in 10 % steps for the River Fonias and the Tsivdoghianni Stream where the native deciduous forest dominates (90 %) in both, followed by the open grass/bush land. For the Tsivdoghianni Stream, some additional land uses (a smaller amount of clear cutting, urban sites and crop land) are present, in contrast to the Fonias River basin where these uses are absent. In conclusion, this situation regarding the minimal environmental pressures affecting the running waters of the Samothraki Island could be characterized as a very rare phenomenon, especially for the Mediterranean Islands.

Table 3.3.2 shows the physico-chemical variables for the Neda River, Fonias River and Tsivdoghianni Stream sites during the summer and winter periods. While all of

---

[6] For exhaustive information and details, see AQEM Consortium, (2002) and www.aqem.de.

Table 3.3.1 The land use practices expressed in 10 % steps in the catchment areas upstream of the sampling sites (NE-1, NE-2 and NE-3) and for the whole catchment area in the investigated rivers Neda (Messinia, Greece) and Fonias (Samothraki Island, Greece), and the Tsivdoghianni Stream (Samothraki Island, Greece).[a] Data taken from two European Research Projects ('STAR' (www.eu-star.at) and 'AQEM' (www.aqen.de)) and from the database of the institute of Inland Water, Hellenic Centre for Marine Research (HCMR), Attica, Greec. Reproduced by permission of Dr. N. Skoulikidis & Dr. T. Koussouris

Land use	Neda River				Fonias River		Tsivdoghianni Stream	
	Subcatchment area (sampling site NE-1)	Subcatchment area (sampling site NE-2)	Subcatchment area (sampling site NE-3)	Total catchment area	Subcatchment area (sampling site PE)	Total catchment area	Subcatchment area (sampling site GV)	Total catchment area
Deciduous native forest	50	40	30	20	90	90	90	90
Coniferous native forest	✓	✓	✓	✓	✓	✓	✓	✓
Mixed native forest	✓	10	10	10	10	10	10	10
Open grass/bushland	✓	✓	✓	✓	10	10	10	10
Reeds		✓	✓	✓	✓	✓	✓	✓
Naturally unvegetated	✓	✓	✓	✓	✓	✓	✓	✓
Non-native forest				✓				
Macchie	50	50	30	40	✓	✓	✓	✓
Crop land	✓	✓	30	30	✓	✓	✓	✓
Pasture	✓	✓	✓	✓	✓	✓	✓	✓
Clear-cutting			✓	✓	✓	✓	✓	✓
Urban sites (residential)	✓	✓	✓	✓	✓		✓	✓
Urban sites (industrial)			✓	✓				
Total	100	100	100	100	100	100	100	100

[a] The symbol '✓' indicates the presence (< 10 %) of the land-use category.

Table 3.3.2 Physico-chemical variables of the sampling sites NE-1, NE-2 and NE-3 of the Neda River and of the sampling sites PE and GV of the Fonias River and Tsivdoghianni Stream, respectively, during the summer and winter periods. Data taken from two European Research Projects ('STAR' (www. eu-star.at) and 'AQEM' (www.aqen.de)) and from the database of the Institute of Inland Water, Hellenic Centre for Marine Research (HCMR), Attica, Greece. Reproduced by permission of Dr. N. Skoulikidis & Dr. T. Koussouris

Physico-chemical variable	Neda River						Fonias River		Tsivdoghianni Stream	
	NE-1 (summer)	NE-2 (summer)	NE-3 (summer)	NE-1 (winter)	NE-2 (winter)	NE-3 (winter)	PE (summer)	PE (winter)	GV (summer)	GV (winter)
Water temperature (°C)	16.2	19.1	19.5	11.8	11.1	14.3	20.3	9.4	21.0	9.8
pH	8.3	8.2	7.7	8.4	8.3	7.5	7.4	7.5	8.5	7.2
Conductivity ($\mu S\,cm^{-1}$)	329	271	368	340	315	407	69	71	112	97
Dissolved oxygen ($mg\,l^{-1}$)	9.1	8.9	7.0	11.1	11.3	6.1	9.1	11.4	8.0	12.0
Oxygen saturation (%)	96.5	99.2	77.0	106.1	106.1	61.5	103.5	102.9	92.4	109.3
Carbonate alkalinity, $[CO_3^{2-}]$ ($mmol\,l^{-1}$)	4.04	3.57	4.14	2.25	1.93	3.43	0.40	0.43	1.30	0.55
Bicarbonate alkalinity, $[HCO_3^-]$ ($mmol\,l^{-1}$)	1.85	1.73	2.11	1.09	0.98	1.88	0.21	0.22	0.58	0.32
Chloride, $[Cl^-]$ ($mg\,l^{-1}$)	2.0	3.0	9.0	4.7	6.0	13.5	8.1	7.0	9.2	6.8
Ammonium, $[NH_4^+]$ ($mg\,l^{-1}$)	0.010	0.020	0.010	0.003	0.004	0.016	0.440	0.015	0.010	0.011
Nitrite, $[NO_2^-]$ ($mg\,l^{-2}$)	0.011	0.011	0.012	0.008	0.008	0.033	0.008	0.012	0.006	0.006
Nitrate, $[NO_3^-]$ ($mg\,l^{-1}$)	1.56	0.95	4.31	1.46	0.89	0.96	2.33	3.93	0.52	3.80
Ortho-phosphate, $[PO_4^{2-}]$ ($\mu g\,l^{-1}$)	120.0	130.0	60.0	100.0	40.0	80.0	680.0	110.0	400.0	70.0
Total Phosphorus, $[T\,P]$ ($\mu g\,l^{-1}$)	42.0	44.0	20.0	37.0	26.0	68.0	283.0	41.0	155.0	42.0
Calcium, $[Ca^{2+}]$ ($mg\,l^{-1}$)	66.80	62.50	75.80	38.90	34.30	67.70	6.60	6.80	17.00	9.40
Magnesium, $[Mg^{2+}]$ ($mg\,l^{-1}$)	4.40	4.10	5.40	3.00	3.10	4.70	1.10	1.30	3.90	2.20
Sodium, $[Na^+]$ ($mg\,l^{-1}$)	4.30	4.60	9.20	3.70	4.10	12.50	7.00	5.90	8.80	5.40
Potassium, $[K^+]$ ($mg\,l^{-1}$)	0.60	0.60	0.90	1.20	1.10	5.50	1.00	0.90	0.70	1.00
Sulfate, $[SO_4^{2-}]$ ($mg\,l^{-1}$)	8.60	10.50	20.10	11.70	14.50	18.10	7.50	6.30	16.30	9.60
Silicate, $[SiO_2]$ ($mg\,l^{-1}$)	6.20	7.00	9.00	5.90	5.90	7.80	10.90	9.00	13.80	8.50
Total hardness, $[CaCO_3]$ ($mg\,l^{-1}$)	186.10	174.08	212.97	110.29	99.23	189.68	21.30	22.65	59.45	33.06

the temperature values lay within a normal range for both seasons and sites, it is interesting to note the colder temperatures of the Neda River due to the existence of karstic sources. The waters deriving from such deep sources maintain characteristically low temperatures. The pH values recorded at the sampling sites NE-1 and NE-3 range between 8.3 and 7.7 for the summer period while for the winter period the range was 8.4 and 7.5, respectively. The conductivities for the PE and GV sites are very low due to the siliceous substrate. Values for the Neda River sites were higher because of limestone in the catchment area. Organic matter originating from anthropogenic impacts, i.e. septic tanks, pasture, livestock, etc., causes the low oxygen saturation values at the NE-3 sampling site, especially during the winter period. Among all of the sites, the highest values regarding oxygen saturation were observed at the upper sites of the Neda River (NE-1 and NE-2) and at the PE and GV sites, due to the absence of pollution sources and due to the high water flow. One site, NE-3, had high sodium and chlorine values because of marine aerosol, but mostly through soil salinization due to irrigation with sea-water-contaminated aquifers. The higher potassium values at the PE and GV sites are due to the existence of granites, while the higher values during the winter period at the NE-3 sampling site is possibly caused by the use of fertilisers in combination with agricultural soil leaching. The dominance of silicate rocks at the PE and GV sampling site basins also contributes to higher potassium levels. As regards the increased recorded values of ammonium, nitrate, orthophosphate and total phosphorus during the summer period at the sampling site PE, this is due to the existence of filamentous algae communities (*Cladophora agg.* mainly) at the upper reaches of the Fonias River. These algae concentrations are found in naturally confined areas – ponds immediately downstream of the cascades and water falls of the river corridor. The substratum of those ponds is dominated more by boulders, cobbles and bedrock mainly, rather than any other material, while they act as leaf collectors and deposits of the adjacent terrestrial vegetation. From the physico-chemical analysis point of view, the environmental status of all sites, and especially that of the PE and GV ones, seems to be at a good level. The NE-3 site shows some signs of environmental degradation (low oxygen saturation, high ammonium and nitrate concentrations resulting from the respiration and the mineralization, respectively, of the aforementioned organic matter).

Macroinvertebrates were collected at all of the above sites by the use of the AQEM method. Beyond the general natural and environmental conditions of the surveyed river basins, the composition of the substrate types found for each sampling period is characterized as satisfactory. As regards aquatic macrophytes in the streams and rivers surveyed, their presence is extremely rare because of the presence of high gradients and current velocity of the waterways. In such streams and rivers, usually the aquatic macrophytes communities are found in depositional zones and in their lowest reaches because of water stagnation (pools). Additionally, the fine abiotic types of substrates (sand, psammal, silt, etc.) appropriate for the development of aquatic macrophyte communities are dominant in such stretches of the running waters.

Regarding the benthic fauna, many of the major macroinvertebrate groups were found in the surveyed areas. The pollution-intolerant Turbellaria *Dugesia* and *Crenobia* were present only at site NE-1 for both seasons, while Gastropoda *Valvata* were found only at the NE-3 site during the summer season. In spite of the calcareous nature of this river basin, the gastropods are very rare, probably due to the almost nonexistent aquatic macrophytes. The pollution-tolerant Oligochaeta taxa *Lumbriculus* and *Stylodrilus* were found only at the NE-2 sampling site during the winter period. At the NE-3 site, the Ephemeroptera taxa showed reduced values for the summer period, while in the winter no mayfly individuals were collected. Species belonging to the Leptophlebiidae and Siphlonuridae families were not collected, while generally the most dominant families were that of Baetidae, Heptageniidae, Caenidae, Ephemerellidae, Ephemeridae and Oligoneuridae. Plecoptera were represented mainly by the Nemouridae (especially during in summer period for all sites) and by a few species of Capniidae and Leuctridae only during winter. Regarding the presence of Trichoptera, in the NE-3 site only a few individuals of *Hydropsyche spp.* were found. Other species belonging to Rhyacophilidae (*Rhyacophila spp.*), Psychomyiidae (*Tinodes spp.*), Polycentropodidae (*Cyrnus sp.*), Philopotamidae (*Philopotamus spp.*), Limnephilidae (*Limnephilus sp.*), Leptoceridae (*Adicella sp.*, *Triaenodes sp.* and *Setodes sp.*) and Hydroptilidae (*Agraylea sp.* and *Hydroptila sp.*) reflected the diversity between the NE-1 and NE-2 sampling sites. Among the Coleoptera, the Elmidae (*Riolus sp.*, *Limnius sp.*, *Elmis sp.*, *Stenelmis sp.* and *Esolus sp.*) were dominant. Dytiscidae, Gyrinidae, Hydraenidae, Helophoridae and Scirtidae were rare, as well as Odonata (*Anax sp.*, *Enallagma sp.*, *Gomphus sp.*, *Calopteryx sp.* and *Onychogomphus sp.*), Amphipoda (*Gammarus spp.*) and Heteroptera (*Velia sp.*, *Gerris spp.*, *Micronecta sp.*, etc.) and were not found in all sites and sampling seasons. In contrast with other taxa, because of the lack of the appropriate taxonomic keys, the identification of Diptera was not carried out at the highest possible taxonomic level (genus or species). Chironomidae were present in all samples, while other families (Simuliidae, Muscidae, Psychodidae, Tabanidae, Stratiomyidae, Athericidae, Empididae, Dixidae and Blephariceridae) varied in presence according to site and season.

The Fonias River and the Tsivdoghianni Stream had a satisfactory presence of benthic faunal groups. Hirudinea and Oligochaeta species, characterized as pollution-tolerant, were not found at the PE and GV sampling sites of the Fonias River and Tsivdoghianni Stream, respectively. Richness and diversity of Gastropoda are very poor. Ephemeropterean density and richness at the sites PE and GV were lower than that of the NE-1 and NE-2 sites. Species include Baetidae (*Baetis sp.*, *Procloeon spp.* and *Centroptilum spp.*), Caenidae (*Caenis spp.*), Ephemerellidae (*Torleya sp.* and *Serratella spp.*) and Heptageniidae (*Ecdyonurus sp.*). Generally, the mayfly community is characterized as scant at all sites with the exception of the PE site during the summer period. Finally, regarding the stonefly community, Perlodidae and Perlidae were not found at the Neda River, but *Isoperla sp.* and *Perla sp.* were present in the PE and GV sites. The Nemouridae *Amphinemura* and *Protonemura* were scarce at these sites. Additionally, a few individuals of the Taeniopterygidae *Brachyptera* and *Rhabdiopteryx* were collected at the PE site during winter.

The richness of the Diptera and Amphipoda taxa at the Samothraki sites (PE and GV) is quite low.

The output of the AQEM software (AQEM Consortium, 2002) for both seasons (summer and winter) for the Neda River, Fonias River and Tsivdoghianni Stream revealed important information (scores of various biotic indices, Mayfly Average Score, evenness, species diversity, locomotion types, feeding types, current prefer-ences, etc.) and crucial observations regarding the real environmental status of the investigated areas. The biotic indices are based upon the number of the collected macroinvertebrate taxa, which is crucial for further elaboration. The AQEM software revealed that the number of taxa, although overall satisfying, is seriously reduced during the winter sampling period (especially at the NE-3 site and at the GV site of the Neda and Fonias Rivers, respectively). This fact has serious consequences on the interpretation of the biotic indices (misidentification of the pollution level). Consid-ering the BMWP index values, and especially its Spanish version, pollution levels for all sampling sites did not exhibit homogeneity regarding the season. Additionally, other indices exhibited analogous behaviour, such as the BBI, Indice Biotico Esteso (IBE) (Ghetti, 1997), etc., a phenomenon attributed to seasonality.

Characteristic is the case of the site NE-3 during the winter period, which is classi-fied as extremely polluted by the BMWP (Spanish version), while the BBI classifies the site as excellent, for the same period. Under the assumption that the overall en-vironmental status in the catchment area level remained unchanged throughout the whole study period, a possible inference is that the above-mentioned indices are not suitable for the surveyed areas. This is due to the fact that the applied indices had been developed in ecoregions not related to those under survey.

Another factor in this complex issue is that of the sampling method, as the overall scores of the indices and metrics have been derived by the application of the AQEM method only. This fact should be a reason for 'over'-or 'under'-collection regarding the qualitative or quantitative characteristics of a sample, for many indices. However, the information conveyed by the collected number of taxa remains intact. Among other results, the values of the ASPT, Diversity indices and Evenness showed stabil-ity regarding their interpretation of the classification of the sites for both seasons. Nevertheless, a slightly increasing tendency concerning the period (mostly during the winter) has been observed.

The mayflies, which are a major taxonomic group easily collected and well repre-sented in benthic macroinvertebrate samples, can be considered as a good biological indicator. The Mayfly Average Score (MAS) has been proposed as a simple pre-liminary step for an extensive application of mayfly nymphs as indicators of the biological quality of streams (Buffagni, 1997), but while the MAS results have not revealed significant differentiations for the Samothraki Island sites for both seasons (all values were equal)for the Neda River sites despite the low degradation impacts on the river basin, the rich channel substrates and the detection of many flow types as determined by the use of AQEM, RHS and South Europe RHS protocols, the richness of mayflies remained poor. The taxa richness of the benthic macroinvertebrate com-munity is often used as a simple and reliable water quality indicator. However, taxa richness for Chironomidae (Sc) has shown variations depending on the stream size and seasonality. It has also been demonstrated that Sc does not always decline with increasing pollution and the increase of Sc might be the result of moderate pollution and sedimentation. Consequently, Sc is characterized as a difficult parameter and should be used with caution in environmental assessment (Lenat, 1983). In the case of the surveyed sites, this has been depicted by the majority of stations and periods. In some sites, especially in the GV and NE-3 ones, the observed high variations could have easily derived from the seasonality and the stream size. Another likely cause, regarding the NE-3 site, is the higher sedimentation processes in comparison with the other sites.

Regarding the current preferences metric, in the Neda River stations, the dominant current preference is rheophile (RP) in both seasons, followed by rheo- to limnophile (RL) and rheobiont (RB), reflecting the slope of the catchment. In the case of the PE and GV stations, the same general preferences are observed, with an increase in limnophile (LP) typology reflecting the presence of natural ponds. The dominant feeding types for the Neda River are the grazers-scrapers and the gatherers/collectors, followed by the predators, shedders, etc. This supports the presence of a fully developed and dynamic trophic chain due to the various types of microhabitats. A similar situation is also seen at the Samothraki Island sites, with a difference regarding the ranking (the gatherers/ collectors dominate instead of grazers-scrapers).

Regarding zonation, at the Neda River sites, hyporhithral, epipotamal, litoral and metarhithral zones were dominant, while at the PE site litoral, hyporhithral and epipotamal zones dominated. The dominant microhabitat preference was mainly the litthal type followed by the phytal type. Finally, the algorithm of the BMG index was applied for all sites and seasons in order to calculate a river-type-specific metric, suitable for the conditions of Greek streams and rivers (Skoulikidis *et al.*, 2004).

The BMG index is as follows:

$$BMG = \frac{\sum_{i}^{N} SC_i}{\sum_{i}^{N} \alpha_i}$$

where: SC_i *is the score of the ith taxon*, and α_i is the total number of taxa.

The development of the BMG index is based on the AQEM data and the results show very high correlation with the Nutrient Pollution Metric (NPM), an indicator of nutrient. It is based on an appropriate algorithm and conveys an inclusive picture of the presence of nutrients for each sampling site (Skoulikidis *et al.*, 2002, 2004). Additionally, the BMG index is highly correlated with the NPM, the land use practices and with potassium and chloride concentrations (N. Th. Skoulikidis, personal communication), reflecting fertilizers and wastewaters.

Finally, the sites were post-classified as follows: the NE-1 and NE-2 as 'Good', the NE-3 as 'Moderate', with the PE and GV as 'Good'. Taking into account the pre-classification of the previous sites, it derives that all quality levels remain the same with the exception of the PE site. This site changed to 'Good' from

Table 3.3.3 HMS and HQA scores for all surveyed sites. Data taken from the European Research Project 'Standardization of River Classifications. Framework for Calibrating Different Biological Survey Results against Ecological Quality Classifications to be developed for the Water Framework Directive (STAR www.eu-star.at) 2001–2005

Stream/site	Neda River			Fonias River River/PE	Tsivdoghianni Stream/GV
	NE-1	NE-2	NE-3		
Habitat Modification Score (HMS)	1	3	12	2	3
Habitat Quality Assessment (HQA)	66	64	56	72	62

the Reference level. As the criteria for the establishment of the reference conditions were meticulously applied, this difference could be attributed to the ability of this index to detect even minimal perturbations (see relevant results in Table 3.3.1).

The River Habitat Survey protocol (Raven *et al.*, 1998) was completed for each sampling site. The HMS and HQA scores are illustrated in Table 3.3.3. On the basis of the HMS (Raven *et al.*, 1998), the sites varied from unmodified (NE-2 and GV) to semi-natural (NE-1 and PE) and modified (NE-3). This is also reflected in the HQA scores.

3.3.10 CONCLUSIONS

It is worth mentioning that during the history of Gaia, human society was organized in different levels (individual, tribe, polis, state, nation and union). The latter followed a parallel route of changes and modifications, along with the environmental alterations, as regards its size and structure as well as its level of organization and administration. The transportation of appropriate knowledge and communication has always been a part and parcel of this behaviour to the earth's environments. The behaviour of these social groups towards environment and time gradually enforced and accelerated the relationship 'action–reaction' upon the ecosystem of the earth (Gritzalis, 2002). At this time, many policies are faced with the results stemming from such behaviour and environmental protection and restoration of running waters need appropriate tools for river quality assessment and biomonitoring.

In Europe, and especially in the Mediterranean region, a general stream typology is lacking. This fact has led in many cases to the use of simple methods and approaches in order for stream types to be defined. In general, for a preliminary differentiation regarding stream types, some criteria of the EU WFD have been adopted by many scientific teams. To achieve this, many criteria been used: (a) *Ecoregions* (according to Illies, 1978), (b) *Size classes (based on catchment area)*, (c) *Geology of the catchment area*, and (d) *Altitude classes*. Unquestionably, the use of other elements of the systems A and B of the WFD in different combinations has also been adopted.

Finally, for the Mediterranean region many assessment tools and studies (Alba-Tercedor and Sánchez-Ortega, 1988; Ghetti, 1997; Solimini *et al.*, 2000; Alba-Tercedor and Pujante, 2000; Buffagni *et al.*, 2001, 2004 Jáimez-Cuéllar *et al.*,

2002; Pinto *et al.*, 2004; Skoulikidis *et al.*, 2004, etc.) for biomonitoring purposes have been developed, but many of those regions have not adopted and implemented the use in national or regional levels of such tools. The majority of the freshwater quality assessment methods which use macroinvertebrates are focusing on wade-able streams and rivers. Generally, the vast majority of these assessment tools trace disruptions of the elements composing ecological integrity. Analogously developed assessment methods appropriate for non-wadeable, intermitted (naturally or not) and heavily modified rivers are fewer. For biomonitoring applications, the identification level of macroinvertebrates always depends on the purposes, while it varies among the current assessment methods. Regarding the establishment of reference conditions, a common Mediterranean approach must be rejected, because of the variety of the ecoregions, pressures and the dissimilar nature of the available historical data. Concerning hydromorphology, this is supportive of the biomonitoring purposes, but methods suitable for the Mediterranean area should also be developed.

It is worth noted that the peculiar nature of the streams and rivers of the Mediterranean islands regarding freshwater quality assessment has also been studied during the last few years (in Sicily, Gerecke, 1986; in Malta, Haslam, 1989; in Samothraki, Gritzalis *et al.*, 2002; in Sicily, Cimino *et al.*, 2002; in eight Aegean Islands, Gritzalis and Karaouzas, 2004, etc.), although not as much as the mainland running waters. In conclusion, during recent years many scientific teams have contributed to a better the freshwater regime in the Mediterranean region through studies dealing with the WFD and biomonitoring requirements.

Regarding the case studies (Neda River, Fonias River and Tsivdoghianni Stream, Greece), the use of the BMG shows the ability of this index to detect organic and inorganic pollution, but further data and elaboration are required in order to cover the characteristics of all the stream types found in Greece.

ACKNOWLEDGEMENTS

The author would like to thank all of the colleagues of the Institute of Inland Waters, Hellenic Centre for Marine Research (HCMR) who participated in these studies and everyone who supported the entire attempt for the development of this manuscript. I am also pleased to acknowledge my colleagues across Europe as such studies reflect collaborative research. This paper was made possible by grants from the European Union and the Greek General Secretariat of Research and Technology, Ministry of Development.

REFERENCES

Alba-Tercedor, J. and Pujante, A. M., 2000. 'Running water biomonitoring in Spain: opportunities for a predictive approach'. In: *Assessing the Biological Quality of Fresh Waters: RIVPACS and Other Techniques*. Wright, J. F., Sutcliffe, D. W. and Furse M. T. (Eds). Freshwater Biological Association: Ambleside, UK, pp. 207–216.

Alba-Tercedor, J. and Sánchez-Ortega, A., 1988. 'Un metodo rapido y simple para evoluar le calidad biologica de las aguas corrientes basado en el de Hellawell (1978)'. *Limnetica*, **4**, 51–56.

Allan, D. J., 1995. *Stream Ecology: Structure and Function of Running Waters*. Chapman & Hall: London, UK.

AQEM Consortium, 2002. 'Manual for the application of the AQEM system. A comprehensive method to assess European streams using benthic macroinvertebrates, developed for the purpose of the Water Framework Directive', Version 1.0, February 2002. www.aqem.de.

Armitage, P. D., Moss, D., Wright, J. F. and Furse, M. T., 1983. 'The performance of a new biological water quality score system based on macroinvertebrates over a wide range of unpolluted running-water sites'. *Water Res.*, **17**, 333–347.

Barbour, M. T. and Yoder, C. O., 2000. 'The multimetric approach to bioassessment, as used in the United States of America'. In: *Assessing the Biological Quality of Fresh Waters: RIVPACS and Other Techniques*, Wright, J. F., Sutcliffe, D. W. and Furse M. T. (Eds). Freshwater Biological Association: Ambleside, UK, pp. 281–292.

Bertrand, C., Siauve, V., Fayolle, S. and Cazaubon, A., 2001. 'Effects of hydrological regime on the drift algae in a regulated Mediterranean river (River Verdon, Southeastern France)'. *Regul. Rivers Res. Manage.*, **17**, 407–416.

Bonada, N., Narcís, P., Munné, A., Rieradevall, M., Alba-Tercedor, J., Álvarez, M., Avilés, J., Casas, J., Jáimez-Cuéllar, P., Mellado, A., Moyá, G., Pardo, I., Robles, S., Ramón, G., Suárez, L., Toro, M., Vidal-Abarca, R., Vivas, S. and Zamora- Muñoz, C., 2002. 'Criterios para la selección de condiciones de referencia en los ríos mediterráneos. Resultados del proyecto GUADALMED'. *Limnetica*, **21**, 99–114.

Buffagni, A., 1997. 'Mayfly community composition and the biological quality of streams'. In: *Ephemeroptera and Plecoptera: Biology – Ecology – Systematics*, Landolt, P. and Sartori, M. (Eds). MTL: Fribourg, Switzerland, pp. 235–246.

Buffagni, A. and Comin, E., 2000. 'Secondary production of benthic communities at the habitat scale as a tool to assess ecological integrity in mountain streams'. *Hydrobiologia*, **422/423**, 183–195.

Buffagni, A. and Kemp, J., 2002. 'Looking beyond the shores of the United Kingdom: addenda for the application of River Habitat Survey in Southern European rivers'. *J. Limnol.*, **61**, 199–214.

Buffagni, A., Kemp, J., Erba, S., Belfiore, C., Hering, D. and Moog, O., 2001. 'A Europe-wide system for assessing the quality of rivers using macroinvertebrates: the AQEM Project and its importance for southern Europe (with special emphasis on Italy)'. *J. Limnol.*, **60**(Suppl. 1), 39–48.

Buffagni, A., Erba, S., Cazzola, M. and Kemp, J. L., 2004. 'The AQEM multimetric system for the southern Italian Apennines: assessing the impact of water quality and habitat degradation on pool macroinvertebrates in Mediterranean rivers'. *Hydrobiologia*, **516**, 313–329.

Cairns, J., Jr, 1995. 'Ecological integrity of aquatic systems'. *Regul. Rivers Res. Manage.*, **11**, 313–323.

Cimino, G., Puleio, M. C. and Toscano, G., 2002. 'Quality assessment of freshwater and coastal seawater in the Ionian area of N.E. Sicily, Italy'. *Environ. Monit. Assess.*, **77**, 61–80.

De Pauw, N. and Vanhooren, G., 1983. 'Method for biological quality assessment of watercourses in Belgium'. *Hydrobiologia*, **100**, 153–168.

Ehlert, T., Hering, D., Koenzen, U., Pottgiesser, T., Schuhmacher, H. and Friedrich, G., 2002. 'Typology and type-specific reference conditions for medium-sized and large rivers in North Rhine-Westphalia: Methodical and biological aspects'. *Int. Rev. Hydrobiol.*, **87**, 151–163.

EU, 2000. 'Directive 2000/60/EC of the European Parliament and of the Council of 23 October 2000 establishing a framework for Community action in the field of water policy'. *Official Journal of the European Communities*, **L327**, 22.12.2000, 1–72.

Furse, M. T., 2000. 'The application of RIVPACS procedures in headwater streams – an extensive and important national resource'. In: *Assessing the Biological Quality of Fresh Waters: RIVPACS and Other Techniques*, Wright, J. F., Sutcliffe, D. W. and Furse, M. T. (Eds). Freshwater Biological Association: Ambleside, UK, pp. 79–91.

Furse, M. T., Moss, D., Wright, J. F. and Armitage, P. D., 1984. 'The influence of seasonal and taxonomic factors on the ordination and classification of running-waters sites in Great Britain and on the prediction of their macro-invertebrate communities'. *Freshwater Biol.*, **14**, 257–280.

Gerecke, R., 1986. 'Le acque interne di Sicilia e la loro fauna: Un patrimonio naturale da salvare'. *Animalia*, **13**, 217–245.

Ghetti, P. F., 1997. *Manuale di Applicazione Indice Biotico Esteso (I.B.E.). I Macroinvertebrati nel Controllo della Qualità Degli Ambienti di Acque Correnti. Agenzia* Provinciale per la Protezione dell' Ambiente: Provincia Autonoma di Trento, Trento, Italy.

Gianferrari, L., 1927. 'Diagnosi preliminare di due nuove specie ittiche di Rodi'. *Atti Soc. Ital. Sci. Nat. Mus. Civ. Stor. Nat. Milano*, **LXVI**, 123–125.

Gibson, G. R., Barbour, M. T., Stribling, J. B., Gerritsen, J. and Karr, J. R., 1996. 'Biological Criteria: Technical Guidance for Streams and Small Rivers (revised edition)', EPA 822-B-96-001. Office of Water, US Environmental Protection Agency: Washington, DC, USA.

Graca, M. A. S., Coimbra, C. N. and Santos, L. M., 1995. 'Identification level and comparison of biological indicators in biomonitoring programs'. *Cienc. Biol. (Ecol. Syst.)*, **15**, 9–20.

Gritzalis, K. C., 1998. 'An approach to the main aquatic animal status of the Mediterranean Sea, and its significance to human health, in the third century AD'. *Fresenius' Environ. Bull.*, **7**, 356–361.

Gritzalis, K. C., 2002. 'Environment, communication, knowledge and globalisation'. In: *Proceedings of the 14th International Conference on Greek Philosophy; Polis and Cosmopolis: Problems of a Global Area*. Pythagoreion Samos: Greece, August 2002, pp. 76–77.

Gritzalis, K. C. and Karaouzas, I. D., 2004. 'The ecological quality assessment and status in permanent streams of the Aegean Islands'. In: *Proceedings of the 5th International Symposium: Fauna and Flora of Atlantic Islands*. Dublin, Ireland, August 2004.

Gritzalis, K., Koussouris, T. and Diapoulis, A., 1993. 'Distribution of the invertebrate fauna with relation to pollution and especial hydrological situation in Arachthos River (Greece)'. In: *Proceedings of the 6th International Congress on the Zoogeography and Ecology of Greece and Adjacent Regions*. Hellenic Zoological Society: Thessaloniki, Greece, April 1993.

Gritzalis, K. C., Skoulikidis, N. T. and Kouvarda, T. D., 2002. 'The biodiversity, locomotion types and the current preferences of the aquatic macroinvertebrate fauna at Samothraki Island, NE Aegean Sea, Greece'. In: *Proceedings of the 9th International Congress on the Zoogeography and Ecology of Greece and Adjacent Regions*. Hellenic Zoological Society: Thessaloniki, Greece, May 2002.

Guerold, F., 2000. 'Influence of taxonomic determination level on several community indices'. *Water Res.*, **34**, 487–492.

Harper, D. M., Kemp, J. L., Vogel, B. and Dawson, M. D., 2000. 'Towards the assessment of "ecological integrity" in running waters of the United Kingdom'. *Hydrobiologia*, **422/423**, 133–142.

Haslam, S. M., 1989. 'The influence of climate and man on the water courses of Malta'. *Toxicol. Environ. Chem.*, **20/21**, 85–92.

HAGS, 1969. *Edition* of: Hellenic Army Geographical Service. Sheet: Nisos Samotraki Scale: 1:50000. General purpose map. Athens, Greece.

Hering, D., Buffagni, A., Moog, O., Sandin, L., Sommerhäuser, M., Stubauer, I., Feld, C., Johnson, R. K., Pinto, P., Skoulikidis, N., Verdonschot, P. F. M. and Zahrádková, S., 2003. 'The development of a system to assess the ecological quality of streams based on macroinvertebrates – design of the sampling programme within the AQEM project'. *Int. Rev. Hydrobiol.*, **88**, 345–361.

Hering, D., Moog, O., Sandin, L. and Verdonschot, P. F. M., 2004. 'Overview and application of the AQEM assessment system'. *Hydrobiologia*, **516**, 1–20.

HMGS, 1978. Edition of: Hellenic Military Geographical Service. Sheet: Filiatra. Scale: 1:100.000. General purpose map. Athens, Greece.

Hughes, R. M., 1995. 'Defining acceptable biological status by comparing with reference conditions'. In: *Biological Assessment and Criteria. Tools for Water Resource Planning and Decision Making*, Davis, W. S. and Simon, T. P. (Eds) Lewis Publishers: Boca Raton, FL, USA, pp. 31–47.

Illies, J. (Ed.), 1978. *Limnofauna Europea*. Gustav Fischer Verlag: Stuttgart, Germany.

IGME, 1982. Publication: Institute of Geology and Mineral Exploration. Sheet: Kiparissa. Scale: 1:50.000. General purpose map. Athens, Greece.

Jáimez-Cuéllar, P., Vivas, S., Bonada, N., Robles, S., Mellado, A., Álvarez, M., Avilés, J., Casas, J., Ortega, M., Pardo, I., Prat, N., Rieradevall, M., Sáinz-Cantero, C. E., Sánchez-Ortega, A., Suárez, L., Toro, M., Vidal-Abarca, R., Zamora-Muñoz, C. and Alba-Tercedor, J., 2002. 'Protocolo GUADALMED (PRECE)'. *Limnetica*, **21**, 187–204.

Karaouzas, I. D. and Gritzalis, K. C., 2002. 'The effects of a modified river on the biodiversity and ecological characteristics on the benthic macroinvertebrate fauna (Pamisos River, Peloponnese, Greece)'. In: *Proceedings of the International Conference, Joint Research Center, Sustainability of Aquatic Ecosystems 'Science in Support of European Water Policies'*. Stresa, Lago Maggiore, Italy.

Keivany, Y., Daoulas, Ch., Nelson, J. S. and Economidis, P. S., 1999. 'Threatened fishes of the world: *Pungitius hellenicus*, Stephanidis, 1971 (Gasterosteidae)'. *Environ. Biol. Fish*, **55**, p. 390.

Kolkwitz, R. and Marsson, M., 1902. 'Grundsätze für die biologische Beurteilung des Wassers nach Flora und Fauna'. *Mitt. Kgl. Prüfanstalt Wasserversorgung Abwassrbeseitigung. Berlin-Dahlem*, **1**, 33–72.

Kolkwitz, R. and Marsson, M., 1908. 'Ökologie der planzlichen Saprobien'. *Ber. Dtschen. Bot. Ges*, **26**, 505–519.

Kolkwitz, R. and Marsson, M., 1909. 'Ökologie der tierischen Saprobien'. *Int. Rev. Hydrobiol.*, **2**, 126–519.

Lenat, D. R., 1983. 'Chironomid Taxa Richness: Natural Variation and Use in Pollution Assessment'. *Freshwater Invertebr. Biol.*, **2**, 192–198.

Metcalfe, J., 1989. 'Biological water quality assessment of running waters based on macroinvertebrate communities: History and present status in Europe'. *Environ. Pollut.*, **60**, 101–139.

Moog, O. (Ed.), 1995. *Fauna Aquatica Austriaca. A Comprehensive Species Inventory of Austrian Aquatic Organisms with Ecological Notes*. Wasserwirtschaftskataster, Bundesministerium für Land-und Forstwirtschaft: Vienna, Austria.

Moog, O. and Chovanec, A., 2000. 'Assessing the ecological integrity of rivers: walking the line among ecological, political and administrative interests'. *Hydrobiologia*, **422/423**, 99–109.

NBN, 1984. 'Qualité biologique des cour d' eau: détermination de l' indice biotique se basant sur les macro-invertébrés aquatiques', NBN T92-402. Institut Belge de Normalization (IBN): Brussels, Belgium.

NIGMR, 1973. Published by: National Institute of Geology and Mining Research. Sheet: Fighalia. Scale: 1:50000. Geological map. Athens, Greece.

Nijboer, R. C., Johnson, R. K., Verdonschot, P. F. M., Sommerhäuser, M. and Buffagni, A., 2004. 'Establishing reference conditions for European streams'. *Hydrobiologia*, **516**, 91–105.

Pinto, P., Rosado, J., Morais, M. and Antunes, I., 2004. 'Assessment methodology for southern siliceous basins in Portugal'. *Hydrobiologia*, **516**, 191–214.

Pires, A. M., Cowx, I. G. and Coelho, M. M., 2000. 'Benthic macroinvertebrate communities of intermittent streams in the middle reaches of the Guadiana Basin (Portugal)'. *Hydrobiologia*, **435**, 167–175.

Raven, P. J., Holmes, N. T., Dawson, F. H., Fox, P. J. A., Everard, M., Fozzard, I. R. and Rouer, K. J., 1998. 'River Habitat Quality: The Physical Character of Rivers and Streams in the UK and Isle of Man, River Habitat Survey Report No. 2. Environment Agency: Bristol, UK.

Resh, V. H., Myers, M. J. and Hannaford, M., 1996. 'Macroinvertebrates as biotic indicators of environmental quality. In: *Methods in Stream Ecology*, Hauer, F. R. and Lamberti, G. A. (Eds). Academic Press: San Diego, pp. 647–667.

Reynoldson, T. B. and Wright, J. F., 2000. 'The reference condition: problems and solutions'. In: *Assessing the Biological Quality of Fresh Waters: RIVPACS and Other Techniques*, Wright, J. F., Sutcliffe, D. W. and Furse, M. T. (Eds). *Freshwater Biological Association*: Ambleside, UK, pp. 293–303.

Siligardi, M., Bernabei, S., Cappelletti, C., Chierici, E., Ciutti, F., Egaddi, F., Franceschini, A., Maiolini, B., Mancini, L., Minciardi, M. R., Monauni, C., Rossi, G. L., Sansoni, G., Spaggiari, R. and Zaneti, M., 2000. *IFF. Indice di Funzionalità Fluviale*, Manuale ANPA November 2000. Agenzío Nazionale per la Protezion e dell'Ambiente: Rome, Italy.

Skoulikidis, N. Th., Gritzalis, K. and Kouvarda, Th., 2002. 'Hydrochemical and Ecological Quality Assessment of a Mediterranean River System'. *Global Nest*, **4**, 29– 39.

Skoulikidis, N. Th., Gritzalis, K. C., Kouvarda, Th. and Buffagni, A., 2004. 'The development of an ecological quality assessment and classification system for Greek running waters based on benthic macroinvertebrates'. *Hydrobiologia*, **516**, 149–160.

Skriver, J., Friberg, N. and Kirkegaard, J., 2000. 'Biological assessment of running waters in Denmark: introduction of the Danish Stream Fauna Index (DSFI)'. *Verh. Int. Limnol.*, **27**, 1822–1830.

Solimini, A. G., Gulia, P., Monfrinotti, M. and Carchini, G., 2000. 'Performance of different biotic indices and sampling methods in assessing water quality in the lowland stretch of the Tiber River'. In: *Assessing the Ecological Integrity of Running Waters*, Jungwirth, M., Muhar, S. and Schmutz, S. (Eds). Kluwer Academic Publishers: Dordrecht, The Netherlands, *Hydrobiologia*, **422/423**, 197–208.

Stoumboudi, M., Barbieri, R., Mamuris, Z., Corsini-Foka, M. J. and Economou, A., 2002. 'Threatened fishes of the world: *Ladigesocypris ghigii* (Gianferrari) 1927, (Cyprinidae)'. *Environ. Biol. Fish*, **65**, p. 340.

Verdonschot, P. F. M., 2000. 'Integral ecological assessment methods as a basis for sustainable catchment management'. *Hydrobiologia*, **422/423**, 389–412.

Vourdoumpa, A. S. and Gritzalis, K. C., 2000. 'Influence of engineering works and controlled river discharge to the use of Biotic Indices'. In: *Proceedings of the 6th Hellenic Symposium on Oceanography and* Fisheries, Vol. II. Chios: Greece, pp. 258–260 (in Greek with English abstract).

Wallin, M., Wiederholm, T. and Johnson, R. K., 2003. 'Guidance on establishing reference conditions and ecological status class boundaries for inland surface waters'. CIS Working Group 2.3 – REFCOND 7th Version. CIS: Common Implementation Strategy.

FURTHER READING

Allan, J. D., 1995. *Stream Ecology: Structure and function of running waters*. Chapman & Hall: London, UK.

Barbour, M. T., Gerritsen, J., Snyder, B. D. and Stribling, J. B., 1999. *Rapid Bioassessment Protocols for Use in Streams and Wadeable Rivers: Periphyton, Benthic Macroinvertebrates and Fish*, 2nd Edition, EPA 841-B-99-002. Office of Water, US *Environmental Protection Agency*: Washington, DC, USA.

Bartram, J. and Balance, R. (Eds), 1996. *Water Quality Monitoring. A Practical Guide to the Design and Implementation of Freshwater Quality Studies and Monitoring Programmes.* Published on behalf of UNEP, the WHO and E and FN Spon. E and FN Spon (an imprint of Chapman & Hall): London, UK.

Giller, P. S. and Malmqvist, B., 1998. *The Biology of Stream and Rivers. Biology of Habitats* (Series Editors: Crawley, M. J., Little, C., Southwood, T. R. E. and Ulfstrand, S.). Oxford University Press: New York, NY, USA.

Hauer, F. R. and Lamberti, G. A. (Eds), 1996. *Methods in Stream Ecology*, Academic Press: London, UK.

Mason, C. F., 1996. *Biology of Freshwater Pollution*, 3rd Edition. Longman: London, UK.

Montgomery, C. W., 1995. *Environmental Geology*, William C. Brown, Publishers: Dubuque, IA, USA.

3.4

Biological Monitoring of Running Waters in Eastern and Central European Countries (Former Communist Block)

Jan Helešic

Biological Monitoring of Rivers   Edited by G. Ziglio, M. Siligardi and G. Flaim
© 2006 John Wiley & Sons, Ltd.

3.4.1   INTRODUCTION

The problem of river water quality arose with the development of industry and intensive farming. In the first part of the 19th Century this was especially the intensive development of the food industry, in particular sugar factories. In the first part of the 20th Century the development of heavy industry in Silesia and other parts of Central Europe was the key factor. Life in large rivers was totally destroyed; very long stretches were lacking fish completely (e.g. Labe – Elbe, Odra – Oder and Morava – March rivers). Agriculture was an source of pollution through soil erosion and runoff from fertilized soils to rivers. Other factors contributing to degraded water quality were the regulation and damming of long stretches of rivers and navigation.

3.4.2   TRADITION OF ENVIRONMENTAL MONITORING AND HISTORY IN THE DIFFERENT CENTRAL AND EAST EUROPEAN COUNTRIES (INCLUDING FORMER EAST GERMANY)

The first act of river management in Central Europe were the miller's regulations issued during the rule of Vladislav of Jagiello (King of Bohemia, Poland and Hungary) in the 15th Century. Millers were responsible for discharge and fish stock in the river stretch assigned to them. This law, with small changes and/or local/national alternations, was valid until the 19th Century. In the 19th Century the Austro–Hungarian Empire adopted the German Imperial Water Act.

After disintegration of the Empire in 1918, the old regulations stayed valid in the emerging states. In Germany a new method of water quality observation was developed. The method was saprobity (Kolkwitz and Marsson, 1908, 1909), which was based on the knowledge on valences of indicator organisms to dissolved oxygen content. The content of oxygen is a measure of fouling processes, by which dissolved oxygen is consumed. This theory was very attractive for water quality monitoring, especially in Central Europe, where the main source of pollution was organic wastes.

This was the time when the first national institutes of applied research in the field of hydrology and technical hydrobiology (wastewater treatment and waterworks) were established. For instance, the Water Research Institute was among the first scientific institutes established in the independent republic of the Czechs and the Slovaks. This priority was the logical consequence of the high attention that water management and use had had in the Czech lands for centuries. The institute was founded in 1919 as the State Hydrological Institute. In 1930, its name was amended to the T. G. Masaryk State Hydrological and Hydrotechnical Institute – now it is called the T.G. Masaryk Water Research Institute. Bohuslav and Záviš Cyrus after World War 1 and Rudolf Šrámek-Hušek after World War 2, were the founders of applied hydrobiology in Czechoslovakia.

Large changes occurred after the Second World War. The major part of Central European countries became the Eastern communistic block, dominated by the Soviet Union. In the 1950s, a new version of saprobic theory was published by Pantle and Buck (1955). This method of water quality evaluation was better and relatively understandable for the technical community. In Germany, Liebmann (1951) published a fundamental work in saprobiology and later a large compendium (1149 pages) (Liebmann, 1958–1960). In Czechoslovakia, the first comparative and methodological works were published by Šrámek-Hušek (1950, 1956), Hanuška *et al.* (1956) and Zelinka *et al.* (1959). Zelinka and Marvan (1961) published a new formula for the saprobic index with indication weight, saprobial valence and abundance of indicator taxon. In 1964, Zelinka and Sládeček published a textbook of technical hydrobiology. In 1965 and 1966, the 'Integral Methods of Biological Analysis of Water' were published (Cyrus and Sládeček, 1965–1966).

A key researcher in the field was Vladimír Sládeček, who worked intensively in saprobiology and published many studies and articles. His efforts culminated in a modern fundamental monograph with detailed sheets of saprobic valences and indication values of more then a thousand taxa of protozoans, algae, macroinvertebrates and fishes (Sládeček, 1973). The saprobity system was accepted in most Eastern and Central European countries. All COMECON (Council for Mutual Economic Aid) member countries had to comply with one common standard (Methods of Assessment, of Water Quality 1977). Besides that, the individual countries had their own national standards (e.g. in Czechoslovakia – ČSN 83 0603 (1967), and STN ČSN 83 0532 (1978/1979), Sládeček *et al.*, 1981; in the German Democratic Republic – Breitig and Tümpling, 1975).

In the 1950s, new water acts were accepted in most Central European countries. One of the most refined acts was prepared in Czechoslovakia (No. 11/1955, later the new and modern No. 138/1973). An integral part of the act was the State Water Management Plan, which proposed all activities in water management for the next five or ten years. The basic act included all terms and regulations regarding pollution, discharge regulation and management. This regulation was supplemented by many decrees and governmental acts. For example, in 1977 there were more then 50 by-laws that negated the originally very good law. This situation was typical for totalitarian regimes, where the secretary of the Communist Party or the government had the first and the last word. Factories were able to operate without sewage treatment plants. According to the Water Act, floodplains with their groundwater supplies were to be protected. Unfortunately, the biggest oil refineries and chemical factories were built on floodplains. For instance, Slovnaft Bratislava was build on the Malý Žitný Island (on island in the Danube channel), with the biggest supply of high-quality groundwater in former Czechoslovakia. The worst situation existed in the 'Black Triangle' between the Czech Republic, Eastern Germany and Poland, with the largest surface coal mines, coal power plants and chemical factories; in Silesia, heavy industry also played a major role. In these areas, together with severe air pollution, the natural hydrological regime and landscape were totally destroyed. The rivers Odra (Oder), Olše (Olsza) and Ostravice in Silesia, as well as the Bílina and

Ohře in West Bohemia, were heavily polluted by industrial and chemical wastes, so that they were without life until the end of the 1970s.

The development of industry and intensive farming since the 1930s has created problems of acid rain and toxic compounds. Soil and water acidification have been destroying spruce forests in the 'Black Triangle'. The main effect of forest 'die-back' has been the change of surface runoff and the entire hydrological regime of streams. The first effects of toxic compounds was shown in Silesian rivers contaminated by phenols.

On the other hand, the COMECON countries attempted to improve the environmental situation as well. In 1962, new regulations for water management were adopted. In 1970, a report about water management, water supplies and water quality in the member countries was presented. Part of this material was a forecast of future development. Typical for these documents was the stress put on the need of water consumption and waste water production as imperative requirements for industrial development. In the 1980s, COMECON issued many recommendations on water management, particularly regarding waste water production and treatment. New guidelines for the protection of water supplies were recommended. For example, in Czechoslovakia several acts were issued (e.g. No. 28/1975 about important water bodies, and 6/1977 about the protection of surface waters against pollution), accompanied by governmental decrees establishing protected areas of natural water accumulation.

Once the Eastern Communist Block fell apart, in practically all countries new water acts and guidelines consistent with the relevant regulations of the EU and OECD were adopted.

3.4.3   STRESSORS/IMPACTS TO RUNNING WATERS AND MONITORING (CASE STUDIES FROM THE CZECH REPUBLIC)

3.4.3.1   Organic pollution (sewage)

This type of pollution is preliminary connected to point sources as outlets from wastewater plants, towns and villages without a central sewage works, food production plants and farms. The worst situation was in the 1960s when long river stretches had highly polluted water (saprobic index higher than 2.5, e.g. alfa or polysaprobity, the worst quality). Especially downstream of pulp, paper and food factories on the Labe and Morava (Czech Republic), the index reached 5, i. e. the worst water quality.

Since the 1980s, the situation has been improving, particularly so after the disintegration of the Soviet Block. All members of the EU have to accept common guidelines or limits for water pollution and construction of sewage treatment plants. Municipal waste waters from the majority of bigger settlements (over 10 000 inhabitants) are treated. National maps of river pollution are published yearly and in 2002 a map of 'Water Quality of the Danube and its tributaries' was published by The International Association for Danube Research (IAD) (http://www.iad-sil.com).

3.4.3.2 Eutrophication

In the 20th Century, eutrophication was caused by intensive farming on large fields. In most former communist countries, e.g. Czechoslovakia, Germany, Hungary and mainly in Russia, individual farming was abolished and big enterprises (state farms or co-operative farms) were farmed on large fields. Runoffs from these sites were the main source of dissolved nutrients in river water.

Euthrophication causes the growth of cyanobacterial phytoplankton, especially in reservoirs. Lower concentrations of phytoplankton are found in running waters, except in stretches downstream of reservoirs. High phytoplankton concentrations, including the occurrence of toxic cyanobacteria, has been found in the Svratka River under the Brno Reservoir, in the Dyje River under the Nové Mlýny Reservoirs, and in the Vltava River (Czech Republic). The working group of Nadatio Flos Aquae (Brno, Czech Republic) produces many handbooks and scientific papers regarding these issues (Gregor and Maršálek 2004; Bláha *et al.*, 2004; Bláha and Maršálek, 2003, etc.).

3.4.3.3 Acidification

Acidification of surface waters by global transportation of pollutants is one of the main effects of human activity on the Earth. The main acidification elements are nitrogen oxides (NO_x) and sulfur dioxide (SO_2). Rainwater contains strong acids, such as HNO_3 and H_2SO_4 (Tables 3.4.1 and 3.4.2.). These acids destroy the hydrogen carbonate balance in water, change the solubility of many elements especially silicon and metals, and have an effect on metal speciation (Foerstner and Wittmann, 1983).

The effect of acidification is realized especially in standing waters – lakes in regions with acid or acidophilous rocks. In the Czech Republic, a very harsh effect of acidification has been observed in frontier mountains – Krušné Mountains, Jizerské Mountains, Krkonoše and Orlické Mountains, because of industrial emissions from the 'Black Triangle' – area in NE Germany, NW Czech Republic and SW Poland (Holoubek, 1990). Acidification can be observed also in other areas of the Czech

Table 3.4.1 Concentrations of sulfur dioxide (mg/m^{-3}) in free air (meteorological station Svratouch – Czech Moravian Highlands, Czech Republic)

Year	1981	1982	1983	1984	1985	1986
Mean SO_2	9.20	20.90	15.20	14.90	21.40	28.20
Maximum SO_2	186.70	203.70	155.20	196.30	203.20	166.80
Year	1988	1989	1990	1991	1992	1993
Mean. SO_2	9.63	20.82	12.73	10.19	14.81	19.81
Maximum SO_2	68.75	180.60	88.80	44.70	107.40	106.70

Table 3.4.2   pH of rainwater (meteorological station Svratouch – Czech Moravian Highlands, Czech Republic)

Year	1989	1990	1991	1992	1993
Maximum pH	6.48	6.39	6.7	6.78	7.01
Mean pH	4.43	4.44	4.44	4.76	4.59
Minimum pH	3.28	3.56	3.29	3.54	3.28

and Slovak Republic, for example, in the Šumava Mountains and the Vysoké Tatry Mountains (Fott *et al.*, 1994; Evans *et al.*, 2001).

Running water ecosystems are commonly less acidified then lentic waters. In areas with significant acidification of soil and only after the natural buffer ability of soil is lost, acidification can be observed in streams as well.

3.4.4   CASE STUDY – CENTRAL PART OF CZECHMORAVIAN HIGHLANDS (CZECH REPUBLIC)[1]

Studied streams are located in the source area of the Svratka River in the Protected Area of Žd'árské vrchy (province of Českomoravská vrchovina – Czech Moravian Highlands). The geology is very diverse and the main rock formations are acid metamorphic rocks of the Svratecké Crystalinic Area. Here, soils are mainly sandy (65 %) and loamy (30 %). Comparison of species richness at individual sites showed that there are two communities, with the lowest diversity at the Řásenský and Holcovský brooks. From the results of pH measurements (Table. 3.4.3), these streams

Table 3.4.3   Minimal, range and median values of pH in sampling streams (1994–1995) – Czech Moravian Highlands, Czech Republic

Site	pH			
	Minimal (acidification episode)	Range[a]	Median	Median[b]
Břímovka brook	3.8	5.3–6.7	5.3	5.6
Řásenský brook	2.8	4.7–6.1	4.8	4.9
Holcovský brook	3.2	4.2–6.4	5.2	5.2
Fryšávka stream (Fryšáva site)	4.8	6.2–7.3	6.3	6.6
Fryšávka stream (kuklik site)	6.0	6.8–7.6	6.8	6.9
Fryšávka stream (Jimramou site)	6.3	7.2–7.7	7.2	7.3

[a] Without acidification episodes.
[b] Without minimal value.

---

[1] For further details, see Scheibová and Helešic, 1999.

seem to be the most influenced by acidic deposits. Lower species diversity in comparison to the Fryšávka Stream in Jimramov is caused especially by acidification but also by a different stream size as well as the characteristics of the river basin. We can compare the influence of acidification on species diversity (Table. 3.4.4) by confronting the Řásenský or Holcovský brooks with the similar non-acidified Fryšava stream at Fryšava; this has twice the number of benthic species than the acidified streams.

Table 3.4.4 Indexes of species richness (Margalef), diversity (Shannon) and (evenness (Shannon)) for six sites

Site	Index		
	Margalef	Shannon	Sheldon
Břímovka brook	15.3	4.2	0.75
Řásenský brook	6.7	2.5	0.56
Holcovský brook	9.4	3.8	0.83
Fryšávka stream (Fryšava site)	17.1	4.4	0.79
Fryšáka stream (Kuklík site)	15.6	3.9	0.71
Fryšáka stream (Jimramov site)	19.8	4.7	0.79

According to Fjellheim and Raddum (1990), each species obtained an acidification index; the average index of species of the macrozoobenthos forming the community is based on the acidification index of the site (0–1, where 0 is the most acidified stream, and 1 is the non-acidified stream).

For the Břímovka brook, the acidification index was 0.20, for the Řásenský brook, 0.14 and for the Holcovský brook, 0.13. For the Fryšávka stream, at Fryšava it was 0.38, at Kuklík, 0.62 and at the Jimramov site, 0.60. From these values, the differences between the followed sites and their zoocenoses are obvious. At the last two sites of the Fryšávka stream, the zoocenosis is composed of species more sensitive to acidification, while the communities at the Řásenský and Holcovský brooks are formed by species tolerant to high acidification.

3.4.4.1 Toxic compounds (metals, PAHs and chlorinated compounds)

Chemical monitoring of water and sediment contamination was developed in the 1990s. Long-term international monitoring programmes have been taken up on the major river basins (International projects, Labe (MKOL), Danube (MKOD) and Odra (MKOO). In the Czech Republic, monitoring is conducted by state organizations such as the Water Research Institute, the River Authorities and the Agriculture Water Management Authority. All activities are co-ordinated by the Ministry of Environment through the subordinate Czech Hydrometeorological Institute. Besides standard analytical procedures, ecotoxicological methods are used (sentinel organisms for bioconcentration studies and toxicity, bioaccumulation, biodegradation tests). The last summarized results are available in Holoubek *et al.* (2004).

3.4.5 CASE STUDY – CONTAMINATION OF SEDIMENTS AND BIOTA (CZECH REPUBLIC)[2]

This study aimed to investigate food web contamination in different types of running water ecosystems in the Czech Republic in Central Europe. The sediment contamination study was presented in 1993–1995 (Holoubek *et al.*, 1994; Fuksa, 1995). In 1996–1997, localities were sampled three times per year; these localities were chosen so that they characterize the basic and most typical types of rivers in the Czech Republic and also rivers which were influenced predominately by global pollution. Research into bioconcentration and bioaccumulation of the chosen pollutants was carried out between 1996 and 1997. To follow the latter, the following pollutants were chosen according to the aims of the grant projects: metals (As, Cr, Cd, Cu, Ni, Pb, Hg, Zn), polychlorinated biphenyls (PCBs; 28, 52, 101, 118, 138, 153, 180), derivatives of DDT (HCB, HCH; alpha, beta, gamma), and DDE, DDD and DDT (Pest-Cls). A bioindicator is defined as an organism which is able to concentrate and accumulate harmful pollutants (Hellawell, 1986).

The highest concentration of metals, PCBs and chlorinated hydrocarbons (Pest-Cls) were found in sediments of the potamal Rivers Berounka, Otava and Labe (Tables 3.4.5–3.4.7, respectively). The concentrations of metals were highest in tissues of water moss (*Fontinalis antipyretica*) and plants (*Batrachium fluitans, Myriophyllum* sp.). As for consumers, Zn accumulated most in filtrators (*Unio* sp.) and predators (*Haemopis sanguisuga*). The concentrations of PCB and Pest-Cls were highest in leeches (*Erpodella* sp. and *Haemopis sanguisuga*) and in larvae of net-spinning caddisflies (*Hydropsyche* sp.). Slight accumulations of PCBs (25, 52, 110), and especially of HCH, were found in congeners.

The best results for accumulation were found in HCH congeners which accumulated in the primary producers and consumers as well. In sediments, the amounts found were non-detectable or near the analytical limits. The 'sums' of the PCBs were used as criteria for selecting useable and significant bioindicators for specific food webs.

3.4.5.1 Morphological and hydrological degradation

Longitudinal and bank regulation – channelization

Construction activities on rivers are very old; millers regulated their own parts of rivers at all times and very old river regulation works can be found inside historic

---

[2] For further details, see Helešic and Scheibová, 2000.

Table 3.4.5 Contents of metals in sediment and biota (mg kg⁻¹)

Sediment/biota	Metal						
	As	Cd	Cu	Ni	Pb	Hg	Zn
Epirhithral zone: Mlýnský Brook, district – Žd'ár nad Sáz							
Sediment	2.0–3.5	0.15–2.0	3.6–13.0	13.0–20.0	8.0–15.9	<0.02–8.0	<50.0–75.0
Algae – *Cladophora* sp.	2.47	0.29	10.18	4.12	3.50	0.015	34.3
Plants – *Fontinalis antipyretica*	8.40–14.11	1.87–4.64	14.94–27.04	32.68–44.70	15.13–21.12	0.055–0.078	171.8–236.7
Macroinvertebrates – Hydropsychidae	0.16–1.53	0.16–1.99	3.38–17.34	0.65–7.04	0.35–3.44	0.115–0.254	33.55–253.1
Metahyporhithral zone: Ředický Brook, district – Pardubice							
Sediment	2.8–3.4	0.64–3.50	13.7–27.0	11.0–15.7	17.0–25.5	<0.02–0.6	268.0–314.0
Algae – *Cladophora* sp.	0.39	0.03	1.55	1.35	2.07	0.016	21.54
Plants – *Batrachium fluitans*	0.18–7.33	0.02–0.85	2.84–33.86	1.07–23.28	0.68–31.48	0.006–0.238	46.38–362.3
Macroinvertebrates – Lymneidae	0.22–2.59	0.04–0.08	4.71–8.88	0.64–2.89	0.59–2.90	0.012–0.028	19.44–33.78
Macroinvertebrates – Hirudinea	1.32	0.09	3.33	0.17	0.37	0.017	107.56
Epipotamal zone: Otava River, district – Starkonice							
Sediment	2.7–13.7	0.57–2.00	4.0–42.0	12.4–27.0	11.3–36.0	<0.02–1.1	42.0–225.0
Mosses *Fontinalis* sp.	9.11	0.45	11.40	13.70	15.85	0.15	87.65
Macroinvertebrates – Lymneidae	1.37	0.27	6.05	1.58	1.44	0.024	23.64
Macroinvertebrates – Hirudinea	—	0.08	2.88	0.20	0.29	0.029	30.28
Macroinvertebrates – *Epeorus sylvicola*	—	0.19	2.84	0.67	0.39	0.014	13.74
Macroinvertebrates – Hydropsychidae	0.51–1.53	0.10–1.99	2.88–17.34	0.91–7.04	0.88–3.44	0.023–0.254	28.79–253.1
Metapotamal zone: Berounka River; district – Praha south; Labe River, district – Pardubice							
Sediment	3.1–5.0	0.55–2.3	5.1–25.0	9.8–22.0	17.0–49.0	<0.02–1.8	63.0–268.0
Algae – *Cladophora* sp.	1.32	0.31	3.60	3.41	5.46	0.022	38.63
Plants – *Myriophyllum* sp.	0.75–4.00	0.27–2.00	4.14–15.96	0.92–13.35	1.54–18.17	0.014–0.089	57.92–215.3
Plants – *Baldingera aerundinacea*	2.48	0.72	33.72	6.42	7.50	0.062	171.80
Macroinvertebrates – *Viviparus viviparus*	0.53	0.11	7.01	7.01	1.57	0.031	36.64
Macroinvertebrates – *Unio* sp.	1.22	0.41	9.28	0.92	0.66	0.027	381.50
Macroinvertebrates – Hirudinea	0.45	0.12	5.02	0.34	0.90	0.031	100.72

Table 3.4.6 Contents of PCBs in sediment and biota (µg kg^{-1})

Variables	Metal						
	PCB 28	PCB 52	PCB 101	PCB 118	PCB 138	PCB 153	PCB 180
Epirhithral zone: Mlýnský Brook, district – Žd'ár nad Sáz							
Sediment	0.35–7.85	<0.2–0.46	0.89–5.10	0.25–0.71	2.28–14.84	2.36–20.17	1.72–14.23
Algae – *Cladophora* sp.	0.44	0.41	0.10	0.37	0.70	0.24	0.12
Plants – *Fontinalis antipyretica*	0.78–1.12	0.59–1.61	0.38–0.41	0.05–0.40	0.69–1.03	0.24–1.68	0.86–1.12
Macroinvertebrates Hydropsychidae	1.61–1.72	0.93–3.43	1.63–7.71	0.51–4.09	4.45–23.29	3.18–23.98	1.44–15.29
Metahyporhithral zone: Ředický Brook, district – Pardubice							
Sediment	1.27–3.45	<0.2–1.02	3.69–22.20	1.19–3.63	13.39–61.41	11.74–93.83	12.18–73.50
Algae – *Cladophora* sp.	2.85	6.93	0.06	0.54	0.31	0.21	0.09
Plants – *Batrachium fluitans*	0.21–0.23	<0.1–0.34	0.35–1.02	<0.1–0.22	0.40–0.76	0.22–1.26	0.19–1.20
Macroinvertebrates Lymneidae	0.25	0.15	0.41	0.24	0.93	1.02	0.63
Macroinvertebrates Hirudinea	0.96	1.01	1.32	0.39	4.78	2.06	0.28
Epipotamal zone: Otava River, district – Starkonice							
Sediment	1.60–2.19	0.2–2.23	4.95–12.72	0.83–3.57	13.46–37.24	21.0–31.82	12.59–31.54
Mosses *Fontinalis* sp.	0.19	2.32	0.31	0.08	2.46	0.82	0.48
Macroinvertebrates Lymneidae	0.14	0.10	0.10	0.10	0.24	0.38	0.15
Macroinvertebrates Hydropsychidae	0.28–1.11	0.10–0.97	0.15–3.74	0.10–0.50	0.32–6.48	0.56–4.41	0.18–1.93
Metapotamal zone: Berounka River; district – Praha south; Labe River, district – Parudubice							
Sediment	1.26–3.45	0.2–1.19	5.52–22.20	0.79–3.63	12.79–61.41	14.66–93.83	12.42–73.50
Algae *Cladophora* sp.	0.51	0.28	0.07	0.08	0.41	0.21	0.08
Plants *Myriophyllum* sp.	1.17–2.20	0.56–2.36	0.38–0.46	0.36–0.38	0.38–3.58	0.45–1.23	0.16–0.76
Plants *Baldingera aerundinacea*	0.45	0.52	1.61	0.10	1.65	3.20	1.17
Macroinvertebrates *Viviparus viviparus*	0.18	0.27	0.44	0.12	0.66	0.77	0.39
Macroinvertebrates *Unio* sp.	0.85	1.97	1.54	0.19	2.72	1.54	0.78
Macroinvertebrates Hirudinea	1.15	1.40	2.25	1.00	6.11	7.61	0.52

Table 3.4.7 Contents of Pest-CIs in sediment and biota (µg kg^{-1})

Variables	Metal						
	α–HCH	β–HCH	γ–HCH	HCB	DDE	DDD	DDT
Epirhithral zone: Mlýnský Brook, district – Žd'ár nad Sáz							
Sediment	<0.01–0.14	<0.01	<0.01–0.41	0.91–1.24	2.17–3.41	<0.5	<0.5–1.75
Algae – *Cladophora* sp.	0.02	0.32	0.21	0.1	0.26	0.07	0.4
Plants – *Fontinalis antipyretica*	0.02–0.85	<0.1–0.71	0.16–1.24	0.45–0.58	0.59–2.26	0.14–0.41	0.80–1.76
Macroinvertebrates Hydropsychidae	0.27–0.90	0.31–0.96	1.08–1.89	0.85–2.11	3.48–10.6	0.38–0.77	2.53–2.69
Metahyporhithral zone: Ředický Brook, district – Pardubice							
Sediment	<0.01–0.08	<0.01	<0.01–0.06	4.92–11.70	10.94–13.39	<0.5	3.26–3.38
Algae – *Cladophora* sp.	0.04	0.32	0.17	0.3	0.24	0.22	0.17
Plants – *Batrachium fluitans*	0.01–0.02	0.01–0.04	0.19–0.71	0.29–0.45	0.52–2.01	0.01–0.33	0.23–0.48
Macroinvertebrates Lymneidae	<0.1	<0.1	0.71	0.14	3.81	0.64	0.67
Macroinvertebrates Hirudinea	0.15	0.41	2.26	1.49	6.83	1.96	2.71
Epipotamal zone: Otava River, district – Starkonice							
Sediment	<0.01–0.44	<0.01	<0.01–0.39	1.06–3.20	2.35–23.80	0.95	1.24–25.55
Mosses *Fontinalis* sp.	0.05	0.9	0.09	0.35	1.12	0.46	1.40
Macroinvertebrates Lymneidae	<0.01	<0.01	0.12	0.07	1.70	0.18	0.05
Macroinvertebrates Hydropsychidae	<0.01–0.05	<0.01–0.81	0.26–1.02	0.1–1.24	1.01–7.76	0.10–0.40	0.21–3.68
Metapotamal zone: Berounka River, district – Praha south; Labe River, district – Pardubice							
Sediment	<0.01–0.1	<0.01	<0.01–0.30	0.83–11.70	1.65–11.70	0.38	<0.5–3.41
Algae *Cladophora* sp.	0.01	0.28	0.07	0.37	0.39	0.19	0.23
Plants *Myriophyllum* sp.	0.04–0.19	0.79–0.92	0.069–0.63	0.69–0.72	0.60–1.01	11689	0.21–2.03
Plants *Baldingera aerundinacea*	0.44	0.24	0.92	0.31	2.24	0.66	0.87
Macroinvertebrates *Viviparus viviparus*	0.04	0.76	0.09	0.12	1.13	0.35	0.38
Macroinvertebrates *Unio* sp.	0.15	0.19	0.17	0.35	2.33	1.14	1.55
Macroinvertebrates Hirudinea	0.13	0.6	1.22	0.79	12.07	2.38	3.47

settlements. Unfortunately, concrete works started after World War 2 in the context of collectivization of private farms and further development of settlements. The majority of wetlands were destroyed and wet meadows were drained. Most lowland streams were regulated. This situation was typical for Czechoslovakia, and partially for Eastern Germany, Hungary, Poland and the Soviet Union. The main aims of these measures were flood protection and land reclamation. In the 1990s, projects of river restoration or revitalization were initiated. Many examples and methods are described in the training handbook by Eiseltová and Biggs (1995), as well as in various national method handbooks.

Damming

A new problem, which appeared before World War 2, was the damming and regulation of rivers. For instance, the Vltava (Moldau) River was dammed for more than 170 km from the town of České Budějovice to Prague (six reservoirs, the biggest being the Orlík Reservoir with an area of 27 km^2, permanent water supply 2.8×10^8 m^3, length 68 km). Hydrological conditions have been changed by the diversion of the watercourse for power production and water supply. These impoundments contributed to the fragmentation of the river continuum. In the 1960s, the construction of the biggest reservoir on the Vltava River, i.e. Lipno (area of about 47 km^2, permanent water supply 2.52×10^8 m^3), was completed. At the end of the 1980s, the Nové Mlýny reservoirs (three reseviors) on the Dyje River in South Moravia were completed (area over 35 km^2, permanent water supply of about 1.3×10^8 m^3).

Ward and Stanford (1983) and Ward and Stanford (1995), respectively who used the serial river discontinuity concept, elaborated the influence of reservoirs in connection with the river continuum concept (Vannote *et al.*, 1980). Other scientists, such as Zwick (1992), use the term 'fragmentation of running waters ecosystems'. There are many studies of reservoir influence on streams biology, for example, in East and Central Europe: Obr (1956, 1963, 1972) – the Orava River in Slovakia, Peňáz *et al.* (1968) – the Svratka River, Helešic and Sedlák (1995) – the Jihlava River, and Helešic *et al.* (1998) and Helešic and Kubíček (1999) – the Dyje River, all in the Czech Republic.

The degree of influence of the discharge by a reservoir depends on the size, type and age of the reservoir, on water manipulation, on the presence or absence of an equalizing reservoir and technical changes of the channel downstream from a reservoir. This is why it is necessary to judge all of these influences on a case-by-case basis.

Generally, water released from a deep reservoir differs from water from a tributary in several basic characteristics:
- in hydrological and sediments discharge regime, and changes of geomorphology of the stream and substratum
- in temperature regime

- in concentration of dissolved gases
- in input of nutrients
- in drifting activity of organisms and migration movements of animals.

3.4.6 CASE STUDY – THE DYJE RIVER[3]

In this work, abundance species composition and trophic structure of the benthic river community in a part of the River Dyje, upstream and downstream of a reservoir, in the Podyji National Park are described. Results obtained from study of the communities of the river were compared with the general model of the river continuum concept (RCC). Altogether, 15 species of aquatic macrophytes, more than 260 species of aquatic invertebrates and 32 species of fish were examined.

The influence of controlled discharges and dependent environmental variables (especially flow velocity, temperature and concentration of dissolved oxygen) in the river downstream from the reservoir creates a specific resilient benthic community which is different from the model of the river continuum concept.

We consider the acute shape of the discharge wave in the river closest to the reservoir and altered temperature regime to be the main stress factors. The absence of an equalizing reservoir, as well as a small capacity of the weir impoundment downstream, does not ensure ecologically negligible minimal discharge, which is only $1.2 \, m^3 s^{-1}$, sometimes declining to only $0.8 \, m^3 s^{-1}$. In times of minimal discharge, about 30% or more of the bottom area are exposed in certain parts of the river below the reservoir. The peak discharge depends on the number of working turbines (one turbine capacity, $15 \, m^3 s^{-1}$, two, $30 \, m^3 s^{-1}$, and three, $45 \, m^3 s^{-1}$); long-term maximal discharges are around $60 \, m^3 s^{-1}$.

Changes in discharge regimes have strongly altered the substrate and bottom permeability. In profiles below the reservoir, the bottom is encrusted by iron and manganese hydroxides. Cobbles and gravel are relatively strongly attached. The hyporheic zone of the bottom is completely different downstream from the reservoir than upstream, not only in substrate granulometry, but in bottom permeability and biota as well.

There is a completely different community of organisms in the River Dyje than might be expected in this type of river. There is a secondary trout zone (epimetarhithral) with some features typical for barbell zone (epipotamal). Tributaries are influenced especially by increased carrying capacity, increased amount of suspended matter and bed-load (significant erosion in the upper parts of the drainage area), as well as built reservoirs. Nonetheless, there are many species typical of subalpine as well as mountain streams (Alps-Carpathian species); on the other side, the occurrence of pannonian fauna individuals was marked. Some remedial measures, especially for minimalizing the negative effects of hydropeaking and minimal flow stress on the bottom community, were recommended.

---

[3] For further details, see Helešic and Kubíček, 1999.

3.4.7   SHORT DESCRIPTION AND REVIEW OF MONITORING METHODS

3.4.7.1   Indication of organic pollution

The pre-eminently used method in Central European countries is the saprobial system of water quality monitoring. This system uses tables with saprobial valence and gives indications of taxa weight. Organisms have to be determined to species level. Higher taxa (genus or families) have low indication values.

The Eastern European Countries use the following formula for the saprobic index:

$$S = \frac{\sum S_i h_i I_i}{\sum h_i I_i}$$

where S_i is the saprobic index of taxon i, h_i is the abundance of taxon i and I_i is the indicative weight of taxon i. S_i is calculated according to the following:

$$S_i = (0x_i + 2b_i + 3a_i + 4p_i)/10$$

The tables of taxon saprobic indexes and indicative weights are in the commentary to the Czechoslovakian Technical Standard (No. 83 0532, Sládeček *et al.*, 1981; new versions, Czech standards No. 75 7716, 1998 and No. 75 7221, 1998).

There are several limits to the practical use of this method. First, the common tables published in fundamental works of Pantle and Buck (1955), Sládeček (1973), etc., are not useable in particular biogeographical regions and/or all types of water bodies. Many countries have own tables of indicator organisms, e.g. Austria (Moog, 1995), Latvia (Cimdins *et al.*, 1995), etc.

The second limit is sampling methodology, including choice of sampling site. The basic idea of saprobity is the affinity of aquatic organisms to absolute content of dissolved oxygen as a measure of fouling processes. This premise proposes that the best indicators are organisms with a small range of tolerance to definite contents of oxygen (stenovalence organisms). The best indicators are reophilic organisms, living in the lotic zone of a stream. Only data taken from riffles in stream produce relevant and comparative results (Marvan *et al.*, 1980).

The third limit is how indicator organisms actually react to the presence of complex factors. Oxygen content is important but it could be covered by the effects of toxic compounds and/or habitat degradation.

The former Index of Pantle and Buck (1955), without indicating the weights of taxa, produces incorrect results. The formula of Zelinka and Marvan (1961) or graphical methods (Rothschein, 1962; Marvan *et al.*, 1980) produce better results.

The other possibility is calculation of the standard deviation (SM), provided that:

$$\sum_{i=1}^{n} h_i \geq 15$$

$$SM = \sqrt{\frac{\sum_{i=1}^{n}(S_i - S)^2 h_i I_i}{(n-1)\sum_{i=1}^{n} h_i I_i}}$$

Zelinka (1978) and Marvan *et al.* (1980) suggested calculation of the saprobic values according to the median based on a set of values Z_j, derived by linear interpolation between the highest negative value of $2 \arcsin Z_j - \pi/2$ and $2 \arcsin Z_j + 1 - \pi/2$, or based on the method of estimating the expected value of a truncated normal distribution. Mrázek (1984) proposed another method of calculation as a modus, furnishing more accurate and precise values.

The best method is a relatively complicated method of index calculations which uses a matrix of indicator values per taxon in individual saprobial degrees (Marvan, 1969; Zelinka and Marvan, 1986; Zahrádka, 1986). The output of this method is a histogram of taxa abundances. Primary values have to have a normal distribution or an asymmetric one with a peak in the upper or lower part of the saprobity axes. Every other type of histogram indicates an incorrect use of this method or the existence of strong enough factors. The saprobic value (S) is calculated as a weighted average, median and modus. If data are normally distributed, differences between S are minimal. In non-normal distributions of data, there are significant differences (Table 3.4.8).

Table 3.4.8 Comparison of values of saprobity index according to different approaches (real examples from monitoring in the Czech Republic)

Type of distribution	Pantle and Buck	Zelinka and Marvan	Median – valence	Modus – saprobity	Modus – valence
Normal distribution	2.13	2.21	2.12	2.02	2.06
Asymmetric	1.58	1.57	1.58	1.99	1.68
Chaotic	1.12	0.93	0.94	1.63	0.81

The fourth problem is in using autotrophic organisms or microbial/protozoan decomposers. Autotrophic organisms, such as cyanoprokaryotes, bacteria, algae and plants, primarily indicate nutrient content, and then light conditions and dissolved oxygen content. This means that these groups are good bioindicators of euthrophication, as can be seen e.g. in Kelly (1996), Marvan and Maršálek (1999) and Pipp (2001). Using decomposers is restricted to the worst quality of natural waters and indicates wastewater status.

In the common guidelines for sampling (Czech Republic), sampling is carried out in accordance with the European standard EN 27 828 (1994). Samples are taken at a homogenous stretch, preferably in a riffle section. Each sample comprises a fixed distance and a defined sampling time. In flowing shallow water, hand-sampling is carried out by disturbing the substratum by hand and picking organisms from stones. Deeper but wadable streams are kick-sampled. In slow-flowing waters, sweep-sampling is applied; substratum is disturbed with the feet and the dislodged fauna is caught by repeated sweeps of the net through the water above the disturbed area. All different types of sampling are made by utilization of a hand-net (meshsize, 500 μm). The Czech approach declares nine saprobic grades and five quality classes (Tables 3.4.9 and 3.4.10).

Table 3.4.9 Saprobic grades according to the water quality balance system

Grade	Saprobic index
1	0–1.0
2	1.01–1.50
3	1.51–2.00
4	2.01–2.50
5	2.51–3.00
6	3.01–3.50
7	3.51–4.00
8	> 4.00

Many countries use a combination of the saprobial system with other methods, for example, with the species-deficit method (Kothé, 1962), or with diversity indices (e.g. Shanon and Weaver, 1949). These approaches can eliminate any inaccuracy in saprobial evaluation. In other countries, the Extended Biotic Index (Woodiwiss, 1978) or its variances are used. The basic condition for the method is the existence of regional tables of indicators. All of these methods are based on the use of bioindicators from an assemblage of macroinvertebrates.

Use of fish as bioindicators is restricted to streams with a natural fish community; unfortunately, in many rivers and brooks only artificial fish communities from stocks exist. If there is a natural fish community, the Index of Biotic Integrity (Karr, 1981) is very applicable.

Table 3.4.10 Approximative conversion to classes according to the czech State Norm 75 7221 (ČSN 75 7221, 1998)

Grade	Class	Saprobic index
1–2	I	< 1.5
3–(4)	II	1.51–2.19
(4)–5	III	2.20–2.99
6	IV	3.00–3.49
7–8	V	≥ 3.5

3.4.7.2   Indication of General Degradation or Ecological
Status of Stream

New methods focus on a wide number of stresses with the aim of evaluating the ecological status of a stream. Many national approaches now exist, together with a first attempt to prepare a common EU norm (AQEM, 2002; http://www.aqem. de).

The **PERLA** prediction system, a new biological method of ecological status assessment of running waters, has been developed in the Czech Republic (Kokeš, 2001, Kokeš *et al.*, 2003; http://www.vuv.cz/perla).

This method is based on the comparison of an observed fauna with a reference fauna and takes into consideration the natural variability of the environment and within biological communities.

The system enables the prediction of macroinvertebrate community composition at a specific site using several environmental variables, and the subsequent comparison of the predicted (target) community with the macroinvertebrate community actually found at the site assessed. The application of PERLA requires the compilation of a reference data set for the given geographical region. The PERLA prediction system is based on the RIVPACS approach.

The sampling procedure is based on the technical norm ČSN EN 27 828 (Water Quality, Methods for biological sampling, Guidance on hand-net sampling of aquatic benthic macroinvertebrates; ISO 7828, 1985).

The first procedure is a description of the stretch (length; sevenfold stream width upstream and downstream; maximum, 100 m); environmental variables, such as slope, character of substrate, aquatic vegetation, degree of shading and riparian vegetation are recorded. In its centre, the smaller sampling stretch of the stream, including all habitats present within the characteristic stretch, is sampled.

Multi-habitat kick-sampling is carried out proportional to the area each habitat covers, using a hand-net of 500 μm mesh size. The total net-sampling time is 3 min. In connection with the sampling of biota, parameters such as stream width, depth, mean velocity, substrates, gradient, bank and bottom characteristics, as well as other additional information, are recorded.

Determination to the species or species groups (genus, family) is recommended (predominantly species level – more precise identification of the families Chironomidae, Ceratopogonidae, and Dolichopodidae is an optional choice). All living biota from the sample must be summarized; the outputs are sheets with abundances of every taxon.

Assessment, calculation, classification and presentation are carried out as follows.

• Assessment is related to the reference conditions based on predictive modelling (see also the RIVPACS approach (Wright *et al.*, 2000), used in the Biological GQA (Environment Agency, 1997a,b)).

- Each group of predicted sites roughly corresponds to one type or group of several types of stream stretches.

- Types of stream stretches not covered by this method: all streams more than 800 m above sea level, peat streams, very large lowland streams and small lowland streams

The calculation method is multivariate: for every species, a reference database and probability of capture at the observed site is computed according to the following formula:

$$C_s = \sum \left(F_{sg} P_g \right)$$

where C_s is the species probability of capture at the observed site, F_{sg} is the frequency of occurrence of the species s in the group g, and P_g is the probability with which the observed site belongs to the group g.

All species are ordered according to their C_s values and the number of species expected at the observed site (lower limit of $C_s = C_{s_l} = 0.5$) is computed as follows:

$$NE = \sum C_s$$

where NE = number of species expected at the observed site. Finally, the index B is computed as follows:

$$B = \frac{NO}{NE}$$

where *NO* is the number of species with $C_s \geq C_{s_l}$ found at the observed site.

There are five quality classes, as shown in Table 3.4.11. This method will be specified in detail during the first phase of the practical application as only a restricted number of sites have been evaluated so far. The conversion formula used at present is preliminary.

The other alternative method is **AQEM Czech** (Brabec *et al.*, 2004; http://www.aqem.de).

Table 3.4.11 Quality classes according to the B index values

Index B	Class
< 0.20	Bad
0.21–0.40	Poor
0.41–0.60	Moderate
0.61–0.80	Good
> 0.80	Reference state

The aims of the system are as follows:

- To classify a stream reach into an Ecological Quality Class from 5 (high) to 1 (bad), based on a macroinvertebrate taxa list, which has been obtained by using a harmonized sampling method.

- To give information about the causes of eventual degradation to help direct future management practices.

AQEM uses an approach that is specifically designed for each stream type: different calculation methods ('metrics') are applied based on comparison with type-specific reference conditions.

Multi-habitat sampling is performed by sampling major habitats according to their proportional distribution within a sampling reach. A total of 20 sampling units is taken from all major habitat types in the reach (e.g. if the habitat in the sampling reach is 50 % sand, then 10(= 50 %) sampling units should be taken in that habitat). Sampling is done by 'kick-sampling' using a hand-net (mesh size, 500 μm; net-opening, $25 \times 25 \, \text{cm}^2$ or $625 \, \text{cm}^2$).

By sieving through a 2000 μm mesh (stony bottom streams) or 1000 μm mesh (sand bottom streams), the sample is split up into two portions, i.e. the coarse and the fine fraction. The coarse fraction must be sorted and identified completely; analysis of the fine fraction is only required if specific stream types are assessed. In this case, *subsampling* is carried out (an estimation which reduces the effort in sorting and identifying the complete fraction), followed by *sorting* and *identification* procedures.

The length of the sampling site depends on stream width and the variability on the habitat (minimum length, 20 m). The selected sampling site must be representative of a minimum *survey area* of 500 m stream length.

Additional environmental data are necessary for multivariable evaluation. The basis set is as follows: height of source, distance from source, stream order, slope, altitude, ecoregion, catchment area, valley form, geology, river continuity (passability), channel form, cross-section of the river bed and/or floodplain, bank and bed fixation, riparian vegetation, degree of shading, land use, temperature, current velocity, discharge, pH, conductivity, dissolved oxygen content, oxygen saturation, TOC (Total Organic Carbon), BOD (Biological Oxygen Demand), nitrite, nitrate, ammonium, phosphorus, salinity, mineral substrates, type and intensity of human impact.

The calculation method is multimetric – a combination of several individual formulae (Tables 3.4.12, and 3.4.13). Each metric is individually converted into a quality class ranging from 5 (high) to 1 (bad). With a multimetric formula the scores are combined to give the final assessment result: 5, high status; 4, good status; 3, moderate status; 2, poor status; 1, bad status (Rolauffs *et al.*, 2004).

Table 3.4.12   Metrics used to assess ecological quality according to stream type[a]

Stream type	Metrics used to assess ecological quality[b]
Mid-sized streams in the central sub-alpine mountains	• Czech saprobic index • ASPT (Average Score Per Taxon) • RETI (Rhithron Feeding Type Index)
Small streams in the Carpathian mountains	• Czech saprobic index • Number of Plecoptera taxa • Number of Ephemeroptera taxa
Mid-sized streams in the Carpathian mountains	• Czech saprobic index • Number of EPT taxa

[a] Each metric is individually converted into a quality class ranging from 5 (high) to 1 (bad). With a multimetric formula, the scores are combined to give the final assessment result: 5, high status; 4, good status; 3, moderate status; 2, poor status; 1, bad status (Rolauffs *et al.*, 2004).
[b] Organic pollution.

Table 3.4.13   Quality status of streams

Stream type	Mid-sized streams in the central sub-alpine mountains	Small streams in the Carpathian mountains	Small-sized streams in the central sub-alpine mountains
High	≤ 1.80	≤ 1.20	≤ 1.70
Good	$1.81-2.10$	$1.21-1.50$	$1.71-2.20$
Moderate	$2.11-2.50$	$1.51-2.00$	$2.21-2.50$
Poor	$2.51-3.00$	$2.01-2.70$	$2.51-3.00$
Bad	> 3.00	> 2.70	> 3.00

3.4.7.3   Bioindication of eutrophication

There are two methods for evaluation of the concentration (development) of phytoplankton as a measure of euthrophication (Kelly and Whitton, 1998; Marvan and Maršálek, 1999, etc.).

The labaratory method is based on cultivation of algae (*Scenedesmus quadricauda*) in water sampled from a water body. The result is an index of trophic potential as a measure of concentration of available nutrients (Žáková, 1980). The field method is based on the measure of chlorophyll concentration in water. Eutrophication is assessed by using chlorophyll-*a* concentration, according to Table 3.4.14.

3.4.7.4   Bioindication of acidification

The results were evaluated by using standard procedures, i.e. calculation of dominance, frequency, species richness (Margalef index), diversity (Shannon index) and

Table 3.4.14 Classes of throphic status

Class	Trophic status	[Chlorophyll-*a*] (μ gl^{-1})
I	Oligotrophic	< 10
II	Mesotrophic	< 25
III	Eutrophic	< 50
IV	Polytrophic	< 100
V	Hypertrophic	> 100

evenness (Sheldon index). The level of acidification is evaluated by the acidification index, according to Fjellheim and Raddum (1990), with the final evaluation achieved by using statistical multivariate method, such as detrended correspondence analysis.

REFERENCES

AQEM, 2002. 'Manual for the application of the AQEM system. A comprehensive method to assess European streams using benthic macroinvertebrates, developed for the purpose of the Water Framework Directive', Version 1.0. http://www.aqem.de/ftp/aqem_manual.zip.

Bláha, L. and Maršálek, B., 2003. 'Contamination of drinking water in the Czech Republic by microcystins'. *Arch. Hydrobiol.*, **158**, 421–429.

Bláha, L., Sabater, S., Babica P., Vilatta, E. and Maršálek, B. 2004. 'Geosmin occurrence in riverine cyanobacterial mats: is it causing a significant health hazard?'. *Water Sci. Technol.*, **49**, 307–312.

Brabec, K., Zahrádková, S., Nemejcová, D., Paril, P., Kokeš, J. and Jarkovský, J., 2004. 'Assessment of organic pollution effect considering differences between lotic and lentic stream habitats'. *Hydrobiologia*, **516**, 331–346.

Breitig, G. and von Tümpling, W. (Eds), 1975. *Ausgewählte Methoden der Wasseruntersuchungen.* VEB Gustav Fischer Verlag: Jena, Germany.

Cimdins, P., Druvietis, I., Liepa, R., Parele, E., Urtane, L. and Urtans, A., 1995: 'A Latvian catalogue of indicator species of freshwater saprobity'. *Proc. Latvian Acad. Sci.*, **570/571**, 122–133.

CSN 83 0603, 1967. Control of Water Quality', Czech Technical State Standard. Czech Standards Institute: Prague, Czech Republic (in Czech).

CSN 75 7221, 1998. 'Water Quality – Classification of Water Quality', Czech Technical State Standard. Czech Standards Institute: Prague, Czech Republic (in Czech).

CSN 75 7716, 1998. 'Water Quality, Biological Analysis, Determination of Saprobic Index', Czech Technical State Standard. Czech Standards Institute: Prague, Czech Republic (in Czech).

Cyrus, Z. and Sládeček, V., 1965–1996. 'Integral methods of biological analysis of water'. Part I, *Vodní Hospodářství*, **15** and **16** (Supplement); Part II, *Vodní Hospodářství*, **19** (Supplement).

Eiseltova, M. and Biggs, J., 1995. *Restoration of Stream Ecosystem: An Integrated Catchment Approach*, IWRP Publication 37. IWRP: Gloucester.

EN 27 828, 1994. 'Water Quality – Methods for biological sampling – Guidance on Hand-net Sampling of Aquatic Benthic Macroinvertebrates', ISO 7828:1985. International Standards Organization: Geneva, Switzerland.

Environment Agency, 1997a. 'Assessing Water Quality – General Quality Assessment (GQA) Scheme for Biology', Fact Sheet. Environment Agency: Bristol, UK.

Environment Agency, 1997b. 'The Quality of Rivers and Canals in England and Wales, 1995'. Environment Agency: Bristol, UK.

Evans, C. D., Cullen, J. M., Alewell, C., Kopacek, J., Marchetto, A., Moldan, F., Prechtel, A., Rogora, M., Vesely, J. and Wright, R., 2001. 'Recovery from acidification in European surface waters'. *Hydrol. Earth System Sci.*, **5**, 283–297.

Fjellheim, A. and Raddum, G. G., 1990. 'Acid precipitation: biological monitoring of streams and lakes'. *Sci. Environ.*, **96**, 57–66.

Foerstner, V. and Wittmann, T. W., 1983. '*Metal Pollution in the Aquatic Environment*. Springer-Verlag: Berlin, Germany.

Fott, J., Pražákova, M., Stuchlík, E. and Stuchlíková, Z., 1994. 'Acidification of lakes in Šumava (Bohemia) and in the High Tatra Mountains (Slovakia)'. *Hydrobiologia*, **274**, 37–47.

Fuksa, J. K., 1995. 'Contamination of metals in the Labe River – differences in longitudinal profile and affinity to sources'. *Vodní hospodářství*, **45**, 251–257 (in Czech).

Gregor, J. and Maršálek, B., 2004. 'Freshwater phytoplankton quantification by chlorophyll alpha: a comparative study of *in vitro, in vivo* and *in situ* methods'. *Water Res.*, **38**, 517–522.

Hanuška, L., Kratochvíl, I., Sládeček, V., Štěpánek, M., Zelinka, K. and Zamoray, I., 1956. '*Biological Methods of the Investigation and Evaluation of the Water Bodies*. Vyd. SAV: Bratislave. (in Slovak).

Helešic, J. and F. Kubíček, F. (Eds), 1999. 'Hydrobiology of the Dyje River in the National Park Podyjí, Czech Republic'. *Folia Fac. Sci. Nat. Masaryk. Brun., Biol.,* **102**.

Helešic, J. and Scheibová, D., 2000. 'Bioaccumulation of harmful pollutants in running water food webs'. *Verh. Int. Verein. Limnol.*, **27**, 3070–3074.

Helešic, J. and Sedlák, E., 1995. 'Downstream effect impoundments on stoneflies: case study of an epipotamal reach of the Jihlava River, Czech Republic'. *Regul. Rivers: Res. Manage.*, **10**, 39–49.

Helešic, J., Kubíček, F. and Zahrádková, S., 1998. 'The impact of regulated flow and altered temperature regime on river bed macroinvertebrates'. In: *Advances in River Ecology*, Bretschko, G. and Helešic, J. (Eds). Backhuys Publishers: Leiden, The Netherlands, pp. 225–243.

Hellawell, J. M., 1986. *Biological Indicators of Freshwater Pollution and Environmental Management*. Elsevier Applied Science: London, UK.

Holoubek, I., 1990. *Chemie a Společnost (Chemistry and Society)*. SPN: Prague, Czech Republic (in Czech).

Holoubek, I., Čáslavský, J., Helešic, J., Vančura, R., Kohoutek, J., Kočan, A., Petrik, J. and Chovancová, J., 1994. 'Project TOCOEN – the Fate of Selected Organic Pollutants in the Environment, Part XXI – The Contents of PAHs, PCBs and PCDDs/Fs in Sediments from the Danube River Catchment Area'. *Toxicol. Environ. Chem.*, **43**, 203–215.

Holoubek, I. (Co-ordinator, Project Manager), Adamec, V., Bartoš, M., Bláha, K., Bláha, L., Budňáková, M., Černá, M., Čupr, P., Demnerová, K., Drápal, J., Hajšlová, J., Hanzálková, M., Holoubková, I., Hrabětová, S., Jech, L., Klánová, J., Kohoutek, J., Kužílek, V., Machálek, P., Matějů, V., Matoušek, J., Matoušek, M., Mejstřík, V., Novák, J., Ocelka, T., Pekárek, V., Petira, K. Petrlík, J., Provazník, O., Punčochář, M., Rieder, M., Ruprich, J., Sáňka, M., Tomaniová, M., Vácha, R., Volka, K. and Zbíral, J., 2004. 'Proposal of National Implementation Plan for the Stockholm Convention Implementation in the Czech Republic'. Project GF/CEH/01/003: 'Enabling activities to facilitate early action on the implementation of the Stockholm convention on persistent organic pollutants (POPS) in the Czech Republic'. TOCOEN, sro, Brno on the behalf of Consortium RECETOX – TOCOEN and Associates, TOCOEN Report No. 252, January/July 2004, + Executive Summary + 9 Annexes. Brno, Czech Republic.

Hungarian Standard MSZ 12 756, 1998. 'Felszíni Vizek Szaprobitásának Meghatározása (Determination of Saprobity Grade of Surface Waters)'. Budapest, Hungary (in Hungarian).

Karr, J. R., 1981. 'Assessment of biotic integrity using fish communities'. *Fisheries*, **6**, 21–27.

Kelly, M. G., 1996. 'The Trophic Diatom Index: a Users' Manual', Research and Development Technical Report E2. Environment Agency: Bristol, UK.

Kelly, M. G. and Whitton, B. A., 1998. 'Biological monitoring of eutrophication in rivers'. *Hydrobiologia*, **384**, 55–67.

Kokeš, J., (Eds), 2001. 'Prediction Models for River Ecosystems', Report of the Grant No. 510/7/99 of the Council of the Government of the Czech Republic for Research and Development, TGM. Water Research Institute: Prague, Czech Republic (in Czech).

Kokeš, J., Zahrádková, S., Hodovský, J. and Němejcová, D., 2003. 'Prediction system PERLA'. In: *Proceedings of the 13th Conference of the Slovak Limnology Society and the Czech Limnology Society*, Banská Stiavnica, June 2003, Bitušik, P. and Novikmec, M. (Eds). *Acta Facul. Ecol.*, **10** (Suppl. 1), 239–242 (in Czech).

Kolkwitz, R. and Marsson, M., 1908. 'Ökologie der pflanzlichen Saprobien'. *Ber. Dtsch. Botan. Ges.*, **26a**, 505–519.

Kolkwitz, R. and Marsson, M., 1909. 'Ökologie der tierischen Saprobien'. *Int. Rev. Ges. Hydrobiol.*, **2**, 126–152.

Kothé, P., 1962. 'Der 'Artenfehlbetrag', ein einfaches Gütekriterium und seine Anwendung bei biologischen Vorfluteruntersuchungen'. *Dtsch. Gewässerkundl.*, **6**, 60–65.

Liebmann, H., 1951. *Handbuch der Frischwasser- und Abwasserbiologie*, Band I. Oldenbourg: Munich, Germany.

Liebmann, H., 1958–1960. *Handbuch der Frischwasser- und Abwasserbiologie. Biologie des Trinkwassers, Badewassers, Fischwassers, Vorfluters und Abwassers*, Band II. G. Fischer Verlag: Jena, Germany.

Marvan, P., 1969. 'Notes on the application of statistical methods in evaluation of saprobiology'. In: *Symposium CMEA on Questions of Saprobity*, Moscow, USSR, pp. 19–43.

Marvan, P. and Maršálek, B., 1999. 'Present monitoring activities in the Czech Republic'. In: *Use of Algae for Monitoring Rivers, III*, Prygiel, J., Whitton, B. A. and Bukowska, J. (Eds). Agence de l'Eau Artois-Picardie: Douai, France, pp. 84–88.

Marvan, P., Rothschein, J. and Zelinka, M., 1980. 'Der diagnostische Wert saprobiologischer Methoden'. *Limnologica*, **12**, 299–312.

Methods of Assessment of Water Quality, 1977. Soviet Ekonomičeskoj Vzaipomšči, Izd. Ot: Moscow, Russia (in Russian).

Moog, O. (Ed), 1995. *Fauna Aquatica Austriaca*. BMLF: Vienna, Austria.

Mrázek, K., 1984. 'Saprobity index from view of statistic and modelling'. *Vodní hospodářství*, **34B**, 314–320 (in Czech).

Obr, S., 1956. 'Hydrobiologische Untersuchung der Fauna des Orava – Flussgebiets mit Hinsicht auf die Wasserreinheit'. *Acta Acad. Sci. Čechosloven. Basis Brun.*, **28**, 337–445 (in Czech with German summary).

Obr, S., 1963. 'Die hydrobiologische Untersuchung des Orava – Flussgebietes in Bezug auf die Wassergüte und die Auswirkung des neuen Stausees auf die Bodenfauna'. *Folia Fac. Sci. Nat. Univ. Purk. Brun.*, **6**, 1–146 (in Czech with German summary).

Obr, S., 1972. 'Die hydrobiologische Erforschung der Fauna im Einzugsgebiet der Orava und deren Entwicklung in Hinblick auf Wassergüte und Auswirkungen des neuen Stausees'. *Folia Fac. Sci. Nat. Univ. Purk. Brun.*, **13**, 1–101.

Pantle, R. and Buck, H., 1955. 'Die biologische Überwachung der Gewässer und die Darstellung der Ergebnisse'. *Gas- und Wasserfach*, **96**.

Peňáz, M., Kubiček, F., Marvan, P. and Zelinka, M., 1968. 'Influence of the Vír river valley reservoir on the hydrobiological and ichthyological conditions in the river Svratka'. *Acta Sci. Nat. Brno*, **2**, 1–60.

Pipp, E., 2001: A regional diatom-based trophic-state indication system for running water sites in Upper Austria and its over regional applicability. *Verh. Internat. Verein. Limnol.* **27**: 3376–3380.

Rolauffs, P., Stubauer, I., Zahrádková, S., Brabec, K. and Moog, O., 2004. 'Integration of the saprobic system into the European Union Water Framework Directive, Case studies in Austria, Germany and Czech Republic'. In: *Integrated Assessment of Running Waters in Europe*, Hering, D., Verdonschot, P. F. M. Moog, O. and Sandin, L. (Eds), Hydrobiologia, **516**, 285–298.

Rothschein, J., 1962. 'Graphical visualization of biological evaluation of water quality'. Výsk. Ústav vodohosp. Bratislava, **9**, 1–63 (in Slovak).

Scheibová, D. and Helešic, J., 1999. 'Hydrobiological assessment of stream acidification in the Czech-Moravian Highlands, Czech Republic'. *Scripta Fac. Sci. Nat. Univ. Masaryk. Brun., Biol.*, **25**, 13–32.

Shannon, C. E. and Weaver, W., 1949. *The Mathematical Theory of Communication*. University of Illinois Press: Urbana, IL, USA.

Simpson, E. H., 1949. 'Measurement of diversity'. *Nature*, **163**, 688.

Sládeček, V., 1973. 'System of water quality from the biological point of view'. *Arch. Hydrobiol. Beih. Ergeb. Limnol.*, **7**, 1–218.

Sládeček, V., Zelinka, M., Rotschein, J. and Moravcová, V., 1981. 'Commentary to CSN 83 0532, No. 6., Assessment of saprobic index'. Vyd. ÚNM: Prague, Czech Republic (in Czech).

Šrámek-Hušek, R., 1956. 'Zur biologischen Charakteristik der höheren Saprobitätstufen'. *Arch. Hydrobiol.*, **51**, 376–390.

Šrámek-hušek, R., 1950. 'Biological control of waste waters'. *Peče o Čistotu*, **1**, 95–109 (in Czech).

STN – CSN 83 0532 – 1 to 8, 1978/1979. 'Biologický rozbor povrchovej vody (Biological analysis of surface water quality)'. Úřad pro Normalizaci a Meřeni: Prague, Czech Republic (in Slovalc).

Vannote, R. L., Minshall, G. W., Cummins, K. W., Sedell, J. R. and Cushing, C. E., 1980. 'The river continuum concept'. *Can. J. Fish Aquat. Sci.*, **37**, 130–137.

Ward, J. V. and Stanford, J. A., 1995. 'The serial discontinuity concept: extending the model to floodplain rivers'. *Regul. Rivers: Res. Manage.*, **10**, 159–168.

Ward, J. V. and Stanford, J. A., 1983. 'The serial discontinuity concept of lotic ecosystems'. In: *Dynamics of Lotic Ecosystems*, Fontaine, T. D. and Bartell, S. H. (Eds). Ann Arbor Scientific Publishers: Ann Arbor, MI, USA, pp. 29–42.

Woodiwiss, F. S., 1978. 'Biological Water Assessment Methods', Second Technical Seminar – Background Information, First Draft Report. Commission of the European Communities: Brussels, Belgium.

Wright, J. F., Sutcliffe, D. W. and Furse, M. T. (Eds), 2000. *Assessing the Biological Quality of Freshwaters. RIVPACS and Similar Techniques*. Freshwater Biological Association: Ambleside, UK.

Zahrádka, J., 1986. 'Using of computer technique in saprobity evaluation of water'. In: *Sb. Biologické Hodnocení Jakosti Povrchových*, vol. 50. MLVH ČSR: Prague, Czech Republic, pp. 38–46 (in Czech).

Zelinka, M., 1978. 'Calculation of saprobic index with aid of median'. *Folia Fac. Sci. Nat. Univ. Purk. Brun.*, **19**, 71–81 (in Czech).

Zelinka, M. and Marvan, P., 1986. 'Saprobic index, its variances and possibility of using'. In: *Sb. Biologické Hodnocení Jakosti Povrchových*, vol. 50. MLVH ČSR: Prague, Czech Republic, pp. 19–37 (in Czech).

Zelinka, M., Marvan, P. and Kubíček, F., 1959. *Surface Water Purity Evaluation.*, Slezský ústav ČSAV: Opava, Czech Republic (in Czech).

Zelinka, M. and Marvan, P., 1961. 'Zur Präzisierung der biologischen Klassifikation der Reinheit fließender Gewässer'. *Arch. Hydrobiol.*, **57**, 389–407.

Zelinka, M. and Sládeček, V., 1964. 'Hydrobiology for water management', SNTL SVTL: Prague, Czech Republic (in Czech).

Zwick, P., 1992. 'Stream habitat fragmentation – a threat to biodiversity'. *Biodiversity Conservat.*, **1**, 80–97.

Žáková, Z., 1980. 'Trophic potential and its application in water management'. Práce a Studie VÚV: Prague, Czech Republic (in Czech with English Summary).

3.5

Key Features of Bioassessment Development in the United States of America

Michael T. Barbour and **Jereon Gerritsen**

3.5.1   THE CLEAN WATER ACT OF THE USA AS A BASIS FOR BIOASSESSMENT

To understand the underpinnings of ecosystem assessment and protection in the USA, one must first be informed of the regulatory statutes behind the development of environmental protection programs (Barbour and Yoder, 2000). The Clean Water Act (CWA) of 1972 identified the restoration and maintenance of physical, chemical

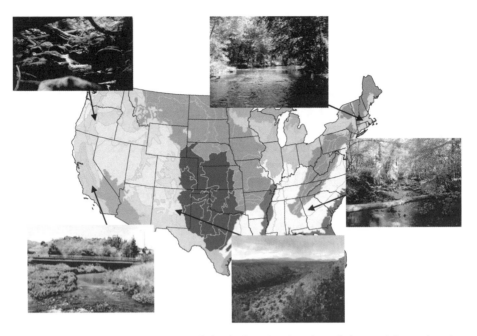

Figure 3.5.1 (Plate 2)   Stream-type variation in the USA: State boundaries are delineated on the map by dark lines, while colours (see plate) denote different ecoregions

and biological integrity as the long-term goal of environmental protection for aquatic resources (Barbour *et al.*, 2000). Under the CWA, the states are required to conduct monitoring and assessment to address the mandates of the law. In addition, First Nations, or Native American tribal authorities, have similar jurisdictional requirements as the states. Therefore, a multitude of water resource agencies exist in the USA to accomplish the stipulated regulations.

The common thread among most agencies conducting bioassessments in the USA is the concept of assessing the biological condition using a regionally defined benchmark, or reference condition, to account for natural variability in dynamic aquatic ecosystems. The biological indicator is then calibrated for the regional classification of the natural streams. Given the wide geographic heterogeneity of streams throughout the USA (Figure 3.5.1), partitioning the natural variability is a first step in establishing a bioassessment program. This geographic diversity of stream types and associated aquatic fauna in the country has resulted in some modification in approaches for bioassessment in various regions. However, the underlying premise and concept is the same throughout the country.

In most bioassessment investigations, there are three kinds of questions that can be addressed with statistical comparisons:

- Asking for a status report on the condition of the aquatic resource at various scales (e.g. catchment, basin, state, region).

- Asking whether a site or waterbody meets, or exceeds, an impairment threshold established by the regulatory entity. This is the standard assessment for assigning sites to a list of impaired waters (in the USA, this practice is to address Section 303d of the Clean Water Act).

- Asking whether a site's biological condition is degraded or improved compared to an earlier time, an upstream site or a nearby site (pairwise comparisons). This question addresses the effects of specific discharges, inputs, Best Management Mitigative actions or restoration, or other site-specific changes, and is typically a trend assessment for the purpose of protecting against further degradation.

In this paper, we examine three advances in bioassessment in the USA and describe the basic ingredients to developing an effective program to address the US Water Law. The three advances are in sampling methodology to obtain some level of consistency in a mosaic of land forms, overall sampling design with increasing adoption of probability-based sample surveys, and evaluation of the variability of biological data and indexes as a result of increasing databases from state and regional sources. We use examples from existing state bioassessment programs to illustrate these premises.

3.5.2 CONSISTENCY IN SAMPLING METHODOLOGY IN DIVERSE REGIONS

The diverse geomorphologic land-forms and climatological regions of the USA underscore the importance of regional specificity in faunal distributions and composition (Barbour and Yoder, 2000). However, the basic premise of a bioassessment approach remains similar across the country. Therefore, a versatile method for sampling, and data interpretation, is needed that will provide some consistency in otherwise disparate areas of rainfall, temperature and geology.

Seven basic climatological regions make up the continental USA (Gabler *et al.*, 1976). For instance, the northwest Pacific coast is cool–temperate rain forest with dense coniferous riparian vegetation (Figure 3.5.1). The streams are generally high gradient with varied geology. The streams in the northeast region (New England) are coldwater, and have silicious-based geology, and dense deciduous riparian vegetation. In contrast, streams in the arid west have little riparian vegetation, are oftentimes intermittent or have high fluctuations in flow, and have high levels of sediment. The streams on the southern California coast are of Mediterranean climate, warmwater, and minimal riparian vegetation. The streams in the southeastern coastal plain are warmwater, sediment laden, but normally with dense riparian vegetation and warm–temperate to subtropical conditions. Climate in the upper midwest and northeast is humid–continental, while the south and southeast are humid and subtropical. As depicted in Figure 3.5.1, the streams in various parts of the country have different geomorphological dimensions. Sampling strategies need to account for this variability and must be representative of the aquatic habitats in those streams. However, a universal sampling strategy is difficult given the wide diversity of stream types in the USA.

Three primary assemblages or biological indicators are used for lotic systems in the USA: periphyton, benthic macroinvertebrates and fish. Benthic macroinvertebrates are surveyed by every state water quality agency. However, many agencies also survey algal and fish assemblages (see Carter *et al.*, 2006 (Chapter 2.5 in this text) for more details on assemblages and sampling strategies in the USA.

While many agencies rely on a single-habitat collection for macroinvertebrates, such as cobble substrate, many streams in the USA lack riffles where this substrate predominates. Several states are using a multihabitat approach (described in Barbour *et al.*, 1999), which advocates sampling the habitat types in proportion to their representation in the stream reach (Table 3.5.1). This strategy is based on the fact that in natural systems the composition of habitat types should be relatively similar throughout each class of streams. The natural characteristics of these streams are the benchmark from which evaluation of habitat alteration is measured. For example, a high-gradient stream in the Cascade Mountains will have an abundance of cobble and boulders, woody debris and pockets of coarse particulate matter. If woody debris is removed, this habitat diminishes and the associated fauna are affected. Sampling the woody debris in reference streams, but missing woody debris in habitat-altered streams provides a basis for evaluating the affects of physical habitat problems. As another example, low-gradient streams of the Central Valley in California have soft, fine sediment, and submerged vegetation predominates in the shore zone. However, there is a distinct absence of cobble substrate and woody debris in these streams.

Table 3.5.1   The multihabitat approach for sampling benthic macroinvertebrates which is used in the USA for different productive and stable habitats (adapted from Barbour *et al.*, 1999)

Benthic habitat type	Description of habitat
Cobble (hard substrate)	Associated with the riffle and run, which are the shallow part of the stream where water flows swiftly over completely or partially submerged pebble to boulder-sized rocks to produce surface agitation. However, riffles are not a common feature of most low-gradient streams
Snags (woody debris)	Submerged woody debris are critical benthic macroinvertebrate habitat in streams, particularly lower-gradient streams. Accumulated woody material in pool areas sometimes provides the only productive habitat for diverse colonization.
Submerged macrophytes	Aquatic plants that are rooted on the bottom of the stream afford a myriad of niches for benthic macroinvertebrates. Submerged macrophytes are seasonal in their occurrence and timing of sampling this habitat is critical
Vegetated bank margins	When the lower portion of banks have roots, plants and snags associated with them, they are a source of extensive colonization by benthic macroinvertebrates. When the banks are of unvegetated or soft soil, few niches exist for organisms
Sand (and other fine sediment)	Usually the least productive macroinvertebrate habitat in streams, this habitat may be the most prevalent in some streams (particularly in low-gradient streams)

The sampling strategy for the algal assemblage is generally to target specific habitats, particularly the cobble, whereas fish are sampled with a multihabitat approach. Regardless, bioassessment in the USA focuses on sampling a stream reach rather than isolated sites or habitat types. Emphasizing the reach enables a consideration of variability that exists in aquatic systems and provides a more robust assessment.

3.5.3 SAMPLING DESIGNS FOR BIOASSESSMENT

The choice of a particular biological assessment design should be based on the monitoring objectives and ultimate decision process incumbent upon the outcome of the study. Any well-designed monitoring and assessment program is inherently anticipatory in that it will provide information for present needs and those not yet determined (Yoder and Rankin, 1995). Measuring several attributes of the biological community ensures that future objectives will have a high probability of being addressed with retroactive data. In the USA, ecological data are collected and analyzed to serve multiple objectives and purposes.

Both state and federal agencies are required to report on the condition of the aquatic resources (our stated first question). The fundamental challenge is that, given our inability to sample every lake and stream in the country, how to select a subset of sites that can be used to infer conditions for all aquatic systems (Hughes *et al.*, 2000). To obtain unbiased estimates of condition, the agencies now often use probability sample surveys. Sites or streams are selected randomly that represent the resource population of interest. Thus, one can extrapolate from the survey results to the entire population. In the USA, streams are identified by resource type, i.e. intermittent, perennial, etc., and further stratified by size to obtain a framework for randomizing the streams to be sampled in the resource population of interest. This design is cost effective in that the entire resource does not have to be sampled – only a representative number of streams. This sampling design was developed by the USEPA's Environmental Monitoring and Assessment Program (EMAP) and has been used to assess the ecological status of waters on different scales of basin, statewide, regional, and national levels (Paulsen and Linthurst, 1994; Hughes *et al.*, 2000). Other designs for determining the condition of aquatic resources are based on intensive monitoring at a smaller scale of the river basin. Examples are the rotating basin approach of state agencies and the basin drainage network used by the National Water Quality Assessment (NAWQA) program of the US Geological Survey (USGS). State agencies use a rotating basin approach to better allocate limited resources each field season and to focus on specific regions of the state for ecological status. This approach allows a longitudinal examination of condition throughout the basin and a characterization of the major impacts and disturbances (Yoder and Rankin, 1995; Bode and Novak, 1995). The basin drainage network design used by the USGS focuses on a set of fixed stations that represent important environmental settings (e.g. relatively homogeneous land use and physiographic conditions) in each basin (Gurtz, 1994). A varying intensity of sampling at the sites is conducted according to different monitoring objectives.

The second question pertains to making a judgment of whether a waterbody meets a prescribed regulation or goal, and the design is more focused on the particular waterbody. In the USA, this sampling design is targeted (i.e. sites are selected in specific locations to determine impairment) and is the most widely used among the state and tribal agencies. The spatial array of the sites is established to determine the extent and magnitude of the perturbation, if present (Yoder and Rankin, 1995). Each agency has its own set of criteria that provide the benchmark for judging impairment (Courtemanch, 1995; Bode and Novak, 1995; Yoder and Rankin, 1995). This design is also useful for teasing out multiple stressors that is critical for making informed decisions on restoration or protection. This approach is not unlike that used in other parts of the world.

Monitoring over time, or trend monitoring, is the essence of the third question. Either a probabilistic or targeted design is used, depending on whether the interest is in the changing condition of the entire resource or the specific waterbody. Paulsen and Linthurst (1994) stated that a probabilistic should enable monitoring to detect temporal patterns for each ecological resource analyses of associations between response indicators and indicators of environmental stress, habitat and exposure, and be adequate for multiple ecological resources. Oftentimes, this question and the resulting design are prescribed to ascertain the effectiveness of some mitigative measures implemented as a result of a previous assessment. Ultimately, the design is necessary to aid in decisions that require a subsequent action for restoration or protection.

3.5.4 REFERENCE CONDITION IS CENTRAL TO BIOASSESSMENTS

The bioassessment models that agencies in the USA employ are based upon comparing test sites to some reference condition. In the multimetric approach, this is usually an average regional reference condition based on the stream classification. For the multivariate predictive model, this is a 'virtual reference' site that is constructed for each site but uses a population of reference sites. In either case, it is necessary to identify a set of reference sites to be used in building these models. In the USA, we stress the importance of establishing strict criteria on the physical characteristics of catchments and streams to serve as reference sites. However, the selection of reference criteria is often a mixture of data analysis and professional judgment. Reference sites ought to be derived from streams that represent minimally disturbed conditions in each region (Bailey *et al.*, 2004). In addition, the reference streams ought to represent typical conditions for a region. Geologically or morphologically atypical sites should be excluded from consideration, as the goal is to define the average and typical regional condition in the absence of human disturbance.

Land use/land cover data, existing water chemistry data, extent of point-source inputs, habitat surveys, presence of impoundments, human population density road density and other existing physical and chemical data are often used as a basis for criteria in identifying regional reference sites (Barbour *et al.*, 1995; Gibson *et al.*, 1996; Bailey *et al.*, 2004). In general, it is not advised to use existing biological data to designate candidate reference sites, because a biological 'benchmark' introduces circularity and confounds the process. However, the experience and knowledge of local professionals, many of whom have sampled biological communities across large regions, is often invaluable for identifying reference sites in any region. In some cases, previous biological sampling could influence their selection of sites. It is best to verify that these sites also meet any physical and chemical reference criteria that may have been selected beforehand.

In the USA, reference criteria are established that explicitly identify desired characteristics of reference sites (Table 3.5.2). Usually sites have to meet all of the reference criteria to be selected as reference sites. Using distinct and quantitative criteria, candidate reference sites can be obtained through probability sample surveys, remote sensing and confirmation surveys of unsampled areas, and from existing sampling programs. In greatly altered systems (e.g. urban watersheds, heavily agricultural areas) minimally impacted sites are usually not available and the least disturbed condition may be heavily disturbed (Gibson *et al.*, 1996). In these cases, historical records or ecological models may have to be used to identify the reference condition (Gibson *et al.*, 1996).

3.5.5 VARIABILITY OF BIOLOGICAL DATA

Routine bioassessment by government agencies in North America now has a history approaching three decades long in some areas. As a result, it is now possible to determine components of spatial and temporal variability, to address concerns raised

Table 3.5.2 Decision criteria or characteristics for identifying reference sites in a tri-state region (Georgia, Alabama, South Carolina) of the southeastern USA

	Stream gradient	
Reference site criteria	High	Low
1. Natural vegetation, i.e. forest/grassland/wetland (% in catchment)	> 65%	> 50%
2. Minimum overall habitat score (% of maximum)	> 70%	> 70%
3. Minimum riparian zone width (m)	> 15 m	> 15 m
4. Riparian zone in catchment (% of stream length)	> 60%	> 60%
5. Agriculture (% in catchment)	< 20%	< 30%
6. Urban land (% in catchment)	< 15%	< 15%
7. Silviculture (active (within 5 years) in catchment)	None	None
8. Road density (length/area of catchment)	?	?
9. Point source discharges (% of flow at 7Q10)	< 5%	< 5%
10. Channel alteration in catchment	None	None

that bioassessment is 'too variable' for reliable water quality monitoring (de Vlaming and Norberg-King, 1999).

Variability has many possible sources. In any sampling program, we desire to (1) minimize variability due to uncontrolled measurement error, and (2) characterize and partition the natural variability. For example, observations are made (e.g. taxa richness) at single points in space and time (riffle, 10 cm depth, 10 AM on 2 July). If the same observation is taken at a different place (pool, 1m depth) or time (30 January), the measured value will be different. These two natural components of variability (space and time in this example) are called sample variability or sampling error (Fore *et al.*, 1994). A third component of variability, called measurement error, refers to our ability to accurately measure the quantity we are interested in. Measurement error can be affected by sampling gear, instrumentation, errors in proper adherence to field and laboratory protocols, variation in taxonomic resolution (Stribling *et al.*, 2003), the choice of methods used in making determinations and small-scale spatial variability at the sample site.

3.5.5.1   Space (sampling error)

Spatial variability of importance to biological assessment of streams exists on all scales from global to microhabitat (Figure 3.5.2). At the continental and regional level, climate, physiographic province (landform and geology), biogeography and ecoregion (Omernik, 1987). All affect the abundance and distribution of species, and hence the values of metrics and indexes. At the level of a watershed or catchment, ecoregional differences may still exist between headwaters and higher-order streams, but the variability is also dominated by local topography and land use. In typical bioassessment programs in the USA, the sampling unit is a defined segment of a

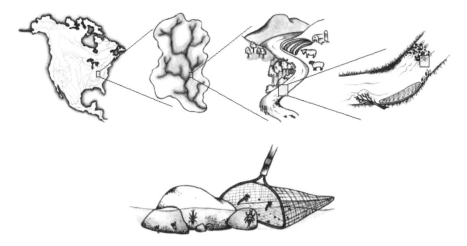

Figure 3.5.2   Scales of spatial heterogeneity in biological sampling

reach, usually either fixed length or a multiple of stream width. Within a sampling unit, there is likewise variability due to riffle-pool sequence within the sample unit, variable distribution of substrates, cover (gravel, cobble, snags) and current (eddies). Some macroinvertebrates colonize specific microhabitats (e.g. suspension feeders in trailing-edge eddies) (Chance and Craig, 1986), whereas predators and some fish may space themselves according to competitive territories. The effect of spatial variability is reduced by stratified sampling in different spatial unity or by spreading sampling effort over multiple habitat units, as in the multihabitat sampling.

3.5.5.2   Time (sampling error)

The second natural component of variability is variation in time. Seasonal change is predictable and results in well-known phenology of species life cycles and seasonal succession. Interannual variation is primarily caused by climate trends and variation, and is less predictable. Short-term variation due to storms and disturbances (natural or anthropogenic) is also unpredictable, and may cause large, short-term disruption. The effect of temporal variability is reduced by sampling during an index period when populations are most stable.

3.5.5.3   Measurement error

Measurement error, is the result of methodological biases and errors, namely gear bias, improper use of gear or improper training, variability in use of gear, laboratory errors (counts, taxa identification (Stribling *et al.*, 2003), chemical analysis) and natural variability that is not taken into account by the sampling design (e.g. by stratification or compositing). Measurement error is minimized with methodological standardization, namely selection of cost-effective, low variability sampling methods, proper training of personnel and quality assurance procedures designed to minimize methodological errors.

If the sampling program develops composite samples from multiple grabs or sweeps of the sampling gear, then small-scale natural spatial variability is also a component of measurement error. There is a random error component regarding which taxa, and how many individuals, are captured in any one grab due to patchiness and variability in the habitat, the ability of some organisms to escape the sampler (gear selectivity) and organisms inadvertently spilling or washing out of the sampler. Combining multiple grabs from several sub-habitats at a site into a single composite sample effectively converts a spatial variance component into the sampling error component. If done randomly (unbiased) and with sufficient effort (area sampled), then compositing of multiple grabs is a cost-effective way to reduce confounding effects of small-scale spatial variability (e.g. Barbour and Gerritsen, 1996).

Multiple observations within a sample unit (e.g. within a reach segment) are used to estimate measurement error, that is, the variability we should expect when a single determination is made.

3.5.5.4 Estimates of variability

We illustrate the variability of derived index values with data from three state programs, Wyoming (sampled by Wyoming DEQ; Zumberge, 2004) Florida (sampled by Florida DEP; data reported in Barbour *et al.*, 1996) and Virginia (sampled by Virginia DEQ; unpublished data). These states re-sampled selected sites within index periods (sample replicates) among seasons in the same year (spring and fall replicates) and multiple years (interannual replicates), allowing estimates of several levels of variability. Streams in the three states differ widely: extreme low-gradient, subtropical (Florida); warm-temperate hills and low mountains (Virginia); high-elevation steppe to cold montane forest and subalpine (Wyoming). Sampling methodology and level of taxonomic identification also differed among the three states.

Figures 3.5.3–3.5.5 illustrate estimated components of variability (as standard deviation) for the state indexes, at three levels: (1) measurement error, based on replicate samples at a site within an index period, often on the same day (Figures 3.5.3 and 3.5.4, Florida and Wyoming only; bars left of dashed vertical line); (2) temporal variability for periods ranging from several years (Wyoming and Florida) to several months (Virginia); (3) spatial variability among streams within ecoregions,

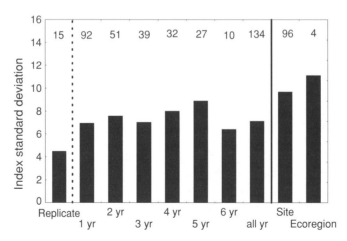

Figure 3.5.3 Variability estimates for the Wyoming benthic macroinvertebrate index in montane streams (1500–2500 m elevation; subregions of the Middle Rockies and Southern Rockies). Numbers across the top are sample sizes. 'Site' and 'Ecoregion' bars are reference sites only; all other bars are both reference and non-reference. Wyoming sampled during a single index period (summer–fall) with eight Surber casts in riffles, all Surbers composited into a single sample and a random subsample of 500 organisms was counted and identified to the lowest practical level

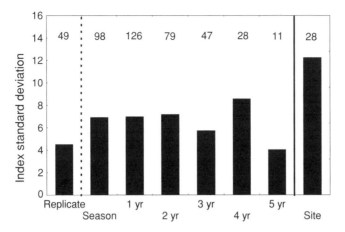

Figure 3.5.4 Variability estimates for the Florida benthic macroinvertebrate index in low-gradient streams of the subtropical Florida peninsula. Florida sampled during two index periods (summer and winter) with 20 D-frame net jabs in all habitats, all net hauls composited into a single sample and a random subsample of 100 organisms was counted and identified to lowest practical level

or among ecoregions (bars to right of solid vertical line). All three biological indexes are on a scale of 0–100, and so no further standardization was necessary.

Index variability in the three widely different regions is remarkably similar (Figures 3.5.3–3.5.5). Replicate samples taken at the same site have standard deviation (sd) of approximately four index units (Figures 3.5.3 and 3.5.4). Replicate

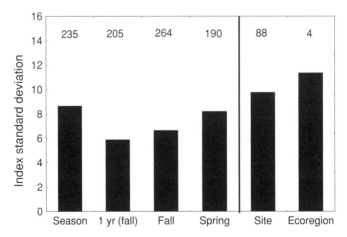

Figure 3.5.5 Variability estimates for the Virginia benthic macroinvertebrate index in upland streams of Virginia (Piedmont, Blue Ridge and Ridge and Valley ecoregions). Virginia sampled during two index periods (spring and fall) with D-frame net jabs in riffles, all net hauls composited into a single sample and a random subsample of 100 to 200 organisms was counted and identified to family level only

variability is an estimate of sampling error, due to field sampling and laboratory analysis. Seasonal variability, due primarily to seasonal cycles of emergence and recruitment, is larger than replicate variability; an sd of 7–9 (Figure 3.5.4, of 100 range) and also larger than interannual variability in temperate Virginia, but the same in subtropical Florida. Variability due to reference sites within ecoregions is an estimate of spatial variability remaining after the fundamental classification (ecoregional classification in these cases: Omernik, 1987, 1995; Hughes and Larsen, 1988; Gerritsen *et al.*, 2000) has been taken into account. An ideal classification of reference sites should reduce the residual spatial variability to that of the seasonal variability within sites (Hawkins *et al.*, 2000); from this consideration it appears that the Florida streams (Barbour *et al.*, 1996) could be classified more finely.

How does the variability of these biological indexes compare to chemical water quality measures? For the data shown here, average index scores for unstressed (reference) sites was in the range 60–75, and thus the coefficient of variation for interannual samples is approximately 10–15 %. Chemical measures in both streams and lakes may vary by an order of magnitude within a single sampling season, especially constituents brought in by runoff during storms such as nutrients, salts or suspended solids (e.g. USGS, 1993; Reckhow and Stow, 1990). The coefficient of variation for these constituents may thus be 50–100 %. As a consequence, chemical monitoring often requires monthly sampling (e.g. Reckhow and Stow, 1990). Biological indexes integrate environmental conditions over a season or longer, and the results shown here suggest sampling intervals of one year or longer are more than adequate for trend monitoring of aquatic biota.

3.5.6   THE FUTURE OF BIOASSESSMENT IN THE USA

Decision-making in environmental management must be based on sound science, which is dependent upon appropriate sampling designs. Environmental decisions with respect to the biological information discussed in this chapter typically concern one or more of the three questions posed earlier, i.e. the ecological condition of the resource, exceedance of regulatory standards or criteria, or detectable degradation of the condition of a waterbody due to a discharge, non-point source, accidental spill or other occurrence. Most decisions are binary: a resource, or portion thereof, is in good condition or not; a site has exceeded criteria or not; a site has been degraded or not. In some cases, a decision requires classifying a site or waterbody into one of several categories, such as exceptional, good, limited resource waters, etc. (e.g. Yoder and Rankin, 1995).

In the USA, thresholds for judging impairment versus unimpairment (also called biocriteria) are determined from a reference condition, which is typically a description of a population of reference sites. We assess a *single* site against this reference distribution (from the sample of reference sites) of the biological indicator or index scores (Figure 3.5.6). The biocriteria test asks whether a single site (represented by a single observation or by multiple observations of the same sample unit) is a member of a population (represented by a sample). Traditional statistical tests (*t*-tests, etc.)

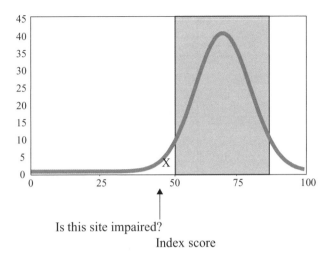

Is this site impaired?

Index score

Figure 3.5.6 Theoretical example of determining an impairment threshold for a site 'X' from a reference population (denoted by shading)

are used to test whether two samples are from two different populations, or if the mean of a sample is different from a predetermined value (say, 0). These tests are fundamentally different.

To provide a consistent framework for environmental practitioners across the USA, a conceptual model was developed to characterize a Biological Condition Gradient (BCG) that could be applied in all waterbody types (USEPA, Unpublished data). The BCG provides a common basis for establishing an anchor for the expectations of the CWA mandates, and a gradient of condition that corresponds to the range of human disturbance and stresses to the aquatic ecosystem (Figure 3.5.7).

Narratives of quality can be assigned to tiers on the BCG based on the range of index values among all reference sites. In Figure 3.5.7, relative positions on the BCG, or tiers, can be described in terms of the natural condition or expectations as communities are exposed to stressors. These narratives can be transformed to an ordinal scale of condition, such as 'very good', 'good', 'fair', 'moderate', 'poor' or 'very poor'. This use of the BCG provides an estimate of resource status, an ordinal ranking of the condition of one site relative to the reference condition and to other assessed sites, and an indication that more intensive, diagnostic sampling may be necessary. The use of the BCG in the USA is similar in concept to the ecological status classes of the Water Framework Directive (Chovanec *et al.*, 2000).

Our purpose was not to describe all elements of bioassessment approaches used in the USA. We emphasized key technical issues that are the underpinnings of an effective bioassessment program: (1) a sampling method that is capable of capturing the representative aquatic community throughout a mosaic of habitat types; (2) sampling designs that are appropriate for the questions being asked; (3) an examination of the variability inherent in biological sampling. The advancement of bioassessment in the USA is proceeding along the lines of the technical issues presented in

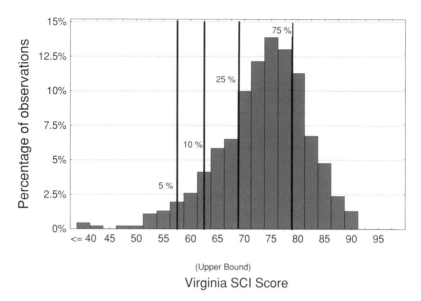

Figure 3.5.7　The conceptual biological condition gradient and descriptions of quality along the continuum

this chapter. Multiple sampling designs are being implemented in various agency programs to improve the overall reporting and documentation of ecological status. The value of combined designs of probabilistic sample survey with targeted site selection for concentrated spatial arrays has been widely accepted throughout the USA. While not fully endorsed by all agencies, the application of multiple designs has increased dramatically over the past decade.

Sampling methods have become more rigorous in that attention to sampling representative habitats and evaluating sampling variability has become a mainstay of the pre-eminent programs. Regional reference conditions have been developed in nearly all regions of the country and continued refinement is ongoing. The development of the Biological Condition Gradient concept and associated ecological attributes provides a common language for all of the state and federal agencies to address both reference condition and impairment classes. We envision that the next decade will be critical for enhancing the scientific underpinnings of all bioassessment programs and developing a scientific framework from which common and consistent assessments can be done across the mosaic of land use and human disturbance in the USA.

REFERENCES

Bailey, R. C., Norris, R. H. and Reynoldson, T. B., 2004. *Bioassessment of Freshwater Ecosystems: Using the Reference Condition Approach*. Kluwer Academic Publishers: Norwell, MA, USA.

Barbour, M. T. and Gerritsen, J., 1996. 'Subsampling of benthic samples: A defense of the fixed-count method'. *J. N. Am. Benthol. Soc.*, **15**, 386–392.

Barbour, M. T. and Yoder, C. O., 2000. 'The multimetric approach to bioassessment, as used in the United States'. In: *Assessing the Biological Quality of Freshwaters: RIVPACS and Other Techniques*, Wright, J. F., Sutcltiffe, D. W. and Furse, M. T. (Eds). Freshwater Biological Association: Ambleside, UK, pp. 281–292

Barbour, M. T., Stribling, J. B. and Karr, J. R., 1995. 'The multimetric approach for establishing biocriteria and measuring biological condition'. In: *Biological Assessment and Criteria: Tools for Water Resource Planning and Decisionmaking*, Davis, W. and Simon, T. (Eds). Lewis Publishers: Ann Arbor, MI, USA, pp. 63–76.

Barbour, M. T., Gerritsen, J., Griffith, G. E., Frydenborg, R., McCarron, E., White, J. S. and Bastian, M. L., 1996. 'A framework for biological criteria for Florida streams using benthic macroinvertebrates'. *J. N. Am. Benthol. Soc.*, **15**, 185–211.

Barbour, M. T., Gerritsen, J., Snyder, B. D. and Stribling, J. B., 1999. *Rapid Bioassessment Protocols for Use in Streams and Wadeable Rivers: Periphyton, Benthic Macroinvertebrates and Fish*, 2nd Edition, EPA/841-B-99-002. Office of Water, US Environmental Protection Agency: Washington, DC, USA.

Barbour, M. T., Swietlik, W. F., Jackson, S. K., Courtemanch, D. L., Davies, S. P. and Yoder, C. O., 2000. 'Measuring the attainment of biological integrity in the USA: A critical element of ecological integrity'. *Hydrobiologia*, **422/423**, 453–464.

Bode, R. W. and Novak, M. A., 1995. 'Development and application of biological impairment criteria for rivers and streams in New York state'. In: *Biological Assessment and Criteria: Tools for Water Resource Planning and Decision Making*, Davis, W. S. and Simon, T. P. (Eds). Lewis Publishers: Boca Raton, FL, USA, pp. 97–107.

Carter, J. L., Resh, V. H., Rosenberg, D. M. and Reynoldson, T. B., 2006. 'Biomonitoring in North American rivers: A comparison of methods used for benthic macroinvertebrates in Canada and the United States'. In: *Biological Monitoring of Rivers: Applications and Perspectives*, Ziglio, G., Siligardi, M. and Flaim, G. (Eds). Wiley: Chichester, UK, pp. 203–228.

Chance, M. M. and Craig, D. A., 1986. 'Hydrodynamics and behaviour of Simuliidae larvae (Diptera)'. *Can. J. Zool.*, **64**, 1295–1309.

Chovanec, A., Jager, P., Jungwirth, M., Koller-Kreimel, V., Moog, O., Muhar, S. and Schmutz, S., 2000. 'The Austrian way of assessing the ecological integrity of running waters: a contribution to the EU Water Framework Directive'. *Hydrobiologia*, **422/423**, 445–452.

Courtemanch, D. L., 1995. 'Merging the science of biological monitoring with water resource management policy: Criteria development'. In: *Biological Assessment and Criteria: Tools for Water Resource Planning and Decision Making*, Davis, W. S. and Simon, T. P. (Eds). Lewis Publishers: Boca Raton, FL, USA, pp. 315–325.

de Vlaming, V. and Norberg-King, T. J., 1999. *A Review of Single Species Toxicity Tests: Are the Tests Reliable Predictors of Aquatic Ecosystem Community Responses?* Technical Report, EPA 600/R-97/11. US Environmental Protection Agency: Duluth, MN, USA.

Fore, L. S., Karr, J. R. and Conquest, L. L., 1994. 'Statistical properties of an index of biological integrity used to evaluate water resources'. *Can. J. Fish. Aquat. Sci.*, **51**, 1077–1087.

Gabler, R. E., Sager, R. J., Brazier, S. and Wise, D. L., 1976. *Essentials of Physical Geography*. CBS College Publishers, Dryden Press: New York, NY, USA.

Gerritsen, J., Barbour, M. T. and King, K., 2000. 'Apples, oranges and ecoregions: On determining pattern in aquatic assemblages.' *J. N. Am. Benthol. Soc.*, **19**, 487–496.

Gibson, G., Barbour, M., Stribling, J., Gerritsen, J. and Karr, J., 1996. *Biological Criteria: Technical Guidance for Streams and Small Rivers* (revised edition), EPA/822/B-96-001. Office of Water, US Environmental Protection Agency: Washington, DC, USA.

Gurtz, M. E., 1994. 'Design considerations for biological components of the National Water Quality Assessment (NAWQA) Program'. In: *Biological Monitoring of Aquatic Systems*, Loeb, S. L. and Spacie, A. (Eds). Lewis Publishers: Boca Raton, FL, USA, pp. 323–354.

Hawkins, C., Norris, R., Gerritsen, J., Hughes, R., Jackson, S. K., Johnson, R. K. and Stevenson, R. J., 2000. 'Evaluation of the use of landscape classifications for the prediction of freshwater biota: synthesis and recommendation'. *J. N. Am. Benthol. Soc.*, **19**, 541–556.

Hughes, R. M. and Larsen, D. P., 1988. 'Ecoregions: an approach to surface water protection'. *J. Water Pollut. Control Fed.*, **60**, 486–493.

Hughes, R. M., Paulsen, S. G. and Stoddard, J. L., 2000. 'EMAP-surface waters: a multiassemblage, probability survey of ecological integrity in the USA'. *Hydrobiologia*, **422/423**, 429–443.

Omernik, J. M., 1987. 'Ecoregions of the conterminous United States'. *Ann. Assoc. Am. Geogr.*, **77**, 118–125.

Omernik, J. M., 1995. 'Ecoregions: A spatial framework for environmental management'. In: *Biological Assessment and Criteria: Tools for Water Resource Planning and Decision Making*, Davis, W. S. and Simon, T. P. (Eds). Lewis Publishers: Boca Raton, FL, USA, pp. 49–62.

Paulsen, S. G. and Linthurst, R. A., 1994. 'Biological monitoring in the environmental monitoring and assessment program'. In: *Biological Monitoring of Aquatic Systems*, Loeb, S. L. and Spacie, A. (Eds). Lewis Publishers: Boca Raton, FL, USA, pp. 297–322.

Reckhow, K. and Stow, C., 1990. 'Monitoring design and data analysis for trend detection'. *Lake Reserv. Manage.*, **6**, 49–60.

Stribling, J. B., Moulton II, S. R. and Lester, G.T., 2003. 'Determining the quality of taxonomic data. *J. N. Am. Benthol. Soc.*, **22**, 621–631.

USGA, 1993. *National Water Summary 1990–1991. Hydrologic Events and Stream Water Quality*, Water-Supply Paper 2400. United States Geological Survey: Reston, VA, USA.

Yoder, C. O. and Rankin, E. T., 1995. 'Biological criteria program development and implementation in Ohio'. In: *Biological Assessment and Criteria: Tools for Water Resource Planning and Decision Making*, Davis, W. S. and Simon, T. P. (Eds). Lewis Publishers: Boca Raton, FL, USA, pp. 109–144.

Zumberge, J. R., 2004. *Water Quality Monitoring Strategy 2004–2008*. Water Quality Division, Wyoming Department of Environmental Quality: Sheridan, WY, USA (http://deq.state.wy.us/wqd/watershed/downloads/monitoring/4-0661doc.pdf).

Section 4
New Tools and Strategies for River Ecology Evaluation

4.1

Monitoring Experiences from Downunder – The Importance of Deciding *a priori* What Constitutes a Significant Environmental Change

Barbara J. Downes

4.1.1   INTRODUCTION

In Australia, as in much of the rest of the world, there is a continuing concern about the damage that human beings wreak on natural environments. That concern is manifest in a multitude of ways, perhaps the most convincing of which is the current focus on restoration of ecosystems deemed to be 'damaged' (Australian State of the Environment Committee, 2001). In Australia, there continues to be declining water quality, over-extraction of water from rivers, increasing salinity and unacceptable frequencies of blue–green algal blooms (Australian State of the Environment Committee, 2001). There has also been considerable effort at developing monitoring programs that might reliably detect and measure human impacts, as well as restoration or rehabilitation projects that might address some of the environmental damage.

There is one large barrier, however, that presently impedes monitoring projects. That barrier is knowing what sizes and what types of changes to the environment should be deemed 'important' for monitoring programmes to detect and for restoration projects to achieve. The natural environment changes over time. Separating out changes caused by humans and those resulting from 'natural' influences is extremely difficult. In Australia and New Zealand (and in many other places in the world), the approach to solving this problem has been to use 'reference conditions' (Norris and Thoms, 1999). Because Australia still has large tracts of relatively undisturbed land, reference rivers are those that are intact, in relatively 'pristine' condition or are thought to have been changed only minimally by human activities. The diversity and abundances of organisms and measures of ecosystem function in these rivers provide information for assessing the condition of similar but putatively impacted rivers.

The reference condition approach is an appealing one. It appears to sidestep the difficult problem of deciding what is 'natural' or 'normal', instead allowing nature to show us what damage has been done. Nevertheless, there are problems that this approach creates, which are not receiving much consideration or debate. In this paper, I explore these problems and offer some alternative thinking. I first set out why assessing river condition has presented barriers and why current thinking on this matter somewhat evades the problem rather than solves it. I then set out a framework that presents a description of the size of the problems in front of us that require tackling. My motivation in writing this chapter is to highlight that I believe we have spent a lot of time discussing what to monitor (invertebrates or fish?) and how to monitor it (fyke nets or electrofishing?) and not enough time debating why we should measure particular things and what can be deduced from such measurements. This is not a review paper and is not intended as a criticism of past work. I offer some thoughts in this area to stimulate different ways of thinking about measuring human impacts.

4.1.2   WHAT TYPES OF ENVIRONMENTAL CHANGES ARE IMPORTANT?

This question seems deceptively simple and straightforward to answer, but it is nothing of the kind. Human activities that create environmental change span a vast array of possibilities. Acute effects – like gross pollution or channelization, where stream channels are converted to simple, straight concrete channels – can cause effects that may be simple to understand in many cases, e.g. mass loss of species diversity, population abundances, ecosystem function, etc. (e.g. Young and Niemi, 1990). Many effects are far subtler and/or represent chronic change rather than acute change. Either way, many changes caused by human activities do not sit starkly outside the 'normal' spectrum of possible changes caused by natural influences in either magnitude or type. For example, natural agents of disturbance (floods, droughts, hurricanes) can also cause wholesale removal of species and lower abundances (Niemi *et al.*, 1990). These events are not abnormal, especially in Australia, where floods and droughts are the norm rather than the exception (Lake, 1995) but change through time and space is common to many places. Indeed, ecological and geographical research over the last few decades has shown, fairly definitively, that ecosystems do not exist in a state of equilibrium or balance, as was largely thought for decades (Perry, 2002).

That finding is not news to ecologists, but can come as a serious surprise to people outside the discipline (even professionals in other fields). The notion that nature exists 'in balance' is a common viewpoint that pervades communities at large and has done so for centuries, largely as a result of philosophical or religious views about nature (Egerton, 1973) that continue to remain uninformed by modern ecology (Botkin, 1990). That viewpoint can sometimes predominate in discussions among environmentalists and managers about 'environmental health'. If people implicitly believe in 'balanced' and equilibrial ecosystems, the usual goal set for monitoring programmes (or restoration projects) is for environments to be kept 'natural' and 'normal'. That target is very difficult to hit given that 'natural' and 'normal' includes a lot of variation through time and space, and not the fixed constancy that many people unconsciously have in mind. A view of nature that presumes it is equilibrial can also be partnered with a presumption that changes caused by humans are inevitably 'bad' and push ecosystems 'out of balance' (Fairweather, 1993). Viewing human beings only as despoilers of the landscape leads us into some further incongruity if we are not careful. Do we view everything that the human species does as 'unnatural' and 'bad'? If so, it can lead us down a slippery slope of deciding, again unconsciously, that the effects of human beings should be removed from the environment altogether. My interest here is not in the philosophical debate about what constitutes 'naturalness', but in the implications such views have when they drive the design of monitoring programmes. If we believe that nature is unchanging and balanced and that the role of humans is simply to destroy that balance, then detecting effects of human beings on the environment looks rather simple. It is only when decision-makers recognize that ecosystems constantly change in space and time, and that human beings can cause subtle changes, that we appreciate the difficulties of quantifying what human beings are doing to the environment (Underwood, 1995).

The nub of the problem, however, is not simply that ecosystems vary so enormously over multiple scales of space and over time but that the majority of that variation is so very poorly understood (e.g. Underwood and Petraitis, 1993; Townsend *et al.*, 2004). Ecology as a science has not yet won an understanding of ecosystems that allows us much predictability, and what predictability we have has been hard won from decades of research conducted over multiple scales of space and time (e.g. Connelly *et al.*, 2001). For most ecosystems, we simply do not have the capacity to make predictions about future species diversity, the abundances of species within them or the rates of ecosystem functions. The recognition of our poor ability to make such predictions highlights the challenges for monitoring programmes. It means that, for the foreseeable future, we have to separate potentially unacceptable changes caused by human beings against a backdrop of continuous, unpredictable, natural change. Contrast the situation with one where there is a good fundamental understanding of the causes of both spatial and temporal variation in environmental variables. The physical processes that change tidal heights, for example, are sufficiently well understood that we can produce Tide Tables that contain predictions of tidal heights four times each day for the next year for most places. I do not mean to suggest that everything is known about predicting tidal heights, but the difference with levels of predictability for almost any ecological process of interest (population abundances, diversity) is utterly stark. Anyone who could produce a Tide Table equivalent to predict fish stocks with that level of accuracy would win the Nobel Prize (if one were awarded for Ecology!).

Nothing I have said above is novel, but it is necessary to describe clearly the problem. We cannot define human-created changes as simply of a different quality or size to those that happen naturally. Nor can we use the simple instance of change – even when associated with some human activity – as strong evidence that the latter has caused it because natural change is common. Measuring changes created by human beings must be done against this backdrop of natural change, most of which is neither understood nor predictable.

4.1.3   THE REFERENCE CONDITION APPROACH

To solve the problem of detecting and measuring human impacts against a constantly changing background, many people have advocated using the reference condition approach (e.g. Barbour *et al.*, 2000; Chovanec *et al.*, 2000). In Australia and other places with tracts of relatively undisturbed land remaining, this approach uses intact rivers in the region to provide reference points (Norris and Thoms, 1999). The logic behind this approach is that we use locations that have not been disturbed by human activities (or only 'minimally disturbed') to provide a picture of what impacted locations ought to look like in the absence of human created change. Typically, we expect reference locations to be in the same general region as the impacted locations, and identifying suitable reference locations may require a formal statistical analysis in some approaches (e.g. Wright, 1995; Linke *et al.*, 1999) or simply the nearest intact streams and rivers and 'professional judgement' (e.g. Barbour *et al.*, 2000). The attraction of the approach is obvious. By using reference rivers, we can avoid the whole messy problem of defining what is 'normal' or 'natural' or 'healthy' because reference locations supply us with the information about what constitutes 'normal' species diversity, abundances or ecosystem function.

Before moving on to discuss some of the limitations of this approach, it is worth pointing out that reference locations are not the same as controls, even though those terms are used interchangeably in the literature. Controls are locations that are comparable to impact sites such that the main difference between them is the instance of human impact of interest (Downes *et al.*, 2002). Hence, if putatively impacted locations suffer an array of human insults but our interest is in one of these in particular, then controls should also suffer all human insults bar the one of interest. Hence controls are not necessarily 'pristine' or 'minimally impacted' – although they could be in circumstances where future impact site is currently close to pristine condition. Control locations are field controls, not laboratory controls where we tightly constrain variables. Their purpose is to isolate a treatment effect (Quinn and Keough, 2002), in this case the effect of a particular human activity. The differentiation between controls and reference locations is critically important to understanding the worth of the comparisons between them and impacted sites. Controls can tell us about changes and about the causes of those changes; references can tell us something about changes but rarely about the causes of them. We can illustrate this difference particularly well by looking at restoration/rehabilitation projects, which require both control and reference locations. The difference between restored locations and controls (which do not sustain any restoration activities) tells us about the effects that restoration activities have – this allows us the inference of *causality*. The reference locations supply the target for restored locations – the direction for change. They do not tell us about the causes of change, but they can tell us whether restored locations are 'improving' relative to undisturbed locations (Downes *et al.*, 2002). Most of the problems that bedevil the definition and choice of reference locations also occur with the search for control locations.

So, what are the difficulties with using reference locations to supply information about health of rivers? Curiously, while a lot of effort has been put into measuring, selecting and using reference locations, much less has been done regarding defining them clearly or discussing the underlying reasoning behind their use (an exception would be Lancaster, 2000). Some of that reasoning does not bear up well under close scrutiny.

Perhaps the first problem is in the definition of 'pristine' or 'minimally impacted'. It is hard to argue, especially in a time of climate change, that anywhere on Earth could be strictly described as 'pristine'. If we drop that use though and use 'minimally impacted', we still have the difficulty of deciding what 'minimally' means in an operational sense. If we understood perfectly the effects that humans have on rivers and streams, then we would know what 'minimal' impact constitutes – but if that were the case we would not actually need reference locations at all anyway. In other words, we need the knowledge that reference locations provide in order to know how to define them properly! The obvious alternative is to use a measure of human activities present as our measure of 'minimally impacted' rather than any putative changes caused by those activities (e.g. Choy *et al.*, 2002). That process operationalizes the 'standard' against which conclusions about condition will be drawn so that is certainly an improvement. Nevertheless, this still requires some knowledge of what constitutes a 'minor' disturbance (so to a degree, we are still putting the cart before the horse) and to operationalize a definition that can then apply anywhere – regardless of the list of human activities in the region – is difficult

(Buffagni *et al.*, 2001). Definitions of reference condition are likely to be regional, not national (or universal) (e.g. Nijboer *et al.*, 2004). The lack of a good working definition can create uncertainty about whether reference sites have been altered by human activities and to what degree. It can lead to instances of reference sites being reclassified as impact sites on the basis of there being lower diversity 'than expected' at those locations (e.g. Marchant *et al.*, 1997). Using the outcome of tests to redefine reference sites is understandable, but unfortunately means we may be creating self-fulfilling predictions, not conducting objective tests about 'health'.

Setting aside the problems of definition *per se* though, we still have the pressing problem that for many situations – and arguably the most urgent – no 'minimally impacted' situations exist (Richardson and Healey, 1996). Even in countries like Australia and New Zealand, with their relatively recent period of European colonization and large landscape change, the most urgent problems occur in areas where all the rivers have undergone the same set of changes (e.g. land clearing, sedimentation, etc.) or have sustained other, equally problematic change (e.g. urbanization). Hence, there are often no rivers left that can provide convincing reference information. The only places left tend to be smaller or upland systems, which are demonstrably unsuitable to supply reference targets for the larger or lowland rivers that have more commonly sustained damage (Thoms *et al.*, 1999). In these situations, we can use historical information if it is available (e.g. Thoms *et al.*, 1999), but history has rarely recorded many of the ecological attributes we value.

Additionally, we sustain a further 'logical blow' if we consider that very few places indeed remain completely undisturbed by human activities if we extend our perspective beyond the last two centuries or so. In Australia, for example, 'human impacts' usually means changes wrought by European occupation over the last two centuries, but there is good evidence that the indigenous peoples of Australia brought about large changes to the landscape. Fire was used extensively to modify the distribution of plant and game species (Fensham, 1997; Yibarbuk *et al.*, 2001), with implications for rivers (Papworth and Lewis, 2003). Do we discount such changes and presume that they were of lesser magnitude or importance or did not create 'degradation'? Perhaps, but assuming that indigenous peoples lived 'in harmony' with their landscape seems to mean we therefore presume that they did not change it in ways we might regard as significant. That view would not seem to be supported by the evidence. As above, I do not wish to enter the philosophical debate: my interest is in the implications for defining reference conditions. We cannot cleanly and logically distinguish all modern changes from those created in the recent past, which makes the simple presumption that it is only modern changes that need concern us illogical. If we continue this theme, and suggest that references are conditions prior to any occupation of the landscape by humans, we are left in an even worse tangle. Indigenous peoples have been present in Australia for tens of thousands of years, not centuries. If we wish to extend our reference back prior to their arrival, we start encountering a continent whose climate was radically different than it is today (e.g. Papworth and Lewis, 2003). Using reference conditions from different climatic regimes is likely to lead to some absurd outcomes. Additionally, we end up defining the human species out of the natural landscape altogether. Unless we are willing to consider the case that humans resulted from an alien invasion from outer space (!), that sort of argument would not seem to be strictly logical either.

The reference-condition approach, on the face of it, seems to solve our problem but leaves us only a little further forward in many circumstances. We may feel comfortable when we have access to places where evidence of human impact is scarce – large tracts of Tasmania, for example, remain as wilderness and are remote from modern human activities. That situation allows us to evade the quagmire of being forced to make *ad hoc* decisions about what level of human disturbance can be considered 'minimal' and deciding what 'reference' really means. In many other circumstances, a definition of 'minimally impacted' may not stand up to scrutiny, and the lack of a formal definition of reference condition, and the logic on which it is based, create difficulties. Indeed, in many places 'reference conditions' may simply be the most intact locations that could be found, meaning that 'reference condition' is a 'moving target' that does not represent a clearly defined and universally consistent measure of health at all. In places where 'the best we could find' still represents fairly disturbed places, we are setting different standards for 'health' than in places where there are intact systems – but the real problem is that we have no idea how much lower we have set the bar for those rivers.

Obviously, improved frequency of debate about defining reference conditions would help, but I suggest that the solution to defining important human-created environmental changes lies in considering the issues more broadly. Before discussing this, I first need to address another problem that we encounter when we have reference (and control) locations: how *large* does an environmental difference have to be for us to consider that it is 'important'?

4.1.4   HOW BIG IS BIG?

Monitoring programmes are set up to detect change, whether it be through an impact deemed to be a negative outcome of some human activity or a positive change sought through restoration activities. For most monitoring, the decision as to whether the programme was 'successful' at detecting some change is driven through the out-come of statistical tests (Downes *et al.*, 2002). If the probability associated with test statistic is sufficiently small, then the difference detected is deemed 'important'. That method of decision-making means that there is no *a priori* discussion about what size of ecological or environmental change might be considered important in light of ecological knowledge (about populations, communities, ecosystems, etc.). Statistical significance is directly equated with ecological significance. The diffi-culty with that approach has been known for a long time: whether a change can be deemed 'statistically significant' is dependent upon the power of our test. Statistical power is a measure of our ability to detect an effect that is real (not due only to sampling variation); it is the inverse of the probability of making a Type II error (concluding that the null hypothesis is correct when it is false, i.e. missing a change)

(Quinn and Keough, 2002). The probability of Type II errors can be quite high (and therefore statistical power quite low) in surveys of ecological variables (Underwood and Chapman, 2003). The larger the sample size and the more invariant our dependent variable, the more likely we are (all else kept constant) that we shall arrive at a statistically significant *P*-value, independent of whether differences are meaningful. This material is not new – the problem has bedeviled the discipline of ecology for some time (e.g. Loehle, 1987).

One proposed solution has been to try to set standards for Type II errors in the same way that we currently set a standard for Type I errors at < 5 %. In ecology for example, a convention of setting the probability of Type II errors at 20 % or less is developing (Di Stefano, 2003). Even if we agree with such conventions, there are two difficulties with this approach for monitoring programmes. First, it sets conventions that will retain Type II error rates as being considered acceptably higher than Type I error rates, i.e. we will be accepting a higher risk of missing a real difference than in falsely concluding there was a difference when there was none. If we seek to balance risks of making wrong decisions, then we need to balance decision errors with respect to stakeholders and the perceived significance of environmental change to the public (Mapstone, 1995). Second, monitoring programmes to detect human-caused changes are, by definition, pitched at large scales of space (multiple locations) and times (usually multiple years). Replication to improve power hence requires replication of whole locations and years and is often obviated by cost or, more significantly, by incapacity (e.g. inability to locate replicate controls or references). Hence, improving monitoring programmes by simply demanding that they increase power to some conventional level is unlikely to be achievable.

So, the problem is that even if we have reference locations we can compare to putatively impacted locations, we cannot simply deem any difference detected (or not detected) to be important (or unimportant) because such decisions are at the mercy of the statistical power of the monitoring programme and that is often weak. Nor can we just ignore this issue. The consequences of making a wrong decision about human impacts can be dire, especially in the case of missed impacts. The latter might result in large changes to the environment that result in prosecution and fines, and environmental restoration or rehabilitation attempted to fix the damage is a great deal more expensive (and chancy) than prevention. We can of course do *a priori* power analysis to settle on an acceptable Type II error rate, but that requires us to decide upon the sizes of environmental changes we deem important to detect.

4.1.5   SETTING *A PRIORI* EFFECT SIZES

The easiest way to think about the sizes of important changes is to think about them in terms of 'effect sizes'. An effect size is the size of difference (say, between two means) that we wish to detect. This can be expressed as the absolute difference between a pair of means (e.g. ten species) or more commonly it is expressed as a percentage increase (or decrease) (e.g. a 20% increase in the number of species) (Quinn and Keough, 2002). How then do we decide what is a 'large difference'? The answer to that question depends upon a number of things. First, we need to realize that effect sizes can be set anywhere along a continuum – there are no universal numbers that can be set for effect sizes across all variables (as will be highlighted below). Secondly, it depends upon the purpose of the monitoring programme – do we need to detect any unacceptable change very early so that we can take action before serious environmental damage is done? In this case, we might set minimum effect sizes smaller than if the consequences of change are less serious and/or are easier to rectify (Downes *et al.*, 2002).

For illustration, I consider first how we might proceed if we had access to perfect information about how important environmental variables change with increasing magnitude of human impact (Figure 4.1.1). That information would include at least three pieces of data: (1) the precise nature of the relation between increasing magnitude of human activity and environmental response – which means we have access to

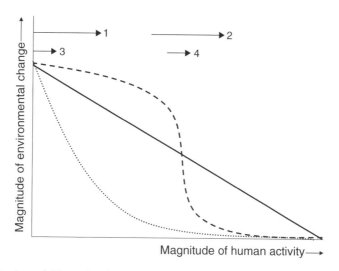

Figure 4.1.1   A graph illustrating the relations between a human activity that creates environmental change and the environmental response. For example, we could consider increasing amounts of sediment moving into rivers as the human impact and examine the effect on species richness of macroinvertebrates. Three possible relations (linear (continuous line), threshold (dashed line) and asymptotic (dotted line)) of many possibilities are illustrated. The arrows indicate increases in human impact in four different situations: '1' and '2' indicate relatively large increases in the magnitude of human activity, whereas '3' and '4' show relatively smaller changes. Arrows '1' and '3' are for locations where there is little human activity of this kind present, whereas arrows '2' and '4' indicate increases in the magnitude of human activity above what is already present. Collectively, the arrows illustrate that we need access to all three items of information (starting condition of site, magnitude of human activity, nature of the relation between the human activity and the environmental response) to understand the likely consequences of any particular human action

curves such as those illustrated in Figure 4.1.1, (2) the base condition of the location – does it have some level of human impact present or not?, and (3) the size of any further change in human activity that is planned. Equipped with these three pieces of information, we would be in a good position to make authoritative predictions about the likely magnitude of environmental change for past human activity as well as future planned ones. Thus, in Figure 4.1.1, we would regard the changes brought about by actions (1) and (3) differently depending on the nature of the relation – we would be less concerned in the case of a threshold relation than in the case of an asymptotic one. For the threshold relation, we would see no large change in the environmental variable, but for the latter relation we would see relatively large changes for both, and the larger change ('arrow 1') would be of particular concern. The situation is different for actions '2' and '4'. For an asymptotic relation, the damage has already been done and arguably little overall change would be seen, whereas in the threshold relation, both actions will cause the magnitude of human impact to tip across the threshold, so that even the small action creates very large changes. It should be clear then that we cannot make any fixed statements about the sizes of changes of concern because this is contingent upon the individual circumstances. Theoretically, we can gather two of the three items of information that we need: the starting condition of the locale to be impacted and the size of the human activity planned. Nevertheless, it is the relation between human activity and environmental response that is critical, and in the vast majority of cases these relations are unknown. Partly, that lack of knowledge is due to the great difficulty in obtaining such curves, because it is very difficult to isolate the precise effect of any one human action on the environment from all others (Downes *et al.*, 2002). This is also due to the lack of consistency in measuring human impacts. Reviewers of the literature usually discover that researchers have used different methods for measuring human impact and different methods for measuring environmental responses; this prevents us from combining the results of different studies to examine relations between human impact and environmental responses (King *et al.*, 2003). We can only hope that some standards of measurement that allow results from different studies to be combined (e.g. Calabrese and Baldwin, 1997) will eventuate.

To summarize, we need to set *a priori* effect sizes for monitoring programmes so that they may then be designed to provide us with adequate power. Reference locations provide us with some information about the condition of rivers (degree of impact or health) when they are available and when we can defend our selections. Regardless, we need to determine ahead of time the size of difference we want our monitoring programme to detect, because we cannot simply compare impact and reference rivers and leave the significance of any differences between them to be determined by the outcome of statistical tests alone, and when we have no reference rivers at all, we have no way of proceeding.

The curves presented in Figure 4.1.1 serve to illustrate mostly the difficulty of the problems in front of us, rather than solutions, given that we are unlikely to have access to such information for most impacts any time soon. Nevertheless, they do illustrate why reference rivers alone cannot provide us with definitive information about 'health'. We still need to know the shape of the curve and where any river location sits along it to draw conclusions about the seriousness of the latter's condition. The curves illustrate the need to think carefully about effect sizes, whether we have 'reference rivers' or not.

Finally, we need to remember that setting effect sizes – deciding what constitutes an 'important difference' – is ultimately a social decision, not a scientific one. Indeed, one can see the tacit acceptance of this in places where the luxury of access to intact, 'undisturbed' rivers is not possible, and reference conditions are based on what is theoretically possible, as well as socially desirable riverine attributes (e.g. Jungwirth *et al.*, 2002). Such effect sizes will represent the value the community places on various environmental attributes – abundances of rare species, species diversity and functioning ecosystems. Such decisions should be informed by the best possible science regarding the consequences to the environment of such changes, but in the end they will reflect the weight placed on protecting the environment by the community via its elected representatives and managers.

REFERENCES

Australian State of the Environment Committee, 2001. *Australia State of the Environment 2001*, Independent report to the Commonwealth Minister for the Environment and Heritage. CSIRO Publishing (on behalf the Department of Environment and Heritage): Canberra, Australia.

Barbour, M. T., Swietlik, W. F., Jackson, S. K., Courtemanch, D. L., Davies, S. P. and Yoder, C. O., 2000. 'Measuring the attainment of biological integrity in the USA: a critical element of ecological integrity'. *Hydrobiologia*, **422/423**, 453–464.

Botkin, D. B., 1990. *Discordant Harmonies: A New Ecology for the Twenty-First Century.* Oxford University Press: New York, NY, USA.

Buffagni, A., Kemp, J. L., Erba, S., Belfiore, C., Hering, D. and Moog, O., 2001. 'A Europe-wide system for assessing the quality of rivers using macroinvertebrates: The AQEM Project and its importance for southern Europe (with special emphasis on Italy)'. *Journal of Limnology*, **60**(Suppl. 1), 39–48.

Calabrese, E. J. and Baldwin, L. A., 1997. 'A quantitatively based methodology for the evaluation of chemical hormesis'. *Human and Ecological Risk Assessment*, **3**, 545–554.

Chovanec, A., Jäger, P., Jungwirth, M., Koller-Kreimel, V., Moog, O., Muhar, S., and Schmutz, S., 2000. 'The Austrian way of assessing the ecological integrity of running waters: a contribution to the EU Water Framework Directive'. *Hydrobiologia*, **422/423**, 445–452.

Choy, S. C., Thomson, C. B. and Marshall, J. C., 2002. 'Ecological condition of central Australian arid-zone rivers'. *Water Science and Technology*, **45**, 225–232.

Connolly, S. R., Menge, B. A. and Roughgarden, J., 2001. 'A latitudinal gradient in recruitment of intertidal invertebrates in the northeast Pacific Ocean'. *Ecology*, **82**, 1799–1813.

Di Stefano, J., 2003. 'How much power is enough? Against the development of an arbitrary convention for statistical power calculations'. *Functional Ecology,* **17**, 707–709.

Downes, B. J., Barmuta, L. A., Fairweather, P. G., Faith, D. P., Keough, M. J., Lake, P. S., Mapstone, B. D. and Quinn, G. P., 2002. *Monitoring Ecological Impacts: Concepts and Practice in Flowing Waters.* Cambridge University Press: Cambridge, UK.

Egerton, F. N., 1973. 'Changing concepts of the balance of nature'. *Quarterly Review of Biology*, **48**, 322–350.

Fairweather, P. G., 1993. 'Links between ecology and ecophilosophy, ethics and the requirements of environmental management'. *Australian Journal of Ecology*, **18**, 3–19.

Fensham, R. J., 1997. 'Aboriginal fire regimes in Queensland, Australia: analysis of the explorers' record'. *Journal of Biogeography*, **24**, 11–22.

Jungwirth, M., Muhar, S. and Schmutz, S., 2002. 'Re-establishing and assessing ecological integrity in riverine landscapes'. *Freshwater Biology*, **47**, 867–887.

King, A., Brooks, J., Quinn, G. P., Sharpe, A. K. and McKay, S., 2003. 'Monitoring programmes for environmental flows in Australia – a literature review'. Department of Sustainability and Environment, Sinclair Knight Merz, CRC for Freshwater Ecology: Melbourne, Australia.

Lake, P. S., 1995. 'Of floods and droughts: river and stream ecosystems of Australia'. In: *Ecosystems of the World*, Vol. 22, *River and Stream Ecosystems*, Cushing, C. E., Cummins, K. W. and Minshall, G. W. (Eds). Elsevier: Amsterdam, The Netherlands, pp. 659–694.

Lancaster, J., 2000. 'The ridiculous notion of assessing ecological health and identifying a useful concept underneath'. *Human and Ecological Risk Assessment*, **6**, 213–222.

Linke, S., Bailey, R. C. and Schwindt, J., 1999. 'Temporal variability of stream bioassessments using benthic macroinvertebrates'. *Freshwater Biology*, **42**, 575–584.

Loehle, C., 1987. 'Hypothesis testing in ecology: psychological aspects and the importance of theory maturation'. *Quarterly Review of Biology*, **62**, 397–409.

Mapstone, B. D., 1995. 'Scaleable decision rules for environmental impact studies: effect size, Type I and Type II errors'. *Ecological Applications*, **5**, 401–410.

Marchant, R., Hirst, R., Norris, R., Butcher, R., Metzeling, L. and Tiller, D., 1997. 'Classification and prediction of macroinvertebrate assemblages from running waters in Victoria, Australia'. *Journal of the North American Benthological Society*, **16**, 664–681.

Niemi, G. J., DeVore, P., Detenbeck, N., Taylor, D., Lima, A., Pastor, J., Yount, J. D. and Naiman, R. J., 1990. 'Overview of case studies on recovery of aquatic systems from disturbance'. *Environmental Management*, **14**, 571–587.

Nijboer, R. C., Johnson, R. K., Verdonschot, P. F. M., Sommerhäuser, M. and Buffagni, A., 2004. 'Establishing reference conditions for European streams'. *Hydrobiologia*, **516**, 91–105.

Norris, R. H. and Thoms, M. C., 1999. 'What is river health?'. *Freshwater Biology*, **41**, 197–209.

Papworth, M. P. and Lewis, B., 2003. 'The development of an historical baseline of water balance and environmental flows'. *Water Science and Technology*, **48**, 139–147.

Perry, G. L. W., 2002. 'Landscapes, space and equilibrium: shifting viewpoints. *Progress in Physical Geography*, **26**, 339–359.

Quinn, G. P. and Keough, M. J., 2002. *Experimental Design and Analysis for Biologists*. Cambridge University Press: Cambridge, UK.

Richardson, J. S. and Healey, M. C., 1996. A healthy Fraser River? How will we know when we achieve this state?'. *Journal of Aquatic Ecosystem Health*, **5**, 107–115.

Thoms, M. C., Ogden, R. W. and Reid, M. A., 1999. 'Establishing the condition of lowland floodplain rivers: a palaeo-ecological approach'. *Freshwater Biology*, **41**, 407–423.

Townsend, C. R., Downes, B. J., Peacock, K. and Arbuckle, C. J., 2004. 'Scale and the detection of land-use effects on morphology, vegetation and macroinvertebrate communities of grassland streams'. *Freshwater Biology*, **49**, 448–462.

Underwood, A. J., 1995. 'Detection and measurement of environmental impacts'. In: *Coastal Marine Ecology of Temperate Australia*, Underwood, A. J. and Chapman, M. G. (Eds). University of NSW Press: Sydney, Australia, pp. 311–324.

Underwood, A. J. and Chapman, M. G., 2003. 'Power, precaution, Type II error and sampling design in assessment of environmental impacts'. *Journal of Experimental Marine Ecology and Biology*, **296**, 49–70.

Underwood, A. J. and Petraitis, P. S., 1993. 'Structure of intertidal assemblages in different locations: how can local processes be compared?'. In: *Species Diversity in Ecological Communities*, Ricklefs, R. and Schluter, D. (Eds). University of Chicago Press: Chicago, IL, USA, pp. 38–51.

Wright, J. F., 1995. 'Development and use of a system for predicting the macroinvertebrate fauna in flowing waters'. *Australian Journal of Ecology*, **20**, 181–197.

Yibarbuk, D., Whitehead, P. J., Russell-Smith, J., Jackson, D., Godjuwa, C., Fisher, A., Cooke, P., Choquenot, D. and Bowman, D. M. J. S., 2001. 'Fire ecology and Aboriginal land management in central Arnhem Land, northern Australia: a tradition of ecosystem management'. *Journal of Biogeography*, **28**, 325–343.

Yount, J. D. and Niemi, G. J., 1990. 'Recover of lotic communities and ecosystems from disturbance – a narrative review of case studies'. *Environmental Management*, **14**, 547–569.

4.2

The Predictive Modelling Approach to Biomonitoring: Taking River Quality Assessment Forward

John F. Murphy and **John Davy-Bowker**

Biological Monitoring of Rivers   Edited by G. Ziglio, M. Siligardi and G. Flaim
© 2006 John Wiley & Sons, Ltd.

4.2.1   INTRODUCTION

The importance of incorporating biological assessment in the monitoring of fresh-water quality is reflected in recent legislation such as the European Union Water Framework Directive (WFD) (European Commission, 2000) and the 1972 US Clean Water Act (most recently amended in 2002). The legislative change in emphasis in recent years, from the sole use of chemical measures to the inclusion of biological information, has provided the impetus for the development of novel approaches to interpreting the information conveyed by the biological community. Frequently, this has involved the use of biological indicators to quantify and simplify the multifaceted ecological status of the water body (Norris and Hawkins, 2000). A biological indicator is defined as a feature of the biota that has particular requirements with regard to a known set of physico-chemical variables such that change in some measure of the indicator implies that current or recent environmental conditions have been altered. A biological indicator can be the rate of a physiological process within individuals, a population or community structural response or variation in the rate of an ecosystem process (Johnson *et al.*, 1993).

Among the most widely used types of indicators in freshwater biomonitoring are diagnostic, relative abundance and diversity indices derived from biotic community structure. Some biomonitoring programmes use a selection of indices, integrated to provide a single measure of ecological condition (the multimetric index) (Karr, 1999). However, natural variation in the values of these indices in different types of rivers or longitudinally down the natural gradient of a river catchment (Vannote *et al.*, 1980) complicates their interpretation. There was a realization that the judgement of river water quality based on indices needed to be qualified by the type of river being assessed. Estimates of what is considered a high quality index score for particular sections of river need to be defined. This was the basis for the development of the predictive modelling approach to bioassessment, whereby an expected index value is predicted for a particular river stretch (were it not subject to anthropogenic stresses) based on its physical and perhaps chemical features. This prediction is derived by following a prescribed methodology of data collection and multivariate analysis and has also been termed the 'Reference Condition Approach' (Bailey *et al.*, 2004).

This chapter will provide an overview of the generic approach used to develop and test predictive bioassessment models. It will be based on the methods used in the development of existing predictive bioassessment models in the UK, Australia, Canada and elsewhere (Wright *et al.*, 1984; Davies, 1994; Reynoldson *et al.*, 1995).

4.2.2   REFERENCE SITES

The crucial first step in the predictive modelling approach is the selection of high quality, unstressed or minimally disturbed reference sites upon which the prediction model will be based. A reference site is defined as a length of river with no, or only very minor, anthropogenic alterations to its hydrochemistry and hydromorphology and with biota usually associated with such undisturbed or minimally disturbed conditions (European Commission, 2000; Bailey *et al.*, 2004). Site selection is a critical phase and must ensure that an adequate number of reference sites, covering the full range of rivers and geological formations in the region, are included. Care must also be taken to encompass a sequence of locations from headwaters to downstream reaches. In the development of the River Invertebrate Prediction and Classification System (RIVPACS) in the UK, reference sites were initially chosen subjectively in consultation with local experienced biologists (Wright *et al.*, 1984). Subsequently, this selection was refined through an iterative process whereby some reference sites, found to be mildly polluted, were removed at later stages of model development and others were added to improve the representation of some stream types (see Section 4.2.8).

At each reference site, biological data and environmental variables representing the perceived environmental drivers, or at least correlates of the biota, are collected. The environmental variables used in the predictive bioassessment model, and measured at each site, should only include those that are unaffected by the stresses whose ecological impacts are ultimately being assessed. Each site may be sampled on a number of occasions, e.g. different seasons, to ensure a complete representation of the biotic assemblage is acquired (Furse *et al.*, 1984). The biota should be identified to the lowest practical taxonomic level to maximize the discriminatory power of the model.

The development of the predictive modelling approach led to the concept of a site-specific (or at least stream-type-specific) expected fauna, termed the reference condition (Reynoldson *et al.*, 1997), and to the idea of comparing the observed fauna at the site with its expected fauna. These principles are now at the heart of the prescribed methodology within the WFD (European Commission, 2000).

4.2.3   STANDARDIZED SAMPLING AND PROCESSING

It is important that the biotic community at all reference sites is sampled using standard protocols. The measurement of environmental variables must also be carried out using a standard methodology. This is to ensure that sampling variation and measurement error are minimized or at least are uniform across all sites.

For the Australian River Assessment Scheme (AUSRIVAS), different in-stream habitats at the reference sites were sampled separately and discrete predictive bioassessment models were developed for each habitat type, e.g. riffle, macrophyte bed and stream edge. A 10 m transect through each habitat was sampled using a 250 μm mesh pond net. Samples were taken twice a year and were processed by live-picking or by sorting of preserved sub-samples until 200 individuals were removed and identified to family level (Simpson and Norris, 2000). In the Fraser River catchment in Canada a single 3 min kick sample was used at each reference site, with all in-stream flow and substrate types at the site being sampled in proportion to their occurrence. In the laboratory, the material collected was sub-sampled using a fixed count method similar to that developed in Australia (Rosenberg *et al.*, 2000). In the development of a predictive bioassessment model for the Adour-Garonne catchment in south western France, macroinvertebrate samples were taken in spring and winter from the various substratum types at each reference site using a 0.1 m² Surber sampler with a 300 µm mesh (Céréghino *et al.*, 2003).

Ultimately, the method used should be that proven to be most effective and efficient at obtaining a representative sample of the biotic community in the region of interest. Crucially, the biological sampling and measurement of environmental variables at any new test site must also conform to that used for the reference sites, thus ensuring that the predicted fauna can be legitimately compared to the observed fauna.

4.2.4   CLASSIFICATION OF REFERENCE SITES

Classifying the reference sites into groups based solely on the similarity of their biotic assemblages provides the basis for the prediction of the expected biological communities for sites in the absence of anthropogenic stress. Several different methods are available for forming a biological classification (see Moss *et al.*, 1999; Giraudel and Lek, 2001). In New Zealand, the UK and Sweden, TWINSPAN (Two-way indicator species analysis) (Hill, 1979) was considered the most appropriate classification method (Moss *et al.*, 1999; Joy and Death, 2002, 2003; Johnson, 2003). TWINSPAN has been used successfully in many branches of community ecology. This repeatedly divides the reference sites into groups in a hierarchical manner based on the similarity of their biota. At each stage, the reference sites in a particular group and their taxa are simultaneously ordered so that sites with similar taxa are placed close together on the ordination axis. The position of sites along the axis is used to split the group into further end-groups (Figure 4.2.1). While the evenness of classification end-group size should be considered, it is perhaps more important that end-groups contain sufficient number of reference sites to generate reliable predictions of taxon occurrence (Moss *et al.*, 1999). In other regions, Bray–Curtis unweighted pair-group mean arithmetic averaging (UPGMA) has been used to form a biological classification of the reference sites (Davies, 1994; Reynoldson *et al.*, 1995; Hawkins *et al.*, 2000) and more recently the applicability of artificial intelligence techniques e.g. Kohonen self-organizing maps has been demonstrated (Giraudel and Lek, 2001; Walley and O'Connor, 2001; Céréghino *et al.*, 2003; Gervey *et al.* 2004). It needs to be emphasized that biological variation across a region is a continuum and that sites do not naturally fall into completely distinct biological types. The biological classification of reference sites into groups is an essential step in the prediction process but does not constitute a definitive assignment of sites to particular end-groups, with a uniform biota wholly distinct from that of other end-groups.

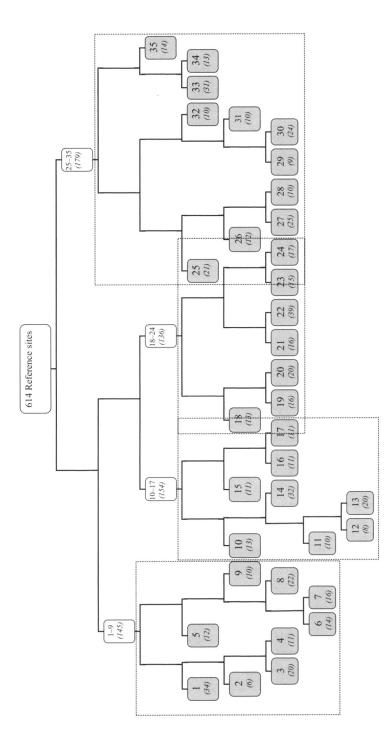

Figure 4.2.1 TWINSPAN classification of 614 reference sites for mainland Great Britain split into 35 end-groups based on their macroinvertebrate assemblage (the number of sites in an end-group is given in parentheses): end-groups 1–9, small streams throughout GB; end-groups 10–17, upland streams and rivers in northern GB; end-groups 18–24, intermediate streams and rivers in northern, mid-and south western GB; end-groups 25–35, lowland streams and river in south eastern Britain (adapted from Rosenberg *et al.*, 2000). Reproduced by permission of the Freshwater Biological Association from Wright, J. F., 2000, "An introduction to RIVPACS" in *Assessing the Biological Quality of Fresh Waters: RIVPACS and Other Techniques*, Wright, J. F., Sutcliffe, D. W. and Furse, M. T. (Eds), Freshwater Biological Association, Ambleside, UK, pp. 1–24

4.2.5 LINKING BIOLOGICAL GROUPS WITH ENVIRONMENTAL VARIABLES

Having classified the reference sites into groups on the basis of their biotic communities, the next step is to derive predictive functions that use the measured environmental characteristics to predict a list of taxa expected at a site in the absence of any disturbance (the reference condition). The occurrence and abundance of taxa at reference sites depends on the availability of suitable habitats, energy sources and the regional species pool. However, it is impractical to gather such detailed information for every site at which an ecological assessment is needed. In the predictive modelling approach, the aim is to select a set of environmental predictor variables that can be easily measured at any type of site. Therefore, while the variables chosen may not necessarily be causal determinants of the community composition, they should be robust correlates of the biota at the site. One of the issues in selecting predictor variables is the need to avoid the inclusion of variables that are themselves affected by the anthropogenic stress whose impact is being assessed. As a result, variables such as nutrient concentrations, macrophyte habitat and recent flow regime are typically unsuitable. The most appropriate predictive bioassessment model may not therefore be the one that gives the best possible statistical fit. The ideal, long-term aim is to have a single fixed prediction of the site-specific fauna expected at any site, based predominantly on time-invariant characteristics, e.g. altitude, distance from source, latitude, etc. (Céréghino *et al.*, 2003).

The process of choosing an optimal set of predictor variables is usually supported by a stepwise selection procedure to eliminate those variables that do not make a statistically significant contribution to the explanatory power of the model. Johnson (2003) used canonical correspondence analysis with forward selection, followed by multiple discriminant analysis (MDA) to define the best set of predictor variables. In the development of RIVPACS, an initial list of 28 physico-chemical variables was modified to remove redundant variables using stepwise MDA, leaving an optimal sub-set of 11 variables in the prediction model (Moss *et al.*, 1987) (Table 4.2.1). MDA is used to derive predictive equations, referred to as discriminant axes, which represent those aspects of the environmental variation that differ most between the classification end-groups. These predictive equations are then used to estimate the probability of a test site belonging to each of the end-groups, based on its environmental characteristics (Figure 4.2.2). Typically, in RIVPACS a test site will have a predicted probability (P_i) of > 1 % of belonging to between one and five of the 35 TWINSPAN end-groups (Figures 4.2.1 and 4.2.2). The probabilistic assignment of sites to biological end-groups is a powerful attribute of the predictive model approach. The predicted community is based on several end-groups and is weighted in proportion to the probability of end-group membership (Figure 4.2.2). This makes the predictions robust with respect to the method of classifying the reference sites (Clarke *et al.*, 2003). Furthermore, this approach provides a means of automatically identifying test sites that are inadequately represented by the reference sites. Test

Table 4.2.1 The optimal sub-set of predictor variables used by RIVPACS to generate the expected fauna at a test site

---

Time invariant map-derived variables

Altitude at site (m)
Latitude
Longitude
Distance from source (km)
Slope (m km^{-1})

Long-term historical data

Mean air temperature
Air temperature range
Discharge category (1–10)
(1 = < 0.31, 2 = 0.31–0.62, 3 = 0.62–1.25, 4 = 1.25–2.5, 5 = 2.5–5.0, 6 = 5–10, 7 = 10–20,
   8 = 20–40, 9 = 40–80, 10 = 80–160 (m^3 s^{-1})

Measured during site visits and averaged over the year

Stream width (m) (mean of three seasonal measurements)

Stream depth (cm) (mean of three seasonal measurements)

Substratum composition (% cover of clay/silt, sand, gravel/pebbles, cobbles/boulders converted to a mean particle size phi score)

Alkalinity (mg l^{-1} CaCO$_3$) (mean of twelve monthly measurements)

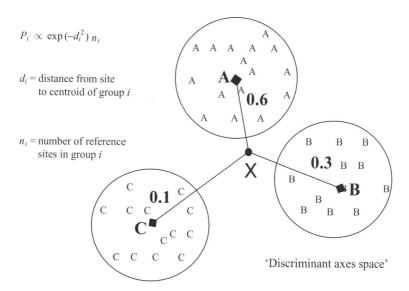

P_i = probability of test site 'X' belonging to group i

$P_i \propto \exp(-d_i^2)\, n_i$

d_i = distance from site to centroid of group i

n_i = number of reference sites in group i

'Discriminant axes space'

Figure 4.2.2 Probabilistic assignment of a test site to biological classification end-groups, where the circles represent the variability between reference sites in each of three end-groups (A, B and C) on the discriminant axes derived from the environmental variables. On the basis of its environmental features, the new site is placed at point 'X'. The distance in the discriminant axes space of site 'X' from a group's mean position is used to calculate its probability of belonging to that group; test site 'X' has a probability of 0.6 of belonging to group A, 0.3 to B and 0.1 to group C

Table 4.2.2   Calculating the expected probability of capture of a taxon at a test site (see Figure 4.2.2)

Site classification end-group	Probability site belongs to (P_i)	% of reference sites in end-group with taxon i present	Contribution of the end-group end-group to the likelihood of capture of taxon i
A	0.6	40	0.24
B	0.3	50	0.15
C	0.1	70	0.07
Expected likelihood of capture (P_E) of taxon i $= 0.46$			

sites with environmental characteristics that give them a low probability of belonging to any of the end-groups can be identified and unreliable predictions can be prevented (Clarke *et al.*, 2003).

4.2.6   PREDICTING THE REFERENCE CONDITION

The probabilities of the test site belonging to each biological classification end-group are used to calculate the expected fauna (Table 4.2.2 and Figure 4.2.2). The probabilities (P_i) may or may not be weighted by the number of reference sites in the end-group (n_i) (Figure 4.2.2). In the example illustrated in Table 4.2.2 and Figure 4.2.2, the test site has a 60 % chance of belonging to group A and taxon *i* occurs in 40 % of Group A reference sites, giving a contribution of 24 % (0.6 × 40) to the expected probability. Using the same process, Groups B and C contribute 15 % (0.3 × 50) and 7 % (0.1 × 0.70) to the expected probability, respectively, giving any overall expected probability of taxon *i* occurring at the test site (P_E) of 46 % (Table 4.2.2). This calculation is done for each taxon in turn. The expected abundance of a taxon at a site could also be calculated in a similar manner, using the average of observed abundances of the taxon for the reference sites in each site group. Using the same approach and site group probabilities ensures consistency in the predictions of expected probability of occurrence and abundance of taxa (Moss *et al.*, 1987; Clarke *et al.*, 2003). As part of the basic output from the RIVPACS software, the macroinvertebrate taxa are listed in decreasing order of expected probability of occurrence (Table 4.2.3). The taxa actually observed in the sample are highlighted with an asterisk. If a site were in reference condition, most taxa with a high expected-likelihood of being present would be expected to occur.

4.2.7   CALCULATING INDICES

The expected probabilities of occurrence and expected abundances can be used to derive estimates of expected values for a wide range of indices. The simplest example is the expected number of taxa, which is the sum of the site-specific expected

Table 4.2.3 An example of *part* of the basic output from the RIVPACS software showing the expected taxa list with associated probabilities of occurrence and expected abundances (River Frome at East Stoke, 'Spring sample'). The taxa actually observed at the test site are 'asterisked' and their actual abundances given; taxa are listed in decreasing order of expected probability of capture (P_E), stopping at 47.5 %, for illustration only

Observed	P_E	Expected $\log_{10}$ abundance	Observed $\log_{10}$ abundance	Taxon
*	1	2.64	2	Chironomidae
*	1	2.37	1	Elmidae
	0.985	1.67	0	Tipulidae
*	0.951	2.28	1	Baetidae
*	0.931	2.44	2	Gammaridae
	0.927	2.11	0	Ephemerellidae
*	0.908	1.88	3	Oligochaeta
*	0.902	1.91	1	Simuliidae
*	0.891	1.51	2	Rhyacophilidae
*	0.846	2.90	1	Leuctridae
*	0.830	1.46	1	Heptageniidae
	0.784	1.20	0	Limnephilidae
*	0.750	1.08	1	Dytiscidae
	0.784	0.96	0	Hydrophilidae
	0.623	0.86	0	Sericostomatidae
*	0.604	1.40	4	Hydrobiidae
*	0.596	0.88	1	Ancylidae
	0.486	0.82	0	Hydropsychidae
*	0.475	0.84	1	Sphaeriidae

probabilities of occurrence (P_E) of all the individual taxa (Table 4.2.3). There is potential for calculating expected values of indices that are proven to be diagnostic of impacts such as organic pollution, e.g BMWP (National Water Council, 1981; Armitage *et al.*, 1983), acidity, e.g. AWIC (Davy-Bowker *et al.*, 2005) and metal toxicity, e.g. SIGNAL-MET (Chessman and McEvoy, 1998). In addition, relative abundance indices such as % Ephemeroptera, Plecoptera and Trichoptera (EPT) or abundance of Baetidae/abundance of Ephemeroptera (Reynoldson *et al.*, 1997;) and diversity or dominance indices such as Simpson or Berger-Parker (see Magurran, 1988) could be considered where they are shown *a priori* to express a significant relationship with a known stressor gradient. Most existing predictive bioassessment models incorporate taxon richness but some systems also include additional indices specifically developed for their region (Armitage *et al.*, 1983; Chessman, 1995).

Existing predictive bioassessment models adopt a range of different threshold P_E values for including taxa in index calculation, e.g. RIVPACS includes all taxa with $P_E > 0$, while AUSRIVAS excludes taxa with $P_E < 0.5$. There is a lack of consensus on the most appropriate threshold P_E value to adopt (Cao *et al.*, 1998; Marchant, 2002). Some workers have found that by excluding rare taxa from the index calculation, the sensitivity of the bioassessment is increased (Hawkins *et al.*,

2000; Simpson and Norris, 2000; Johnson, 2003) while others contend that there may be a loss of information when rare taxa are excluded (Cao *et al.*, 1998, 2001). The threshold P_E value chosen for a particular model is usually a compromise between increasing the power to detect change by increasing the potential taxon list and lowering the variance associated with the predictions (Johnson, 2003). The most appropriate cut-off point will be different for individual models, depending on the regional taxon diversity and field sampling method. There are also suggestions that the effect of the sample processing method may influence the ability to discriminate assemblages, in particular where fixed-count (< 500 individuals) sample processing is undertaken (Cao *et al.*, 2002).

Uniquely, the Benthic Assessment of Sediment system (BEAST) developed in Canada adopts a different approach to comparing test and reference sites (Reynoldson *et al.*, 1995). Rather than predicting index values for comparison with the observed state, the location of the test site (based on its community composition) is compared in ordination space to the distribution of appropriate reference sites (Reynoldson *et al.*, 1995). Such an approach has also been applied to the Fraser River catchment in British Columbia (Rosenberg *et al.*, 2000).

4.2.8   VALIDATING THE MODEL

It is vitally important that the accuracy of the model predictions is determined. A number of techniques for fulfilling this purpose are available. Cross-validation (also termed jack-knifing) is a method that is used to assess the accuracy of the model at an early stage. The process of constructing the predictive bioassessment model is followed as detailed above but one reference site is omitted from the data set and used as a test site. Assessing whether the test site has been assigned to the same end-group as in the original classification, including all reference sites, then tests the MDA predictive functions (Moss *et al.*, 1999). A different test site is omitted each time and the model reconstructed. The percentage of sites assigned correctly to their biologically predetermined end-groups is a measure of the accuracy of the model. Most models achieve a predictive accuracy $\geq 50\%$ (Johnson, 2003; Joy and Death, 2003; Gevrey *et al.*, 2004). In practice this is a very severe, and potentially misleading, test of predictive accuracy because it ignores the fact that in most predictive bioassessment models the expected fauna for a test site is based on the likelihood of the site belonging to all end-groups, not just the most probable.

An alternative approach to model validation is to calculate the agreement between observed and predicted values of indices derived from the lists of observed and predicted taxa. Ideally, model validation should be on an independent data set of undisturbed sites but these independent sites still need to be within the same range of environmental conditions and high qualities as the reference sites themselves. Additionally, the ability of the model to detect deterioration in ecological quality can be assessed by comparing the observed fauna at a range of sites along a known gradient of impact with the predicted reference condition.

There is also a very important iterative step whereby the reference sites themselves are passed through the predictive bioassessment model to quantify the distribution of observed/expected (O/E) ratio values for a given index (see Section 4.2.9). In theory, the O/E ratios should vary around unity with approximately half of the reference sites having O/E values > 1 (and potentially better than average reference conditions) and half having O/E values < 1 (and potentially poorer than average reference conditions). This distribution can be described by the mean ± 95 percentile-based confidence interval. The quality of the model and the reference site data set upon which it is based can be assessed in terms of the degree of dispersion in the range of O/E values (Clarke, 2000; Johnson, 2003). Sites that lie at the extremes of the O/E frequency distribution must contain uncharacteristically diverse or depauperate fauna. These sites can be identified by simple rules, e.g. those having $< 75\%$ of the taxa that the model predicts, and should be removed from the reference site data set and the model reconstructed.

4.2.9 ECOLOGICAL QUALITY CLASSES

The comparison of the observed fauna with that predicted by the model is usually achieved by calculating O/E ratios for a given index, e.g. taxon richness. O/E ratios provide a means of standardizing indices, so that a particular O/E value implies the same ecological quality, regardless of stream-type. It is largely because of the success of the predictive modelling approach using O/E ratios as a robust standardization tool, that the WFD prescribes the use of O/E (termed Ecological Quality Ratios by the WFD) for reporting biomonitoring results (European Commission, 2000).

To simplify the interpretation of numerical O/E values, they are usually banded into ecological quality classes. The number and width of the classes is usually based on the distribution of O/E values for the reference site data set. In AUSRIVAS, sites that fall between the 10th percentile and the 90th percentile of the reference site O/E distribution are considered to be equivalent to the reference condition. Those with O/E values > 90th percentile are considered better than reference condition, i.e. sites with exceptionally diverse fauna. It may also indicate sites where the taxon richness is enhanced by mild organic enrichment. Sites with O/E values below the 10th percentile are considered to be below the reference condition. The range of O/E values below the 10th percentile is divided into classes with a width equivalent to the central 80 % of the reference site O/E distribution (Simpson and Norris, 2000).

The WFD stipulates that a five-class reporting system, ranging from 'High' to 'Bad' Status, is adopted by member states (European Commission, 2000). Intercalibration studies at a standardized network of sites across the continent will be used to set the boundary O/E values. At the time of writing, however, there is no objective approach to determining the location of ecological quality class boundaries across the range of O/E values. Such a scheme is required to ensure that the presentation and interpretation of bioassessment results is seen as rigorous and unbiased.

4.2.10   QUANTIFYING UNCERTAINTY
IN THE ASSESSMENT

It is essential to be able to assess whether there is a statistically significant or eco-logically real difference in the quality estimates obtained for two samples, whether from the same site at two points in time or from two different sites. Therefore, it is necessary to be able to estimate the confidence of the assignment of a site to a partic-ular ecological quality class, i.e. the probability of 'misclassification'. The decision on how many classes to have and where to place the class boundaries in terms of the O/E values should take account of the fact that having more classes gives finer apparent discrimination of quality but greater actual misclassification rates (Clarke *et al.*, 1996).

The actual uncertainty of estimates of biological condition of a test site must de-pend on errors in generating a predicted fauna as well as uncertainty in deriving the observed fauna. There are a number of potential sources of error in the predictions and measures should be taken to minimize their occurrence. It is critical that the model is based on an adequate set of reference sites with all site types well repre-sented. It needs to be remembered that although the reference sites are unstressed or minimally impacted, they will still vary considerable in quality. Additionally, there will be variation associated with the biological sampling and measurement of environmental variables at each reference site. While it is clear that the choice of reference sites, environmental variables and prediction method all contribute uncer-tainty to the model, these can be considered as an integral part of the definition of the quality index. Therefore, the only errors in estimating the expected fauna for test sites can be assumed to arise from error and variation in obtaining values for the environmental predictor variables (Clarke, 2000).

Sources of uncertainty in estimating the observed fauna at a test site include field sampling, sample processing (including taxonomic identification) and temporal vari-ation in community composition. Information on the magnitude of the uncertainty can be obtained from a variety of sources. Spatial sampling variation can be quantified by replicate sampling at the same site and calculating the proportion of taxa cap-tured in one replicate relative to all replicate samples, e.g. a single 3-min RIVPACS sample has been shown to contain 50 % of the species found in six replicates (Furse *et al.,* 1981). Sample-processing-biases errors can also be measured from quality control procedures. Quantifying the natural temporal variation of communities can be achieved by using time series data. Spatial and temporal variation due to field sampling and sample processing biases and errors need to be quantified for a wide range of types and qualities of site.

The RIVPACS software system for river sites in the UK has a powerful function whereby it generates statistical information on the reliability of classification (or risk of misclassification) using an uncertainty simulation model that integrates estimates of error from a variety of sources to generate simulated values of O/E (see Clarke *et al.*, 1996). Five hundred or more simulated O/E values are generated to estimate the distribution within which the actual O/E value for the site probably lies, thus enabling the statistical likelihood and extent of real change in ecological quality between the two sites or two surveys at the same site to be assessed (Clarke, 2000).

4.2.11 APPLYING THE PREDICTIVE BIOASSESSMENT MODEL

When a fully validated predictive bioassessment model has been developed for a region, it can then be used to assess the ecological status of new sites of unknown quality. At these test sites, a biological sample is collected and the relevant physico-chemical variables measured by using the same standard protocols as were used for the reference sites. The predictive bioassessment model then generates a probabilistic list of the expected fauna from the measured physico-chemical variables from which expected values of biotic indices can be calculated. The biological sample taken at the test site provides an observed fauna from which the same biotic indices are also calculated. The expected and observed values for various biotic indices are then compared using O/E ratios. The latter can be banded into quality classes to aid interpretation of the data across the entire region or country.

4.2.12 FUTURE RESEARCH

Throughout this chapter, we have drawn on examples from existing predictive bioassessment models, with a bias towards those developed using stream macroinvertebrate communities but the approach has also been successfully applied to lotic fish (Joy and Death, 2002) and diatom (John, 2000; Gevrey *et al.*, 2004) communities as well as lentic macroinvertebrate communities (Reynoldson *et al.*, 1995; Johnson, 2003). The WFD prescribes that information from benthic macroinvertebrates, fish, macrophytes and phytobenthos communities is considered in the overall assessment of ecological quality (European Commission, 2000). However this multi-assemblage approach needs to acknowledge that the different communities respond to stresses acting at different spatial and temporal scales. Further research is required to determine how best to integrate and interpret the information from the different biotic groups.

There is considerable scope for improving the diagnostic capabilities of predictive bioassessment models. Most existing models calculate taxon richness and perhaps one or two other diagnostic indices. For national assessments of the biological quality of UK river sites, RIVPACS calculates taxon richness and BMWP-related indices that indicate general degradation and organic pollution impacts, respectively, but it is not constrained to calculate only these indices. By incorporation of *a priori* tested peer-reviewed indices and the development and testing of new ones where necessary, predictive bioassessment models could calculate expected values for any index and therefore offer accurate diagnosis of the probable causes of deviation from the reference condition.

In many ways this represents a merging with some elements of the multimetric approach. This approach has similar aims to the predictive modelling approach in that it seeks to quantify ecological quality by measuring biological attributes of a community and comparing them to that expected given the location of the test site and its physical and chemical characteristics (Karr, 1999). However, it differs fundamentally from predictive modelling in the way it classifies reference sites and generates a predicted fauna for comparison with the observed (Norris and Hawkins, 2000). Reference sites are categorized primarily based on their geophysical features and subsequently based on their biotic assemblage. The reference condition is derived as the mean of only those reference sites in the same category as the test site (Reynoldson *et al.*, 1997). More detailed discussions of the relative merits of predictive modelling and multimetric approaches to freshwater bioassessment are available elsewhere (Reynoldson *et al.*, 1997; Norris and Hawkins, 2000; Karr and Chu, 2000). The addition of more diagnostic indices to predictive bioassessment models probably provides the best of both worlds, whereby diagnostic power is maximized by the use of a predetermined range of robust indices, but the target value for each index is standardized by site-specific predictions.

Incorporation of functional indicators, such as decomposition or primary production rates or nutrient retention, is also compatible to a predictive modelling approach, e.g. deriving a site-specific reference condition for the rate of primary production. Such indicators would detect deleterious changes to key ecosystem functions beyond that provided by community structure indices, e.g. taxon richness. However, functional measures should not be interpreted in isolation given that the structural integrity of a community could be altered from reference condition without a significant change in the rates of indicator processes, due to functional redundancy in community composition (Hooper *et al.*, 2002).

There is also potential to enhance the core capabilities of predictive bioassessment models. Most of these models are currently static, i.e. they are derived from a spatially explicit reference site data set. While these models make extremely robust predictions of reference state communities, there is a growing awareness that biotic communities may fluctuate in response to climatic cycles and climate change (Bradley and Ormerod, 2001). Incorporating knowledge of the climatic conditions prevailing at the time the reference site data set was collected, together with long-term temporal data sets, may allow climatic cycles to be integrated into predictive bioassessment models. Inclusion of temporally explicit data may also allow the effects of common stresses to be modelled. For example, biological and environmental data collected from sites in a catchment before and after improvements to sewage treatment may allow predictive bioassessment models to be built to forecast the ecological consequences of changes in the levels of organic pollution. Currently, the static nature of predictive bioassessment models precludes their use to reliably forecast the biological response to proposed changes in the environment (Armitage, 2000). Dynamic predictive bioassessment models would be beneficial to regulatory authorities in helping to prioritize improvement works to achieve the best overall improvement to the aquatic environment.

In conclusion, this chapter describes the generic approach to the development, testing and application of predictive bioassessment models. Such models are routinely used by environment protection agencies in the UK and Australia and have been successfully applied in regional and national bioassessment programmes in Canada, the USA, Sweden, New Zealand, the Czech Republic, France and Spain. The predictive modelling approach has many advantages over other approaches to bioassessment (Norris and Hawkins, 2000). Further refinement and development of predictive bioassessment models will ensure that biomonitoring of water quality remains reliable, accurate and ecologically meaningful.

ACKNOWLEDGEMENTS

The authors would like to thank Patrick Armitage, Ralph Clarke, Mike Furse, John Hilton and John Wright for their helpful discussions and improving comments on this chapter. The Natural Environment Research Council, UK, supported this work.

REFERENCES

Armitage, P. D., 2000. 'The potential of RIVPACS for predicting the effects of environmental change'. In: *Assessing the Biological Quality of Fresh Waters: RIVPACS and Other Techniques*, Wright, J. F., Sutcliffe, D. W. and Furse, M. T. (Eds). Freshwater Biological Association: Ambleside, UK, pp. 93–111.

Armitage P. D., Moss, D., Wright, J. F. and Furse, M. T., 1983. 'The performance of a new biological water quality score system based on macroinvertebrates over a wide range of unpolluted running water sites'. *Water Research*, **17**, 333–347.

Bailey, R. C., Norris, R. H. and Reynoldson, T. B., 2004. *Bioassessment of Freshwater Ecosystems Using the Reference Condition Approach.* Kluwer Academic Publishers: Boston, MA, USA.

Bradley, D. C. and Ormerod, S. J., 2001. 'Community persistence among stream invertebrates track the North Atlantic Oscillation'. *Journal of Animal Ecology*, **70**, 987–997.

Cao, Y., Williams, D. D. and Williams, N. E., 1998. 'How important are rare species in aquatic community ecology and bioassessment'. *Limnology and Oceanography*, **43**, 1403–1409.

Cao, Y., Larsen, D. P. and Thorne, R. St -J., 2001. 'Rare species in multivariate analysis for bioassessment: some considerations'. *Journal of the North American Benthological Society*, **20**, 144–153.

Cao, Y., Larsen, D. P., Hughes, R. M., Angermeier, P. L. and Patton, T. M., 2002. 'Sampling effort affects multivariate comparisons of stream assemblages'. *Journal of the North American Benthological Society*, **21**, 701–714.

Céréghino, R., Park, Y. -S., Compin, A. and Lek, S., 2003. 'Predicting the species richness of aquatic insects in streams using a limited number of environmental variables'. *Journal of the North American Benthological Society*, **22**, 442–456.

Chessman, B. C., 1995. 'Rapid assessment of rivers using macroinvertebrates: a procedure based on habitat-specific sampling, family level identification and a biotic index'. *Australian Journal of Ecology*, **20**, 122–129.

Chessman, B. C. and McEvoy, P. K., 1998 'Towards diagnostic biotic indices for river macroin-vertebrates'. *Hydrobiologia*, **364**, 169–182.

Clarke, R. T., 2000. 'Uncertainty in estimates of river quality'. In: *Assessing the Biological Quality of Fresh Waters: RIVPACS and Other Techniques*, Wright, J. F., Sutcliffe, D. W. and Furse, M. T. (Eds). Freshwater Biological Association: Ambleside, UK, pp. 39–54.

Clarke, R. T., Furse, M. T., Wright, J. F. and Moss, D., 1996. 'Derivation of a biological quality index for river sites: comparison of the observed with the expected fauna'. *Journal of Applied Statistics*, **23**, 311–332.

Clarke, R. T., Wright, J. F. and Furse, M. T., 2003. 'RIVPACS models for predicting the expected macroinvertebrate fauna and assessing the ecological quality of rivers'. *Ecological Modelling*, **160**, 219–233.

Davies, P. E., 1994. *River Bioassessment Manual: Monitoring River Health Initiative*. Department of the Environment, Sport and Territories, Land and Water Resources R and D Corporation, Commonwealth Environment Protection Agency: Canberra, Australia.

Davy-Bowker, J., Morphy, J. F., Rutt, G. P., Steel, J. E. C. and Furse, M. T., 2005. 'The development and testing of a macroinvertebrate biotic index for detecting the impact of acidity on streams'. *Archiv für Hydrobiologie*, **163**, 383–403.

European Commission, 2000. *Establishing a Framework for Community Action in the Field of Water Policy*, Directive 2000/60/EC, *European Commission PE-CONS 3639/1/100 Rev. 1*. European Commission: Luxembourg.

Furse, M. T., Wright, J. F., Armitage, P. D. and Moss, D., 1981. 'An appraisal of pondnet samples for biological monitoring of lotic macroinvertebrates'. *Water Research*, **15**, 679–689.

Furse, M. T., Moss, D., Wright, J. F. and Armitage, P. D., 1984. 'The influence of seasonal and taxonomic factors on the ordination and classification of running water sites in Great Britain and on the prediction of their macroinvertebrate communities'. *Freshwater Biology*, **14**, 257–280.

Gevrey, M., Rimet, F., Park, Y. -S., Giraudel, J. -L., Ector, L. and Lek, S., 2004. 'Water quality assessment using diatom assemblages and advanced modelling techniques'. *Freshwater Biology*, **49**, 208–220.

Giraudel, J. L. and Lek, S., 2001. 'A comparison of self-organizing map algorithm and some conventional statistical methods for ecological community ordination'. *Ecological Modelling*, **146**, 329–339.

Hawkins, C. P., Norris, R. H., Hogue, J. N. and Feminella, J. W. 2000. 'Development and evaluation of predictive models for measuring the biological integrity of streams'. *Ecological Applications*, **10**, 1456–1477.

Hill, M. O., 1979. '*TWINSPAN – a FORTRAN program for arranging multivariate data in an ordered two-way table by classification of the individuals and attributes*'. Cornell University, Ithaca, New York, NY, pp. 90.

Hooper, D. U., Solan, M., Symstad, A., Diaz, S., Gessner, M. O., Buchmann, N., Degrange, V., Grime, P., Hulot, F., Mermillod-Blondin, F., Roy, J., Spehn, E. and van Peer, L., 2002. 'Species diversity, functional diversity and ecosystem functioning'. In *Biodiversity and Ecosystem Functioning*, Loreau, M., Naeem, S. and Inchausti, P. (Eds). Oxford University Press: Oxford, UK, pp. 195–208.

John, J., 2000. *Diatom Prediction and Classification System for Urban Streams: A Model from Perth, Western Australia*, Urban Subprogram, Report No. 6, National River Health Program: Project S3/UCU, Land and Water Resources Research and Development Corporation Occasional Paper 13/99. Canberra, pp. 150.

Johnson, R. K., 2003. 'Development of a prediction system for lake stony-bottom littoral macroin-vertebrate communities'. *Archiv für Hydrobiologie*, **158**, 517–540.

Johnson, R. K., Wiederholm, T. and Rosenberg, D. M., 1993. 'Freshwater biomonitoring using individual organisms, populations and species assemblages of benthic macroinvertebrates'.

In: *Freshwater Biomonitoring and Benthic Macroinvertebrates*, Rosenberg, D. M. and Resh, V. H. (Eds). Chapman and Hall: New York, NY, USA, pp. 40–158.

Joy, M. K. and Death, R. G., 2002. 'Predicative modelling of freshwater fish as a biomonitoring tool in New Zealand'. *Freshwater Biology*, **47**, 2261–2275.

Joy, M. K. and Death, R. G., 2003. 'Biological assessment of rivers in the Manawatu–Wanganui region of New Zealand using a predicative macroinvertebrate model'. *New Zealand Journal of Marine and Freshwater Research*, **37**, 367–379.

Karr, J. R., 1999. 'Defining and measuring river health'. *Freshwater Biology*, **41**, 221–234.

Karr, J. R. and Chu, E. W., 2000. 'Sustaining life in rivers'. *Hydrobiologia*, **422/423**, 1–14.

Magurran, A. E., 1988. *Ecological Diversity and its Measurement*. Cambridge University Press: Cambridge, UK.

Marchant, R., 2002. 'Do rare species have any place in multivariate analysis for bioassessment?'. *Journal of the North American Benthological Society*, **21**, 311–313.

Moss, D., Furse, M. T., Wright, J. F. and Armitage, P. D., 1987. The prediction of the macroinvertebrate fauna of unpolluted running-water sites in Great Britain using environmental data'. *Freshwater Biology*, **17**, 41–52.

Moss, D., Wright, J. F., Furse, M. T. and Clarke, R. T., 1999. 'A comparison of alternative techniques for the prediction of the fauna of running water sites in Great Britain'. *Freshwater Biology*, **41**, 167–181.

National Water Council, 1981. *River Quality: The 1980 Survey and Future Outlook*. National Water Council: London, UK.

Norris, R. H. and Hawkins, C. P., 2000. 'Monitoring river health'. *Hydrobiologia*, **435**: 5–17.

Reynoldson, T. B., Bailey, R. C., Day, K. E. and Norris, R. H., 1995. 'Biological guidelines for freshwater sediment based on BEnthic Assessment of SedimenT (the BEAST) using a multivariate approach for predicting biological state'. *Australian Journal of Ecology*, **20**, 198–219.

Reynoldson, T. B., Norris, R. H., Resh, V. H., Day, K. E. and Rosenberg, D. M. 1997. 'The reference condition: a comparison of multimetric and multivariate approaches to assess water-quality impairment using benthic macroinvertebrates'. *Journal of the North American Benthological Society*, **16**, 833–852.

Rosenberg, D. M., Reynoldson, T. B. and Resh, V. H., 2000. 'Establishing reference conditions in the Fraser River catchment, British Columbia, Canada, using the BEAST (BEnthic Assessment of SedimenT) predictive model'. In: *Assessing the Biological Quality of Fresh Waters: RIVPACS and Other Techniques*, Wright, J. F., Sutcliffe, D. W. and Furse, M. T. (Eds). Freshwater Biological Association: Ambleside, UK. pp. 181–194.

Simpson, J. C. and Norris, R. H., 2000. 'Biological assessment of river quality: development of AusRivAS models and outputs'. In: *Assessing the Biological Quality of Fresh Waters: RIVPACS and Other Techniques*, Wright, D. W., Sutcliffe, D. W. and Furse, M. T. (Eds). Freshwater Biological Association: Ambleside, UK, pp. 125–142.

Vannote, R. L., Minshall, G. W., Cummins, K. W., Sedell, J. R. and Cushing, C. E., 1980. 'The river continuum concept, *Canadian Journal of Fisheries and Aquatic Sciences*, **37**, 130–137.

Walley, W. J. and O'Connor, M. A., 2001. 'Unsupervised pattern recognition for the interpretation of ecological data'. *Ecological Modelling*, **146**, 219–230.

Wright, J. F., Moss, D., Armitage, P. D. and Furse, M. T., 1984. 'A preliminary classification of running water sites in Great Britain based on macroinvertebrate species and prediction of community type using environmental data'. *Freshwater Biology*, **14**, 221–256.

4.3

A New Approach to Evaluating Fluvial Functioning (FFI): Towards a Landscape Ecology

Maurizio Siligardi and **Cristina Cappelletti**

4.3.1   INTRODUCTION

The survey methods, which are used to evaluate the quality of watercourses at a given moment in time, are not the mere reflection of the technological level of the analytical instruments, but are above all an answer to the cultural and economic needs of the social context.

Rapid and continuous growth in production in the post-war period and the mirage of affluence have influenced the cultural and social frame of reference. The conservation of the environment and the landscape was only demanded by a few enlightened souls, whereas it was a totally unimportant issue for civil society. The consequences for the environment were an unprecedented urbanization of coasts and alluvial plains, as well as water, air and soil pollution. In full consistency with such a socio-economic context, regulations to prevent water pollution were almost non-existent.

The extent of the environmental damage and 'dis-economies' led to the passing of laws centred on protecting water *from pollution*, which directed control activities for more than twenty years. The main cultural limits of this system lay in the attention paid to *waste* instead of the receptor; this led to a weakening in the fight against pollution, which was the basic goal of the regulations.

By an increasingly frequent application of 'biotic indexes', these patterns of thought experienced a rupture. The deterministic approach of chemical and bacteriological methods was halted. For the first time, evaluation was no longer limited to water, but was extended to the presence of microhabitats, periphyton, aquatic vegetation, environmental variety and the hydraulic regime; for the first time, the impact of the cementification of the river bed and the uniformization of the fluvial environment could be mapped by using an appropriate survey method.

The increasing use of biotic indexes and the transfer of the relevant judgements onto coloured maps of the biologic quality of watercourses (making them intuitively and immediately understandable to laymen as well) also enabled the general public to comprehend and to acquire an environmental awareness that until then was the preserve of a narrow circle of experts with a biological-naturalistic background.

The attention paid to the wet river bed (instead of to water only) made it possible to quickly widen the scientific horizons. It was thus possible to recognize the close functional interrelations between a watercourse and its surrounding area, and the essential importance of – above all – the riparian vegetation strips. At a cultural level, a great stride was made: at last, scientific interest moved from the survey of a single water drop at a given instant to considering the entire fluvial environment, thus giving rise to an increasingly pressing need for survey methods which allowed a comprehensive evaluation of the *functioning* of watercourses.

Such a technical need was reinforced by the growth of an equally pressing social need. In fact, in the meantime sustainable development had become one of the goals of the European Union and the governments of Member States, and the restoration of watercourses, the minimum vital flow, the right of future generations to inherit an unspoilt environment and the conservation of biodiversity began to figure largely in legislation. This is the case of European Directive No. 2000/60, which mainly concerns the biological matrix, with strong ecological-functional connotations that focus on various biotic components of the aquatic environment, such as macroinvertebrates, periphyton, macrophytes and fish.

4.3.2   AN ECOSYSTEM VISION

The original alluvial plains are areas that are composed of various combinations of herbage, shrubbery and arboreal vegetation, which have an important hydraulic and ecological function. They give shelter from overflows, erosion and sedimentation, improve the quality of the aquatic environment and form a natural habitat for animals and plants (Brooks, 1988). Man's determination to regulate rivers and control their flow has influenced territorial planning, thus causing significant problems for aquatic environments and – in many instances – compromising the functioning and the ecological quality of rivers.

A river has a close relationship with its surrounding environment since – in most cases – it is the sole receptor of the outflows of a basin, including the pressure and impact agents; in fact, through the biogeochemical mechanisms of resilience (Niemi *et al.*, 1990), the river ecosystem gives rise to a process of cutting down and controlling of pressures. However, quite often this activity is destroyed by a management of the river system and the territory that is unaware, and possibly ignorant, of the existing environmental complexity.

It is true that a river can absorb pressures through its own biogeochemical mechanisms – such as the beakdown of organic substances (Cummins, 1974), the retention (Vought *et al.*, 1991), the spiralling of nutrients (Newbold, 1994) and others – but it must be enabled to put them into effect by ensuring the best possible morpho- and biodiversity of the aquatic environment. In addition, it is necessary to preserve and improve the buffer action of perifluvial areas on those nutrients, such as nitrogen and phosphorus, that come about mainly as a result of diffuse pollution of agricultural origin (Moss, 1980; Peterjohn and Correl, 1984; Lowrance *et al.*, 1984; Johnston and Naiman, 1990; Calow and Petts, 1994; Haycock *et al.*, 1996).

It is therefore strongly desirable that specific action be taken, including the creation of *buffer strips* along river banks and especially in spring areas and by watercourses of lesser importance (Vennix and Northcott, 2004); this would not only lead to the cutting down of pollutants but would also bring an economic advantage by reconstituting that landscape element – riparian vegetation – that is the indication of a balanced functioning of the ecosystem (Lamb *et al.*, 2003).

4.3.3   THE FLUVIAL FUNCTIONING INDEX (FFI) METHOD

4.3.3.1   Historical outline

The Fluvial Functioning Index (FFI) method is based on the **RCE–I** (Riparian Channel Environmental Inventory). This method – which was devised by R. C. Petersen (1992) at the Limnology Institute of Lund University (Sweden) and published in 1992 – was based on a card containing 16 questions, each with 4 predefined answers. The main goal was to gather information on the main ecological characteristics of watercourses in order to produce an inventory on the state of the riverbeds and riparian zones of Swedish rivers. Within such utilization, the formulation of environmental evaluations that could be based on the scores assigned to the single features was more a 'byproduct' rather than an explicit target of the survey.

In 1990, the card was applied to 480 stretches of the main alpine watercourses in Trentino, North Italy (Siligardi and Maiolini, 1990). The critical analysis of the collected data highlighted the need to make some important changes in the original method in order to adjust the methodology to the morpho-ecological features of Italian watercourses, above all Alpine and pre-Alpine ones. In the course of its many applications, it became increasingly evident just how important the methodology was, not only as a support to an inventory of environmental features, but also as a model to define environmental quality. The **RCE-2** index, with a new evaluation card, was therefore proposed (Siligardi and Maiolini, 1993).

In the meantime, while the importance of the specific information content achieved through well-established biological, microbiological and chemical indexes was fully appreciated, the need for new instruments to evaluate the ecosystem became strongly felt within the world of hydrobiology, as shown by the rapid spread of the application of the new RCE-2 index in Italy.

Such a proliferation of applications proved the notable significance of this method and its correspondence to various needs, but at the same time showed its inadequate adjustment to the wide range of types of Italian watercourses and made more than tangible the worry that the generic name of RCE would begin to include a heterogeneous family of indexes with divergent contents and objectives. Hence, the need to update the method in order to make it more generalizable (covering the various fluvial typologies), to define its aims with greater exactitude and ensure the comparability of its results by drafting guidelines and precise instructions for users (Siligardi, 1997).

To this end, a working party was sponsored by the Italian environmental protection agency APAT (*Agenzia per la Protezione dell'Ambiente e Tutela del Territorio*), which after in-depth deliberation and confrontation produced various changes in the questions and the answers on the card, in their meaning and their weight. All of these changes – often apparently slight, but actually substantial – turned out to be so notable that a new name for the index was requested. The new name given to the index, the **Fluvial Functioning Index (FFI)** (Siligardi *et al.*, 2000), effectively stresses the new interpretation given to each question on the survey card.

4.3.3.2   Aims of the method

The main aim of the index is to evaluate the comprehensive status of the fluvial environment and its functioning, intended as the result of the synergy and the integration of an important series of biotic and abiotic factors present in the aquatic ecosystem and the terrestrial ecosystem connected to it.

The description of the morphological, structural and biotic parameters of an ecosystem in the light of fluvial ecology principles makes it possible to survey the function connected to the parameters as well as their possible deviation from the condition of maximum functioning. Thus, a critical and integrated interpretation of environmental features enables a comprehensive functioning index to be defined.

This methodology, precisely because of its holistic approach, enables specific data to be gathered that can differ – often noticeably – from those obtained using other indexes or methods that restrict the survey to a more limited number of aspects and/or environmental sectors (see, for example, Biotic Indexes, chemical, microbiologic analyses, etc.). Therefore, it is not a question of alternative or competing methods, but rather of complementary methods, which contribute to a more in-depth knowledge of the various hierarchical levels of a fluvial system. When applied to easily understandable maps, the FFI enables the functioning of individual river stretches to be grasped straightaway; it can therefore be a useful instrument for planning the reclamation of the fluvial environment and for supporting a policy aimed at the conservation of the most unspoilt environments.

4.3.3.3   Sphere of application

The FFI can only be applied to environments of flowing water. Therefore, it cannot be applied to ecosystems of stagnant water (lakes, lagoons, ponds, relict water, etc.).

Some environments, due to their specific features, are not suitable for this methodology: this is the case, for example, of transition and mouth areas where salt cones flow up the river on the tide. In other cases, the results obtained must be interpreted with great care to avoid incorrect estimates. In fact, there are running water environments for which the 'expected' functioning level is naturally low: for example, the areas near oligotrophic springs situated above the tree line (for that biogeographic area). Such areas are characterized by an ecologic-functional 'fragility' and are particularly vulnerable. It is the task of the operator to correctly evaluate the results and suitably interpret the data provided by the maps of the fluvial functioning levels. The most suitable period for a survey to be correctly carried out is the one falling between the moderate flow and minimum flow regimes and, in any case, it must be a period of vegetation growth.

Before starting the field study, it is very important to carry out an accurate analysis of the aerial photos of the watercourse in order to plan the practical work of the on-site survey and to become familiar with the general situation of the study area.

4.3.3.4   Structure of the card

The FFI card (Table 4.3.1) comprises an introductory part containing general environmental information and 14 questions concerning the main ecological features of a watercourse; for each question, only one of the four given answers is possible. It is clear that not all the possible situations arising from the survey can be fitted onto the

Table 4.3.1 The Fluvial Functioning Index – an example of an FFI card

Basin.................... Watercourse Name ....

Location.........................

Stretch (metres)............. width (metres).......... altitude (metres)...........

Date.......... Record no.......... Photograph no............Code...........

	Left	Side	Right
(1) Land use pattern of the surrounding area			
Undisturbed, consisting of forests, woods and/or natural wetlands	25		25
Meadows, pasture, woods, a few areas of arable and uncultivated land	20		20
Mainly seasonal cultivation and/or mixed arable and/or permanent cultivation	5		5
Urbanized area	1		1
(2) Vegetation of primary perifluvial zone (fluvial zone around watercourse)			
Arboreal riparian formations	30		30
Shrub riparian formations (shrubby willow thicket) and/or cane thicket	25		25
Non-riparian arboreal formations	10		10
Made up of non-riparian or herbaceous or absent shrub species	1		1
(2b) Vegetation of secondary perifluvial zone			
Arboreal riparian formations	20		20
Shrub riparian formations (shrubby willow thicket) and/or cane thicket	15		15
Non-riparian arboreal formations	5		5
Made up of non-riparian or herbaceous or absent shrub species	1		1
(3) Extent of the perifluvial vegetation zone			
Perifluvial vegetation zone > 30 m	20		20
Perifluvial vegetation zone 5–30 m	10		10
Perifluvial vegetation zone 1–5 m	5		5
Perifluvial vegetation zone absent	1		1

(4) Continuity of the perifluvial vegetation zone		
Perifluvial vegetation zone intact without breaks in vegetation	20	20
Perifluvial vegetation zone with breaks in vegetation	10	10
Frequent breaks or only continuous and consolidated herbaceous growth	5	5
Soil without or with thin herbaceous vegetation	1	1
(5) Water conditions of the river bed		
Width of the annual peak flow bed less than three times that of the wet river bed	20	
Width of the annual peak flow bed more than three time that of the wet river bed with discharge fluctuations with seasonal return	15	
Width of the annual peak flow bed more than three times that of the wet river bed with discharge fluctuations with frequent return	5	
Wet river bed non-existent or almost non-existent or presence of impermeabilization of the cross-section	1	
(6) Structure of the river bank		
Bank with arboreal vegetation and/or stones	25	25
Bank with grass and shrubs	15	15
Bank with a fine grassy layer	5	5
Bare banks	1	1
(7) Retention structures of trophic matter		
River bed with large boulders and/or old trunks firmly embanked or presence of cane thicket or hydrophyte bands	25	25
Presence of boulders, cobbles and/or branches with depositing of sediment or scarce and non-extensive cane thicket or hydrophyte	15	15
Retention structures free and mobile during flooding or absence of cane thicket or hydrophyte	5	5
River bed with sandy sediment without algae or smooth artificial profile with uniform current	1	1
(8) Erosion		
Scarcely evident and not important	20	20
Only at bends and/or narrow passages	15	15
Frequent, with cutting of the banks and of roots	5	5
Very evident, with undercutting of banks and landslips or presence of artificial intervention	1	1

(Continued)

Table 4.3.1   The Fluvial functioning Index – an example of an FFI card (*Continued*)

	Left	Side	Right
(9) Cross-section			
Natural		15	
Natural with some artificial intervention		10	
Artificial with some natural elements		5	
Artificial		1	
(10) Watercourse bottom			
Diversified and stable		25	
Moveable in stretches		15	
Easily moveable		5	
Cemented		1	
(11) Riffles, pools or meanders			
Clearly distinguished and recurrent		25	
Present at different distances and at irregular intervals		20	
Long pools which separate short riffles or vice versa, few meanders		5	
Meanders, riffles and pools absent, straightened path		1	
(12) Vegetation in the wet river bed			
Periphyton only noticeable on touching and/or low covering of macrophytes		15	
Periphyton visible and/or limited covering of macrophytes		10	
Periphyton fair, presence of filamentous algae and/or monotonous macrophytes		5	
Periphyton thick and/or macrophytes relatively unvaried		1	

	Score
(12b) Vegetation in the river bed in low-flowing sections	
Periphyton scarcely developed and low presence of tolerant riverine vegetation	15
Periphytion fair with low low presence of tolerant riverine vegetation or scarcely developed with limited presence of tolerant riverine vegetation	10
Periphytion fair, or scarcely developed with relevant presence of tolerant riverine vegetation	5
Periphyton thick, and/or relevant presence of tolerant riverine vegetation	1
(13) Detritus	
Presence of leaves and woods, vegetable fragments recognisable and fibrous	15
Leaves and woods scarce, vegetable fragments fibrous and pulpy	10
Pulpy fragments	5
Anaerobic detritus	1
(14) Macrobenthonic community	
Well structured and diversified, appropriate to the fluvial type	20
Sufficiently diversified but with altered structure compared to that expected	10
Poorly balanced and diversified with a prevalence of taxa tolerant of pollution	5
Absence of a structured community, presence of a few taxa all relatively tolerant of pollution	1
Total Score	
Fluvial Functioning Level – FFI	

card, and so the operator must choose the answer that comes closest to the situation observed.

The general data required are those concerning the basin, the watercourse, the locality, the width of riverbed at moderate flow, the length of the homogeneous strip under examination and the altitude.

In the space provided at the bottom of the card, it can be useful to note – in addition to the clearness of the water and the specific conditions of the survey and the environment – the kind of substratum (limestone or siliceous or mixed), which, since it conditions the amount of dissolved acids, plays an important role in the biota.

The questions can be arranged into four functional groups:

• Questions 1–4 concern the *vegetational conditions of the banks and the land surrounding the watercourse* and analyse the different structural typologies influencing the fluvial environment, as, for example, the use of the land or the width of the natural riparian zone.

• Questions 5 and 6 refer to the *relative width of the wet river bed* and the *physical and morphological structure of the banks*, which give information on hydraulic characteristics.

• Questions 7–11 concern the *structure of the river bed* by identifying the typologies that favour environmental diversity and the self-treatment of a watercourse.

• Questions 12–14 aim at identifying *biological features* by means of the structural analysis of macrobenthonic and macrophytic communities and the conformation of the detritus.

All answers are given numerical weights which are grouped into four classes (from a minimum of 1 to a maximum of 30) and which express the functional differences between the individual answers. The attribution of numerical weights to individual answers is not based on mathematical formulae but on weighing up the whole of the functional processes that are influenced by the features that are the object of each answer.

The FFI value obtained by adding up the partial score of each question can range from a minimum of 14 to a maximum of 300.

Two questions (Question 2 and Question 12) offer two alternative versions and must be approached by answering only the version pertaining to the study situation:

• Question 2 (primary perifluvial zone) and Question 2 (b) (secondary perifluvial zone).

• Question 12 (turbulent flow) and Question 12 (b) (laminar flow).

4.3.3.5 Levels and maps of functioning

The FFI values are translated into five Levels of Functioning (LF) – expressed in Roman numerals (from I which indicates the best situation to V which indicates the worst) – corresponding to the evaluation of functioning; intermediate levels are also provided in order to gradate the change from one level to the next (Table 4.3.2).

Each Functioning Level is associated with a conventional colour on the map; intermediate levels are represented by two colours with a bar separating them.

The graph is drawn with two lines corresponding to the colours of the Functioning Levels and distinguishing the two banks of the watercourse. It can be drawn on 1:10 000 or 1:25 000 scale maps for a detailed representation and on 1:100 000 scale maps for an overall representation (Figure 4.3.1).

For an operative and accurate use of the achieved results it is important not to stop at an interpretation of the overall cartography, but to examine in detail the FFI values relating to the various stretches of the watercourse as well. If needed, it is possible to analyse the scores given to the individual questions or the various groups of questions. This more detailed analysis can be represented cartographically to highlight aspects of the environment that are most in jeopardy and, consequently, devise more precisely aimed programmes for environmental restoration.

4.3.3.6 The operator's competence

The essential prerequisite for the application of the FFI is an adequate knowledge of fluvial ecology and related functional dynamics. In fact, although the FFI card allows for an objective surveying of the fluvial characteristics under examination, a critical reading of the environment and strong analytic skills are required to fill it in. By filling in the card in a superficial, almost mechanical way, it is possible to make wrong assessment that may even be very distant from a correct estimate of the functioning.

It is therefore necessary, at least at the stage of the first application of the index, to operate under the guidance of experts or to attend special training courses.

Table 4.3.2 Functioning levels, evaluations and reference colours

Score	Functioning level	Evaluation	Colour map
261–300	I	Excellent	
251–260	I–II	Excellent–good	
201–250	II	Good	
181–200	II–III	Good–fair	
121–180	III	Fair	
101–120	III–IV	Fair–poor	
61–100	IV	Poor	
51–60	IV–V	Poor–very poor	
14–50	V	Very poor	

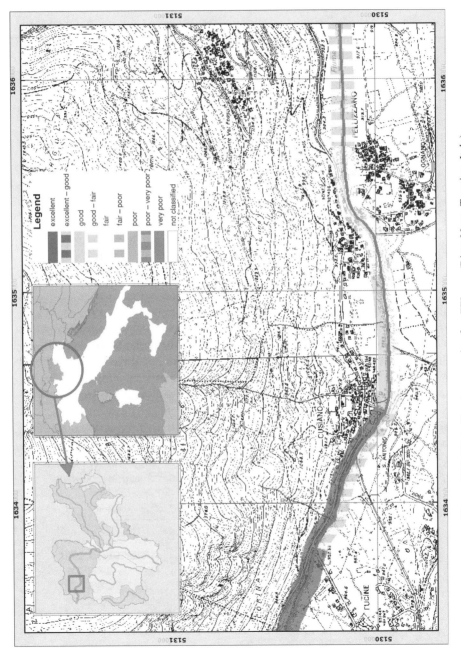

Figure 4.3.1 (Plate 3)   An example of an FFI map (River Noce, Trentino, Italy)

4.3.4 LANDSCAPE ECOLOGY AND THE FFI METHOD

Landscape Ecology is a relatively modern discipline that is aimed at studying landscape patterns, the interactions existing between the various levels of organization and the relationship between the various patches of an ecomosaic (O'Neill *et al.*, 1986; Ingegnoli, 1993; Nakagoshi and Otha, 1992).

From a thermodynamic point of view, an ecosystem is an open system, which is far from balanced and therefore dissipative. Everything going in or going out has the same importance as the exchanges taking place within the system (Prigogine *et al.*, 1972; O'Neill *et al.*, 1986). This applies both to effectively definable entities such as a wood or a city, and to a complex and open ecosystem, such as a river.

Landscape ecology focuses mainly on three aspects:

- Structure – spatial relations between distinct ecosystems, i.e. the distribution of energy and matter, relations between species, size and quantity of the biocenosis.

- Function – interaction between spatial elements such as energy flows, materials and individuals, in ecosystemic components.

- Change – alteration in the structure and function of an ecomosaic.

Landscape Ecology thus allows the ecofunctional aspects of a large ecosystem to be analysed and defined. A fluvial system is a mixed ecosystemic condition, composed of a wet part, which counts as a real ecosystem, and a riparian part, which can be considered as a transitional system between two ecosystems, in other words an ecotope (Haber, 1990).

According to landscape ecology, this condition can lead to a global vision of the river system. The river basin can be seen as a structure and a connecting grating of a system of ecosystems (Hibbs and Chan, 2001).

For this reason, great emphasis has been placed on methods of analysis using Geographic Information Systems (GISs) and on methods for the survey of the relationship between a watercourse and the surrounding territory in a functional sense. The applications of these methods can be of great help as an instrument of analysis at a watershed level, both for the drafting of plans for the protection of the territory and as an instrument for town planning.

4.3.4.1 The FFI as an instrument for the scale management of a river stretch

By applying this method, it is possible to obtain a series of data about the functioning capacity of a watercourse involving various sections of an ecosystem with specific functions. In other words, the FFI highlights the positive and negative implications related to the fluvial functioning, thus making it easier to locate the possibilities of intervention aimed at the improvement of the environment.

In fact, if the data obtained by an FFI survey on a given river stretch show a situation that deviates from the optimum, it is possible to hypothesize ameliorative measures capable of increasing the functioning level. Let's imagine, for example, a river stretch with bare banks, without any vegetation, or with a river bed that is scarcely diversified or that has been channellized: it appears evident that if its banks were revegetated or if the bed roughness or the meandering were increased, the FFI score would rise.

By means of this method, it is therefore possible to devise the course of action to be followed in order to improve the eco-functional conditions of a watercourse and to undertake programmes of river restoration (Negri *et al.*, 2004).

Similarly, the method can also be used in programmes for the evaluation of the environmental impact of planned infrastructural works (roads, motorways, railways, etc.) or works for the hydraulic settlement of a watercourse and the cutting of vegetation. In this case it is possible to highlight – by designing virtual scenarios – the loss of functioning and the possible effects not only on the river itself, but also on the fluvial territory, with the inhibition of the ecological connectivity that is made possible by the river corridor.

4.3.4.2   The FFI as an instrument for the scale management of a basin

The results of the FFI at the river stretch level, that is, on a small scale, make it possible to work at the local level to mitigate impact or to requalify the environment. At the basin scale, in contrast, this method can be used as a planning means for land management as part of town and land planning schemes.

The application of the FFI on all the main watercourses in Trentino (an Alpine region in the north of Italy) has given plausible support to deliberations on land planning and river management as provided by the *Piano Generale di Utilizzo delle Acque Pubbliche* (PGUAP) (General Plan for the Use of Public Water). This plan has priority over town planning legislation, i.e. any plan for land use is subject to the PGUAP regulations.

In this specific case, the operators' main concern has been to try and satisfy the following aspects and objectives:

- to ensure the greatest possible integrity of the transversal and longitudinal dimensions of the watercourses in Trentino;

- to increase the efficiency of the riparian zones as 'buffer zones' against diffuse pollution;

- to contribute to enhancing the landscape of 'Fluvial Environments'.

Starting from the FFI data of the main rivers of Trentino and, above all, taking only the first six questions on the card into consideration, it was possible to define a functioning condition for the riparian zone only. The data obtained from the research in the field made it possible to locate environmentally critical and delicate situations, in which the functioning ability and the resilience of a river can be jeopardized by ecological emergencies in the absence of a town planning scheme that is integrated both in its concepts and decisions.

Thanks to this work it was possible to define delimitation models for areas adjoining a watercourse by means of differentiated protective regulations, so as to objectively support planning choices. The three theoretical and practical criteria for the definition of such areas are as follows.

(a) *A zone of 'adequate ecologic quality', made up of well-established arboreal and shrubby riparian formations, that must be protected and correctly maintained.*
This zone – which is green on the map and lies between the river system and the surrounding territory – functions as an 'eco-buffer', since it intercepts and purifies the nutrients and pollutants percolating through the territory before they reach the river. In addition, it ensures the presence of a river corridor for the conservation of biological flows from upstream to downstream and vice versa.

(b) *A fluvial zone that is 'ecologically altered, but with possibilities of restoration', located in scarcely urbanized, agricultural, pasture or uncultivated areas.*
This zone –in yellow – has a width of 30 m from the line of the river bank: this is the limit set by the international literature for an effective buffer action. Accordingly, protection is required also for those zones in which artificial banks are present, in order to create the conditions for a possible restoration of the eco-functions of the perifluvial zone. It stands to reason that the restoration of areas for production or commercial activities or residential use must take the needs of the river into account.

(c) *A zone that is 'highly urbanized' and altered inside areas where urbanization is well established.*
Zones of this kind are represented by red lines only, showing their longitudinal delimitations along the watercourse. In fact, the intense urbanization of the perifluvial zones makes a rigorous cross-delimitation superfluous. Intervention is restricted to the requalification of the river bed to increase environmental morphodiversity and, consequently, to diversify ecological niches.

The three zones adjoining a river, as described and defined here, can be of great assistance in acknowledging the importance of a watercourse as an essential and fundamental element of the complex ecosystem of a basin.

4.3.5   CONCLUSIONS

The above method for the evaluation of fluvial functioning is becoming widespread in Italy, particularly because it gives biological monitoring operators – so far used to considering only the wet part of a river – an instrument capable of providing them with a more complete picture of the situation.

This method has also been used in the European field within the RIP-FOR (Riparian Forestry) project, financed by the European Community, and was included as a *best practice* on the list of Biological Methods in European Directive 2000/60EU.

The approach used in this method obliges the operator to know the foundations of River Ecology, but above all to fulfil three essential duties:

- To operate in the field by moving along the river from downstream to upstream, thus becoming acquainted with the existing situations in a longitudinal and continuous way.

- To try hard to read the signs and signals of the environment concerned from the point of view of environmental semiotics.

- Not to take any measurements, but only estimate quantitative ratios. With this attitude the operator is obliged to trust a normal practice of the human mind, which is used to appraising events more than measuring them, and to take responsibility for his/her evaluation and the relevant answer without relying – sometimes rather absent-mindedly – upon the answers given by measuring instruments.

The achieved results show how the FFI can be effectively proposed as an instrument for evaluating impacts or processes of environmental restoration as well as for land planning. In fact, it enables operators to define the functioning conditions of a watercourse and locate ecological peculiarities that are important in the comprehensive view of the sustainable development of the land.

The restoration of fluvial functions and their conservation as an integrated element of the landscape – and not only as an aesthetic factor or as an environment for dissolved or suspended matter to float in – are important in order to give back to the land one of its elements capable of exercising all those ecological functions that have been demanded by so many for so long.

Thus, town planning has an instrument at its disposal that can be highly effective in drawing up town planning schemes and for land management in general.

This is an enormous step forward. In fact, in processes of hydraulic reorganization it is possible to single out intervention models that respect and comply more fully with the dictates of naturalistic engineering. In land and town planning, watercourses can no longer be left on the fringes of the planners' interests but will have to be considered as a real and integral part of the land with their own function and importance.

REFERENCES

Brookes, A., 1988. *Channelized Rivers: Perspectives for Environmental Management*. Wiley: Chichester, UK.

Calow, P. and Petts, G. E. (Eds), 1994. *The Rivers Handbook – Hydrological and Ecological Principles.*, Vol. 2. Blackwell Scientific Publications: Oxford, UK.

Cummins, K. W., 1974. 'Structure and function of stream ecosystems'. *BioScience*, **24**, 631–641.

Haber, W., 1990. 'Basic concepts of landscape ecology and their application in land management'. *Ecology for Tomorrow. Physiology and Ecology Japan*, **27** (Special Issue), Kawanabe, H., Ohgushi, T. and Higashi, (Eds), 131–146.

Haycock, N. E., Burt, T. P., Goulding, K. W. T and Pinay, G. (Eds), 1996. *Buffer Zones: Their Processes and Potential in Water Protection*, Proceedings of the International Conference on Buffer Zones, September 1996. Quest Environmental: Harpenden, UK.

Hibbs, D. E. and Chan, S., 2001. 'Developing management strategies for riparian areas'. In: *Proceedings of the 22nd Annual Forest Vegetation Management Conference: Water, Aquatic Resources and Vegetation Management*, USDA, Redding, CA, USA 2001, pp. 84–92.

Ingegnoli, V., 1993. *Fondamenti di Ecologia del Paesaggio*. Città Studi Edizioni: Milano, Italy.

Johnston, C. A. and Naiman, R. J., 1990. 'Browse selection by beaver: effects on riparian forest composition'. *Canadian Journal of Forest Research*, **20**, 1036–1043.

Lamb, E. G., Azim, U. M. and Mackereth, M. W., 2003. 'The early impact of adjacent clearcutting and forest fire on riparian zone vegetation in northwestern Ontario'. *Forest Ecology and Management*, **177**, 529–538.

Lowrance, R., Todd, R., Fail, J. Jr, Hendrickson, O., Leonard, R. and Asmussen, L., 1984. 'Riparian forests as nutrient filters in agricultural watersheds'. *BioScience*, **34**, 374–377.

Moss, B., 1980. *Ecology of Freshwaters*. Blackwell Scientific Publications, Oxford, UK.

Nakagoshi, N. and Otha, I., 1992. 'Factors affecting the dynamics of vegetation in the landscape of Shimokamagari Islands, south western Japan'. *Landscape Ecology*, **7**, 111–119.

Negri, P., Siligardi, M., Francescon, M., Fuganti, A., Monauni, C. and Pozzi, S., 2004. 'The fluvial functioning index: an ecological assessment applied for river restoration'. In: *Proceedings of the 3rd European Conference on River Restoration*, May 2004, Zagreb, Croatia, pp. 221–227.

Newbold, J. D., 1994. 'Cycles and spirals of nutrients'. In: *The Rivers Handbook*, *1*, Calow, P. and Petts, G. E. (Eds). Blackwell Scientific Publications: Oxford, UK, pp. 379–410.

Niemi, G. J., DeVore, P., Detenbeck, N., Taylor, D., Lima, A., Pastor, L., Yount, J. D. and Naiman R. J., 1990. 'Overview of case studies on recovery of aquatic ecosystem from disturbance'. *Environmental Management*, **14**, 571–587.

O'Neill, R. V., De Angelis, D. L., Waide, J. B. and Allen, T. H. F., 1986. *A Hierarchical Concept of Ecosystems*. Princeton University Press: Princeton, NJ, USA.

Peterjohn, W. T. and Correl, D. L., 1984. 'Nutrient dynamics in an agricultural watershed: observations on the role of a riparian forest'. *Ecology*, **65**, 1466–1475.

Petersen, R. C., 1992. 'The RCE: a Riparian, Channel, and Environmental inventory for small streams in agricultural landscape'. *Freshwater Biology*, **27**, 295–306.

Prigogine, I., Nicolis, G. and Babloyatz, A., 1972. 'Thermodynamics of evolution'. *Physics Today*, **25**(11), 23–28; **25**(12), 38–44.

Siligardi, M., 1997. 'Ecologia del paesaggio e sistemi fluviali'. In: *Esercizi di Ecologia del Paesaggio*, Ingegnoli, V. (Ed.) Città Studi Edizioni: Milano, Italy, pp. 73–103.

Siligardi, M. and Maiolini, B., 1990. 'Prima applicazione di un nuovo indice di qualità dell'ambiente fluviale'. In: *Atti del Convegno 'AMBIENTE '91'. Terme di Comano (TN)*, October 1990, La Spada, P. (Ed.) Provincia Autonoma di Trento, Servizio Ripristino e Valorizzazione Ambientale: Trento, Italy, 147–177.

Siligardi, M. and Maiolini B., 1993. 'L'inventario delle caratteristiche ambientali dei corsi d'acqua alpini: guida all'uso della scheda RCE-2'. *Biologia Ambientale*, **7**(2), 18–24.

Siligardi, M., Bernabei, S., Cappelletti, C., Chierici, E., Ciutti, F., Egaddi, F., Franceschini, A., Maiolini, B., Mancini, L., Minciardi, M.R., Monauni, C., Rossi, G.L., Sansoni, G., Spaggiari, R. and Zanetti, M., 2000. IFF *Indice di Funzionalità Fluviale*, Manuale ANPA/November 2000. Agenzia Nazionale per la Protezione dell'Ambiente: Rome, Italy.

Vennix, S. and Northcott, W., 2004. 'Prioritizing vegetative buffer strip placement in an agricultural watershed'. *Journal of Spatial Hydrology*, **4**, 1–18.

Vought L., Petersen, M. and Petersen, R. C., 1991. 'Short term retention properties of channelized and natural streams'. *Verhand lungen der Internationalen Vereinigung für Theoretische und Angewandte Limnologie*, **24**, 678–685.

4.4

Planning to Integrate Urban and Ecological Processes: A Case Involving Fluvial Functionality

Corrado Diamantini

4.4.1   AN OLD PROBLEM

The paradigm of sustainability faces town planners with the problem of the relations between the city and the natural ecosystems. This is an old problem that emerged at the very beginnings of modern town planning, i.e. the second half of the 19th Century, and has accompanied it ever since. The diverse solutions originally proposed – and which in certain respects are still proposed – refer on the one hand to the separation of functions, including urban ones, and, on the other, to the integration of functions.

Biological Monitoring of Rivers   Edited by G. Ziglio, M. Siligardi and G. Flaim
© 2006 John Wiley & Sons, Ltd.

The first solution, which has managed to impose itself mainly by virtue of the economies obtained through agglomeration, is amply represented by the Ville Industrielle study undertaken by Toni Garnier: an industrial area, with a steel mill, is situated on a plain surrounded by hills among which there is a dam spanning a riverbed. The dam, although used to produce electrical energy, becomes a metaphor for the separation between modern city and nature, while the latter – in its turn – succumbs to the emergent needs of industrial production. In this town-planning scheme, green areas are essentially ornamental, delimiting urban activities that, nevertheless, dominate physical space.

The study in question is an explicitly unreal image. Toni Garnier does not refer to a well-defined place or to a specific project, but to an ideal-type industrial city, whose rules can be reproduced in a variety of places. However, this unreal image also implies the de-contextualization of urban design from physical space and its specific features, and not only from nature but also from history. It also attributes aseptic features to urban functions, which are not at all realistically represented. Such features are at odds with the daily ones of the British or German industrial cities at the turn of the last century, whose slag heaps rising up within and around steel mills depict a ghastly landscape of black hills, with dense smoke rising from chimney stacks, blackening all the buildings, and watercourses and canals turned into dark industrial sewers.

The second solution, originally proposed by Ebenezer Howard in his Garden City, has found expressive force in the city (another ideal-type construction) designed by Frank Lloyd Wright at the beginning of the 1930s, i.e. Broadacre City. This is a 'landing-place' of the critical thinking on urban planning that developed in the second half of the 19th Century in response to the urban decay produced by expanding industry. Unlike Ebenezer Howard's Garden City, which is 'set in the countryside', Broadacre City is 'a countryside converted into a city'. According to Wright, this was to be build in harmony with an omnipresent nature and with forms that respect and blend into the landscape. This meant the renunciation of major engineering works such as those needed by heavy industry and their replacement by light engineering and, furthermore, by the intermixing of functions to establish a balance between them rather than the dominance of industrial production over other functions. The result is a city extended over the territory whose forms are barely perceivable, that foregoes two features that have practically become the hallmarks of the modern city: the concentration of functions and centrality. Contrary to Ville Industrielle, Broadacre City is based on the principle of the contextualization of urban design and its adaptation to real places by modelling itself on the latter's form, also as concerns the buildings forms.

Wright's proposal, like other proposals of critical thinking, was rebuffed as utopian. The modern city has taken form in response to centripetal forces and successive densifications, which were not countered by the first expressions of the suburbanization process. Thus, most people have considered an urban form alternative to agglomeration impossible, especially as the spatial organization of the productive cycles made concentration clearly necessary. However, it can, in any case, be observed that the dematerialization of the economy, emerging only a few decades after Wright's hypothesis, now provides the conditions required to realize his vision.

4.4.2   TOWARDS A REFORMULATION OF THE RELATIONS BETWEEN THE CITY AND THE NATURAL ECOSYSTEMS

The modern city has developed in an antithetical manner with respect to its physical settings and negated every form of integration with nature. However, this only took place after a certain moment, through the progressive loss of the city's original compact form, which made it immediately recognizable and 'separate' with respect to the surrounding territory. Such a loss of compactness in Europe began to take shape around the middle of the last century, when the productive activities and the residences, followed by commercial activities, gradually abandoned the cities and occupied suburban spaces at ever greater distances from them. The roads, as also the railways and later the communication networks, first constituted 'umbilical' cords, to allow any activity, wherever localized in the territory, to main functional relations with the city but later a network, which developing, allowed other relations to come into being, thus creating the conditions for increasing autonomy with respect to the metropolises. Sprawltown (Ingesoll, 2004) is not only and not so much a modern city that expands its functions often in a discontinuous manner in space but primarily a territory that becomes a city. However, the transformation takes place in a disorderly manner because it does not happen in response to the impulses of a particular core area and thus conforming to a preceding urban order: Cacciari (2004) in this regard speaks of a kind of 'nervous collapse' of the metropolis.

This observation indicates that the large-scale urban fragmentation within the territory has nothing to do with the hypotheses of integration between city and country, which were still proposed – for example in Italy – in the second half of the 20th Century. The 'urbanized countryside', for example, proposed during the 1960s in the Trentino territorial plan (Provincia Autonoma di Trento, 1968), is a conscious construction that make explicit use of urban and landscape elements. Sprawltown, on the contrary, pays no attention to the nature, the environment or the landscape, except in the best of cases as decorative elements, in the manner proposed by Toni Garnier.

In a wholly changed context, with respect to the past, the sustainability paradigm still puts planners before the old problem of the relationship between cities and natural ecosystems. However, while in the past an attempt was made to give a form to the city in a world still dominated by the countryside, today we are attempting to give form to the territory in a world dominated by the city. Furthermore, while once it was a matter of guaranteeing better living conditions and social relations to urbanized populations, now it is a matter of conserving resources and life on the planet.

The attention of some territorial planners has consequently been attracted by the environmental emergences reported during the Rio de Janeiro Conference, and first of all by climatic change and biodiversity loss (Kirkby, *et al.*, 1995). Traffic in the industrialized countries is today one of the main sources of emissions of carbon dioxide and the increase in traffic is also the result of the urban agglomerations that present, in addition to a large territorial extension, an ever more complex and efficient road network as well as a good supply of parking places. This, in other words, is the consequence of urban sprawl (Newman and Kenworthy, 1999). In addition, when we consider another important source of emissions, i.e. domestic heating, it appears that carbon dioxide emissions increase with the growth in the heat-dispersing surfaces of buildings with respect to their volume, which is a fact associated, above all, with urban sprawl. This is the reason for the need not only to stop urban sprawl (Banister *et al.*, 1997; O'Meara Sheehan, 2002) but also to review the manner in which the city is used.

The loss of biodiversity, another of the urban emergencies indicated by the Rio de Janeiro conference, was related by some town planners (Maciocco, 2003) to the dysfunctions of marginal territories, that is, the relations established between urban and environmental processes in suburban areas, which are no longer agricultural and still not urbanized. Here, urban fragmentation has emerged, multiplying the settlements and augmenting the infrastructural networks and hence causing an equivalent fragmentation of the ecosystems, especially the agricultural ecosystems, which end up as residual elements, whose functions are impoverished or are reduced to residual areas available for building growth. Realizing a relationship of continuity between cities and agricultural areas has thus become a significant theme in urban planning. The objectives are multiple and range from safeguarding natural resources to the protection of the agrarian landscape and from environmental synergies with the urban system to the rehabilitation of the identity of physical places. Similarly, different scales of intervention are deployed, ranging from measures with a major territorial scope to residual, marginal or interstitial spaces immediately outside or inside the city, along the motorways or railway tracks or between two buildings (Palazzo, 2004).

Generally speaking, for the solution of the problem planners revert to a time-honoured invention, the cultivated garden, or in more recent times, the municipal park, with its innumerable variations over time. Thus we can cite the Parc André Citroên in Paris, where agriculture is used with recreational and didactic functions, or the Jardin Botanique in Bordeaux, that solves the problem of an urban connection between the new quarter, La Bastide, and the city's historic centre. Alternatively, with reference to solutions to overcome the separation from the natural environment, which in this case refers to the river Hudson, we can mention the creation of the Hudson River Park in Manhattan, along the West Side Water Front.

Some important architects have moved in the same direction. In addition, in this sense it is interesting to consider the critical reflection on the ways in which modern architecture established a relationship with the natural landscape, by denying it; hence today it would be matter of operating a transition from the compact to the landscape city. Renzo Piano at Trento, in a context of public and private sector co-operation, concentrates constructions alongside the railway, thereby creating a park that naturally slopes down towards the river, which in this manner re-establishes a relation with the city.

However, there are even more telling examples in this framework, comprising projects that create green areas from nothing and commencing from sites that were severely compromised on account of the use to which they had been put: landfills for industrial waste, contaminated sludge or industrial sewers. The projects refer, respectively, to Prosper Park created at Bottrop, in Ruhrgebiet, in an area occupied by a decommissioned mine, to Dania Park, along the bank of the Øresund Straits, that divide Sweden from Denmark, set up on a landfill, and to the renaturalization of a small watercourse, the Deininghauser Bach, at Essen, again in Ruhrgebiet, that had become artificial and functioned as a trunk sewer.

4.4.3 THE CHANGE IN THE SCALE OF THE RELATIONS BETWEEN URBAN AND NATURAL PROCESSES

The twin objective of recompacting the city and opening it up to agricultural green areas or, as suggested by some projects, the natural landscape, is, for the foregoing reasons, fully acceptable. However, it seems inadequate to handle the problems of the contemporary city, assuming that the latter is not only the result of the ample fragmentation caused by the explosion of the metropolises but also of the transformation, as discussed above, of the entire territory into a city (Bonomi and Abruzzese, 2004). A transformation due certainly to a process generated by the metropolises and their connective networks, but also to a widespread local growth whereby vast territories, hitherto excluded from significant urbanization processes, take on the features and functions of the city. There are many examples of this dispersed settlement pattern in many parts of Europe, but it is particularly evident in some Italian regions (Indovina *et al.*, 1990; Secchi, 1996, Munarin and Tosi, 2001) where the transformation of previously agricultural territories is all too apparent following the growth and transformation of many small and medium settlements, that end up by creating a tightly knit connective fabric in which the headquarters of large-scale multifunctional manufacturing firms intermingle with farmhouses, software houses, small factories scattered along country roads and huge amusement parks (Figure 4.4.1).

If dispersion is the primary characteristic of the contemporary city, the problem of its containment or compaction appears to have been overtaken by events. Adapting an observation by Huber (Tjallingii, 2000) on the ineluctability of industrial society, we can say that there are no alternatives to urban sprawl, only within it. Naturally, this does not mean that attempts to rationalize urban agglomerations in terms of the use of resources and the conservation of environment should be abandoned but that the same attention should be given to the problems created by the territorial scale of the settlement sprawl.

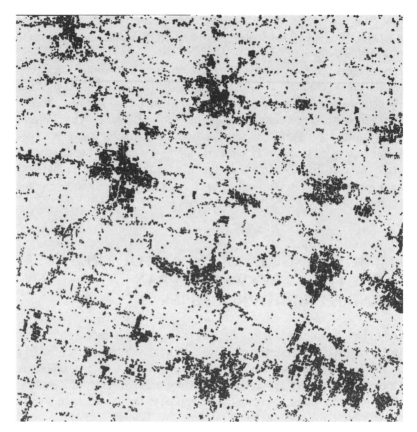

Figure 4.4.1 Dispersion features: small centres growth is accompanied by the scattering of single buildings in the countryside, made possible by the presence of a thick network of rural roads (Veneto Foothills, Italy)

As already pointed out (Diamantini, 2004), this dispersion of the built environment is not only redesigning the territory, posing wholly new problems for urban planners – problems of cohesion and identity in a large area – but also multiplying and scaling up the problems raised by the interactions between settlements and natural processes, insofar as these are now problems which are refearable no longer only to the city and its edge, that is, to an urban growth that interferes with the surrounding countryside. In other words, the edge cannot so much be identified with the edge of the city as with a multiple and multiform edge of a growing built environment that ramifies over a vast territory and interferes more pervasively – virtually from inside – with environmental and ecological processes.

In terms of sustainable development, the fact that the edge of the built environment coincides with a vast territory calls for a wholly new approach to planning practices *vis-à-vis* the natural environment. The purpose of this new approach is not so much that of simple guaranteeing green areas or pleasant landscapes for the urban population, but to guarantee a real dialogue between urban and natural processes.

This calls for the following:

• Effective co-operation between planners and other professional figures, especially ecologists, to avoid (in line with the tradition that has constantly addressed the problem of the form to be given the city), that the ways of integrating urban and natural processes once again revert to the form to be given the territory (Nassauer, 2002).

• A much greater flexibility on the part of planners, who are accustomed to propose territorial plans based on forces mainly brought into being by economic factors, to deal with building processes, which should also adjust themselves to meeting, if not to increasing, the functional integrity of ecosystems, in order that equal status be attributed to urban and natural processes (Congiu and Serreli, 2001).

Naturally, the same commitment should be requested of ecologists, who must indicate the means whereby the integration between settlement forces and natural processes can be feasibly achieved.

From this point of view, an interesting working approach is provided by Tjallingii (1996, 2002), who in distancing himself from the 'still dominant approach based on the polarity of urban and rural worlds' and the widespread idea that everything that is built represents an irreparable damage for nature, indicates that the design of multiple situations of contrast between the built and the natural environment could be a feasible approach, but on a territorial scale and not at the level of the neighbourhood, to avoid the prospect of 'an endless suburban sea of houses with gardens'. Different territorial forms, hydrography and the historically consolidated models of land use will constitute three design elements to be used in the new territorial management, as an alternative to the disorder represented by the sprawl.

A similar approach has been put forward by some ecologists (Ghetti, 2002) who propose, albeit with different emphases, to make use of ecological networks in territorial planning. Such networks would be able to provide a continuous reinforcement for the entire landscape, where nodes comprise natural or semi-natural areas (natural reserves, woods, lakes) and whose connective elements are represented by ecological corridors performing a mobility function. In addition, akin to the suggestions of Tjallingii, the hydrographic reticule plays a fundamental role in this framework.

4.4.4 PLANNING FROM THE VIEWPOINT OF SUSTAINABLE DEVELOPMENT: THE RESTORATION OF THE BRENTA RIVER

I shall now discuss some planning work that was undertaken assuming this viewpoint. It was conducted in a region, Trentino, which by being situated in the Alps does not appear to have the problem of the integration between urban and natural processes given the seemingly grandiose character of the natural setting (the forests account for 54 % of the entire region surface).

In point of fact, however, along the valleys and in the highland territory, especially close to major built-up areas, settlement sprawl is at work and now characterizes vast plain and foothill territories. The sprawl, as occurs elsewhere, is the result of suburbanization, on the one hand, and the growth of localities, in this case, stimulated by tourist flows, on the other. This takes the form of a slow but inexorable encroachment on agricultural green areas by buildings in the absence, here as elsewhere, of clear criteria for selecting building land. To this should be added the invasive character of the major transport infrastructures that run, for their entire extension, along valley floors, despite the limited spaces for human activities, on account of their morphological and hydrogeological features. It should also be added that as a result of conflicts over land use, the riverbanks often finish up as an ideal alignment for such new infrastructures (Diamantini and Geneletti, 2004).

Furthermore, along valley floors characterized by a less intensive land use, the parts of the territory that have not succumbed to residential or productive forces have been earmarked for residual uses, such as landfills for urban or special waste, quarries and areas for processing aggregates, waste treatment plants, or animal rearing farms that confer the appearance of a wasteland upon sites that still retain some natural traits, such as those adjoining the watercourses.

One of these valley floors, the Valsugana, is the site for the rehabilitation project of the Brenta river (Diamantini, 2003a), and it has given rise to some projects aimed at making the functionality of the ecosystems, environmental quality and the maintenance of the landscape characteristic key factors in its development.

The Valsugana is a narrow valley created by the Brenta river that extends in a long arc which connects the Adige Valley with the Veneto Foothills. In the last thirty years, the valley, hitherto regarded as a marginal area compared to the wellbeing degree achieved by other areas of the Trentino region, has experienced an unexpected transformation following the localization of some industrial activities. This industrial development has stimulated a growth in the settlements that, on the one hand, has mainly concerned alluvial fans on which small centres grew up for reasons of safety and, on the other, the valley floor, where the valley centre widens – Borgo Valsugana. The settlements growth presents most of the aspects mentioned earlier, including the encroachment on agricultural land as a result of the lack of any coherent plan, and on the other the proliferation along the riverside of industrial enterprises and ancillary activities such as quarries and aggregate dumps, which came into being with the building of the main road axis. Even agriculture in the valley bottom has changed insofar as it is now mainly based on fodder crops. The edge areas, in this context, are made up of parts of the territory that have not yet undergone the settlement process on account of morphological difficulties yet which, by being near to urbanized areas or principal roads, or – on the contrary – on account of their marginality with respect to the settlement processes, have become a 'no-man's land' characterized, here as elsewhere, by spontaneous, unregulated activities rather like the degraded or abandoned periphery spaces of cites (Maciocco, 2003).

Valsugana exhibits an alternation of recurrently critical situations and similarly recurrent situations of considerable environmental and landscape quality. The latter refer, on the one hand, to the persistence of a traditional type of settlement characteristic of mountain agriculture found along the sides of the mountain sloping down to Borgo Valsugana, and, on the other, to the presence – where the riverbed of the Brenta liberates itself from constrictions of hydraulic structures – of an ecological floodplain area whose natural characteristics remain more or less intact.

The disassociation between these constitutive elements of the territory represents the starting point of the design proposal. The measures take account of the relevance or frequency with which the problematic parts of the valley floor are found, as also of the relevance and frequency with which quality elements are to be found. Critical areas and quality contribute, therefore, to the construction of a design grid to generate measures whose priority is dictated by the degree of intensity with which critical situations or the valley's potentialities appear. The design proposals therefore take on the meaning of actions designed to overcome or reduce the critical aspects of the territory or to exploit its potentialities without harming its sensitiveness.

In this framework, an equal level of attention is given to problems posed by urban development and problems posed by the presence of natural ecosystems. In the first case, the objective is to reconfigure the settlement forms and thus contain urban sprawl. In the second case, the objective is to guarantee the continuity of the natural ecosystems along the valley floor by recreating the missing links in the ecosystem network and, therefore, by reconstituting nature, whenever necessary.

For such reconstitution work, the design criteria to be used will not be the traditional ones that make reference to the form of places and thus of the landscape. They must, instead, be inferred from the paradigm of sustainability and, in particular, from the indicators that emerge from studies and local sustainable development plans, such as those that signal the state of the ecosystems or fluvial functionality.

4.4.5 THE RECOURSE TO NEW DESIGN CRITERIA: THE ECOLOGICAL FUNCTIONALITY OF THE WATERCOURSE

One of the territorial contexts to which the project has dedicated most attention is that between the towns of Borgo Valsugana and Roncegno. The area in question is mainly agricultural, reclaimed from marshland and the frequent flooding of the Brenta by two temporally distant measures. The first was the land reclamation carried out at the time of Maria Teresa of Austria and the second is a change made to the riverbed carried out during the 1930s. However, this part of the territory still falls victim to flooding, as the latest flood showed, which also struck the town of Borgo Valsugana. Today, the territory is involved by building processes, hitherto limited to artisan type constructions stretching along the new riverbed of the Brenta, whose pace is likely to pick up following the completion of a large shopping mall near to the centre of Borgo Valsugana, just at the junction of the two riverbeds of the Brenta, the original one and that constructed at the turn of the last century (Figure 4.4.2).

The restoration of the old riverbed – the Brenta Vecchio – has been proposed in synergy with the construction of a storage area – conceived in order to defend the

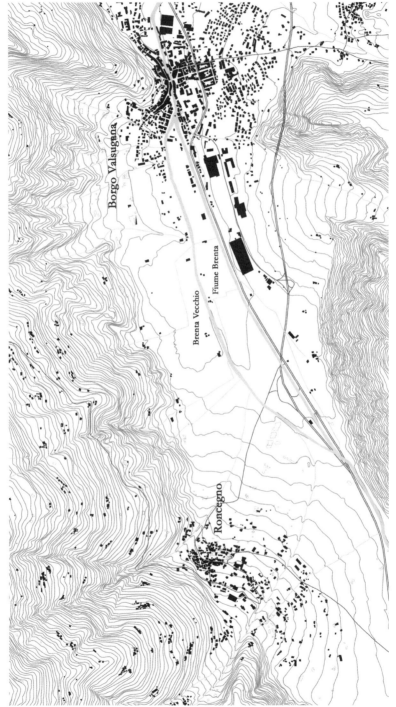

Figure 4.4.2 The area involved by the project, situated between Roncegno and Borgo Valsugana

town of Borgo Valsugana from flooding – which will occupy an ample area between the two riverbeds (Diamantini, 2004). This restoration acquires, in this framework, also the force of an intervention aimed at defining the urban form upstream of the town of Borgo Valsugana and rehabilitating the natural conditions of the area. From this point of view, the shopping mall forms part of a design that limits urban expansion to the west by constituting an interruption between the town and the open space that, while retaining agricultural features, will perform eco-systemic functions.

Such functions have been deduced by eco-mosaic readings obtained, on the one hand, from the cartographic breakdown of the eco-systemic units, whether characterized by high or low anthropogenic pressures (Cantiani, 1997) and, on the other, from the identification of the interrelations between these units, and first of all those represented by the old Brenta riverbed. With regard to the eco-mosaic, the following eco-systemic units were identified.

A first group refers to units characterized by low anthropogenic pressure or conditions close to those pertaining in nature. Among the first, there are, above all, the woods that slope down and almost touch the valley bottom, and are mainly made up, in proximity to the valley bottom, of chestnut, ash, white beech and oak trees. These woods used to constitute an important resource for the rural settlements that are still be found along the valley sides. On the valley floor, there is the biotope, La Palude, which represents one of the last examples of a riparian hydrophilic wood, characterized by the presence of rare vegetation combinations that have virtually disappeared from the rest of the valley.

A second group is represented by units characterized by high anthropogenic pressure constituted by farmland obtained from the land reclamation work, which, organized into cadastral lots with their characteristic geometrical forms, is a distinguishing feature along most of the valley floor. Once essential for the poor local economy, these fields are today intensively farmed (chiefly maize) for either animal breeding or residual family-based purposes.

Linking units are also found, in the form of coppices on farmland or near to housing.

The hydrographic reticule, as a potential medium for ecological connections, was also duly studied. As stated, most attention was given to Brenta Vecchio, given that it was impossible to take measures of any consequence designed to connect up the two sides of the valley on account of the combined presence, along the valley floor, of a road, a railway, and, for a long stretch, an industrial area.

For purposes of re-establishing the ecological functionality of the Brenta Vecchio, the Fluvial Functionality Index was adopted (ANPA, 2000; Siligardi, *et al.*, 1997), already included among the indicators of sustainability chosen for the Sustainable Development Project for the Trentino region (Diamantini, 2003b). In other words, the Fluvial Functionality Index has not been exclusively used to assess the functionality of Brenta Vecchio but also as a design criterion, able to act as a guideline for the work of renaturalizing the watercourse. The watercourse, in question, for reasons of both analysis and later project design, was divided into five sections, which returned values that ranked it between the second and the third level of functionality. The elements that were mainly responsible for returning such a poor assessment are summarized below (Tajoli, 2003).

The riverside is highly deteriorated and often reduced to grass overgrowth or is colonized by exotic plants (black locust). This is the result of the poor maintenance of the banks, in turn the result of hydraulic risk prevention, but also the consequence of the fact that intensively farmed fields extend up to the boundaries of the watercourse. This undermines the efficacy of the buffering action of various nutriments obtained from the surrounding territory. In addition, the sparse coverage afforded by shrubbery cannot provide the river with the necessary biomass, which would otherwise constitute useful *pabulum* for the macroinvertebrate community.

Furthermore, the absence of structures, such as large rocks, to retain particulate organic matter within the riverbed, which as stated earlier, is the result of precautions against hydraulic risk, means that the river environment has a limited capacity to capture leaves and the result is monotonous macrobenthic community, mainly comprised of plankton feeders. In addition, in these conditions bacterial demolition processes are set in motion that consumes much of the oxygen dissolved in the water.

The functional impoverishment of the riverbed, confined during the bank remediation measures which oblige it to follow a virtually rectilinear course, leads, on the one hand, to a loss of morphological diversity and, on the other, to the loss of the micro-habitat that contributes towards biodiversity. Furthermore, the corrections to the riverbanks have the effect of abbreviating the watercourse and thus eroding the river bottom in which the benthic fauna perform the recycling of organic substances.

The restoration of the river's old riverbed will transform it into an essential element acting as a matrix of ecological links, which, in their turn, are functional to the restoration of more widespread natural conditions. By commencing from the biotope, the creation of an ecological corridor along the watercourse will enable a transversal link to be made on which minor corridors can develop to give leverage to the agricultural territory itself in terms of biodiversity. These minor corridors can also establish links with the new Brenta riverbed, and in so doing may render its renaturalization possible, albeit only partially.

The restoration of Brenta Vecchio, leaving out here the question of the estimate of the minimum vital flow, is based upon a two-pronged measure: one aimed at remodelling the morphology of the riverbed and the other at reconstructing the riverside (Figures 4.4.3 and 4.4.4). As concerns the morphology of the riverbed, all

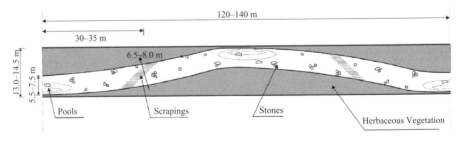

Figure 4.4.3 The layout of the riverbed follows the present one, maintaining or increasing the bending radius of the present four curves, thus allowing the formation of pools and scrapings

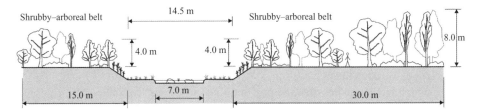

Figure 4.4.4 The transversal structure of the riparian belt aims to maximize the filter effect, where the planimetric disposition of plants keeps into account the requirement of alternating open and closed spaces

arbitrary solutions were avoided, where possible, by establishing the course layout that the river would have taken over time if left to evolve. A sufficiently diversified morphology was achieved by reprofiling the deepest part of the riverbed with a series of alternate riffles that cause the water flow to develop a sinuous course functional to the increment of the habitats. Similarly, it was deemed important to place large rocks in the shallow riverbed to achieve two functions: to guarantee shelter for fish fauna and to act as retention structures for organic material.

As concerns the riverside stretch, it was decided to link up the various and demarcated stretches of different vegetation into one continuous extension and to let a pair of trees grow at the foot of the bank each backwater along the watercourse in order to increase the shade and also to cover the banks with willow cuttings.

In the light of these design solutions, the parameters of the Fluvial Functionality Index undergo a significant change insofar as presenting a good riverbed quality (Class II), which becomes excellent (Class I) in the section adjacent to the biotope.

4.4.6 CONCLUSIONS

The sustainability paradigm, pointing out the value of biodiversity (Murray, 1995), raises the question how to safeguard, in front of an increasing urban sprawl, the natural processes and, in, particular the ecosystems functionality.

Up until now, the answer has consisted most of all in separating urban and natural processes: the former have occurred without limitations, apart from those suggested by more convenient land uses; the latter have been segregated in the protected areas, in a way not much dissimilar from that which occurred to Native Americans.

The significance of this separation has been attenuated by the presence of large ecosystems, like the agricultural ones, even though characterized by high human pressure. Now, in many parts of the world the pervasiveness of the building processes makes more and more evident the contraction of such 'natural areas' and reveals the artificial character of many natural components.

Another, alternative answer consists in operating to integrate urban and natural processes, giving both these processes almost the same attention. In the case of watercourses, this means to safeguard or, in the case, to reestablish fluvial functionality wherever possible, pursuing, on a large scale, the goal of creating networks of ecosystems.

The river restoration experience described above, carried out in collaboration by planners, ecologists and engineers, goes in this direction. A key factor, in this experience, has been the assumption, as design criterion, of the Fluvial Functionality Index which has made possible the prevention of solutions from the arbitrariness.

REFERENCES

ANPA, 2000. *Indice di Funzionalità Fluviale*. Agenzia Nazionale per la Protezione dell'Ambiente (ANPA), Agenzia Provinciale per la Protezione dell'Ambiente (APPA): Trento, Italy.

Banister, D., Watson, S. and Wood, C., 1997. 'Sustainable Cities: Transport, Energy and Urban Form'. *Environment and Planning*, **24**, 124–143.

Bonomi, A. and Abruzzese, A. (Eds), 2004. *La Città Infinita*. Bruno Mondadori: Milan, Italy.

Cacciari, M., 2004. 'Nomadi in prigione'. In: *La Città Infinita*, Bonomi, A. and Abruzzese, A. (Eds). Bruno Mondadori: Milan, Italy, pp. 51–58.

Cantiani, G., 1997. 'Le componenti naturali del territorio'. In: *La Riqualificazione dell'Asta Fluviale del Brenta*. Dipartimento di Ingegneria Civile e Ambientale (DICA): Trento, Italy, pp. 82–99.

Congiu, T. and Serreli, S., 2001. 'Funzioni ambientali e funzioni urbane: prospettive di integrazione'. In: *La Città Latente*, Maciocco, G. and Pittaluga, P. (Eds). Angeli: Milan, Italy, pp. 25–41.

Diamantini, C., 2003a. 'L'approccio interdisciplinare al progetto ambientale: la riqualificazione dell'asta fluviale del Brenta'. In: *Territorio e Progetto*, Maciocco, G. and Pittaluga, P. (Eds). Angeli: Milan, Italy, pp. 270–287.

Diamantini, C., 2003b. 'Selezione d'indicatori di sostenibilità per il territorio alpino: indicazioni dal Progetto per lo sviluppo sostenibile del Trentino'. *Atlas*, **25**, 63–70.

Diamantini, C., 2004. 'Integrating infrastructures and natural processes: the occasion given by protecting settlements from natural hazards. In: *Sustainable Urban Infrastructures*, Zanon, B. (Ed.). TEMI: Trento, Italy, pp. 224–233.

Diamantini, C. and Geneletti, D., 2004. 'Reviewing the application of SEA to sectoral plans in Italy. The case of the mobility plan of an Alpine region'. *European Environment*, **14**, 123–133.

Ghetti, P. F., 2002. 'Verso una sostenibilità ambientale'. In: *Atti del Seminario di Studi 'Nuovi Orizzonti dell'Ecologia'*. Baldacchini, G. M. and Sansoni, G. (Eds). APPA: Trento, Italy, pp. 1–9.

Indovina, F., Matassoni, F., Savino, M., Sernini, M. Torres, M. and Vettoretto L., 1990. *La Città Diffusa*. Daest: Venice, Italy.

Ingesoll. R., 2004. *Sprawlton*. Meltemi: Rome, Italy.

Kirby, J., O'Keefe, P. and Timberlake, L. (Eds), 1995. *Sustainable Development*. Earthscane: London, UK.

Maciocco, G., 2003. 'Territorio e progetto. Prospettive di ricerca orientate in senso ambientale'. In: *Territorio e Progetto*, Maciocco, G. and Pittaluga, P. (Eds). Angeli: Milan, Italy, pp. 21–29.

Munarin, S. and Tosi, C., 2001. *Tracce di Città. Esplorazioni di un Territorio Abitato: l'Area Veneta*. Angeli: Milan, Italy.

Murray, M., 1995. 'The Value of Biodiversity'. In: *Sustainable Development*, Kirby, J., O'Keefe, P. and Timberlake, L. (Eds). Earthscane: London, UK, pp. 17–29.

Nassauer, J. I., 2002. 'Ecological science and landscape design: a necessary relationship in changing landscapes.' In: *Ecology and Design*, Johnson, B. R. and Hill, K. (Eds). Island Press: Washington, DC, USA, pp. 217–230.

Newman, P. and Kenworthy, J., 1999. *Sustainability and Cities: Overcoming Automobile Dependence*. Island Press: Washtington, DC, USA.

O'Mehara Sheenan, M., 2001. *City Limits: Putting the Brakes on Sprawl*, Worldwatch Paper, 156 Worldwatch Institute: Washington, DC, USA.

Palazzo, V., 2004. 'Ecosistemi urbano e agricolo. Un'ibridazione possibile?'. *Areavasta*, **6/7**, 13–20.

Provincia Autonoma di Trento, 1968. *Il Piano Urbanistico del Trentino*. Marsilio: Venice, Italy.

Secchi, B., 1996. 'Un'interpretazione delle fasi pi recenti dello sviluppo italiano: la formazione della città diffusa e il ruolo delle infrastrutture'. In: *Infrastrutture e Piani Urbanistici*, Clementi, A. (Ed.). Palombi: Rome, Italy, pp. 27–44.

Siligardi, E., Monauni, C. and Cappelletti, C., 1997. 'Analisi della qualità dell'ambiente acquatico del fiume Brenta attraverso l'uso della scheda RCE 2'. In *La Riqualificazione dell'Asta Fluviale del Brenta*. Dipartimento di Ingegneria Civile e Ambientale (DICA): Trento, Italy, pp. 100–109.

Tajoli, D., 2003. *Criteri di Rinaturalizzazione dell'Ambiente Fluviale: Applicazione al Caso della Brenta Vecchia*, Master's Thesis. Corso di laurea in Ingegneria dell' Ambiente e del Territorio: Trento, Italy.

Tjallingii, S. T., 1996. *Ecological Conditions. Strategies and Structures in Environmental Planning*, PhD. Thesis. Technical University of Delft: Delft, The Netherlands.

Tjallingii, S. T., 2000. 'Ecology on the edge: Landscape and ecology between town and country'. *Landscape and Urban Planning*, **48**, 103–119.

4.5

Beyond Biological Monitoring: An Integrated Approach

Piet F. M. Verdonschot

4.5.1   INTRODUCTION

During the 20th Century, environmental problems have increased from sewage discharge in the first decennia towards climate change today. An increase in scale of threats implies an increase in scale of management and assessment. Successively, physico-chemical, biological and ecological assessment evolved but failed to stop deterioration. The 5-S-Model, a frame that divides the stream ecosystem into five major components, i.e. System conditions, Stream hydrology, Structures, Substances and Species (Verdonschot *et al.*, 1998), is a first attempt to comprise theoretical knowledge in 'best-practice' integrated management and assessment. Integrated ecological management and assessment comprises an ecological typology, an ecological catchment approach and a societal approach. Integrated assessment using ecological parameters includes the following.

• *Environmental parameters that are relevant for the structure and functioning of the ecosystem.* In order to make the proper choices in integrated stream and catchment management, one has to understand the functioning and interactions (dominance and feed back) of the controlling factors. The conceptual basis for integrated assessment should therefore be embedded into a landscape ecological frame. To simplify the ecological complexity of catchment ecology the 5-S-Model, a conceptual model that provides guidelines for assessment and management, was formulated (Verdonschot *et al.*, 1998). The main structure of this model is shown in Figure 4.5.1. The five key components are as follows:

1. System conditions comprise the processes related to climate (temperature, rainfall), geology and geomorphology (such as slope, soil composition). System

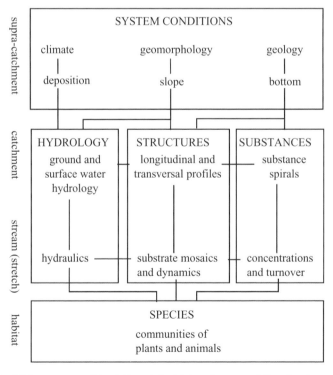

Figure 4.5.1 Main structure of the 5-S-model with key factors and functional aspects (adapted from Verdonschot *et al.*, 1998)

conditions are composed of ultimate controlling factors and are boundary conditions for a stream. The system conditions set the possibilities and limits for stream ecosystem functioning. Ultimate controlling factors continuously interact with a stream at a high hierarchical scale level in space (the catchment), as well as in time (± 100 years). Generally, system conditions cannot be changed by management. Human activities influence this level through, for example, atmospheric deposition and climate change. Stream rehabilitation does not focus on these factors but one has to consider the effects of these boundary conditions as well as the long-term effects of change.

2. Stream hydrology characteristics are set by the system conditions. Stream hydrology comprises, at the scale level of catchment, the hydrological processes, such as infiltration, ground water flow, seepage, run off and discharge. At the level of stream and habitat, stream hydrology comprises hydraulic processes, such as current velocity and turbulence. Stream hydrology refers to the water quantity parameters. The direction of the water flow strongly influences the direction of all other parameters in the system. The two main directions of flow are one running from the boundary of the catchment towards the stream (lateral) and one running from source to mouth of the stream (longitudinal).

3. Structures of the stream valley and the stream itself are strongly determined by the hydrological and hydraulic processes of stream hydrology. Structures imply the morphological features of the longitudinal and transversal shape of the stream bottom, banks and bed, as well as the substrate patterns within. Structures also refer to cut off meanders, wetlands, sand deposits and others in the stream valley. The dynamics of these structures directly relate to the dynamics in hydrology and hydraulics.

4. Substances comprise the dissolved components, such as nutrients, organic matter, oxygen, major ions and contaminants. Substances directly follow the water flow. From catchment boundary towards the stream the amount of dissolved substances increases. In addition, from source to mouth this increase is visible. Substances refer to the water quality parameters.

5. Stream hydrology, structures and substances together compose the group of controlling factors that directly determine how the stream community functions. These controlling factors take an intermediate position in between the high-scale and low-scale levels, and include the latter. Species are the response to the functioning of all above-mentioned groups of controlling factors. Species and their communities are the actual goal of ecological stream management and rehabilitation.

Controlling and response characteristics are not solely related to one of the mentioned groups of factors. There are mutual interactions. Structures, for example, can respond to the action of stream hydrology but can also reduce discharge fluctuations. Alternatively, species can be adapted to stream hydrology but, for example, trees can operate on stream hydrology and morphology. Despite a dominant hierarchical effect, a feed back is always present. Thus, factors interact on different hierarchical scale levels and with different intensity.

Knowledge of the hierarchy in factors and processes acting in space and time in streams, allows us also to infer the direction and magnitude of potential changes due to human activities (Naiman *et al.*, 1992): changes which refer to disturbance as well as to restoration, and the time involved/needed (Niemi *et al.*, 1990). Human disturbances can be seen as a sixth 'S'; the 'S' of Steering. The disturbance and restoration of streams is steered in a negative or positive direction. Integrated ecological assessment includes these aspects.

- *Biological parameters that are indicative for the ecological quality state of the eosystem.* Within the use of biological parameters two approaches dominate assessment:

  1. Indicators;

  2. Communities or species assemblages.

4.5.2 INTEGRATED ASSESSMENT AT THE COMMUNITY LEVEL: A CASE EXAMPLE

An example of a typological study to develop a reference framework based on the community approach is illustrated by the results of the project 'Ecological characterization of surface waters in the province of Overijssel (The Netherlands)' (Verdonschot, 1990). This study was situated in the province of Overijssel. In this region, a number of physical-geomorphological water types are present, such as canals, ditches, pools, streams, rivers and small lakes. This diversity in water types is quite representative for most waters present in The Netherlands. In total, we sampled 664 sites, hereby including all major environmental variability relevant to this region – thus, as natural as possible waters, as well as waters under all kinds and intensities of disturbances.

Macroinvertebrates were chosen as the biotic parameter. There are several reason to select macroinvertebrates for this purpose (Armitage *et al.*, 1992), as follows:

1. The wide diversity and abundance of species in almost all freshwater habitats.

2. Their relatively sedentary habit that allows the presence of most taxa to be related directly to conditions at their place of capture.

3. The length of many species life-cycles that allows for an overview of conditions at a site over several months.

4. Their effective dispersal and colonization mechanisms which mean that species are mostly within their potential geographical range.

5. The ability of macroinvertebrate communities to integrate and to respond to a range of environmental stresses simultaneously.

6. Many species are important accumulators and concentrators of toxic substances.

7. Qualitative sampling is easy and inexpensive.

The macroinvertebrates were collected with a pond net, a micro–macrofauna shovel or an Ekman-Birdge grab. After sorting in the laboratory if possible they were identified to species level.

About 70 abiotic parameters that were considered physically, chemically or biologically relevant, were measured at each sampling site. Finally, 853 macroinvertebrate taxa, plus 70 environmental variables at each of the 664 sampling sites were collected. This huge amount of data was elaborated by multivariate analysis techniques.

The processing of data consisted of the following main steps.

Step 1. All macroinvertebrates collected were identified to species level if possible. However, identifying macroinvertebrates still has to deal with taxonomical shortcomings. Sometimes individuals cannot be identified further then genus or even family level. These taxonomical higher units mostly have a wider ecological amplitude and thus occur in a wider range of water types. This would influence the final result of the analysis. Therefore, each taxon in the final list was screened on its status, on overlap with other taxonomical units and it was then decided whether or not to take it in the analysis (e.g. Nijboer and Schmidt-Kloiber, 2004).

A second problem deals with the abundances of the species. Three problems arise, as follows:

1. First, when a macroinvertebrate deposits its eggs, often a very high number of young individuals will appear. Through predation, starvation and other causes, their number will be reduced through the life stages. Furthermore, the number of predators will often be much smaller then that of the prey. For example, often you will find many more small worms in a sample and only one or two large predating water beetles.

2. Second, the samples were taken by means of a pond net. This is a semi-quantitative technique, by which abundances are often less accurate. During sampling, fast swimming individuals can escape while slow bottom inhabitants will all be caught more easily.

3. Third, the multivariate analysis techniques used are sensitive to high abundances – a mathematical problem. If one is interested in dominance this won't cause a problem but if one is interested in the more characteristic species, which are more often less dominant, then abundance should be transformed.

For the reasons listed above, species abundances were transformed several techniques are available, such as standardization, transformation or weighting of data. These

Table 4.5.1   Features of standardization, transformation and weighting of data

Technique	Approach	Data		Emphasis	
		Biotic	Abiotic	Dominant taxa	Rare taxa
Standardization	Centering	+			+
	Sites/pecies total	+			+
	Site/species maximum	+			+
	Zero mean/unit variance		+		
Transformation	Logarithm	+	+		+
	Square root	+	+		+
	Exponential	+		+	
	Scaling	+	+	+	+
Weighting	Rare species[a]	+		+	
	Arbitrary	+		+	+

[a] Down weighting as the main option.

are listed in the left-hand columns of Table 4.5.1. The data columns in the middle indicate the main objective of the technique, i.e. biotic or abiotic. The columns on the right-hand side indicate the main emphasis in the biotic data.

Standardization is used to make objects comparable, for example, abundance of species among samples, or values of different environmental factors. Within standardization of biotic data, one can use 'centering', 'standardizing to site or species total' and 'standardizing to site or species maximum'. These standardizations strongly overweight the rare species and downweight the common ones. They are recommended when, for example, different trophic levels are present in the species list. One should only use these options if diversity of sites in a set do not differ too much.

'Standardizing to zero mean and unit variance' implies centering but makes environmental variables with different units of measurement mutually comparable.

'Transformation' is used to ensure a better fit of the values in a regression model. Generally, 'logarithmic and square root transformations' are used to obtain statistically attractive properties. Both give less weight to dominant species, in other words, give more weight to qualitative aspects. The 'exponential transformation' is used to emphasize dominant species. 'Scaling' is used to transform values to an ordinal or nominal scale. Dependent on the class limits, one can influence the results in all possible ways, although mostly a logarithmic scale is used. Scaling always includes loss of information. Therefore, if continuous data are available, any other transformation is to be preferred.

Within weighting, 'downweighting of rare species' is the most commonly used option. A lower weight dependent on the species frequency is assigned to rare species to underemphasise the final result. This is advised if these species occur merely by chance and if the multivariate technique to be used is sensitive to rare species. Weighting can also be carried out arbitrarily for ecologically less relevant species or if there is doubt on the reliability of species identification or sampling procedure.

In this study the most commonly used transformation, i.e. the logarithmic transformation is chosen. In addition, in order to reduce extreme values of environmental parameters, transformation is advisable. Many of the variables have a skew distribution and should be transformed to better adapt to the algorithms in the analysis techniques being used.

Step 2. Secondly, the sites were clustered. Clustering is often carried out by using the program 'TWINSPAN' (Hill, 1979). TWINSPAN is a commonly used, hierarchical, divisive method of classification. However, this type of technique is sensitive to aberrant sites and to dominant species. Both of the latter strongly influence the results. Therefore, it is advisable to use non-hierarchical, agglomerative techniques.

In cluster analysis, groups of sites are calculated based on the similarity between the species compositions of the sites. The more that sites resemble each other, the better they are grouped together. An example of the results of clustering are often different kinds of ordered tables. A very high number of clustering techniques is available. Most techniques are based on calculations of similarity or dissimilarity indices. Here, the program FLEXCLUS (Van Tongeren, 1986) was used. This program is based on an initial, non-hierarchical clustering, following the algorithm of Sørensen for a site-by-site matrix based on a similarity ratio.

Step 3. Thirdly, the sites were ordinated by detrended canonical correspondence analysis (DCCA), using the program 'CANOCO', an ordination technique based on reciprocal averaging.

Since the development of canonical correspondence analysis (CCA) by Cajo ter Braak in 1986 and its implementation in his computer program CANOCO, a number of articles have been published using this program package. Birks and Austin (1992) have published an annotated bibliography on canonical or constrained ordination methods. They showed a strong increase in the number of publications in the 1990s; worldwide, almost 80 % of the publications referred to European studies at that time. This is surprising, considering the widespread use of methods such as detrended correspondence analysis (DCA) in the USA. Out of 168 publications, 35 concerned freshwater ecology.

A basic assumption in ordination is that species, sites and environmental variables that are grouped together have similar requirements with respect to environmental circumstances. The most important question is 'Is multivariate analysis needed therefore?'. Often, the answer should be 'no', although it can be useful in processing complex data sets, for example:

- to obtain a rough overview of the data;
- for communication;
- for executing overall statistical tests.

Ecological studies often involve the sampling of many sites, species and environmental variables. Multivariate techniques are adopted and adapted to best organize these numerous data. A first condition is the number of species and sites. Small data sets do not need to be treated with multivariate techniques – this is comparable to 'shooting a mosquito with a cannon'! The number of objects should be in hundreds and not in tens. A second condition should be that the data are spread over more then one environmental gradient. Multivariate techniques are useful to analyse the generally non-linear, non-monotonic response of species assemblages to a complex of environmental circumstances.

Therefore, multivariate analyses often comes down to association analysis (Figure 4.5.2). Multivariate analysis includes indirect techniques, thus classification and ordination, and direct techniques, thus regression and constrained ordination. Which to use often depends on the research question and the data available. If environmental data are available together with biological data, then direct techniques should be used.

The computer program CANOCO is designed to make several multivariate techniques available. These are summarized in Figure 4.5.3. This program uses a linear or an unimodal model. The choice of the model can be decided on the basis of a simple rule. Run a detrended correspondence analysis (DCA) program and if the gradient length is smaller than 2 sd (standard deviation units) use a linear model and if the gradient length is larger than 3 sd use a unimodal model. The linear model, furthermore, requires 'neat' quantitative data while the unimodal model 'accepts' many zeros.

Detrending is only used if an 'arch effect', or in principle component analysis (PCA), a 'horse shoe' is observed. Jackson and Somers (1991) warn against the use

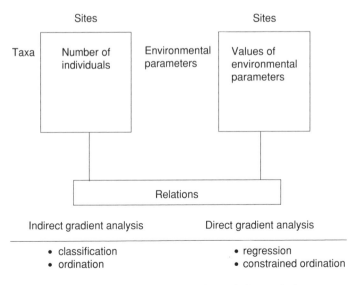

Figure 4.5.2   Main structure of association analysis

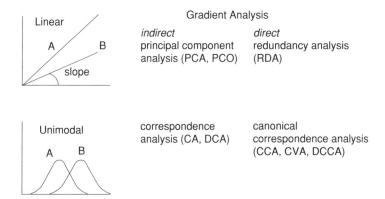

Figure 4.5.3 Different multivariate techniques available in the CANOCO package (after Braak & Smilaver, 2002)

of detrending because the results may be unreliable, especially when small data sets are processed.

In the 35 publications on the use of direct gradient analysis in freshwater ecology using CANOCO, 14 % used redundancy analysis (RDA), a linear technique, while 77 % used CCA and 23 % DCCA, both unimodal techniques (Birks and Austin, 1992).

The most important feature in ordination is the graphical arrangement of species and/or sites and/or environmental variables in a diagram.

- At first, one can arrange sites in a diagram such that sites similar in species composition are close together while sites that are dissimilar lie far apart.

- Secondly, one can arrange sites and species together in a diagram such that it allows the species compositions of the groups of sites to be derived from it as well as possible.

- Thirdly, one can arrange sites and species and environmental variables in a diagram such that it allows us to relate the sites and species to the environmental variables.

These kind of approaches are all of a quite descriptive nature. To find out which variables best explain the data, CANOCO offers the option of forward selection of environmental variables which can be combined with the option of testing whether these variables are statistically significant by means of a Monte Carlo permutation test. This option can be used to select the minimal number of explanatory variables that can explain the largest amount of variation in the data. The Monte Carlo permutation test can also be used to test the significance of the axes of a constrained ordination. Under the null hypothesis, the sites are exchangeable among the set. Especially with small data sets, this test may warn you that the relationships found may be due to 'chance effects' only. It is noticeable that only about 17 % of the publications using CANOCO in freshwater ecological studies utilized this option (Birks and Austin, 1992).

Partial ordination, another option in CANOCO, is used to control for nuisance variation, for example, for particular spatial or temporal variation, and to focus the ordination on other, underlying variables. In CANOCO, partial ordination can be carried out by specifying co-variables. This yields an ordination of the residual variation in the data after the co-variables are factored out by multivariate regression. Verdonschot and Ter Braak (1994) demonstrated the use of this option by carrying out a study on the effect of time in experimental ditches. The data were ordinated by RDA. The resulting diagram showed the relation between oligochaetes collected in five experimental ditches and the environmental variables. Sampling date is an important variable and splitted the diagram into three parts, in fact into three seasons. Differences due to sampling date hamper the interpretation of the experiment. Partial ordination, by handling sampling date as a co-variable, produces a diagram which represents the remaining variation that is attributable to the other variables; in this case, the simple relationship between species and the control experimental ditches as variables. Furthermore, it was tested as to whether the indicated remaining variables, the ditches, in the diagram are really significantly related to the data. Therefore, the Monte Carlo significance test was carried out and appeared in this case to be significant. Thus, the remaining variation was related to the ditches. In this experiment, the control ditches appeared to be quite different from each other.

Another approach was chosen for the analysis of the large data set obtained for Overijssel, as discussed above. For this data set, a strategy based on the removal of groups of sites along identified gradients and subsequent re-ordination was used. This procedure was advised by Peet (1980). He argued that the efficiency of information recovery decreases by an increasing complexity of data. The reason is that the species–environment relations do not correspond to one consistent model like our unimodal model. Thus, ordination results beyond two or three dimensions are difficult to interpret in terms of the underlying environmental variables. Peet stated that ordination may reveal the identification of the most important and conspicuous trends in a data set with a simple underlying environmental structure, but these would also have been deduced by a competent field ecologist. Thus, in this study we choose this method of progressive removal of groups of sites. In the first ordination two distinctive groups were recognized, acid moorland pools and helocrene springs and small streams. These groups were removed and the remaining sites were re-ordinated. Hereby, the impact of the originally observed variables is greatly reduced. In the re-ordination process, other groups and other variables were recognized and subsequently removed. This process was repeated four times.

Multivariate analysis results in ideas about the relationships between species, sites and environmental variables. These relationships can be statistically tested. One can test the individual relationship between species, which is an important factor, or test the differences between groups of sites. Some of these tests are already part of the CANOCO program.

Both the results of clustering and ordination are often used to recognize and describe groups of species and sites in relation to environmental factors. However, the direct results of clustering and ordination should be further examined. By looking at the species composition of an individual site and its environmental condition one can compare these with the overall composition of its group or of related groups. Furthermore, knowledge about the ecology of species, at least of the more characteristic ones, supports the ecological validity of presented groups. Thus, existing ecological knowledge is used to adjust the composition of groups. The verification of the results of multivariate analysis is maybe the most important step for an ecologist.

An example of a technique used for ordering is provided by the NODES program (Verdonschot, 1990). Taxa can be ordered according to their typifying weight in a matrix of taxa versus site groups. The numbers of sites, averages and standard deviations of the quantitative environmental variables, and the relative frequency of the nominal variables per site group, can be calculated.

4.5.2.1   Web of cenotypes

Let us return to the example of the Overijssel data. The typology of surface waters in the province of Overijssel (The Netherlands) resulted in the description of 42 site groups. These groups meet the demands earlier posed for the recognition of water types. On the basis of environmental variables and the abundance of organisms, the entities can be recognized. Hereby, no clear boundaries were described between the entities, only a recognizable centroid, and each entity shows a limited internal variation. These entities are referred to as 'cenotypes', meaning community types.

To show all 42 cenotypes and their environmental relationships, a apparently complex figure was constructed (Figure 4.5.4). The graphical result of the first ordination run of all data on axes 1 and 2 was used as a basis. The 'white contour line' indicates the variation, in macrofauna composition and environmental conditions, present between all 664 sites. All sites together form a continuum within this contour. The ecological entities recognized in the continuum are presented by the centroids of the respective cenotypes, the green dots. These centroids of the cenotypes are arranged along four major environmental gradients, the large arrows in the in-set. The four most dominant key factors for these gradients are stream character, acidity, duration of drought and dimensions. The spatial configuration of cenotypes more or less corresponds to their biological similarity. Other, still significant but less dominant, environmental relations between cenotypes are indicated by arrows. They mutually connect individual cenotypes and represent short environmental gradients.

The 42 cenotypes, as shown in Figure 4.5.4, with their mutual relationships, constitute a web. The web looks quite complex, but can be explained in detail by means of a simple example. Figure 4.5.5 shows a small part of the web. Three cenotypes, all middle reaches of streams, from the web are shown, the coded circles, as well as the most important environmental relationships, the arrows, and some profile shapes. Cenotype S5, at the top, represents polysaprobic or organically polluted streams. The

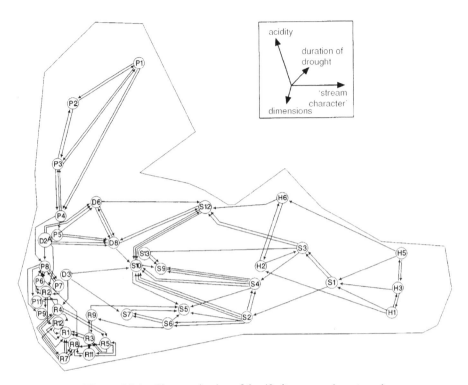

Figure 4.5.4   Characterization of the 42 site groups (cenotypes)

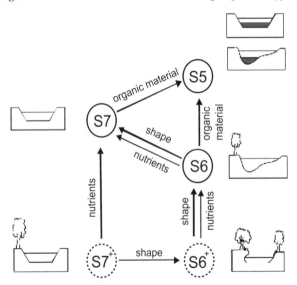

Figure 4.5.5 Five cenotypes (circle with code) with their mutual relation-ships (arrows). S7: α-mesosaprobic middle reaches of regulated streams, S5: polysaprobic upper and middle reaches of natural and regulated streams, S6: α-mesosaprobic middle reaches of semi-natural streams. S6+: β-mesosaprobic middle reaches of natural streams, S7+ β-mesosaprobic middle reaches of regulated streams. For further explanation see the text.

relationships between this type and both cenotypes S7 on the left and S6 on the right are illustrated by the arrows, both related to the amount of organic material. Cenotype S7 represents β-mesosaprobic-regulated streams and is related to the cenotype S6, β-mesosaprobic seminatural streams. These are related by the parameters profile shape, thus morphology and hydrology of the stream and the catchment and nutrient concentration, and hence the intensity of agricultural activities in the watershed. A general feature in this region is the combination of intensive agricultural activity and increased discharge fluctuations by stream canalization and land drainage. Through these human activities, streams belonging to cenotype S6 can shift towards those of S7. The construction of a sewage treatment plant which discharges in a stream belonging to cenotype S6 or S7 will cause a shift towards cenotype S5.

4.5.2.2 Reference conditions and reference framework

For water quality assessment or nature valuation, one tries to compare the actual state of a water with that of another, preferably an anthropogenical less or undisturbed water. Most known assessment systems use a singular succession series, as the series under A of white open circles, with one static endpoint as the undisturbed state or the optimum in ecosystem development; the endpoint R. However, it is difficult to give objective criteria for the definition of this static point R and the singular series is not flexible.

In Figure 4.5.6, a web is illustrated of more and less well defined states, the dots, and their relationships, the arrows. Within such a web it is possible to indicate different potential developmental directions from actual states (A) towards ecologically more optimal states (T and R). The ecological optimal states are defined as the conditions whereby an ecosystems under the given climatological, geomorphological and

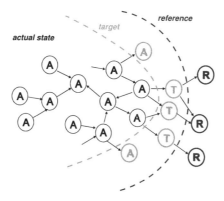

Figure 4.5.6 An example of cenotypes with actual cenotypes (A), target cenotypes (T) and reference (R): the arrows indicate the most important environmental relationships (adapted from Verdonschot, 1990)

geological conditions function as self-maintaining. The directional process is defined as ecosystem development. The degree of ecosystem development informs us about the actual state of the aquatic system and about its potential development. This means that different more or less defined states and different directions of potential development from actual states can be indicated. Such a web can serve as a reference framework for water quality assessment and nature valuation. The web is open and flexible. The choices which direct the development are up to water management. Our web of cenotypes can become such a reference framework.

4.5.2.3   The web as a reference framework

The example of a part of the web which I showed earlier with typs S5, S6 and S7, is extended with two developmental states towards the ecological optimum (Figure 4.5.5). These states were not included in the web of cenotypes shown earlier and are indicated by the extension of plus signs. The potential states S6+, oligosaprobic natural streams, and S7+, oligosaprobic regulated streams, are shown. The relationship between cenotypes S6 and S6+ is mainly due to the parameters profile shape, again morphology and hydrology, and nutrient concentration. The latter is also important between cenotypes S7 and S7+. Streams which belong to cenotypes S6 or S7 can be managed in the direction of more optimal states S6+ and S7+, respectively. This web and its extension can also be used for assessment and valuation by adding a valuation scale to the different cenotypes.

4.5.2.4   The web's application

As an example to show an application of the web, I take the crayfish, *Astacus astacus*, as an umbrella species for the whole stream. The crayfish is an endangered species in The Netherlands. It is named in the Dutch Nature Policy Plan as one of the special taxa of which the population should be conserved and where possible be restored. The deterioration of the crayfish population is not only due to the crayfish pest but also to pollution, regulation and canalization of streams and rivers. To improve the crayfish population and thus the streams, one has to deal with three scale levels.

The crayfish has specific habitat requirements. It looks for food at night and hides at day. Therefore, it needs specific structures to hide, consisting of hollow banks, stones, branches or holes in a loamy bank. These structures are necessary to avoid predation.

Looking at a section of the stream, the reach system, *Astacus astacus* needs clear water rich in oxygen. This species does not tolerate any pollution. These conditions are only met in a natural stream valley. For the aquatic compartment, the stream should belong to the formerly mentioned cenotype S6+, as illustrated in Figure 4.5.7. For the terrestrial compartment, the natural riparian zone should be present because the animals leave the water. This riparian zone offers moisture, often also shaded places where both the males in the mating period, as well as the females when they carry eggs, are protected against desiccation.

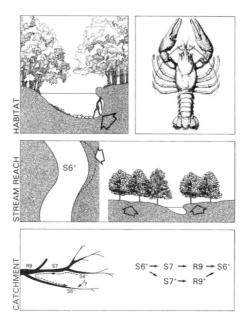

Figure 4.5.7 The habitat, stream reach and catchment distribution of *Astacus astacus* in the Netherlands. For further explanation see the text.

Furthermore, the crayfish must be able to distribute from one stream to another – some routes are indicated in Figure 4.5.7. For such population expansion, the ecological infrastructure should be suited. There may not be any barriers, specifically no weirs and dams but also the streams sections themselves should be passable. This implies requirements in water quality and habitat structures. So, if stream restoration leads to a habitat improvement, for example, cenotype S6 improves towards type S6+, the population can stabilize at that place. However, for the distribution also the stream sections belonging to the cenotypes S7 and R9 should be suited to pass and reach another section of cenotype S6+. Perhaps, therefore the conditions of the cenotypes S7+ and R9+ are necessary – thus catchment restoration.

In conclusion; the web of cenotypes offers a basis for the daily practice of regional water and nature management. The web supports the development of water quality objectives and standards, it supports the methods to monitor and assess waters, it indicates the potentials of waters and it informs about the management and restoration of waters.

4.5.3 INTEGRATED ASSESSMENT AT THE SPECIES LEVEL: A CASE EXAMPLE

Each index or metric, in general, is limited to a single impact factor. The disadvantage of an index reflecting a single aspect of the stream is that it may fail to reveal the effects of other or of combined impact factors (Fore *et al.*, 1994; Barbour *et al.*, 1996). Karr (1981) introduced the multimetric index for fish in the United States. A multimetric index consists of a combination of different individual metrics, each of which provides different ecological information about the observed community. The combination of these individual metrics act as an overall indicator of the biological quality state of a water. The strength of the multimetric index is its ability to integrate information from individual, population, community and ecosystem level (Karr and Chu, 1999). A multimetric index provides detection capability over a broad range of stressors, and provides a more complete picture of the ecosystem than single biological indicators are able to do (Intergovernmental Task Force on Monitoring Water Quality, 1993).

EBEOSWA (ecological assessment of running waters) is a system for the biological assessment of Dutch streams (Peeters *et al.*, 1994).

EBEOSWA is intended to assess more than one impact factor, and as such it can be qualified as a multimetric index. The system considers metrics related to stream velocity, saproby, trophy, functional feeding-groups and substrate. The disadvantages of the system are that it gives separate scores for each metric instead of one final classification for a location, and the ecological status of a water body is not determined by comparing the actual status of a water body with near-natural reference conditions. Furthermore, EBEOSWA is based on data collected in the 1980s. These data comprise mainly impacted sites, and collection and identification was not carried out in a standardized manner (Vlek *et al.*, 2004).

The Water Framework Directive (WFD) has led to a demand for a 'European-wide' usable assessment system. The criteria set by the WFD, to which assessment should comply, are as follows (European Commission, 2000):

- the use of different ecological quality elements – benthic invertebrate fauna, phytoplankton, fish fauna and aquatic flora;

- the ecological status of a water body is determined by comparing the biological community composition of the investigated water body with near-natural reference conditions;

- the method is based on a stream-type specific approach;

- the final quality classification of water bodies ranges from 5 ('high' status) to 1 ('bad' status).

4.5.3.1   The development of Dutch multimetrics

For the development of Dutch multimetrics, two independent data sets were used, as follows:

1. A set composed of existing data collected by water manager in the period 1990–2000. Such a data set composes information on macroinvertebrates, macrophytes and environmental variables from 949 samples taken in streams in the Netherlands. An 'a posteriori' quality classification was used to select metrics.

This quality classification was based on the results of a multivariate analysis using data on macroinvertebrates and environmental variables (Verdonschot and Nijboer, 2000). The resulting groups of sites with similar taxon composition and ranges of environmental variables were evaluated on the basis of using 'expert-judgement'.

2. A collection of a smaller data set for validation of the multimetric index.

A summary of the steps taken in the process of multimetric development is shown in Figure 4.5.8 (Vlek *et al.*, 2004).

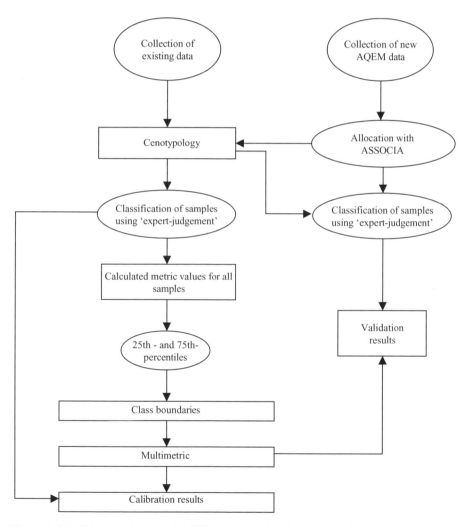

Figure 4.5.8 Diagram showing the different steps taken in multimetric development: 'ovals' represent applied techniques, while 'squares' represent accomplished results (Vlek *et al.*, 2004). Reproduced by permission of Kluwer Academic Press

A number of national multimetric systems with stressor-specific approaches for running waters have been developed (AQEM Consortium 2002; Hering *et al.*, 2004; Brabec *et al.*, 2004; Buffagni *et al.*, 2004; Ofenböck *et al.*, 2004; Sandin and Hering, 2004). Parallel to these studies, a large number of metrics representing different ecosystem components have been selected and tested. To assess whether all metrics have the ability to discriminate between the different ecological quality classes, a graphical analysis using 'box-and-whisker plots' was applied (Barbour *et al.*, 1996; Karr and Chu, 1999; Vlek *et al.*, 2004). This graphical approach provides insight in the response of macroinvertebrates to degradation, and one can easily determine the range over which a metric is sensitive and produce a metric response curve (linear, unimodal or discontinuous at a threshold level).

For each sample, the values calculated for each metric were plotted against the quality class classification and were plotted as a 'box-and whisker plot'. The suitable metrics were selected if no interquartile overlap between one or more quality classes in the 'box-and-whisker plot' occurred. To determine a possible overlap in the interquartile range, the 25th and 75th percentile were therefore calculated all metrics for each quality class (Fore *et al.*, 1996; Barbour *et al.*, 1996; Royer *et al.*, 2001). If none of the metrics showed no interquartile overlap between all classes, only the ones that showed no interquartile overlap between one class and all other classes were selected. Where possible, only those metrics reflecting different quality aspects of the ecosystem were selected – in order to be able to reveal the effect of multiple stressors. Finally, two to four metrics were selected per quality class. Tables 4.5.2 and 4.5.3 show the metrics selected to compose the multimetric for slow- and fast-running streams, respectively (Vlek *et al.*, 2004).

After selection of the metrics, the class boundaries were established either for single or combination of metrics (Tables 4.5.2 and 4.5.3). The class boundaries were set at the 25th percentile and/or the 75th percentile of the metric values. With these class boundaries, scores can be assigned to the individual metrics. Suppose a metric, value for a sample scores within the class boundaries (for a combination metric, the values for both metrics have to score within the class boundaries), this score is equal to the class that metric indicates (equal to the values shown in columns six of Tables 4.5.2 or 4.5.3) . For example, a sample taken from a slow-running stream with a metric value of 0.43 for hypopotamal (%) scores 4 for this metric (Table 4.5.2). When a metric value scores outside the class range, the site scores 0 for the respective metric. Next, the scores for the individual metrics were combined into the following multimetric indices.

Slow-running streams
    S: final score
    T-1: sum of scores for the individual metrics indicating Class 1
    T2:   sum of scores for the individual metrics indicating Class 2
    T3:   sum of scores for the individual metrics indicating Class 3
    T4:   sum of scores for the individual metrics indicating Class 4
    n1:   number of indices indicating Class 1

Table 4.5.2 Metrics that met the test criteria, metrics included in the multimetric index and their class boundaries for the slow-running streams (adapted from Vlek *et al.*, 2004)

Metric	Meets test criteria				Class for which the metric is selected as indicator[a]	Class boundaries
	Class 4	Class 3	Class 2	Class 1		
German Saprobic Index (DIN 38 410)	Yes	No	No	No	—	
Saprobic Index (Zelinka and Marvan, 1961)	Yes	No	No	No	4	< 2.12
Metapotamal (%)	Yes	No	No	No	—	
Hypopotamal (%)	Yes	No	No	No	4	< 0.55
Metarhithral (%)	Yes	No	No	No	—	
Hyporhithral (%)	Yes	No	No	No	—	
Shredders (%)	Yes	No	No	No	—	
Type Pel (%)	Yes	No	No	No	4	< 8.4
Type Lit (%)	Yes	No	No	No	—	
Type Aka (%)	Yes	No	No	No	—	
Type RP (%)	Yes	No	No	No	4	> 29.4
Type IN (%)	Yes	No	No	No	—	
Gastropoda hypopotamal (%) -EPT/OL (%)	Yes	No	No	—	3	< 3.22 – > 0.91[b] and > 1.3[c]
Gastropoda-EPT/OL (%)					3	≤ 6 – ≥ 2[b] and > 1.3[d]
No. of EPT/OL taxa	No	No	Yes	No	2	< 0.67
EPT/OL (%)	No	No	Yes	No	2	< 0.51
Grazers + scrapers/gatherers + filter feeders	No	No	No	Yes	1	> 2
Gastropoda (%)	No	No	No	Yes	1	> 9.92

[a] Where no class is shown, this indicates that the metric was not included in the multimetric index.
[b] Class boundary for the metric hypopotamal (%).
[c] Class boundary for the metric EPT/OL (%).
[d] Class boundary for the metric EPT/OL (%).

    n2: number of indices indicating Class 2
    n3: number of indices indicating Class 3
    n4: number of indices indicating Class 4

Fast-running streams
    S: final score
    T-1: sum of scores for the individual metrics indicating Class 1

Table 4.5.3 Metrics that met the test criteria, metrics included in the multimetric index and their class boundaries for the fast-running streams (adapted from Vlek *et al.*, 2004)

Metric	Meets test criteria				Class for which the metric is selected as indicator[a]	Class boundaries
	Class 4	Class 3	Class 2	Class 1		
Saprobic Index (Zelinka and Marvan, 1961)	Yes	No	Yes	Yes	4	< 2.02
					2	≥ 2.46
					1	> 2.27 – < 2.46
Metapotamal [%]	Yes	No	No	No	—	

[a] Where no class is shown, this indicates that the metric was not included in the multimetric index.

T2: sum of scores for the individual metrics indicating Class 2
T3: sum of scores for the individual metrics indicating Class 3
T4: sum of scores for the individual metrics indicating Class 4
n1: number of indices indicating Class 1
n2: number of indices indicating Class 2
n3: number of indices indicating Class 3
n4: number of indices indicating Class 4

To correct for this disproportional distribution in number of metrics included, the final score was corrected for this number. Table 4.5.4 shows the conversion of the final quality class into the multimetric index quality class.

Table 4.5.4 Class boundaries for transformation of the multimetric index scores into the final quality classes.

Quality class	Score
5 ('high' status)	Not applicable
4 ('good; status)	> 3.5 – < 4
3 ('moderate' status)	> 2.5 – < 3.5
2 ('poor' status)	> 1.5 – < 2.5
1 ('bad' status)	< 1.5

4.5.3.2 Calibration and validation

Calibration of the multimetric index took place by comparing the quality class derived through a posteriori quality class classification with the calculated one. Two possible types of errors can occur during these calculations:

(a) *Type I error*: the calculated quality class for a sample is lower than the quality class derived through a posteriori classification.

(b) *Type II error*: the calculated quality class for a sample is higher than the quality class derived through a posteriori classification.

The percentage of correctly classified samples varied between 48 and 100 %, depending on the quality class of a steam type. Only very low percentages of the samples from the slow-running streams (8 %) and from the fast-running streams (9 %) deviated more than one class from the 'post-classification'. Overall, 67 % of the slow-running streams and 65 % of the fast-running streams were classified in the same class. The Type I and Type II errors varied between 15 and 19 %. Most errors occurred with the classification of samples that received a quality Class 3 during a posteriori classification (Vlek *et al.*, 2004).

The multimetric index was validated with the above-mentioned smaller data set. This set was newly collected and processed according to standardized techniques (AQEM Consortium, 2002). Therefore, this data set was independent from the first one. Furthermore, the process of validation was the same as the one used for calibration.

Overall, 54 % of the samples were classified correctly during validation. Most of the samples that were classified in a different class, only deviated one quality class from the a posteriori classification. The Type I and Type II errors for the total data set were 32 and 14 %, respectively.

4.5.4 INTEGRATIVE TOXICITY ASSESSMENT: AN EXAMPLE

Den Besten *et al.* (1995) assessed the quality of the sediment on 46 sites in the delta of the rivers Rhine and Meuse (The Netherlands). They used an integrative approach and combined physico-chemical analysis, field observations on the macrobenthic community structure, accumulation studies and bioassays using *Chironomus riparius, Daphnia magna* and *Photobacterium phosphoreum*. The analyses results were classified either by using national criteria for sediment quality by chemical analyses or by using criteria derived from reference conditions (field studies or literature data) of both field studies and bioassays. Two methods of risk assessment were used: a sediment quality triad and a multi-criteria analysis. The sediment quality triad demonstrated causal relations between the effects on the macroinvertebrate community structure, the effects derived from the bioassays and from sediment pollution data. The multi-criteria analysis ranked all sites according to their relative risk. These authors concluded that the sediment quality triad and the multi-criteria analysis provide additional information that can be used to establish priorities for remedial action.

4.5.5   CONCLUDING REMARKS ON COMMUNITY AND INDICATOR ASSESSMENT

Over the last century, stream assessment has developed from physico-chemical, physical and biological approaches towards ecological approaches. There has been a change in water management from, in the early phases, responding to problems in the past, later on towards detecting problems in an early stage, and recently to approaches of predicting and preventing problems expected to occur in the near future. Along this chronological line a, comparable development in ecological tools has taken place. Up until the 1970s, assessment mainly dealt with either organic pollution or intrinsic natural values – the latter used to assess major anthropogenic impacts such as dam construction, channelization and mining. With the increase in impact of human activities on streams, and thus the loss of biological values, the need for nature conservation and nature development grew. The increase in stream types and scale of threats has led to an increase in management objectives and refinement in ecological tools.

In general, three major developments can be deduced from the foregoing (Verdonschot *et al.*, 1998). In Verdonschot (2000), first, integrated ecological assessment needs a frame where scale, hierarchy and concepts of the catchment and diversity are integrated: an *ecological catchment approach*. Secondly, the abiota and biota should be made operational in an *ecological typology approach*; entities which can be applied in the daily practice of water management. Thirdly, actual demands or consequences of society should be included: a *societal approach*. Overall, this also implies that assessment is just one aspect of sustainable catchment management.

Using biological entities in assessment, the two major approaches, the community and the indicator approach, sometimes seem opposing. Communities reflect the influence of all stressors on their environment (Karr, 1999; Karr and Chu, 1999). Such a multiple stress condition exerts its influence on the whole ecosystem. Under such conditions, a specific 'cause-and-effect' assessment may be difficult. For example, in The Netherlands habitat degradation and organic pollution often coincide, whereby the individual role of each of both stressors is difficult to determine. Community assessment results in an overall evaluation of the ecosystem, without referring to the cause or causes. Several authors have also indicated that community assessment is more precise and accurate than multimetrics (Reynoldson *et al.*, 1997; Bailey *et al.*, 1998; Milner and Oswood, 2000). However, multivariate techniques are complex, contain a number of arbitrary choices, need high expertise, and are difficult to communicate to policy makers.

Indicator assessment is designed to either asses the overall ecosystem condition or to assess individual stressors. Assessment of a specific stressor asks for knowledge about the indicator stressor cause–effect relationship. Autecological databases that contain such information are in development (among others, see Hering *et al.*, 2004), but on the other hand a large amount of information is still lacking. Furthermore, quantified knowledge on cause–effect relationships is yet scarce.

A community related indicator approach is based on combined use of the following:

- positive and negative dominance taxa

- positive and negative indicator (characteristic) taxa

- rare taxa

A combination of community analysis and multimetrics is favourable in future assessment development.

A five-step procedure will lead to an appropriate ecological assessment, as follows:

1. *Establish the stream type* – to be able to choose the appropriate assessment system, one has to identify the stream type.

2. *Identify the reference conditions* – for the stream type identified, one has to describe the reference conditions, including the relevant organism groups and key environmental parameters.

3. *Assess the actual biological condition* – apply a combined community and indicator assessment to obtain maximum information on the state of the stream under study.

4. *Assess the actual environmental condition* – apply an assessment of the physico-chemical and hydromorphological condition of the stream under study and check if these conditions support the biological valuation.

5. *Set the final ecological quality score* – combine the foregoing steps into a final valuation accompanied by an identification of the causes of disturbance and the measures to be taken.

REFERENCES

AQEM Consortium, 2002. 'Manual for the application of the AQEM system. A comprehensive method to assess European streams using benthic macro-invertebrates', developed for the purpose of the Water Framework Directive, Version 1.0, February 2002. pp. 198.

Armitage, P. D., Furse, M. T. and Wright, J. F., 1992. 'Environmental quality and biological assessment in British rivers – past and future perspectives'. In: *Jornadas sobre la Gestion Integral de Ecosistemas Acuaticos, Direccion de Investigacion y Formacion Agropesqueras*, Gobierno Vasco, Spain, pp. 477–511.

Bailey, R. C., Kennedy, M. G., Derbis, M. Z. and Taylor, R. M., 1998. 'Biological assessment of freshwater ecosystems using a referente condition approach: comparing predicted and actual benthic invertebrate communities in Yukon streams'. *Freshwater Biology*, **39**, 765–774.

Barbour, M. T., Gerritsen, J., Griffith, G. E., Frydenborg, R., McCarron, E., White, J. S. and Bastian, M. L., 1996. 'A framework for biological criteria for Florida streams using benthic macroinvertebrates'. *Journal of the North American Benthological Society*, **15**, 185–211.

den Besten, P. J. C., Schmidt, A., Ohm, M., Ruys, M. M., van Berghem, J. W. and van de Guchte, C., 1995. 'Sediment quality assessment in the delta of rivers Rhine and Meuse based on field observations, bioassays and food chain implications'. *Journal of Aquatic Ecosystem Health*, **4**, 257–270.

ter Braak, C. J. F., 1986. 'Canonical correspondence analysis: a new eigenvector technique for multivariate direct gradient analysis'. *Ecology*, **67**, 1167–1179.

ter Braak, C. J. F. and Smilaver, P., 2002. 'CANOCO Reference Manual and Users Guide to CANOCO for Windows. Software for Canonical Community Ordination (version 4.5)'. Centre for Biometry: Wageningen, Wageningen, The Netherlands.

Brabec, K., Kokes, J., Nemejcova, D. and Zahradkova, S., 2004. 'Assessment of organic pollution effect considering differences between lotic and lentic stream habitats'. *Hydrobiologia*, **516**, 331–346.

Birks, H. J. B. and Austin, H. A., 1992. *'An Annotated Bibliography of Canonical Correspondence Analysis and Related Constrained Ordination Methods, 1986–1991*. Botanical Institute: Bergen, Norway.

Buffagni, A., Erba, S., Cazzola, M. and Kemp, J. L., 2004. 'The AQEM southern Apennines module (South Italy): pool communities of rivers to assess water quality and habitat degradation'. *Hydrobiologia*, **516**, 313–329.

European Commission, 2000. 'Directive 2000/60/EC of the European Parliament and of the Council – Establishing a Framework for Community Action in the Field of Water Policy', 23 October 2000. European Commission: Brussels, Belgium.

Fore, L. S., Karr, J. R. and Conquest, L. L., 1994. 'Statistical properties of an index of biological integrity used to evaluate water resources'. *Canadian Journal of Fisheries and Aquatic Science*, **51**, 1077–1087.

Fore, L. S., Karr, J. R. and Wisseman, R. W., 1996. 'Assessing invertebrate responses to human activities: evaluating alternative approaches'. *Journal of the North American Benthological Society*, **15**, 212–231.

Hering, D., Moog, O., Sandin, L. and Verdonschot, P. F. M., 2004. 'Overview and application of the AQEM assessment system'. *Hydrobiologia*, **516**, 1–20.

Hill, M. O., 1979. 'TWINSPAN. A FORTRAN program for arranging multivariate data in an ordered two-way table by classification of the individuals and attributes'. Cornell University Press: Ithica, New York, NY, USA.

Intergovernmental Task Force on Monitoring Water Quality, 1993. 'The ecoregion concept, reference conditions and index calibration'. ITFM Environmental Indicators Task Group, Position Paper 1, August 27, 1993. pp. 34.

Jackson, D. A. and Somers, K. M., 1991. 'Putting things in order: the ups and downs of detrended correspondence analyis'. *American Naturalist*, **137**, 704–712.

Karr, J. R., 1981. 'Assessment of biotic integrity using fish communities'. *Fisheries*, **6**, 21–27.

Karr, J. R. and Chu, E. W., 1999. *Restoring Life in Running Waters: Better Biological Monitoring*. Island Press: Washington, DC, USA.

Milner, A. M. and Oswood, M. W., 2000. 'Urbanization gradients in streams of Anchorage, Alaska: a comparison of multivariate and multimetric approaches to classification'. *Hydrobiologia*, **422/423**, 209–223.

Naiman, R. J., Lonzarich, D. G., Beechie, T. J. and Ralph, S. C. 1992. 'General principles of classification and the assessment of conservation potential in rivers'. In: *River Conservation and Management*, Boon, P. J., Calow, P. and Petts, G. E. (Eds). Wiley: Chichester, UK, pp. 93–124.

Niemi, G. J., DeVore, P., Detenbeck, N., Taylor, D., Lima, A., Pastor, J., Yount, J. D. and Naiman, R. J. 1990. 'An overview of case studies on recovery of aquatic systems from disturbance', *Journal of Environmental Management*, **14**, 571–587.

Nijboer, R. C. and Schmidt-Kloiber, A., 2004. 'The effect of excluding rare taxa on the ecological quality assessment of running waters'. *Hydrobiologia*, **516**, 347–363.

Ofenböck, T., Moog, O., Gerritsen, J. and Barbour, M. T., 2004. 'The development of a macroinvertebrate-based multimetric index for monitoring the ecological status of running waters in Austria'. *Hydrobiologia*, **516**, 251–268.

Peet, R. K., 1980. 'Ordination as a tool for analyzing complex data sets'. *Vegetatio*, **42**, 171–174.

Peeters, E. T. H. M., Gardeniers, J. J. P., and Tolkamp, H. T., 1994. 'New methods to assess the ecological status of surface waters in the Netherlands. Part 1: Running waters'. *Verhandlungen der Internationalen Vereinigung für Theoretische und Angewandte Limnologie*, **25**(3), 1914–1916.

Reynoldson, T. B., Norris, R. H., Resh, V. H., Day, K. E. and Rosenberg, D. M., 1997. 'The reference condition: a comparison of multimetric and multivariate approaches to assess water-quality impairment using benthic macroinvertebrates'. *Journal of the North American Benthological Society*, **16**, 833–852.

Royer, T. V., Robinson, C. T. and Minshall, G. W., 2001. 'Development of macroinvertebrate-based index for bioassessment of Idaho rivers'. *Environmental Management*, **27**, 627–636.

Sandin, L. and Hering, D., 2004. 'Comparing macroinvertebrate indices to detect organic pollution across Europe: a contribution to the EC Water Framework Directive intercalibration'. *Hydrobiologia*, **516**, 55–68.

van Tongeren, O., 1986. 'FLEXCLUS, an interactive flexible cluster program'. *Acta Botanica Netherlands*, **35**, 137–142.

Verdonschot, P. F. M., 1990. *Ecological Characterization of Surface Waters in the Province of Overijssel (The Netherlands)*, PhD Thesis. University of Wageningen: Wageningen, The Netherlands.

Verdonschot, P. F. M., 2000. 'Integrated ecological assessment methods as a basis for sustainable catchment management'. In: *Proceedings of the International Conference, Assessing the Ecological Integrity of Running Waters*, Jungwirth, M., Muhar, S. and Schmutz, S. (Eds), Vienna, Austria; 'Developments in Hydrobiology 149'. *Hydrobiologia*, **422/423**, 389–412.

Verdonschot, P. F. M. and ter Braak, C. J. F., 1994. 'An experimental manipulation of oligochaete communities in mesocosms treated with chlorpyrifos or nutrient additions: multivariate analyses with Monte Carlo permutation tests'. *Hydrobiologia*, **278**, 251–266.

Verdonschot, P. F. M. and Nijboer Rebi, C., 2000. 'Typology of macrofaunal assemblages applied to water and nature management: a Dutch approach'. In: *Assessing the Biological Quality of Fresh Waters: RIVPACS and Other Techniques*, The RIVPACS International Workshop, 16–18 September, 1997, Oxford, UK, Wright, J. F., Sutcliffe, D. W. and Furse, M. T. (Eds). Freshwater Biological Association: Ambleside, UK, pp. 241–262.

Verdonschot, P. F. M., Driessen, J. M. C., Mosterdijk, H. G. and Schot, J. A., 1998. 'The 5-S-Model, an integrated approach for stream rehabilitation'. In: *River Restoration '96, Session lectures Proceedings*, Hansen, H. O. and Madsen, B. L. (Eds). International Conference arranged by the European Centre for River Restoration. National Environmental Research Institute, Copenhagen, Denmark.

Vlek, H. E., Verdonschot, P. F. M. and Nijboer, R. C., 2004. 'The development of a Dutch asessment system based on macroinvertebrates: a multimetric system'. *Hydrobiologia*, **516**, 173–189.

Zelinka, M. and Marvan, P., 1961. 'Zur Präzisierung der biologischen Klassification er Reinheit fliessender Gewässer'. *Archive für Hydrobiologie*, **57**, 389–407.

Index

Note: page numbers in **bold** refer to tables and those in *italic* to figures.

With kind thanks to W. F. Farrington for creation of this index.